NASA SP-4016

I0495848

ASTRONAUTICS AND AERONAUTICS, 1971

Chronology on Science, Technology, and Policy

Text by
Science and Technology Division
Library of Congress

Sponsored by
NASA Historical Office

Scientific and Technical Information Office   1972
NATIONAL AERONAUTICS AND SPACE ADMINISTRATION
Washington, D.C.

Foreword

This volume is the eleventh in NASA's series of annual chronologies of astronautics and aeronautics. The intent is to provide a preliminary historical reference which at least identifies, orders, and offers partial documentation for key events of the year.

One of the curious aspects of these chronology volumes is that the very process of documenting NASA's portion of them reveals not only continuities but also distinctive patterns in the unfolding events. Each year has seemed to emerge with characteristics of its own. The year 1971 was no exception. As NASA's record unfolds in these pages, it can clearly be seen as a transitional or bridging year. On the one hand many on-going programs demonstrated their value in solid accomplishments: *Apollo 14* and *Apollo 15* brought manned lunar scientific exploration to dramatic maturity; *Mariner 9* went into orbit around Mars after its 400-million-kilometer journey and began its extensive photography of that fascinating nearby planet; *OSO 7* discovered "polar caps" on the sun; in aeronautics jet aircraft equipped with the experimental supercritical wing made 27 successful research flights and the quiet jet engine program was successful in its initial tests.

On the other hand, the shape of the next decade of the space program took much firmer shape. It became clear that after a decade in which the moon was the major focus of the U.S. space program, the next ten years would emphasize earth-orbital programs geared to intensive study of our homeland in the cosmos, the planet Earth. In manned spaceflight, even as the final preparations were underway for the final two Apollo lunar missions, all flight systems for Skylab were in final manufacturing stages or in checkout; another year's intensive study by NASA and industry had refined and hardened the design of the space shuttle to the point that we could ask Administration approval to proceed, confident that a reusable, cost-effective space transportation system could be built.

In unmanned space programs, the first Earth Resources Technology Satellite was readied for its launch in 1972, while aircraft flights were testing its remote sensing devices. In aeronautics, the joint Department of Transportation-NASA study of national civil aviation R&D requirements led to the contract for design of the experimental STOL transport aircraft. Administratively, the quintupling of funds for space technology transfer to the civilian economy and the establishment of the NASA Office of Applications were other indications of the increased priority for practical benefits from space research. While all of the foregoing was accomplished within the constraints of the lowest space budget since 1962, congressional approval of an FY 1972 budget that was slightly larger ended the trend of progressively smaller space budgets that had begun in 1967. This was taken in NASA as endorsement for the new stable realignment of the space program toward the needs of man on earth for the decade ahead.

I commend to you, in the pages that follow, 1971, the year of transition.

James C. Fletcher
Administrator
National Aeronautics and Space Administration

October 31, 1972

Contents

	PAGE
Foreword	iii
NASA Administrator James C. Fletcher	
Illustrations	vii
Preface	ix
January	1
February	35
March	59
April	93
May	119
June	151
July	185
August	215
September	245
October	275
November	305
December	337
Summary	365
Appendix A: Satellites, Space Probes, and Manned Space Flights, a Chronicle for 1971	371
Appendix B: Chronology of Major NASA Launches, 1971	407
Appendix C: Chronology of Manned Space Flight, 1971	411
Appendix D: Abbreviations of References	415
Index and List of Abbreviations and Acronyms	419

Illustrations

	PAGE
Dr. George M. Low and Dr. Wernher von Braun watch Apollo 14 launch activities in Firing Room 2	26
Saturn V propels *Apollo 14* toward the moon	26
The modularized equipment transporter leaves tracks across the lunar surface	28
Scientists examine a seismic reading of the S–IVB impact on the moon	28
Apollo 14 astronauts relax in the recovery raft after splashdown	31
Dr. Richard T. Whitcomb congratulates test pilot Thomas McMurtry after the first flight of the supercritical wing on the TF–8A	66
Cosmonaut Yuri A. Gagarin and U.S.S.R. Premier Nikita Khrushchev respond to crowds celebrating the first manned space flight—10 years ago	100
Dr. James C. Fletcher is sworn in as NASA Administrator by Judge James A. Belson as President Nixon, Mrs. Fletcher, and Dr. George M. Low watch	114
Astronaut Alan B. Shepard, Jr., the first American in space, stands by the plaque marking Launch Complex 5/6 in ceremonies 10 years after the *Freedom 7* flight	123
President Kennedy—May 25, 1961—urges Congress to commit the United States to landing a man on the moon	143
The *Mariner 9* spacecraft and its flight path to Mars	149
Interior of the U.S.S.R.'s *Salyut 1*	155
Lee Scherer, Director of NASA's Apollo Lunar Exploration Office, exchanges lunar samples with Soviet Academy of Sciences President Aleksander P. Vinogradov	159
President Nixon and Secretary of Agriculture Clifford M. Hardin are briefed on the Corn Blight Watch Experiment	169
Apollo 15 prelaunch activities: NASA Administrator James C. Fletcher talks with NASC Executive Secretary William A. Anders, L/G Samuel C. Phillips (USAF), and NASA Deputy Associate Administrator for Planning Wernher von Braun	202
The lunar roving vehicle Rover on the moon near Hadley Rille, with Astronaut David R. Scott	204
Astronaut James B. Irwin scoops a trench in the lunar surface	205
The *Apollo 15* command and service modules with the SIM bay orbit the moon	206

	PAGE
Apollo 15 CM splashes down on two parachutes	208
NASA's experimental, quieter jet aircraft engine is checked out just before noise tests	241
The "Genesis rock" collected on the moon by the *Apollo 15* mission	258
A white light coronagraph of solar streamers with inset of the *Oso 7* Orbiting Solar Observatory	270
Interior of the 1-g training model of Skylab's Orbital Workshop	285
Mariner 9 photo of a sinuous rille on the surface of Mars	316
One of man's first closeup photos of the Martian moon Phobos	317
Nix Olympica, gigantic volcanic mountain on Mars	317
The descent apparatus of the U.S.S.R.'s *Mars 3*, which landed a capsule on the surface of Mars	338
Graph—U.S.–U.S.S.R. Space Activity, 1957–1971	366
Table—Summary of Soviet and United States Payloads by Mission Category	367

Preface

A chronicle is not a history explaining "whys" and "wherefores" in refined literary form. But a reference of immediate usefulness is provided by a chronology attempting to be comprehensive, with documentation to available sources. *Astronautics and Aeronautics* for 1971, the eleventh annual volume so compiled, also helps give future historians a running start on events in aerospace science and technology gleaned from the vast contemporary literature. More detailed chronology will be derived from historical research in the primary document sources as well as from oral interviews concerning pivotal events not necessarily committed to paper form. This volume is thus but a first step in the historical process.

One of the intrinsic features of this annual chronology is the inclusion of policy statements, some commentary on the impact of technology and social concerns, and biographical notes which amplify the central story of hardware, programs, and science activities. It covers worldwide aerospace events. As with earlier volumes the reader is advised to make use of the detailed index for subjects of particular interest. It has been somewhat surprising that a few readers find it of value to read through the entire volume. A concise chronology does not seek high literary marks other than brevity, accuracy, and source citation to assist additional research.

General editor of this volume was the Deputy NASA Historian, Frank W. Anderson, Jr., and the technical editor was Mrs. Carrie E. Karegeannes. The entire NASA Historical Office participated in source selection, review, and publication. Archivist Lee D. Saegesser collects current documentation. The Science and Technology Division of the Library of Congress, under an exchange of funds agreement, drafts monthly segments in comment edition form, which are circulated for corrections and use. At the Library Mrs. Patricia D. Davis, Mrs. Carmen B. Brock-Smith, and Mrs. Shirley M. Singleton carry principal responsibility. At the end of the year, the entire manuscript is reworked to include comments received and recent additional information. Arthur G. Renstrom of the Library of Congress prepared the extensive index, which is indispensable to the usefulness of a chronology. The index serves also as a glossary of abbreviations and acronyms used in the chronology.

Appendix A, "Satellites, Space Probes, and Manned Space Flights, 1971," Appendix B, "Chronology of Major NASA Launches, 1971," and Appendix C, "Chronology of Manned Space Flight, 1971," were prepared by Leonard C. Bruno of the Library staff. Appendix D, "Abbreviations of References," was prepared by Mrs. Brock-Smith.

Without the validation throughout NASA and other Federal agencies, the content of this volume would be less reliable and less nearly complete. Comments, additions, and criticisms are always welcomed by the NASA Historical Office.

Eugene M. Emme
NASA Historian

January 1971

January 1: Discovery of galaxies Maffei 1 and Maffei 2 about 3 million light years from earth was reported in *Astrophysical Journal* by team of Univ. of California at Berkeley, Cal Tech, and Carnegie Institution astronomers headed by Robert Landau of Univ. of California at Berkeley. Galaxies previously had been obscured by interstellar dust in Milky Way. Team had begun study after Landau read 1968 report of Italian astronomer Paolo Maffei that two strange objects had been observed on infrared photo Maffei made of region between constellations Perseus and Cassiopeia. Using Mount Palomar and Lick Observatory telescopes and advanced astronomical tools and research techniques, team had located galaxies twice as far from earth as Andromeda, nearest galaxy to Milky Way. Team believed brighter of new galaxies might be larger than either Milky Way or Andromeda and thus possibly largest member of local group of galaxies, measuring 50 000 to 100 000 light years in diameter. (*Astrophysical Journal*, 1/1/71, L25–31)

- *Washington Post* editorial commented on "verbal violence" at Dec. 26–31, 1970, AAAS convention in Chicago: "It should not be beyond the power of scientists to restore reason to its normal throne at their conventions. They had better set about doing so without further paltering or delay—by calling in the police if necessary to eject disrupters and impose order. It is a scientific fact, we believe, that only a single speaker can be heard at a particular time in a particular place. Those who want to hear him should be free to do so; those who do not should be free to go away. This is not alone the basis of science; it is also the essence of freedom." (*W Post*, 1/1/71, A18)

January 2: GAO sent report to Congress that North American Rockwell Corp. stood to receive extra $1.5 million for beating target costs on NASA contract to develop F–1 rocket while, actually, target cost had been overstated originally by $5 million. NR had not used most current cost data available. GAO recommended that NASA seek adjustment in fees to NR and ensure that cost estimates in existing and future contracts were based on accurate, complete data. Report included NR statement of disagreement with findings and recommendations. (Beckler, *W Post*, 1/3/71, A8)

- USAF long-range planners were advocating DOD development of pilotless combat aircraft, *Business Week* reported. DOD had consulted USAF, USN, and NASA about operational robots and had talked with RAND Corp., Aerospace Corp., and Mitre Corp. Ryan Aeronautical Div. of Teledyne, Inc., Boeing Co., Northrop Corp., and Cessna Aircraft Co. also had given DOD their views. Experts felt existing technology used in unmanned spy aircraft made pilotless combat aircraft possible. DOD had approved experimental hardware and flight tests. (*Bus Wk*, 1/2/71)

January 4: NASA was looking at air bases in the Carolinas, south Florida, and the Bahamas as possible landing site for space shuttle rocket launched from KSC, UPI said in *Philadelphia Inquirer*. NASA engineers also were considering refueling shuttle booster's 1st stage in flight so that it could return to launch site for landing, like aircraft. (*P Inq*, 1/4/71)

- Discovery of cloud of hydrogen gas 80 billion km (50 billion mi) long and traveling through space at 210 000 km per hr (130 000 mph) in different direction and at 60° angle from solar system was reported in *Washington Post*. In interview Dr. Gary Thomas, head of Univ. of Colorado team of astronomers that discovered cloud, had said it would take cloud at least 50 yrs to pass by solar system and could take forever. "For all we know . . . it's been there forever." Cloud was discovered first by *Ogo 5* in September 1969 and later confirmed by same satellite three times. (O'Toole, *W Post*, 1/4/71, A11)

- New York Assemblyman Andrew J. Stein told Washington, D.C., press conference he had 76 cosponsors for bill which would ban supersonic transports from operating at New York's John F. Kennedy International Airport. Bill, if passed, would make it illegal to operate any aircraft producing noise at levels above that of existing subsonic jets. (Sehlstedt, B *Sun*, 1/5/71, A6)

January 5: Mayor Carl B. Stokes of Cleveland, Ohio, and LeRC Director Bruce T. Lundin announced that LeRC would assist Cleveland's Div. of Air Pollution Control by identifying and cataloging trace elements and compounds in air, determining their concentrations, and finding trends or patterns related to weather, season, and geographical area. (LeRC Release 71-1)

- Oklahoma Governor Dewey H. Bartlett told state legislature in Oklahoma City there was good possibility that Oklahoma could become launch site for NASA space shuttle. "Based on the launch azimuth and the orbital inclinations . . . Oklahoma provides both a desirable launch and recovery location." (*Daily Oklahoman*, 1/6/71, 1)

January 6: NASA announced award of $4-million, cost-plus-fixed-fee contract modification to Ball Brothers Research Corp., Inc., for changes in OSO-H Orbiting Solar Observatory to increase experiment capacity. Change orders definitized by contract included larger physical volume to handle larger experiment packages, improved command system security, and increased power for instruments. OSO-H would be launched in mid-1971. (NASA Release 71-2)

- Lockheed Aircraft Corp. had rejected Dec. 30, 1970, DOD proposal that firm accept $200-million loss on disputed C-5A contract, Lockheed Board Chairman Daniel J. Haughton announced at New York press conference. Lockheed had chosen to take matter to court. Haughton estimated litigation would take two to five years. (Getler, *W Post*, 1/7/71, A1)

- Permanent appointment of Gustav E. Lundquist as Associate Administrator for Engineering and Development was announced by John H. Shaffer, FAA Administrator. Lundquist had been in acting capacity since January 1970. David R. Israel, former Deputy Director of Defense Communications Planning Group in DOD, had been named Director of newly created Office of Systems Engineering Management and Spencer S. Hunn had been named Director of National Airspace System Program Office in FAA. (FAA Release 71-2)

- NOAA said scientific team that studied 23 310 000-hectare (90 000-sq-mi) region east of Barbados in summer of 1969 had reached preliminary conclusion that dust blown off North African deserts into upper air ended in atmosphere over South Atlantic and influenced North American weather. Scientists, who had hoped to find atmosphere relatively free of particulate matter, had discovered instead African dust in Barbados air at levels to 5000 m (16 000 ft). (O'Toole, *W Post,* 1/7/71, A15)

January 7: NASA launched series of three Nike-Cajun sounding rockets carrying GSFC experiments to obtain temperature, pressure, density, and wind data in upper atmosphere by detonating grenades and recording their sound arrivals on ground. Rocket launched from Point Barrow, Alaska, reached 122-km (75.8-mi) altitude, with all 19 grenades exploding as planned. Rocket launched from Wallops Station reached 130.9-km (81.3-mi) altitude, with 16 of 19 explosions confirmed by ground. Third rocket was launched from Churchill Research Range. (NASA Rpt SRL)

- Release of "Statement of Government Policy on Satellite Telecommunications for International Civil Aviation Operations" was announced by Clay T. Whitehead, Director of Telecommunications Policy for Executive Office of the President. Statement said U.S. would promote preoperational deployment of satellite communications in Pacific in 1973 and in Atlantic in 1975. DOT and FAA would assume program management responsibility for preoperational and operational systems and services. Dept. of State, with DOT, would seek international use of preoperational system and initiate cooperative efforts with other nations to establish operational system by 1980. U.S. would use commercial communications facilities to maximum. U.S. Government would use UHF frequency band near 1600 mhz in both preoperational and operational satellite air-traffic-control communications. Experimental evaluation of independent surveillance by satellite should begin with system deployment in Pacific and be followed by preoperational evaluation in air-traffic-control environment after 1975. Unified program to satisfy Government and airline requirements should be adopted to provide economic benefits of single program. (EOP Release)

January 8: Three aerospace teams, including seven European firms interested in support roles, submitted proposals to MSFC on preliminary design of research and applications module (RAM) proposed for space shuttle and space station. Proposals were submitted by McDonnell Douglas Corp., General Dynamics Corp., and General Electric Co. Team headed by General Dynamics would include MATRA of France, ERNO of Germany, SAAB of Sweden, Hawker Siddeley of U.K., and Fiat of Italy. General Electric team would include Messerschmitt-Boelkow-Blohm of Germany and Thompson-CSF of France. (MSFC Release 71-10)

- HUD had evolved $4-million pilot project to put 1500 unemployed space industry technicians and former servicemen to work in model cities projects, *Washington Post* reported. Plan, developed by HUD Assistant Secretary Floyd Hyde, would recruit unemployed and give them fast orientation courses on urban problems on assumption that aerospace skills could be transferred to urban improvement. Hyde proposed to start project in March with up to 350 participants and estimated cost to Government of about $2000 per person placed in new job. (*W Post*, 1/8/71)

- *St. Louis Post-Dispatch* editorial commented on Dec. 23, 1970, DOD announcement that Sprint ABM had successfully intercepted ICBM nosecone over Pacific: Announcement "proved only two things: the military establishment recognizes that the war in Congress over full deployment of the ABM system is far from over; consequently the real target of the test firing was not a nose cone but Congressional critics whose arguments the Pentagon has yet to refute." Paper wondered "why the secondary Sprints are needed in the allegedly foolproof Safeguard system if the Spartan is as successful as the military claims it to be." More puzzling was "how the Pentagon can call the Sprint a success when the crew that fired it from a South Pacific atoll was told in advance exactly when the nose cone target would be fired." (*St Louis P–D*, 1/8/71)

January 9: Apollo 14 Astronauts Alan B. Shepard, Jr., Edgar D. Mitchell, and Stuart A. Roosa held press conference at MSC. Astronauts described plans for Jan. 31 lunar landing mission and said they were eager to fly on what Mitchell called "a more mature spacecraft" in which "residual risks go down." (Lannan, *W Star*, 1/11/71, A4)

- *Economist* editorial discussed lunar science investigations: "The growing importance of all this peering at, prodding, pounding, baking and growing plants upon moon soil is that the results are contradicting most of the moon theories of the past 15 years. It is not often that scientific concepts are set by the ears, and the piquancy this time lies in the way that the bulk of serious scientists originally believed that putting men on the moon was an unnecessary extravagance; instruments and unmanned probes could do the job, they felt, for a fraction of the cost and none of the risk. But, as it turns out, they could not. The instruments that were sent to the moon ahead of the astronauts did not produce the same results that the actual collection of rock samples has done, and it will be interesting to see if the Russians have done any better out of the robot shovelling of Luna 16." (*Economist*, 1/9/71)

January 10: U.S.S.R.'s *Lunokhod 1* lunar rover, energized by solar cells after surviving second lunar night on moon's Sea of Rains, completed third exploration operation, moving about 128 m (140 yds) during 4½-hr period. Vehicle had been inoperative since beginning of lunar night Dec. 23, 1970, and had been opened to receive solar energy Jan. 8. It had been released on surface by *Luna 17*, which had soft-landed on moon Nov. 17, 1970. (*SBD*, 1/11/71, 28; 1/12/71, 35)

- Newspapers reviewed *Of a Fire on the Moon* by Norman Mailer. In *New York Times Book Review* Morris Dickstein said book aimed at confrontation with "brave new world of science and technology." Mailer found himself "frustrated at every turn. The event seems packaged, distant, unapproachable. . . ." Mailer gave "rather depressed personal account of his attempt . . . to make an approach to the mission, especially to the astronauts and their machines." Mailer felt "betrayed by the gap between his romantic expectations and the gray but immense realities. Seeking to understand, he turns the book into an account of the other America, the Wasp hinterland." In end Dickstein concluded, "This is not perhaps the book on the impact of technology that we needed, but it is important nonetheless, and offers much to ponder and prey on." (*NYT Book Review*, 1/10/71, 1, 42–45)

In Washington *Sunday Star,* Day Thorpe said: "Mailer wasn't a passenger on Apollo 11, but in his motel rooms in Houston and Canaveral, and in his home in Provincetown . . . he has put together a long and excellent book about the flight, the story not only of what the adventure meant to the three astronauts who lived it but, no less interesting, what it meant to Mailer himself. The bureaucrats Mailer encountered first in hermetic Houston and then in hermetic Canaveral were unlike all other bureaucrats. They were infallibly courteous, considerate and helpful, and they invariably told the truth." Mailer showed *Apollo 11* astronauts as "three human beings, as similar as necessary for their mission, but as unlike one another as experimental prototypes should be." (W *Star*, 1/10/71, D 6)

- M/G John B. Medaris (USA, Ret.)—first Commanding General of Army Ballistics Missile Agency (ABMA) at Redstone Arsenal, Huntsville, Ala., in 1955 and head of team that launched *Explorer 1* Jan. 31, 1958—had become an Episcopal priest, *National Enquirer* reported. He was curate of Church of Good Shepherd in Maitland, Fla. (*Natl Enq*, 1/10/71)

January 11: Apollo 14 Astronauts Alan B. Shepard, Jr., Edgar D. Mitchell, and Stuart A. Roosa began preflight semiquarantine at KSC to minimize their exposure to disease or illness that could delay Jan. 31 liftoff toward moon. Astronauts would be restricted to three buildings and contacts would be limited to 160 persons, all of whom had undergone comprehensive medical examinations. (UPI, W *Star*, 1/11/71, A9)

- Nike-Cajun sounding rocket launched by NASA from Point Barrow, Alaska, carried GSFC payload to 115.9-km (72-mi) altitude to obtain temperature, pressure, density, and wind data in upper atmosphere by detonating grenades and recording their sound arrivals on ground. Rocket and instruments functioned satisfactorily, with all 19 grenades exploded and recorded as planned. Launch was part of series of similar experiments at three sites [see Jan. 7]. (NASA Rpt SRL)

January 11–14: Second annual Lunar Science Conference in Houston attracted 750 scientists who had been studying 55.8 kg (123 lbs) of lunar material gathered by *Apollo 11* and *Apollo 12* missions.

ARC team of scientists headed by Dr. Charles P. Sonett announced that first measurements of moon's interior indicated moon had basalt-like mantle 338 km (210 mi) deep surrounding "cool core of primordial olivine-like rock." Measurements suggested moon's outer layer was melted during first billion years of its 4.5-billion-yr history. Measurements had been made with ARC magnetometer placed on moon by *Apollo 12* astronauts and also made by *Explorer 35* moon-orbiting satellite launched July 19, 1967. (ARC Release 71-2)

Cornell Univ. astrophysicist Dr. Thomas Gold said stratifications found in *Apollo 12* lunar surface samples were caused by earth in same process that made near side of moon different from far side. Moon passed behind earth with face turned earthward for four days during regular monthly cycle and was bombarded by high-energy electrons in earth's magnetic field. Dr. Gold said he and coworkers had reproduced lunar surface markings inside vacuum chamber by bombarding finely ground earth material and small amounts of lunar soil with high-energy electrons up to 800 ev. Dr. Gold's theory of "electrostatic motion" was disputed at meeting by Univ. of California physicist Dr. Harold C. Urey. (Lannan, W Star, 1/12/71, A1)

NASA announced plan to continue lunar exploration with unmanned robots after last Apollo manned lunar landing in 1972. Plan would cost minimum $1 billion. Robot vehicles would travel 965 km (600 mi) across lunar surface and return to earth with samples or would rendezvous with other robot spacecraft that could return samples to earth. NASA spokesman said, "We sincerely hope that collaboration with the Russians takes place in space. A common program to explore the moon with instrumented spacecraft remotely controlled from earth might be the best way to help that collaboration take place." (O'Toole, W Post, 1/14/71, A1)

Dr. Gerald R. Taylor, MSC scientist, described laboratory experiment in which three highly resistant strains of bacteria were killed within 10 hrs of contact with one core sample from a number of centimeters below surface of moon's Sea of Tranquility. Micro-organisms in protein soup had been exposed to lunar soil.

Dr. Narenda Bhandari of Tata Institute of Fundamental Research in Bombay, India, reported that analysis of cosmic ray tracks on moon had provided "conclusive proof" of existence in lunar soil of fossils from chemical elements far heavier than any previously discovered in nature or in laboratory. If verified, discovery would greatly extend periodic table of elements. (Wilford, NYT, 1/14/71, 9)

Dr. P. Buford Price of Univ. of California at Berkeley, said during interview that *Apollo 12* lunar samples might contain first real evidence of "magnetic monopoles," basic force units that were to magnetism what electrons were to electricity. If their existence could be confirmed, it would mean rewriting books on magnetism and electrodynamics. Dr. Price's group of scientists appeared to have found two fossil tracks of nuclear type of particles which, theoretically, could have been made by monopoles in two lunar cysts. (Lannan, W Star, 1/14/71, A3)

Dr. Aleksander P. Vinogradov, Soviet academician and Director of Vernadsky Institute for Analytical Chemistry in Moscow, told conference U.S.S.R. would continue to explore moon with unmanned spacecraft and would attempt bolder missions in future. He believed U.S.S.R. would exchange lunar samples with U.S. beginning with samples of Sea of Fertility returned by Soviet *Luna 16*. He said main task for *Lunokhod 1*, Soviet moonwalker still on moon, was to determine how eight-wheeled vehicle could best move about moon and how it could best be used in future. "We discovered that the moonwalker can move in very large angles and can go down into craters at angles of 20 degrees. The mechanical qualities of this car are very good." Dr. Vinogradov said *Lunokhod 1* carried instrument to analyze soil, but device had not done so thus far. He described *Luna 16* samples as bearing no evidence of water or life on moon. Drill used to dig samples had been stopped at 330 mm (13 in) below lunar surface because it hit rock. "We were afraid to send it any deeper." Dr. Vinogradov described Sea of Fertility as "blackish powder" sprinkled with "cosmic beads." *Luna 16* sample was similar to soil brought back by *Apollo 12* from Ocean of Storms and not very different from samples returned by *Apollo 11* from Sea of Tranquility. Age of Sea of Fertility had been calculated at 4.6 billion yrs, older than some rocks returned by Apollo missions. (O'Toole, *W Post*, 1/15/71, A1)

Dr. Paul W. Gast, chief of MSC Earth and Lunar Sciences Div., reported discovery by MSC scientists of "exotic components" in lunar soil believed to be fragments from primitive crust that covered moon 4.5 billion yrs ago. Fragments differed from other Apollo lunar samples and were thought to be pieces of crustal bedrock scattered by impact of meteorites pounding more ancient lunar highlands. Discovery, through chemical analysis, supported theory that moon once had hot, molten surface and that crust had hardened as it began cooling off. Discovery also helped explain why lunar soil was mostly one billion years older than lunar rocks.

Dr. John A. Wood of Smithsonian Astrophysical Observatory described norite found in chemical analysis of five soil samples. Norite was lighter in color than most Apollo samples and did not appear to be chemically native to landing sites. Most norite represented "near-surface material in an ancient lunar crust." Coarser grained relative of norite might have come from "somewhat deeper in the moon—perhaps several miles." If discovery proved true, samples would go back in age to beginnings of solar system.

Dr. Gerald J. Wasserburg—Cal Tech physicist who had dated *Apollo 12* Sample 13, oldest rock yet found, at 4.5 billion yrs (other *Apollo 12* rocks were about 3.4 billion yrs old)—said Sample 13's oldest parts were granite that crystallized early in lunar history and could be some of original crustal material and some of "magic component" that accounted for difference in ages of lunar rocks and soil. Dr. Wasserburg believed that fact there was always some missing ingredients in lunar soil meant soil contained debris from granites and other materials originating elsewhere on moon.

Dr. Gary V. Latham, Apollo program chief seismic investigator, reported *Apollo 12* seismometer was recording weak but frequent moonquakes along rilles near landing site. He suggested rilles might be "fault zones," where slight movements in lunar surface were occurring. (Wilford, *NYT*, 1/17/71, 4:7; Gast, MSC)

January 12: U.S.S.R. launched *Cosmos 390* from Baykonur into orbit with 270-km (167.8-mi) apogee, 202-km (125.5-mi) perigee, 89.2-min period, and 65.0° inclination. Satellite reentered Jan. 25. (GSFC *SSR*, 1/31/71; *SBD*, 1/13/71, 40)

• Library of Congress published *United States and Soviet Progress in Space: Some New Contrasts*, report by Dr. Charles S. Sheldon II, Chief of Congressional Research Service Science Policy Research Div. Report summarized "how far these two major space powers have come in the last 13 years," gave "answers to frequently raised questions about the comparative aspects of the two programs," and looked at possible future developments.

Since U.S.S.R. did not disclose information on its space budget and manpower, exact comparison with U.S. could not be made. Dr. Sheldon speculated, however, that Soviet budget was about 2% of GNP—more than twice as much as U.S. percentage—and that Soviet space work force was close to 600 000 persons. Greatest number of successful launches from a single site to earth orbit and beyond between 1957 and 1970 had been 311 from Vandenberg AFB, Calif. U.S.S.R.'s Baykonur was second with 282, followed by KSC with 189 and Plesetsk, U.S.S.R., with 147. There seemed to be little difference between U.S. and U.S.S.R. space programs in general purpose and direction. Both were broadly based with elements of scientific exploration, technology development, national image building, practical applications, and military support services. Although military missions made up about two thirds of launches by both countries, number of presumptively specialized U.S. military missions had declined from 34 missions in 1966 to 16 in 1970, while U.S.S.R. total had increased from 28 missions in 1966 to 57 in 1970.

Although U.S. and U.S.S.R. were competing in space they were also cooperating. Cooperation already under way included exchange of information at scientific meetings, negotiation of treaties, trading of space-collected weather pictures, joint efforts to write textbook on space biology, coordination of efforts on geomagnetism, and negotiation on possible common designs for docking attachments to facilitate space rescue or joint projects. (Text)

- Nike-Cajun sounding rocket launched by NASA from Wallops Station carried GSFC experiment to 132.5-km (82.3-mi) altitude to obtain temperature, pressure, density, and wind data in upper atmosphere by detonating grenades and recording their sound arrivals on ground. Rocket and instruments functioned satisfactorily with 14 of 19 grenades confirmed by sound. Launch was part of series of similar experiments from three sites [see Jan. 7]. (NASA Rpt SRL)
- MSFC announced it was checking out extreme UV coronal spectroheliograph, first flight instrument to be delivered for ATM solar observatory. Instrument was one of two designed by Ball Brothers Research Corp. from NRL concepts. Second NRL flight instrument—extreme UV spectrograph—would be shipped to MSFC in March 1971. (MSFC Release 71–7)
- AFSC announced plans to launch 1500-w FRUSA (flexible rolled-up solar array) in fall 1971. Two panels would roll into cylinder 25 cm (10 in) in diameter for launch and would unroll in space to overall length of 10 m (32 ft), oriented to sun and held rigid by parallel extension arms. FRUSA would be placed in 740-km (460-mi) polar orbit by Thor-Agena booster for six-month flight to test electrical and dynamic characteristics. (AFSC Release 309.70)
- DOD announced award of $148 399 036 to Boeing Co. for full production of SRAM (short-range attack missile). Award brought total obligated to date under fixed-price-incentive-fee USAF contract to $183 599 036. (DOD Release 30–71)

January 13: U.S.S.R.'s *Lunokhod 1* unmanned lunar rover explored lunar surface for six hours, testing navigation and steering techniques. Vehicle covered 553 m (1814 ft), bringing total distance traveled since landing Nov. 17, 1971, to 2930 m (9610 ft). *Lunokhod 1* would be parked until Jan. 15 because high angle of sun during lunar noon made contrast range for TV pictures unacceptably low. (*SBD*, 1/14/71, 47)

- Calibration had been completed at MSFC on new tube wind tunnel capable of test speeds up to twice speed of sound, *Marshall Star* reported. Tube was 130 cm (52 in) in diameter and was 119 m (390 ft) long. It would be used in wind-tunnel test program for space shuttle. (*Marshall Star*, 1/13/71, 1)
- NASA launched two Nike-Cajun sounding rockets. Rocket launched from WSMR carried MSC experiment to study atmospheric composition. Mission was partial success. Rocket launched from Churchill Research Range carried GSFC grenade experiment and functioned satisfactorily. (SR list)
- Delivery of first operational automated radar terminal system (ARTS III) to new control tower at Chicago's O'Hare International Airport was announced by Secretary of Transportation John A. Volpe. Computer-based system would be operational by mid-1971. Components, when added to existing airport surveillance radars, would permit display of vital flight information on radar scopes used by air traffic controllers. Alpha-numeric data tags would be attached to related aircraft target or "blip." (FAA Release 71-5)

January 14: U.S.S.R. launched *Cosmos 391* from Plesetsk into orbit with 796-km (494.6-mi) apogee, 266-km (165.3-mi) perigee, 95.2-min period, and 70.9° inclination. Satellite reentered Feb. 21, 1972. (GSFC *SSR*, 1/31/71; 2/29/72; *SBD*, 1/15/71, 56)

January 16: Tokyo Univ. Institute of Space and Aeronautical Science launched two rockets from Uchinoura to collect data on oxygen density, electron density, and temperature in preparation for launch of Japanese satellite Feb. 12. S-210 observation rocket launched at 11:00 am local time reached 115-km (72-mi) altitude; Kappa 9-M launched at 4:45 pm local time reached 357-km (222-mi) altitude in 303 sec. (*SBD*, 1/20/71, 73)
- Soviet Academician Dr. Anatoly A. Blagonravov said in broadcast that Soviet lunar vehicle Lunokhod's safety margin and "perfect design" had "surpassed all expectations." He said at this stage of space technology man could not have stayed on moon for as long a time as Lunokhod had. In future it would be possible to assign to automatic devices such tasks as studying meteorites, exploring volcanoes, and studying radiation in near-moon space. What was more important "is that we now have an almost ideal means of conveyance on the moon, a means independent of super-rigorous conditions of vacuum and sharp changes of temperature. We can load such a selenomobile with different scientific apparatus." (*AF Mag*, 6/71, 59-9)
- Richard Hirsch, author and retired White House and NASC staff member, died of heart attack at age 58. Hirsch had been writing book on NASA in Praeger Press U.S. Government Departments and Agencies series. (*W Star*, 1/17/71, E19)

January 17: U.S.S.R.'s *Lunokhod 1* lunar rover returned to *Luna 17* landing stage, completing one of planned experiments for *Luna 17*. Tass said vehicle "turned in its tracks about half a kilometer [about one third mile] from the landing area. For some time the old track showed clearly on the television screen and the controllers . . . used it for guidance. For the first time in the history of cosmonautics there was a solution to a practical navigation problem: The return of the self-propelled apparatus to a preplanned point at a preplanned time on the surface of a different celestial body." Vehicle had traveled 3173 m (10 410 ft) since it landed on moon's Sea of Rains Nov. 17, 1970. (*SBD*, 1/19/71, 68)

- Ornithologists working with FAA had attached midget radios to migratory whistling swans to trace migration patterns and help avoid collisions with aircraft, *New York Times* reported. In joint project of Johns Hopkins Univ. and Queens Univ. in Kingston, Ontario, small aircraft pursued migrating flock between U.S. and Canada to monitor radios. FAA spokesman had said aircraft collisions with birds occurred "fairly frequently" though fatalities were "very rare." (Devlin, *NYT*, 1/17/71, 41)

January 18: Widow of Astronaut Virgil I. Grissom, one of three astronauts killed in Jan. 27, 1967, Apollo spacecraft fire, filed $10-million suit in Brevard County Circuit Court against North American Rockwell Corp. and subsidiaries North American Aviation, Inc., Rockwell Standard Corp., and Rockwell Standard Co. Mrs. Betty Grissom charged negligence in fire which killed her husband and Astronauts Roger B. Chaffee and Edward H. White II. Suit asserted spacecraft did not have proper fire extinguisher system, had no emergency egress, and was "defective" because electrical wiring permitted electrical arc to flash in cabin. (AP, *W Star*, 1/19/71, B10)

January 18-21: Dr. George M. Low, Acting NASA Administrator, and Mstislav V. Keldysh, President of Soviet Academy of Sciences, met in Moscow with representatives of other agencies to exchange views for increased U.S.-U.S.S.R. cooperation in exploration and use of outer space for peaceful purposes. At close of meetings, joint communiqué was released by Soviet Academy and U.S. Embassy in Moscow. Academy and NASA officials had found discussions "useful" and had agreed to exchange lunar surface samples. Procedures were established to produce recommendations for joint consideration of results of space research, improvement of existing weather data exchanges, research with meteorological rockets, techniques for studying natural environment, expanded exchange of data on space biology and medicine, and work in other fields. Preliminary document was initialed on Jan. 21. (NASA Release 71-9)

January 19: Apollo 14 Astronauts Alan B. Shepard, Jr., Edgar D. Mitchell, and Stuart A. Roosa completed final dress rehearsal for scheduled Jan. 31 launch toward moon. Countdown began at 9:23 am EST, with simulated liftoff on time at 3:23 pm EST. (*P Inq*, 1/20/71, 3)

- NASA announced it was requesting proposals from potential U.S. and foreign experimenters for investigations of data to be acquired from earth resources experiment package (EREP) to fly on manned Skylab spacecraft in late 1972. Data would be used to appraise value and direct applications of space observations in agriculture, geography, forestry, geology, hydrology, oceanography, and cartography. Objectives of EREP were to extend use of sensors; use man to observe, discriminate, and select study areas; and provide early source of unique research data for analysis. Three-man experimental space station would be manned by three different crews for total five months during eight-month mission in circular orbit of 435 km (270 mi) with $50°$ inclination. Astronauts would investigate solar astronomy, space medicine, space physics, bioscience, and material processing in addition to operating EREP sensors, including six high-precision, 70-mm cameras, infrared spectrometer, multispectral scanner, microwave radiometer/scatterometer and altimeter, and passive radiometer. (NASA Release 71-5)

- Return of *Apollo 11* Astronaut Edwin E. Aldrin, Jr., to active service with USAF in July 1971 was announced by Gen. John D. Ryan, USAF Chief of Staff. Col. Aldrin would assume command of AFSC's Aerospace Research Pilot School at Edwards AFB, Calif. He had joined NASA in January 1964 as astronaut trainee after serving with USAF in Germany and Korea. He held B.S. degree from U.S. Military Academy and Sc.D. degree from MIT. (NASA Special Release; DOD Release 45-71)

January 20: U.S.S.R. launched *Meteor 7* meteorological satellite from Plesetsk into orbit with 655-km (407.0-mi) apogee, 629-km (390.8-mi) perigee, 97.5-min period, and 81.2° inclination. (GSFC *SSR*, 1/31/71; *SBD*, 1/26/71, 96)

- President Nixon announced intention to nominate James H. Wakelin, Jr., as Assistant Secretary of Commerce for Science and Technology succeeding Myron Tribus, who had resigned in November. Nomination was submitted to Senate Jan. 25. (*PD*, 1/25/71, 84; 2/1/71, 141)

- U.S. Geological Survey published *The National Atlas of the United States of America,* first official national atlas produced in U.S. Preface noted that revised editions might be "greatly expedited" by "repetitive coverage at short time-lapse intervals, possibly from orbiting satellite systems." Systems might "expedite the production of more up-to-date and less costly regional and state as well as national atlases or even make it possible to obtain directly from storage data banks and electronic computers the type of information now gained in part from maps and atlases." (DOI PIO; Text)

January 21: USAF launched unidentified satellite by Titan IIIB-Agena booster from Vandenberg AFB into orbit with 244-mi (392.7-km) apogee, 130.4-km (81-mi) perigee, 89.2-min period, and 110.8° inclination. Satellite reentered Feb. 9. (Pres Rpt 72)

- *Cosmos 392* was launched by U.S.S.R. into orbit with 276-km (171.5-mi) apogee, 203-km (126.1-mi) perigee, 89.2-min period, and 64.9° inclination. Satellite reentered Feb. 2. (GSFC *SSR*, 1/31/71; 2/28/71)

- USAF X-24A lifting-body vehicle, piloted by NASA test pilot John A. Manke in joint program, completed 21st flight from FRC. Objectives of powered flight were to expand flight envelope to mach 1.5, determine lateral-directional derivatives, and obtain longitudinal trim and lift-to-drag data with 40° upper flap at 0° rudder bias. Two of four engines were shut down because of sticky angle-of-attack indicator; alternate two-engine flight plan was followed and completed successfully. (NASA Proj Off)

- ATM acceptance checkout at MSFC was nearing completion and ATM would be ready for prototype acceptance testing in early March, MSFC announced. ATM would fly on Skylab cluster in 1972. (MSFC Release 71-11)

- Grumman Aerospace Corp. confirmed that investigators of Dec. 30, 1970, crash of first prototype F-14A Tomcat fighter aircraft had found probable cause to be crack in titanium hydraulic lines caused by vibrations in hydraulic pumps. Grumman and USN investigators had recom-

mended replacing titanium hydraulic lines with heavier stainless steel and installing mufflers on hydraulic pumps. (*Newsday*, 1/22/71, 15)
- AFSC announced award of $8.2-million contract to Bendix Corp. Instrument and Life Support Div. for 7053 improved aircraft altimeters. Improved altimeters—to be delivered to USAF and USN beginning in early 1972—had two operating modes and plus-minus accuracy factor of no greater than 76 m (250 ft). Some previously used systems had errors in excess of 305 m (1000 ft). (AFSC Release 5.71)
- President Nixon nominated R/A Don A. Jones as Director of National Ocean Survey of NOAA and R/A Harley D. Nygren as Director of the Commissioned Officer Corps of NOAA. Both had held these positions in acting capacity since October 1970. (*PD*, 1/25/70, 85)

January 21–22: Mockup of 12-man space station concept being studied by McDonnell Douglas Astronautics Co. for NASA was inspected by 100 Government and industry representatives attending quarterly review at MSFC. Mockup was 10 m (33 ft) in diameter and 15 m (50 ft) tall, with four decks and large simulated power section. Concept called for tunnel section in center with 3-m (10-ft) diameter. (MSFC Release 71–13; MSFC PIO)

January 22: NASA announced reorganization of Office of Manned Space Flight to reflect new management requirements for future manned flight programs. Reorganization established two new technical support offices, Engineering and Operations Office and Quality and Safety Office; established two functional offices, Administration Office and Program Analysis Office; and abolished Mission Operations, Field Center Development, Institutional Operations, and Space Medicine organizations.

M/G John D. Stevenson (USAF, Ret.), Director of Mission Operations, became Deputy Director of Engineering and Operations. Capt. Robert F. Freitag (USN, Ret.), Director of Field Center Development, became Special Assistant to Associate Administrator for Manned Space Flight. Space Medicine Office had been combined with other life science activities under Dr. James W. Humphreys, Director of Life Sciences, as announced Dec. 3, 1970. (NASA Release 71–10)
- New world distance record in heavyweight turboprop class was claimed by USN for P–3C Orion aircraft piloted 11 280 km (7010 mi) by Cmdr. Donald H. Lilienthal in nonstop flight from Atsugi Naval Air Station, Japan, to Patuxent Naval Air Station, Md., in 15 hrs 21 min. Previous record of 7662 km (4761 mi) had been set by Soviet Il–18 aircraft in 1967. (DOD Newsfilm Release 60–71)
- Plastic-coated solar cells were being developed by LeRC and Lockheed Missiles & Space Co. scientists to reduce cost of solar cell power systems. Cells, which produced abundant power when formed into panels or arrays, were lightweight, flexible, insulated, completely sealed, and resistant to breakage. (LeRC Release 71–5)
- Nike-Tomahawk sounding rocket was launched by NASA from Andoeya, Norway, carrying Norwegian experiment for auroral studies. Rocket and instruments functioned satisfactorily. (SR list)
- *Apollo 11* and *Apollo 12* lunar rock groups differed in potassium-uranium abundance systematics, JPL scientists Dr. Fraser P. Fanale and Douglas B. Nash reported in *Science*. Difference indicated "that relatively little exchange of regolith material has occurred between

Mare Tranquillitatis [Sea of Tranquility] and Oceanus Procellarum [Ocean of Storms]. Two suites appear to have been derived from materials of identical potassium and uranium content." It appeared unlikely "that bulk lunar material has the ratio of potassium to uranium found in chondrites. However, systematic differences in the potassium-uranium ratio between Apollo samples and crustal rocks of the earth do not preclude a common potassium-uranium ratio for bulk earth and lunar material." (*Science*, 1/22/71, 282–4)

- NSF released report on industrial R&D spending in 1969. Funds had totaled $18.4 billion, 6% above 1968 level and about 50% more than $12.6 billion in 1963. Growth was 1% from 1968 to 1969. Deflated 1969 level was 22% above that of 1963. Industrial firms accounted for about 70% of R&D in U.S. Federal financing of R&D in industry amounted to $8.6 billion in 1969, same as in 1968. Although Federal support was primarily responsible for increases in industrial R&D between 1953 and 1966, funds had leveled off, rising less than 3% between 1966 and 1969. In real dollars, Federal support showed a decline. (NSF Highlights, 1/22/71, 1)
- President Nixon announced intention to nominate Thomas E. Carroll to be Assistant Administrator for Planning and Management of Environmental Protection Agency. Nomination was submitted to Senate Jan. 25. (*PD*, 1/25/70, 88)
- Harry F. Guggenheim—financier, philanthropist, sportsman, and publisher—died in Sands Point, N.Y., at age 80. His enthusiasm for aviation as pilot in World War I had led him to persuade his father, Daniel Guggenheim, to establish Guggenheim School of Aeronautics at New York Univ., first university school of aeronautics in U.S.

    Daniel Guggenheim Fund for the Promotion of Aeronautics, Inc., of which Harry Guggenheim was president, helped establish aeronautical engineering schools at MIT, Georgia Institute of Technology, Cal Tech, Univ. of Washington, Stanford Univ., and Univ. of Michigan. Fund aided Charles A. Lindberg after first transatlantic crossing. Daniel and Florence Guggenheim Foundation, established in 1924, sponsored much of pioneer rocket research by Dr. Robert H. Goddard. (*NYT*, 1/23/71, 1)

January 23: Apollo 14 backup pilot Eugene A. Cernan narrowly escaped death when helicopter in which he was practicing lunar landing maneuvers crashed and burned in Indian River near KSC. After leaping from helicopter and swimming toward shore Cernan was rescued and taken to Patrick AFB for treatment of minor cuts and bruises and singed eyelids and eyebrows. (UPI, *W Post*, 1/24/71, A5)

- Baltimore *Sun* editorial praised U.S.–U.S.S.R. agreement to exchange lunar samples: "The scientists and engineers are unmistakably pointing the way which soon or late must be taken if men are to make the most of opportunities offered them by the opening of vast new extraterrestrial horizons of inquiry and adventure." (*B Sun*, 1/23/71)

January 24: Aerobee 150 sounding rocket was launched by NASA from WSMR carrying GSFC experiment to study stellar spectra. Rocket and instruments functioned satisfactorily. (SR list)

- Lovell Lawrence, Jr., chief research engineer for Chrysler Corp. since 1964, died at age 55. Lawrence had been President of American Rocket Society, recipient of 1950 Goddard Memorial Lecture Award,

President of Reaction Motors, Inc., and key figure in development of of USA Redstone ballistic missile. (*NYT*, 1/25/71, 39)

January 25–26: *Intelsat-IV F–2* comsat was launched by NASA for ComSatCorp on behalf of INTELSAT. Satellite, launched from ETR at 7:36 pm EST by Atlas-Centaur booster, entered elliptical transfer orbit. Primary objective was to place satellite into transfer orbit accurate enough for onboard propulsion systems to place spacecraft in planned synchronous orbit.

Apogee kick motor was fired at 10:44 pm EST, Jan. 26, about 11 hrs earlier than planned because of slight temperature fluctuations in apogee motor shaft. *Intelsat-IV F–2* entered orbit with 36 410.4 km (22 624.4 mi) apogee, 35 740.0 km (22 207.8 mi) perigee, 1450-min period, and 0.59° inclination. Satellite would drift from position at 124° east longitude over Pacific to permanent station over Atlantic at 25.5° west longitude by about March 19.

Intelsat-IV F–2 was first in Intelsat IV series of advanced comsats. Satellite—largest commercial comsat launched—was 5.4 m (17.6 ft) high with 237.5-cm (93.5-in) diameter and weighed 1397 kg (3080 lbs) at launch. It had 12 transponders, providing 12 TV channels and 3000–9000 telephone circuits, and was capable of multiple-access and simultaneous transmissions. Expected lifetime was seven years—two years more than Intelsat III comsats and over five years more than *Intelsat I*. (NASA Proj Off; ComSatCorp Release 71–4; ComSatCorp PIO)

January 25: Dr. Charles A. Berry, Director of Medical Research and Operations at MSC, directed four-hour physical examination of Apollo 14 astronauts and declared them "certified and ready to fly."

Countdown for launch began on schedule at 9:00 am EST in preparation for liftoff toward moon at 3:23 pm EST Jan. 31. (AP, B *Sun*, 1/26/71, A5)

- NASA launched four sounding rockets. Two Nike-Apaches, launched from Churchill Research Range, carried GSFC experiments to study energetic particles and fields. Aerobee 150, launched from WSMR, carried Johns Hopkins Univ. experiment to study Venus UV. Aerobee 170, launched from WSMR, carried Johns Hopkins Univ. experiment to study dawn airglow. All rockets and experiments functioned satisfactorily. (SR list)

- Dr. Robert R. Gilruth, MSC Director, named to five-man board chaired by Astronaut James A. Lovell, Jr., to investigate Jan. 23 crash of NASA helicopter piloted by Astronaut Eugene A. Cernan. (MSC Release 71–03)

- *Newsweek* commented on "shrinking space shuttle." U.S. space program had been "hit with another budgetary broadside." Space shuttle to service planned space station "will get no more than a third of the money its designers insist they need to build it." Original request for $300 million had been cut by NASA to $225 million. "Now White House budgeteers have shrunk it to $105 million—and it has yet to face the anti-technology faction in Congress." (*Newsweek*, 1/25/71)

- Steel stress plate 2.5 cm (1 in) thick and 61 cm (2 ft) wide on F–111A fighter cracked during ground testing at McClellan AFB, Calif. Later USAF reported crack came at subzero temperatures under loads equal to 60% of maximum load aircraft was designed to withstand. (*W Post,* 1/30/71, A4)

January 25–27: Ninth annual AIAA Aerospace Sciences Meeting was held in New York. Dr. Coleman duP. Donaldson, President of Aeronautical Research Associates, delivered Hugh Dryden Research Lecture, "Calculation of Turbulent Shear Flows for Atmospheric and Vortex Motions."

Sylvanus Albert Reed Award was presented to Ira G. Hedrick, Vice President of Engineering for Grumman Aerospace Corp., for technical excellence, design innovation and leadership in the structural development of several generations of aerospace vehicles."

Space Science Award was presented to Univ. of California physicist William I. Axford for "outstanding contributions to the fluid dynamics of the sun-earth environment and to the theory of cosmic ray interaction with the solar wind."

Robert M. Losey Award went to Verner E. Suomi of Univ. of Wisconsin for "creativity and ingenuity in designing advanced meteorological sensors for satellite applications as exemplified by the Spin-Scan camera which has made it possible to view the earth's atmosphere as an entity."

G. Edward Pendray Award was presented to Dr. Nicholas J. Hoff of Stanford Univ. for "outstanding contributions in the fields of structures and structural mechanics through his many books and publications, as well as direct consultation for government agencies and industry."

Martin Goland, President of Southwest Research Institute, was named new President of AIAA. (AIAA Release, 1/8/71, 1/27/71)

January 26: U.S.S.R.'s *Venus 7* (launched Aug. 17, 1970) had landed intact on planet Venus Dec. 15 and "relayed scientific information from the surface of another planet for the first time," Tass announced. Spacecraft, equipped with "highly stable frequency generators," transmitted signals for 23 min, recording surface temperature at 748 K ± 20° (475°C ± 20°) and pressure, at 90 ± 15 atmospheres. Tass said continuous measurements of the atmospheric parameters were made all the way to surface and it was established "that Venus possesses an unusually hot atmosphere the density of which, at the surface, is about 60 times greater than the density of the atmosphere on the earth's surface." (*Sov Rpt,* 3/2/71, 4; Gwertzman, *NYT,* 1/27/71, 1)

- U.S.S.R. launched *Cosmos 393* into orbit with 417-km (259.1-mi) apogee, 281-km (174.6-mi) perigee, 91.3-min period, and 70.9° inclination. Satellite reentered June 16. (GSFC *SSR,* 1/31/71; 6/30/71)

- International aspects of U.S. space program were discussed by Dr. George M. Low, Acting NASA Administrator, in luncheon speech before National Space Club in Washington, D.C.: Cooperation with Western nations in past decade had included 70 countries, 250 specific agreements, and more than 24 joint satellite projects at run-out cost in excess of $400 million. More than 50% of total had been funded by other nations. In 1970s, "opportunities exist for cooperation on a significantly larger scale." U.S. wished "to build a foundation for important benefit- and cost-sharing in the major space programs of the future." Canada, Japan, and Australia—"countries having the most obvious potential for post-Apollo work-sharing"—had been given "every chance to become fully acquainted with our plans and studies" to enable them to decide whether they wished to commit their own resources.

European Space Conference had put several million dollars into studies of post-Apollo program possibilities and some member countries had invested still larger amounts. British, French, and West German firms were working with NASA prime contractors on space shuttle design studies. Europe would have to choose between significant participation in shuttle funding and funding independent European rocket program. Decision would be necessary "even if Europe funds only 10 per cent of the shuttle's price tab." U.S. had said it would "sell launch services for projects consistent with peaceful purposes and existing international agreements . . . [and] that general technical access to the entire program would be available, but that technology at the level of commercial know-how would be transferred in either direction only where one side required it to complete its commitments to the other." Since Europe was talking of only 10% share of shuttle program, "we said that we would retain decision-making responsibility except where European costs were directly affected, in which case decisions would be joint."

In U.S.–U.S.S.R. cooperation in space, "it is particularly important that we give clear recognition to *both* . . . competition and cooperation." U.S. was competing with U.S.S.R. "because accomplishments in space represent a measure of the state of our technology; and in today's world, the level of our technology is of first importance, strategically, economically, and politically." U.S. was cooperating with U.S.S.R. "because we both live in a vast universe that must be explored, where important new knowledge is to be gained, for the benefit of all men, everywhere."

While U.S. retained demonstrated lead over U.S.S.R., "it will not be an enduring lead without major new initiatives on our part—initiatives like the space shuttle and the exploration of the outer planets. . . ."

In Jan. 18–21 Moscow talks with Soviet Academy of Sciences, "we specifically agreed, initially, to exchange three grams [0.11 oz] of sample obtained from Luna 16, for three grams each from Apollo 11 and 12. These relatively small amounts are sufficient for detailed scientific examination of a comparative nature."

Additional meetings on compatible docking arrangements had not been held because "docking discussions are well underway, and the planned exchange of information is taking place."

On visit to Soviet cosmonaut training area at Star City, "we were welcomed by Cosmonauts Beregovoy, Nikolayev, Leonov, Shatalov, and Valentina Tereshkova. We were shown the Soyuz mission trainer, and were invited to attempt to dock two Soyuz spacecraft in their docking trainer. While there, I also presented a plaque in memory of Yuri Gagarin [first man in space] from the U.S. astronauts." (Text)

- President Nixon announced appointments of members of President's Science Advisory Committee for terms expiring Dec. 31, 1974: Dr. Lee A. DuBridge, former Presidential Science Adviser; Dr. Herbert Friedman of NRL; Dr. Daniel P. Moynahan, former Counsellor to the President; Kenneth H. Olsen, President of Digital Equipment Corp.;

and Dr. John G. Truxal of Brooklyn Polytechnic Institute. (*PD*, 2/1/71, 120)
- Sen. Edward M. Kennedy (D-Mass.) introduced S. 302 to authorize National Park Service acquisition of historic site in Auburn, Mass., where first liquid-fuel rocket flight was launched by Robert H. Goddard in 1926. Bill was referred to Committee on Interior and Insular Affairs. (*CR*, 1/26/71, S222)
- *Philadelphia Inquirer* editorial commented on Apollo 14: "With $400 million of taxpayers' money behind it and the dubious future of manned spaceflight depending on it, the Apollo 14 mission to the moon . . . is grimly earnest business." If mission was successful there would be "another Apollo shot in midsummer, two more next year." If it was not, there would be "a whale of a battle in Congress and elsewhere about appropriations and the wisdom of the whole idea." (*P Inq*, 1/26/71)
- Sylvia Porter analyzed Federal budget in Washington *Evening Star* to show part individual taxpayers played in financing Federal projects. Only small segment of tax dollars was taken by "controversial" programs. If "you eliminated space entirely, you'd have only 2c more left; if you cut out foreign aid entirely, this would give you only another 2c extra." (*W Star*, 1/26/71, A11)

January 26–28: House Committee on Science and Astronautics held 12th annual seminar with Panel on Science and Technology to discuss international science policy. In keynote address Secretary of State William P. Rogers said: "Our basic goal is to put science and technology at the service of human—and humane—ends." U.S. was increasing "emphasis on science and technology in our aid to developing countries. We are encouraging an international effort to preserve the quality of the world's environment. And we are seeking greater international cooperation to enhance the benefits of technology and to curb its dangers. . . . We now want to . . . make the exploration of space a truly international endeavor. Space should not be the exclusive preserve of a small number of countries. Our post-Apollo program aims at the development on a multilateral basis of a new generation of reusable space vehicles designed to make the exploration and use of space easier, and more economical." (Text; Transcript)

January 27: Apollo 14 astronauts completed test of ability to react to inflight emergencies in KSC simulators. During four-hour session, engineers simulated malfunctions that could occur during flight and monitored astronaut reactions. (AP, *B Sun*, 1/28/71, A10)
- USN P-3C Orion turboprop antisubmarine aircraft established world record for nonrefueled, long-distance flight for heavyweight turboprop aircraft. P-3C Orion, piloted by Cdr. Donald H. Lilienthal (USN), attained speed of 808 km per hr (502 mph). Previous record of 727 km per hr (452 mph) had been established by Soviet Il-18 in May 1968. (*AF Mag*, 4/71, 15)
- President Nixon announced recipients of 1970 National Medal of Science, Federal Government's highest award for distinguished achievement in science, mathematics, and engineering. Recipients included Dr. George E. Mueller, Senior Vice President of General Dynamics Corp. and former NASA Associate Administrator for Manned Space Flight. Dr. Mueller received award for "his many individual contributions to the design of the Apollo System, including the planning and

interpretation of a large array of advanced experiments necessary to insure the success of this venture into a new and little known environment."

Other recipients were: Dr. Richard D. Brauer, professor of mathematics, Harvard Univ.; Dr. Robert H. Dicke, professor of physics, Princeton Univ.; Barbara McClintock, Carnegie Institution of Washington geneticist; Dr. Albert B. Sabin, president of Weizmann Institute of Science in Rehovoth, Israel; Dr. Allan R. Sandage, Cal Tech astronomer; John C. Slater, professor of physics and chemistry, Univ. of Florida; Dr. John A. Wheeler, professor of physics, Princeton Univ.; and Dr. Saul Winstein, deceased, formerly professor of chemistry at UCLA. (PD, 2/1/71, 124)

- Dr. Edward E. David, Jr., OST Director and President's Science Adviser, addressed AIAA luncheon held in conjunction with House Committee on Science and Astronautics' 1971 international science panel: "One of the key challenges facing this nation, and other nations as well, is finding means by which technology can be advanced, shared, and applied to common problems without restricting our individual national ability to compete for markets with the products of this technology." This could be done by application of comsats for global point-to-point communication. "Here we have seen the initial research blossom into a demonstrated capability and then into an international cooperative venture." Intelsat story "indicates that we should have no illusion about the readiness of nations to give up competitive advantages without compensating return. At the same time it is fair to say that we have seen there a willingness to accept indirect as well as direct benefits . . . such things as enhanced international security resulting from international stability, an enriched quality of life, and better understanding of man as a biological entity. This is a hopeful sign and it is a point of view which I think needs to be emphasized and cultivated." (CR, 2/1/71, E320-3)

- National Religious Broadcasters, at annual convention in Washington, D.C., celebrated 50th anniversary of first religious broadcast on commercially licensed station by beaming their closing program around world in first international religious broadcast transmitted by satellite. (Janson, NYT, 1/28/71, 1)

- USN plans to maintain offensive capability without exorbitant investment were described by Adm. Elmo R. Zumwalt, Jr., Chief of Naval Operations, in speech before Cleveland, Ohio, chapter of American Ordnance Assn.: "We intend to develop small, high speed platforms. These will include escorts having a hard punch but which do not individually constitute a high percentage of our total capability. These will probably be...hydrofoils, air cushion platforms and surface effect ships. We have placed a very high priority on the surface effect ship on the premise that this offers the potential for destroyer-size ships to conduct anti-submarine warfare on top of the water at speeds 2 or 3 times greater than the fastest submarines." With addition of V/STOL aircraft, "one can envision a fleet of highly mobile craft which would be a potent striking force when concentrated, or a number of unique and flexible weapons systems when operating as individual units. By supplementing this force with small high speed craft as missile-launching platforms we will add a new dimension to our present powerful aircraft carriers, frigates, destroyers, and guided missile ships." (Text)

January 28: NASA launched two Nike-Apache sounding rockets from TERLS carrying Indian experiments to study ionosphere. Rockets and instruments functioned satisfactorily. (SR list)

NASA announced award to Lockheed Electronics Co. of $800 000, cost-plus-fixed-fee contract for scientific and technical support services for MTF Earth Resources Laboratory. Contract covered one year with provision for two one-year extensions. Company would support laboratory in research on application of remote-sensing data to Mississippi-Louisiana-Gulf area. Data would be obtained from aircraft, ERTS, Skylab, Gemini, and Apollo experiments. (NASA Release 71-13)

- ESRO awarded six-month study contracts for European regional comsat to three European consortiums—MESH (S.A. Engins M.A.T.R.A. of France, ERNO-Raumfahrttechnik GmbH of West Germany, SAAB Aktiebolag of Sweden, and Hawker Siddeley Dynamics Ltd. of U.K.), STAR (Satellites for Telecommunications, Applications, and Research), and COSMOS, led by Marconi Space and Defence Systems in U.K. (*SF*, 4/71, 121)
- Senate adopted S.R. 15, modifying Senate rules on total number of Senators assigned to certain standing committees. Committee on Aeronautical and Space Sciences was set at 11 members.

    Senate also adopted S.R. 16, electing majority members of Senate Committees. Members elected to Committee on Aeronautical and Space Sciences were Sen. Clinton P. Anderson (D-N. Mex.), Chairman; Sen. Warren G. Magnuson (D-Wash.); Sen. Stuart Symington (D-Mo.); Sen. John C. Stennis (D-Miss.); and Sen. Howard W. Cannon (D-Nev.). (*CR*, 1/28/71, S406)
- President Nixon submitted to Senate nominations of Robert M. White as Administrator of NOAA, Howard W. Pollock as Deputy Administrator, and John W. Townsend, Jr., as Associate Administrator. White had served as Acting Administrator of NOAA since his appointment as Acting Deputy Administrator Oct. 9, 1970. Before establishment of NOAA, he was Administrator of ESSA. (*PD*, 2/1/71, 127, 141)
- Award of $23 800 contract to American Airlines to determine feasibility of using STOL aircraft for short-haul runs was announced by Secretary of Transportation John A. Volpe. Six-month study would cover certification and safety, avionics, STOL service categories, system development requirements, user and community acceptance criteria, metroflight feasibility demonstration, New York area STOLport sites, and STOLport acceptance. (DOT Release 1771)
- European Organization for Nuclear Research announced in Geneva that it had effected first head-on collision between intense beams of high-energy protons. Beams of protons were injected into intersecting storage rings by 28-bev particle accelerator at nearly speed of light. Collisions occurred at intersection points of two rings in which very high vacuum was maintained. Experiment would enable scientists to study fundamental reactions that took place in interior of stars or in upper atmosphere. (*NYT*, 1/29/71, 7)
- *New York Times* editorial commented: "The important steps recently taken toward great Soviet-American space cooperation are still small, but they provide a beginning for a much wider and mutually advan-

tageous effort. This recent progress also suggests that there is now a political interest in cooperation in Moscow as well as in Washington, a welcome change from the situation in earlier years." (*NYT*, 1/28/71)

- *Christian Science Monitor* editorial praised Soviet *Venus 7* softlanding on Venus, "another major first in the conquest of space." Apollo 14 would be "bigger attention-getter" because there was "more glamour and excitement in manned space flight." But U.S.S.R. had achieved "remarkable technological performances." U.S.S.R. could point "with justifiable pride" to fact that, while *Venus 7* was transmitting signals from Venus to earth, moonrover *Lunokhod 1* was also sending information from moon. (*CSM*, 1/28/71)

January 29: President Nixon in message transmitting FY 1972 budget to Congress said: "Science and technology can make major contributions to the public by developing creative solutions to a wide range of national problems.... I am proposing in this budget a substantial expansion in outlays for federally supported research and development." President recommended increased funding to "make greater use of our scientific and engineering capabilities and resources to cope with ... pollution, crime, health, transportation, and other environmental and social problems; strengthen research essential to the advancement of our technology and economic productivity; and increase our investment in fundamental science which leads to the long-term progress of our society." President recommended increase in NSF budget from $506 million in 1971 to $622 million in 1972.

In 1972, defense spending would increase but, even with increase, would "drop from 36% of total spending in 1971 to 34% in 1972." Human resources programs' share would be 42% of total. Budget proposed "sharing Federal revenues with States and communities," "national strategies to improve health care" and to ensure "an income floor for every family," and efforts toward "full-employment in peacetime with relative price stability." (*PD*, 2/1/71, 130–9)

- President Nixon sent $229.2-billion FY 1972 budget request to Congress, including total science R&D budget of $16.7 billion. Total request for NASA new obligational authority (NOA) of $3.271 billion (1.4% of total U.S. budget) was $27 million less than FY 1971 NOA of $3.298 billion. NASA expenditures were budgeted to decline $216.5 million against decline of $486.4 million in FY 1971. Reduction to $3.152 billion placed proposed NASA FY 1972 funding at lowest level since 1962. Of budget request, $2.518 billion would go for R&D, $56.3 million for construction of facilities, and $697.4 million for research and program management.

Apollo funding decrease of $302.2 million, to $612.2 million—reflecting cancellation of Apollo 18 and 19—would be partially offset by increase from $515.2 million in FY 1971 to $672.8 million in FY 1972 for manned space flight operations. Manned space flight operations would include $535.4 million for Skylab, $100.0 million for space shuttle, and $37.4 million for orbital systems and experiments. Advanced missions would receive $1.5 million, to bring total for manned space flight (including Apollo) to $1.286 billion, down $144.6 million from FY 1971.

Funding for space science and applications programs would increase $184.7 million, from $565.7 million in FY 1971 to $740.4 million in FY 1972. Increases would go to physics and astronomy program for solar observatory ($19 million in FY 1972), High Energy Astronomical Observatory ($13.4 million), and space telescope study ($500 000). Funding for lunar and planetary exploration would increase $166.6 million, to $311.5 million, with $38.3 million going toward Mariner-Venus/Mercury 1973, $180.4 million toward Viking project, and $18.8 million toward supporting research and technology advanced studies. Funding for Pioneer program of planetary exploration would decrease from $35.5 million in FY 1971 to $17.6 million. Funding for bioscience program, including Biosatellites, was eliminated from NASA budget for FY 1972. Increase of $15.5 million in space applications programs, to $182.5 million in FY 1972, included $48.5 million for Earth Resources Survey, $13 million for Synchronous Meteorological Satellite, and $60.3 million for ATS.

Total for advanced research and technology programs decreased from $264.2 million in FY 1971 to $212.8 million. STOL technology program would receive $15 million, against $1.9 million in FY 1971. NERVA rocket funding would decrease from $32 million in FY 1971 to $9.9 million.

Tracking and data acquisition funding would fall $26 million, to $264 million.

DOD's FY 1972 budget of $76 billion for outlays included $1.55-billion space funding, $500 000 less than previous year [see April 2]. Budget also included $370 million for B-1 intercontinental bomber, up from $75 million in FY 1971; $39 million for LAMPS antisubmarine-warfare helicopter, up from $4 million, $145 million for air defense aircraft system (AWACS), up from $87 million, $47 million for USAF AX close-support aircraft, up from $28 million, $409 million for Poseidon missile system, and $926 million for Minuteman III program, up from $720 million in FY 1971.

AEC funding for NERVA and space electric power would decrease from $84 million in FY 1971 to $39.7 million.

Increase of $437 million in total DOT FY 1972 budget of $7.8 billion would go mostly to expanded aviation and urban mass transit spending. DOT funding for reduction of aircraft noise and study of environmental effects of SST would increase from $27 million in FY 1971 to $54 million. These funds also would be used for pollution reduction from Coast Guard installations. NOA of $235 million was requested for SST prototype program, up from $210 million in FY 1971. (US Budget for FY 1972; BOB Special Analysis; NASA budget briefing transcript; DOD budget briefing transcript; AIAA release 71-2; *CR*, 1/29/71, H265; DDR&E Dir Foster Testimony, 4/2/71)

- NASA released Jan. 28 briefing on FY 1972 budget request, in which Dr. George M. Low, NASA Acting Administrator, said: "For the past several years we have reported ... a further decrease in budget authority. This year we can report a clear halt in this trend: the budget authority for FY 1972 is at the same level as the funds appropriated for FY 1971. The exact figure for FY 1972 is $3.271 billion.

NASA would complete Apollo program with four previously planned missions but would delay Apollo 17 five months to ensure that mission carried experiments originally scheduled for canceled Apollo 18 and

19. Skylab missions would be delayed about four months, to follow as soon as practical after last Apollo flight. Plans for unmanned planetary exploration included two Mars orbiters in 1971, Jupiter flybys in 1972 and 1973, launch in 1973 for Venus-Mercury flyby in 1974, and two Viking launches in 1975 to land on Mars in 1976.

Dr. Low said NASA had asked for $190 million for space shuttle in FY 1972. "We have included in our budget $100 million. This reduction means essentially that we will move out with the engine development exactly as we had planned; that we are still in a position to make a decision after we have completed the studies on the air frame development, whether to continue with additional design or whether to proceed with the development of the air frames."

Near-earth science programs included Orbiting Solar Observatories in 1971, 1973, 1974, and 1975 and last orbiting Astronomical Observatory in 1972.

Space applications program would progress as planned, including two ERTS, two Nimbus experimental weather satellites, two Synchronous Weather Satellites, and two advanced communications experiments with ATS-F and ATS-G.

New FY 1972 projects would include start of work on High Energy Astronomy Observatory (HEAO) and program for exploration of outer planets. "To do this we will develop a versatile new spacecraft, and we will plan Grand Tour missions to Jupiter, Saturn, and Pluto in 1976–1977; and to Jupiter, Uranus, and Neptune, in 1979." NASA also would study mission to orbit Jupiter as alternative to first Grand Tour.

Major new initiative in aeronautics in FY 1972 would be development of experimental STOL aircraft.

Downward trend in aerospace industry employment on NASA programs would be halted during FY 1972. "Although aerospace jobs will continue to decline in FY 1971, we expect employment to start increasing by the middle of FY 1972, with the end-of-year level being about equal to that at the beginning of the year."

NASA would reduce civil service work force by 1500 positions by end of FY 1972, bringing total decrease to 6800 since July 1967.

Work on NERVA nuclear rocket engine would be limited to "essentially long-lead time items," preserving ability "to move forward ... when the need arises," without large expenditures in FY 1972.

Program included "no plans at this time for a manned Mars landing mission." (Transcript)

- NASA launched two sounding rockets to obtain data on ionosphere. Nike-Apache, launched from Univ. of Michigan's Keweenaw Rocket Launch Site (KRLS), carried GSFC payload to 121.9-km (76.2-mi) altitude to obtain data during magnetic activity in ionosphere while radiofrequency absorption in 2- and 10-mhz region of spectrum was high. Rocket and instruments functioned satisfactorily.

  Boosted Arcas II, launched from Wallops Station, carried GSFC experiment to 122-km (75.8-mi) altitude to study enhanced ionization of lower ionosphere and its relationship to precipitation of energetic electrons. One telemetry channel with one RF probe was lost; all other instruments functioned satisfactorily. (NASA Rpts SRL)

- Lunar landing trainer crashed and exploded at Ellington AFB, Tex. NASA pilot Stuart M. Present parachuted to safety. Trainer was one of two $1.9-million craft used by Apollo 14 commander Alan B. Shepard, Jr., in training. Preliminary investigation indicated cause of crash was loss of electrical power. MSC Deputy Director Christopher C. Kraft, Jr., later said accident should not affect Apollo 14 mission since LLTV electrical system was "totally different from the one in the lunar module." (MSC Release 71-05)

- Selected fire departments across U.S. would be testing new fire-resistant outfits developed by MSC for use by NASA rescue crews, NASA announced. Prototype outfit had been displayed at symposium of International Assn. of Fire Fighters. Clothing included thermal underwear, coverall, chaps, two kinds of trousers, two jackets, and two proximity suits that permitted fire fighters to move close to fire or even enter flames. Nonflammable materials in garments had been developed to ensure maximum safety of Apollo crews in oxygen-rich atmosphere. (NASA Release 71-12)

- Micrometer-sized silicate grains from all depths in *Apollo 11* and *Apollo 12* lunar core samples showed unusually high density of nuclear tracks, Univ. of California physicists reported in *Science*. Densities exceeding 100 billion per sq cm in several percent of grains indicated possibility that grains had been irradiated in space as extralunar dust orbiting sun before being caught by moon to build lunar soil—a possible extralunar origin of part of lunar soil. (Barber, Hutcheon, Price, *Science*, 1/29/71, 372-4)

- Evidence obtained from UV photometers aboard two NRL Aerobee 150 sounding rockets—one launched from White Sands into night sky in August 1967 and one Oct. 13, 1969—had indicated earth was immersed in magnetoglow, NRL scientists reported in *Science*. Glow, from large volume of glowing helium ions, at 304 Å, was similar to geocoronal hydrogen glow "in that it extends to very high altitudes, but it is unique in that it is largely confined to the closed field line portion of the magnetosphere." Because of magnetic containment, radiation had been termed "magnetoglow." Observations of magnetoglow from inside and outside magnetosphere cavity promised to provide "valuable means of studying the structural dynamics of the magnetosphere." (Johnson, Young, Holmes, *Science*, 1/29/71, 379-81; NRL PAO)

- European Broadcasting Union said in Geneva that it would provide live TV coverage of Apollo 14 for 21 countries in Western Europe and North Africa and coverage for TV stations in U.S.S.R. and Eastern Europe. Fourteen multilateral broadcasts would be transmitted via *Intelsat-III F-6* to earth station at Raisting, West Germany. (*NYT*, 1/31/71)

- NASA and ARPA announced selection of ARC as site for ILLIAC IV computer. ILLIAC IV program was ARPA-sponsored R&D effort to develop, test, and evaluate parallel array processing concept by constructing and operating large-capacity computer. ARC would perform final system integration and testing, operate computer for DOD, and perform NASA research in aeronautic fluid mechanics. Program had started at Univ. of Illinois in 1966 and was approaching manufacturing phase. ILLIAC IV computer was being built by Burroughs Corp. and would be moved to ARC after completion in summer of 1971. (NASA Release 71-14)

- Senate adopted S.R. 20, electing minority members to Senate Committees for 92nd Congress. Elected to Committee on Aeronautical and Space Sciences were Sen. Carl T. Curtis (R-Neb.), Sen. Margaret C. Smith (R-Me.), Sen. Barry M. Goldwater (R-Ariz.), Sen. Lowell P. Weicker (R-Conn.), and Sen. James L. Buckley (R-N.Y.). (*CR*, 1/29/71, S505)

January 30: President Nixon sent individual messages from Caneel Bay, Virgin Islands, wishing success to Apollo 14 astronauts Alan B. Shepard, Jr., Edgar D. Mitchell, and Stuart A. Roosa on eve of launch. Message to Shepard said: "The courage and determination you have displayed in coming back against such great odds to fly again have earned our unanimous admiration at home and abroad. That this important mission should be entrusted to our Nation's first man in space makes us especially proud and confident in its success." (*PD*, 2/8/71, 151)

- LRV had run into development difficulties that had boosted estimated cost more than 60%, Thomas O'Toole said in *Washington Post*. Cost was now estimated at $41 million, $12 million more than $19 million stipulated in October 1969 contract let to Boeing Co. and General Motors Corp. Overrun was due "almost entirely to overtime paid to the two companies to make sure the Rover will be ready to fly with Apollo 15 when it takes off for the moon July 25." Trouble with electronic controls and independent harmonic drive and other problems had delayed testing of LRV until Dec. 14, 1970, almost three months late. (*W Post*, 1/30/71, A3)

- Threat of military superiority was "inherent" in Soviet space program, *New York Daily News* said. "For it was the the development of a Russian rocket for intercontinental bombardment that led the way to the technological race in space in the first place. We certainly had nothing like it on Oct. 4, 1957, when the Russians startled the world with the orbiting Sputnik and came along a month later to launch Sputnik II, a 1200-pound [500-kg] vehicle carrying a live dog...." U.S. Saturn rocket had no military usefulness; "one doubts that the Russians can say the same about their spacecraft launching vehicles." (Greene, *NY News*, 1/31/71)

- Smithsonian Institution commissioned at Fort Pierce, Fla., the *Johnson-Sea-Link*, five-man submersible research vessel designed to operate in water to depth of 900 m (3000 ft). First of its kind, 8-m (25-ft) vessel could remain under water for up to 48 hrs. (AP, W *Star*, 1/30/71, A9)

January 31–February 9: NASA's *Apollo 14* (AS–509) carried three-man crew on successful lunar landing mission. LM-8 *Antares* landed on moon's Fra Mauro and two astronauts conducted experiments and explored lunar surface for 9 hrs 24 min before rejoining orbiting CSM-110 and returning safely to earth with lunar samples in CM *Kitty Hawk*.

January 31–February 3: Spacecraft, carrying Astronauts Alan B. Shepard, Jr. (commander), Stuart A. Roosa (CM pilot), and Edgar D. Mitchell (LM pilot), was launched from KSC Launch Complex 39, Pad A, at 4:03 pm EST Jan. 31 by Saturn V booster. Launch was 40 min 3 sec later than planned because of high overcast clouds and rain. Delay caused revision of flight azimuth from 72° to 75.6°. Launch was watched by about 2500 invited guests, including Vice

President Spiro T. Agnew, Prince Juan Carlos and Princess Sophia of Spain, and *Apollo 11* Astronaut Neil A. Armstrong, Jr.

Spacecraft and S-IVB combination entered parking orbit with 188.9-km (117.4-mi) apogee and 183.1-km (113.8-mi) perigee. Midcourse correction (MCC-1) modified trajectory so spacecraft arrival time at moon would be same as if launch had been on time. CSM separated from LM/S-IVB/IU at 3:02 GET. Onboard TV was initiated to cover CSM docking with LM, but difficulty in docking was encountered. First five docking attempts by CM pilot Roosa were unsuccessful. Mission Control in Houston, watching on TV, studied problem that could prevent extraction of LM and make lunar landing impossible. Alternate missions also were under study. Astronauts noted catches seemed properly cocked but did not release. Sixth try however, was successful; docking was achieved at 4:57 GET, 1 hr 54 min later than planned.

Separated S-IVB/IU impacted lunar surface at 83:18 GET (2:41 am EST Feb. 4) at 7°49′ south latitude and 26° west longitude 174 km (108 mi) southeast of planned impact point. *Apollo 12* seismometer detected impact and showed vibrations for about 2 hrs.

Unscheduled 2-hr 22-min TV transmission from spacecraft was initiated at 10:00 GET to troubleshoot probe and drogue of docking mechanism. Crew removed probe and drogue but found no foreign material or abnormal damage. Capture latch assembly was actuated and system performed nominally. Crew and ground were unable to determine why CSM and LM had failed to dock properly, but they found no indication that systems would not work normally again. Officials decided mission would continue as planned [see Feb. 1].

Second midcourse correction, with 10.1-sec SPS burn at 30:36 GET, increased velocity by 21.7 m per sec (71.1 fps). Following maneuver, crew completed about nine frames of dim-light photography on earth's dark side and continued to check out systems. GET update was performed at 55:40 GET to add the 40 min 3 sec lost by launch delay. MCC-3, scheduled for 60:38 GET, was not necessary. Third TV transmission, for 42 min beginning at 60:40 GET, showed Shepard and Mitchell transferring into LM and checking out LM systems. MCC-4, at 77:38 GET with 0.6-sec SPS burn, changed velocity 1.1 m per sec (3.5 fps).

February 4–6: *Apollo 14* entered lunar orbit with 314.1-km (195.2-mi) apolune and 108.2-km (67.2-mi) perilune after LOI burn at 82:37 GET, including 40-min update (1:55 am EST Feb. 4). Roosa called to Mission Control: ". . . you're not going to believe this—it looks just like the map." Shepard described the moon and near-moon space as "really quite a sight. No atmosphere at all. Everything is clear up here. Really fantastic." Mitchell said moon looked like "a plaster mold that somebody has dusted with grays and browns." Region where daylight stopped and darkness began had "some fairly high crater walls and high country, with those long shadows, it really looks rugged. It looks like you could walk along that surface into the darkness and fall into nothing."

Shepard and Mitchell reentered LM and checked out systems before firing RCS thrusters at 104:28 GET to separate CSM and LM. Spurious bit in LM guidance computer was corrected to preclude inadvertent

January 31–February 9: *Propelled into space by the Saturn V, Apollo 14 carried Astronauts Alan B. Shepard, Jr., Stuart A. Roosa, and Edgar D. Mitchell to the moon on the third successful United States lunar landing mission. After Feb. 5 touchdown, Shepard and Mitchell explored the Fra Mauro area for 9 hours 24 minutes in two extravehicular periods before the LM Antares lifted them off to rejoin Roosa in the orbiting CM Kitty Hawk for the return to earth. Prelaunch activities were monitored by Dr. George M. Low (at left below), Acting NASA Administrator, and Dr. Wernher von Braun, Deputy Associate Administrator for Planning, in Firing Room 2 at KSC.*

abort during powered descent, and Shepard controlled descent manually.

LM *Antares* landed on moon's Fra Mauro at 108:54 GET (4:17 am EST Feb. 5) on 8° slope 9–18 m (30–60 ft) short of planned landing point. "We seem to be sitting in a bowl," Mitchell said. "It's choppy, undulating. There is a large depression to our right to the north of us, which forms another bowl. And I can see several ridges and rolling hills of perhaps 35 to 40 feet (10 to 12 meters) in height."

First EVA began 49 min later than planned because of intermittent PLSS communications. Shepard started down ladder, deployed MESA containing camera which recorded descent, and stepped on lunar surface at 114:30 GET (9:53 am EST Feb. 5). Mitchell stepped down six minutes later. "Not bad for an old man," Mission Control told Shepard as he touched surface. "You're right," Shepard replied "I'm on the surface. It's been a long way and I'm here. The surface on which the forward footpad landed is extremely soft. . . . it's in a small depression. The soil is so soft that it comes all the way to the top of the footpads, it even folded over the sides to some degree. . . . we can see the boulders on the rim. It looks as though we have a good traverse route up to the top of the Cone."

Crew collected 19.5-kg (43-lb) contingency sample; deployed TV, S-band antenna, American flag, and solar wind composition experiment; and photographed LM, lunar surface, and experiments. ALSEP was deployed about 152 m (500 ft) west of LM and laser-ranging retroreflector was deployed about 30 m (100 ft) west of ALSEP. Mitchell conducted active seismic experiment, firing 13 thumper shots into lunar surface. Signals were recorded by geophones and transmitted to earth. Eight of the 21 cartridges misfired, but the 13 fired sent vibrations 18–21 m (60–70 ft) into moon. Crew was granted 30-min extension on EVA and reentered LM after 4 hrs 49 min.

After resting inside LM and checking plans for second EVA period, astronauts left LM at 131:48 GET (3:11 am EST Feb. 6)—2 hrs 27 min earlier than planned. Astronauts loaded MET with photographic equipment and lunar portable magnetometer (LPM) and proceeded toward Cone Crater. MET bounced slightly during traverse but appeared stable. Astronauts' main problem was lunar dust which stuck to their space suits up to their knees. "We're filthy as pigs," Mitchell said. "Everything's going to be covered with dust before long." LPM site measurement was made at first stop and LPM traverse measurement was later made near Cone Crater. Geological features were described and materials collected. Crew was unable to reach rim of Cone Crater because slopes were steeper than expected and climb would have required running behind time line by about 30 min. Samples were collected in blocky field near rim. On return leg of traverse crew dug 0.5-m (1.5-ft) trench and collected samples. Unsuccessful triple-core-tube attempt was made and other containerized samples were collected. Alignment adjustment was made to ALSEP central station's antenna to improve signal strength being received at MSFN ground stations. Crew was unable to perform a number of secondary tasks because of timeline and minor problems.

Just before reentering LM Shepard made surprise announcement: "You might recognize what I have in my hand as the handle for the

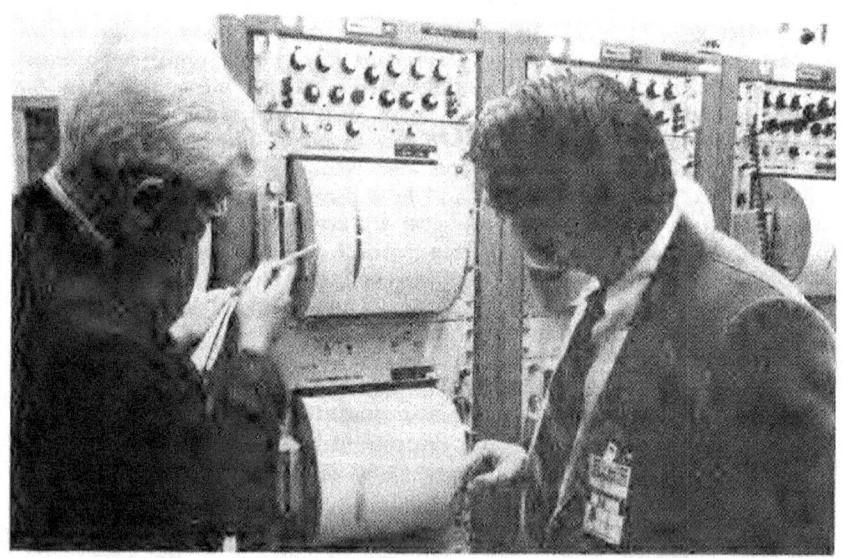

February 4–6: *Tracks of the modularized equipment transporter, used for the first time on the Apollo 14 mission, were photographed (at top left) leading across the lunar surface from the LM Antares. The inverted umbrella of the S-band antenna was positioned by the astronauts to the left of the LM. Lunar dust clung to the legs of Astronaut Edgar D. Mitchell (above) as he moved across the surface during extravehicular activity. Meanwhile, in the Manned Spacecraft Center at Houston, Texas, a seismic reading of the Feb. 4 impact on the moon of the Saturn launch vehicle's S-IVB stage was examined by Dr. Maurice Ewing (bottom left), Director of Columbia University's Lamont-Doherty Geological Observatory, and Columbia graduate student David Lammlein. The impact released energy comparable to 11 tons of TNT.*

contingency sample return and it just so happens to have a genuine six iron on the bottom of it." Shepard dropped golf ball on lunar surface and on third attempt drove ball about 366 m (400 yds). Astronauts reentered LM at 136:26 GET, after 4 hrs 35 min, bringing total EVA time to 9 hrs 24 min.

While LM was on moon, Roosa, orbiting moon in CSM, completed bootstrap photography of Descartes landing site and astronomic photography, including Gegenschein experiment, and prepared for shortened rendezvous technique.

Shepard and Mitchell depressurized LM, discarded excess equipment, repressurized LM, and lifted off lunar surface with 43 kg (94.8 lbs) of samples at 142:25 GET—33 hrs 31 min after landing on moon. During braking phase for docking, telemetry indicated that abort guidance system had failed, but *no caution and warning signals were on*. Docking was accomplished successfully at 144:13 GET (3:36 pm Feb. 6) with no probe-drogue problems. Probe was returned to earth for postflight analysis. Excellent TV during rendezvous and docking showed docking maneuver clearly. Crew transferred from LM to CSM with samples, equipment, and film. LM ascent stage was jettisoned and intentionally crashed onto lunar surface at 148:22 GET (7:05 pm EST Feb. 6) at 3° 35' south latitude and 19° 40' west longitude. Impact was recorded by *Apollo 12* and *Apollo 14* ALSEPs. Transearth injection maneuver, on schedule at 149:16 GET, injected CSM into transearth trajectory after 66 hrs 40 min (34 revolutions) in lunar orbit.

February 7–9: MCC–5 at 166:14 GET increased velocity by 152 mm per sec (0.5 fps). MCC–6 and MCC–7 were not required. Inflight demonstrations of composite casting, liquid transfer, and heatflow connection were televised beginning at 172:30 GET. Press briefing at 195:09 GET (6:32 pm EST Feb. 8) was of good quality.

CM *Kitty Hawk* separated from SM at 216:12 GET. Parachute deployment and other reentry events occurred as planned and *Kitty Hawk* splashed down in mid-Pacific about 7 km (4 nm) from recovery ship U.S.S. *New Orleans* at 216:42 GET (4.05 pm EST Feb. 9). Astronauts, wearing flight suits and masks, were carried by helicopter from CM to recovery ship, where they entered mobile quarantine facility with recovery physicians and technician. Crew, physicians, and technician would remain inside MQF until ship neared Samoa and would then be transferred to another MQF for flight to LRL quarantine in Houston. CM was retrieved and mated to MQF transfer tunnel on board recovery ship. From inside MQF–CM containment envelope, MQF engineer removed lunar samples and equipment through decontamination lock and CM was sealed until delivery to LRL. Sample return containers, film, and other data were flown via Pago Pago to Houston for LRL.

Primary *Apollo 14* objectives—to make selenological inspection, survey, and sampling of materials in preselected region of Fra Mauro; deploy and activate ALSEP; develop man's capability to work in lunar environment; and photograph candidate exploration sites—were achieved. Launch vehicle and spacecraft systems performance was near nominal except for docking problems, spurious signals, and an occasional loss of communications. Flight crew performance was excellent. Crew exhibited exceptional poise during CSM/LM docking attempts and while troubleshooting AG computer. Accomplishments included first use of MET, largest payload—32 500 kg (71 650 lbs)—placed in lunar orbit, longest distance traversed on lunar surface, largest payload returned from lunar surface, longest lunar surface stay time, longest total EVA, first use of shortened rendezvous technique, first active seismic experiment, and first inflight technology demonstrations.

Apollo 14 was 11th Apollo mission to date, 8th manned Apollo mission, and 3rd successful lunar landing mission. *Apollo 11* (July 16–24, 1969) and *Apollo 12* (Nov. 14–24, 1969) had proved capability to land men on moon and retrieve lunar samples for study on earth. *Apollo 13* (April 11–17, 1970), aborted by SM tank explosion, had demonstrated capability to perform under emergency situation and return crew safely to earth under alternate plan. Apollo program was directed by NASA Office of Manned Space Flight; MSC was responsible for Apollo spacecraft development, MSFC for Saturn V launch vehicle, and KSC for launch operations. Tracking and data acquisition was managed by GSFC under overall direction of NASA Office of Tracking and Data Acquisition. (NASA Proj Off; NASA Release 71-245; *NYT*, 2/1-10/71; *W Post*, 2/1-10/71; *Av Wk*, 2/15/71, 19)

January 31: President Nixon in Caneel Bay, Virgin Islands, issued statement following successful *Apollo 14* launch: "Today, we take the most amazing things almost for granted. It is difficult to imagine, but some day the discovery of the New World will seem almost inconsequential alongside the explorations of our own time. In a way we are like the people who must have stood on the dock at Palos, Spain, and watched Columbus' ships disappear, believing they were watching the end of a great event, when they were watching the beginning. But where they watched the world being opened, and did not know it, we are watching the heavens being opened and we do know it. While those men went under the patronage of individual monarchs, our men go for the American people and, in a larger sense, for all mankind." (*PD*, 2/8/71, 151-2)

- Vice President Spiro T. Agnew told workers at KSC after *Apollo 14* launch that days of declining space budget were past and "we're going forward together not just to the moon but I'm certain that the American program will continue and press forward into the reaches of interplanetary space." (AP, *W Star*, 2/11/71, A7)
- Nike-Apache sounding rocket, launched by NASA from Univ. of Michigan's Keweenaw Rocket Launch Site (KRLS), carried GSFC payload to 184.8-km (114.8-mi) altitude to obtain data during magnetic activity in ionosphere while radiofrequency absorption in 2- and 10-mhz region

February 9: *Safely back on earth after a nine-day journey to explore the surface of the moon, the* Apollo 14 *astronauts watched a pararescueman close their command module hatch following splashdown in the planned Pacific Ocean landing zone.*

of spectrum was high. Rocket and instruments functioned satisfactorily. Data would be compared with data from similar launch Jan. 29. (NASA Rpt SRL; NASA Release 71-1)

- Tenth anniversary of Mercury-Redstone mission that carried chimpanzee Ham up to 253-km (157-mi) altitude on suborbital flight. Ham was recovered in good health. He was now ward of National Zoo in Washington, D.C. (AP, *W Post*, 1/31/71, A4; Natl Zoo PIO)
- Capabilities of NASA's largest computer were described by Dr. Robert Jastrow, Director of GSFC Institute for Space Studies, in *New York Times Magazine* article: "The 'thinking' of this machine is carried out in a computing unit, containing 100,000 transistors, which performs 10 million elementary actions per second, such as addition and subtraction, comparisons of numbers, logical choices between alternatives. . . ." Numbers and words employed "are stored in its memory, which has a capacity for holding one million separate items of information and can produce . . . items, or ingest new ones, in less than one-millionth of a second. . . . In normal use, 5 per cent of this memory is filled up with the programing instructions which furnish the instinctive reactions of the machine." Machine also received "specialized training" from GSFC staff, "a set of instructions also placed in the machine's memory, and taking up . . . 10 per cent or 15 per cent of the total memory capacity." Remainder of computer's memory contained facts pertinent to problem at hand. Computer executed 24-hr global weather forecast in two hours. Process required "one trillion additions and subtractions, which would take a person working at a desk calculator 10,000 years to complete." (*NYT Mag*, 1/31/71, 14ff)
- Author Norman Mailer said in San Francisco that Americans were as interested in *Apollo 14* as they would be in "border war in Bolivia." He criticized images of astronauts projected by NASA. They were "tough men—daredevils" but NASA presented them as "priests." NASA was "asking this country to love saints and Americans are not noted for that." (AP, *W Post*, 1/31/71, A24)
- *Los Angeles Times* editorial commented on space funding: "If Congress wants to be truly responsible about our national priorities, it will avoid making further cuts which would damage the space program out of proportion to the savings involved." In addition to "advancing the frontiers of knowledge about the origins of the earth and the nature of the universe," space program promised "to pay off in much more concrete ways."

  "We are already reaping huge benefits from weather and communications satellites, as well as from the development of compact, high-speed computers, new metals and fabrics, microminiature circuits and many other technological advances." Space program also promised more efficient use of earth's resources, help in global attack on environmental pollution, possible prediction of earthquakes, and hope for more productive agriculture by use of lunar soil ingredients.

  "The line between true and false economy is hard to draw. But the space program is one area where the time has come to draw it." (*LA Times*, 1/31/71)

During January: Creation of global resources management system was discussed by Dr. Wernher von Braun, NASA Deputy Associate Administrator for Planning, in guest editorial in *Astronautics & Aeronautics*:

"Can we afford to create a global resources management system? I think we cannot afford not to develop it. As the richest nation in the world we have a moral obligation to bring our new technology to bear to help the helpless. But let us talk about our national security. On both well-fed sides of the Iron Curtain, an uncomfortable amount of raw thermonuclear power has been piled up in the name of 'deterring the other side.' In the meantime, the under-fed, over-populated part of the world, too poor to build thermonuclear bombs, has quietly developed a far more terrifying threat to the thermonuclear powers: the Population Bomb. Is is not just common sense to spend at least a small part of our tax money set aside for national security on the task of defusing *that* bomb? It is gratifying indeed to know that, by establishing a global Earth-resources survey and management system, we can combine a project bound to vastly enhance our national security with a noble contribution to the welfare of the world, a charity quite likely to save the lives of millions." (*A&A*, 1/71, 24-5)

- NSF released *Research and Development in Local Governments, Fiscal Years 1968 & 1969* (NSF 71-6): Local government R&D expenditures doubled between 1966 and 1969, from $20 million to $40 million. Approximately 50% of funds spent by local governments for R&D came from Federal Government. Life, social, and engineering sciences were largest fields in local government R&D. (Text)

February 1971

February 1: Capt. Chester M. Lee (USN), Apollo Mission Director, discussed *Apollo 14*'s docking problems with press at MSC and assured press mission would continue toward moon as planned: ". . . first of all we did have a successful docking and every indication is that it was a normal docking." Engineers "have been making extensive analysis here and we can find nothing wrong with the probe. So far as we're concerned it's all operating. We do not fully understand why it did not operate or why the first attempts at docking were not successful. However, we are confident that the drogue . . . is good. As to the cause, it could have been a foreign particle. . . . we did ask the crew to look diligently for any foreign particles and they didn't find any. . . . We did ask them to inspect the probe head very carefully . . . to see if there were any foreign particles. We found none. If that were the problem, it apparently dislodged and is no longer there and is no longer causing a problem . . . and the probe certainly works now." (Transcript)

- Newspaper editorials commented on *Apollo 14* difficulties in docking CSM with LM [see Jan. 31–Feb. 9].

    New York Times said that "as the world waited tensely for the outcome of the struggle to solve the problem, there was a deep sense of disappointment at the possibility that all the effort, training and planning that went into this mission have been imperiled." (*NYT*, 2/1/71, 30)

    Washington *Evening Star* editorial called *Apollo 14* "pressure mission." It had "already qualified for several entries in the record book. It is the first Apollo mission to be crewed entirely by men who have never before flown in earth orbit and commanded by a veteran whose total previous time in space was 15 minutes. The mission is led by the oldest man to go into space, the first American to ride a rocket, the only astronaut to return to active duty after a medically enforced period of idleness. It is the first Apollo shot to be delayed because of weather. It is the first to run into a major problem on its first day out. It is a mission that, in the opinion of many in the top ranks of the space agency, will decide the future of manned space flight." (W *Star*, 2/1/71)

- *Izvestia* published *Apollo 14* story filed by its New York correspondent claiming that cane presented as joke to Alan B. Shepard, Jr., before launch had become "symbolic" of entire mission. Report emphasized mission's technical difficulties. (AP, *W Post*, 2/2/71, A6)

- NASA issued memorandum change to solicit proposals for participation in analysis of space flight data from photography and auxiliary records acquired on Apollo lunar exploration missions. Apollo 15, 16, and 17 would carry three orbital scientific instruments designed to provide photos and supporting data, panoramic camera, mapping camera system, and laser altimeter. Proposals also were solicited for analysis of lunar

photography from *Apollo 14* and for photos of astronomical objects from Apollo missions 14 through 17. (NASA Memo Change 36)
- Lockheed Aircraft Corp. announced it would accept $200-million loss on C–5A transport imposed by DOD as price of resolving contract disputes [see Jan. 27]. Lockheed Board Chairman Daniel J. Haughton said in statement released to press that acceptance meant Lockheed would forfeit $100 million it had already provided toward C–5A costs and would repay second $100 million by securing lien on its investment in Lockheed-Georgia plant. Lockheed said it expected to conclude restructured credit arrangement successfully within a few weeks, providing additional financing for L–1011 TriStar transport and other programs. (Lockheed Release, 2/1/71)
- President Nixon resubmitted to 92nd Congress proposals previously submitted to 91st Congress, including proposed legislation to increase airline passenger ticket tax and departure tax on international flights to cover cost of civil air and ground security program, particularly sky marshals established by Federal Government to combat aerial hijacking. President noted "number of airline hijackings that seemed to be taking place almost daily months ago has been reduced." (*PD*, 2/1/71, 109–20)
- Interpretation of Gum Nebula as fossil of super nova, or exploded star, was reported in *Astrophysical Journal* by GSFC astronomers Dr. John C. Brandt, Theodore P. Stecher, and Dr. Stephen P. Maran and Kitt Peak National Observatory astronomer David L. Crawford. Gum Nebula, largest known object within Milky Way Galaxy, extended 60° across southern sky. Light took some 3000 yrs to cross it at speed of 299 000 km (186 000 mi) per sec. Previous theory had been that nebula glowed visibly because it contained "hot" stars that ionized surrounding hydrogen gas cloud. New interpretation, based on more accurate information on nebula's distance from earth, indicated Gum Nebula was produced with pulsar during supernova explosion some 11 000 yrs ago when burst of radiation ionized surrounding hydrogen. (*Astrophysical Journal*, 2/1/71, L99–104)
- Discovery of rarest nuclear particle anti-omega-minus baryon in photo of nuclear interaction was reported by physicist Dr. Gerson Goldhaber of Univ. of California at Berkeley at American Physical Society meeting in New York. Particle's track had been found in search of more than 500 000 photos and further confirmed that baryon particles had antiparticles. Photos of particle collisions were taken in heavy-hydrogen-filled bubble chamber of particle accelerator at Stanford Linear Accelerator Center (AP, *NYT*, 2/2/71, 19)
- Promotion of Thomas W. Morgan, Manager of KSC Apollo-Skylab Programs, to rank of permanent brigadier general in USAF was noted in *Armed Forces Journal*. (*AFJ*, 2/1/71, 45)

February 1–3: Joint U.S.–U.S.S.R. editorial board for preparation and publication of review of space biology and medicine held second meeting in Washington, D.C. Cochairmen were Professor Melvin Calvin of Univ. of California and Professor Oleg G. Gazenko, Director of Institute of Medical Biological Problems of Soviet Ministry of Health. (NASA Release 71–16)

February 2–4: North Atlantic Treaty Organization's *Natosat 2* (*Nato 2*; NATO–B) military comsat was launched from ETR at 8:41 am EST by

NASA for USAF and NATO by thrust-augmented Thor-Delta booster. Spacecraft entered transfer orbit with 37 712.2-km (23 433.3-mi) apogee, 272.9-km (169.6-mi) perigee, 673.9-min period, and 25.9° inclination. Primary NASA objective was to place spacecraft into transfer orbit accurate enough for onboard propulsion systems to place spacecraft in acceptable synchronous orbit.

Apogee motor was fired at 11:56 pm EST Feb. 4, placing *Natosat 2* into circular orbit with 37 053.0-km (23 023.7-mi) apogee, 34 493.6-km (21 433.3-mi) perigee, and 0° inclination. Spacecraft would drift about 1° per day from 135° east longitude until it reached permanent station over Atlantic at 26° west longitude in late March.

Natosat 2 was second of two NATO comsats launched under DOD–NATO agreement. USAF managed project for DOD and would reimburse NASA for launch services. NATO would reimburse USAF. Initial operation of satellite telemetry and command functions were performed from USAF satellite control facility. Control of orbital operations would be transferred to NATO after spacecraft reached station. (NASA Proj Off)

February 2: President Nixon sent message to Congress transmitting *Aeronautics and Space Report of the President, Transmitted to the Congress January 1971:* "The activities of our space program during the year are consistent with the recommendations I made in March for a balanced space program. Our goals are continued exploration, scientific knowledge and practical applications. The technology acquired through our space programs has many practical applications on earth ranging from communications, meteorology and navigation to agriculture, education and transportation." Specific objectives were continued lunar exploration to increase scientific return on Apollo program investment, exploration of solar system and universe, reduction of space operation costs, expansion of knowledge of "man's ability to perform productively in the hostile environment of space and to relate this knowledge to uses here on earth" and application of space technology to environmental problems. "We should also promote international cooperation in our space program by pursuing joint space ventures, exchanging scientific and technical knowledge, and assisting in the practical application of this knowledge. We are greatly encouraged by European interest in joining us in cooperative post-Apollo planning."

Year 1970 had produced "substantial contributions to continued U.S. pre-eminence in civil aviation, major improvements in aeronautical services, and impressive developments in a sound SST program." New military aeronautics programs initiated in 1970 would "enhance our national security." (Pres Rpt 71; *PD*, 2/8/71, 158)

- White House Office of Management spokesman said FY 1972 spending for public relations would be cut $45 million, or nearly 30%, under Nov. 6 Presidential order to curtail "self-serving and wasteful public relations activities." NASA PR spending would be cut from $13.7 million to $9.7 million, one of three largest cuts. Other agencies most effected were HEW and DOD. (Knap, *W News*, 2/2/71)
- U.K. government agreed to finance half of BAC's contribution to NR's Phase B space shuttle study for NASA. Total financing was about $550 000. (*SF*, 4/71, 121)
- Sen. David H. Gambrell (D-Ga.) was assigned to Senate Committee on Aeronautical and Space Sciences. (NASA *LAR* X/7)

- *New York Times* editorial on NASA decision to continue *Apollo 14* mission: "Space officials know that they have taken on an extra degree of responsibility with the 'go' decision, and have opened the way for massive recrimination should their judgment be proved wrong by later developments. Nevertheless there is some reassurance in the fact that the original docking problem—which for a time seemed certain to abort the flight—was successfully overcome by the astronauts thousands of miles from earth. The nation and the world will be watching this flight with renewed hope—and admiration for the brave men embarked upon it." (*NYT*, 2/2/71, 34)

February 3: Total new obligational authority for U.S. space program including all pertinent agencies in President Nixon's proposed FY 1972 budget was $4.707 billion—down $73.6 million, or 1.1%, from FY 1971 space NOA of $4.780 billion—*Space Business Daily* estimated. Total FY 1972 space expenditures would be $4.633 billion, down 5.2% from $4.888 billion in FY 1971. (*SBD*, 2/3/71, 148)

February 4: Rolls-Royce Ltd. declared bankruptcy. Prestigious British automotive and aircraft engine company blamed collapse on losses incurred in developing RB-211 engine for Lockheed Aircraft Corp. TriStar airbus. Rolls-Royce said it could not proceed with engine under current fixed-price contract. (Lee, *NYT*, 2/5/71, 1)

- USN P-3C Orion turboprop antisubmarine aircraft achieved world's record for altitude in horizontal flight of 13 686 m (44 900 ft). Previous record of 12 990 m (42 618 ft) had been set by Soviet Il-18 in June 1969. (*AF Mag*, 4/71, 15)

- Training model of French-made UV panorama (S-138) experiment—only foreign-made experiment for first Skylab flight—was delivered to McDonnell Douglas Astronautics Co. at Huntington Beach, Calif. Experiment would gather photographic data on massive hot stars. Training model, which weighed 68 kg (150 lbs) and was 1.2 m (3.8 ft) high, would be incorporated with other training equipment at McDonnell Douglas. Flight model would be delivered in May. (MSFC Release 71-19)

- House adopted resolution setting number of members of Committee on Science and Astronautics at 30 and resolution providing for assignment of majority and minority members of Committee. Majority (Democratic) members were: George P. Miller, Calif., Chairman; Olin E. Teague, Tex.; Joseph E. Karth, Minn.; Ken Hechler, W. Va.; John W. Davis, Ga.; Thomas N. Downing, Va.; Don Fuqua, Fla.; Earle Cabell, Tex.; James W. Symington, Mo.; Richard T. Hanna, Calif.; Walter Flowers, Ala.; Robert A. Roe, N.J.; John F. Seiberling, Jr., Ohio; William R. Cotter, Conn.; Charles B. Rangel, N.Y.; Morgan F. Murphy, Ill.; Mike McCormack, Wash. Minority (Republican) members were: James G. Fulton, Pa.; Charles A. Mosher, Ohio; Alphonzo Bell, Calif.; Thomas M. Pelly, Wash.; John W. Wydler, N.Y.; Larry Wynn, Jr., Kans.; Robert D. Price, Tex.; Louis Frey, Jr., Fla.; Barry M. Goldwater, Jr., Calif.; Marvin L. Esch, Mich.; R. Lawrence Coughlin, Pa.; John N. Happy Camp, Okla. (*CR*, 2/4/71, H428, H434)

- Winners of 11th annual Federal Women's Award were announced by Patricia R. Hitt, Assistant Secretary of Health, Education, and Welfare. They included Rita M. Rapp, Subsystems Manager of Apollo Food and Personal Hygiene Items and Head of Tests and Integration Activities for Skylab Food System at MSC. Winners, selected from 119 upper-grade Federal career women nominated, would be honored at Feb. 25 dinner in Washington, D.C. (*Federal Times*, 2/17/71, 3; Dean, W *Star*, 2/4/71, C3)

- U.S.S.R. would spring "Sputnik-like surprises" on U.S. this decade that could make major U.S. military weapon systems technically obsolete, Dr. Eberhardt Rechtin, DOD Principal Deputy Director of Defense Research and Engineering, told Univ. of Southern California audience. U.S. response would have to be doubling of current $7-billion-per-year defense budget. (Text)

February 5: President Nixon telephoned MSC Director of Flight Crew Operations Donald K. Slayton and asked him to send President's congratulations to *Apollo 14* astronauts and ground crew. Later President issued statement: "*Apollo 14*'s successful landing on the moon was a thrilling event for humanity. I congratulate the astronauts and also the thousands of other dedicated people who helped make this awesome achievement possible." (*PD*, 2/8/71, 173)

- USAF's X-24A lifting-body vehicle, piloted by Maj. Cecil Powell (USAF), successfully completed 22nd flight at FRC. Purpose of glide flight was to check out new pilot. (NASA Proj Off)
- Government-industry program for development of experimental jet STOL aircraft was proposed by NASA and DOT at meeting of NAE Aeronautics and Space Engineering Board in Washington, D.C. Program would pool Government and industry funds in partnership to develop low-noise-level jet STOL technology. Development of prototype and operational aircraft for commercial use would be left to competitive industry decision and financing. NASA later said development of experimental aircraft was necessary to complete technology and proof-of-concept work for production of economically and environmentally acceptable STOL air transport system. Roy P. Jackson, NASA Associate Administrator for Advanced Research and Technology, estimated that research aircraft could be flight-tested in approximately two years from inception of cooperative program. (NASA Release 71-20; NAE PIO)
- Collapsed craters formed in terrestrial basalt flows near Grants, N. Mex., exhibited size-frequency distributions similar to those for craters in terraces on inner walls of lunar crater Copernicus, ARC scientists Ronald Greeley and Donald E. Gault reported in *Science*. Similarities suggested Copernicus craters had origin similar to basalt collapse craters. Scientists suggested that interior terraces of Copernicus were lava flows through fractures during postimpact adjustments after formation of Copernicus. (*Science*, 2/5/71, 477-9)
- Acid hydrolysis, vacuum crushing, and vacuum pyrolysis experiments to investigate amount, nature, and isotopic composition of carbon in *Apollo 12* sample 12023 were described in *Science* by ARC exobiologists Sherwood Chang, Keith A. Kvenvolden, James G. Lawless, and Cyril A. Ponnamperuma and UCLA geologist I. R. Kaplan. Results indicated that sample contained indigenous carbide and methane in amounts comparable to those found in *Apollo 11* fines. Hydrolyzable carbide in *Apollo 12* sample was heavier, isotopically, than any known carbide phase in meteorites. Large difference between isotope values for lunar and meteorite carbide suggested "that the former was either native to the moon or deposited by meteorites and subsequently sub-

jected to isotope fractionation processes on the lunar surface." If meteorites were not source of lunar carbide, "then we are left with the possibilities that the carbides represent primordial indigenous lunar carbon, however formed, or the product of interaction between solar wind carbon and metals or metal oxides or minerals on the lunar surface." Evidence existed for presence of carbides in *Apollo 11* rocks but amounts were too low for isotopic measurements. "If the carbon was indeed indigenous to the lunar rocks, the observed range of isotopic composition would be more representative of primordial lunar carbon." (*Science*, 2/5/71, 474–6)

- President Nixon's FY 1972 science budget was analyzed in *Science*: Budget proposed selective increases for R&D which would "start the federal science budget upward again after several years of virtually static financing." Impact in FY 1972 spending would not be dramatic "since there is a lag of as much as 2 or 3 years between the obligation and actual spending" but budget served as "declaration of intentions." Nixon Administration showed "inclination to maintain high-quality capabilities for research but to restrain expansion of research and, particularly, of manpower training except in areas judged to contribute to the solution of major nation problems." Priorities were indicated by choices of agencies and programs for major increases: NSF budget would exceed $600 million for first time, with major portion of increases going to additional support of research grants. New cancer research program would receive $100 million. Substantial increases would provide funds for research on environmental problems, highway safety, air traffic control, and crime reduction. DOD R&D spending would be increased with additional funds split between work on strategic and tactical weapons.

  Obligational authority for total Federal science budget would rise to $16.7 billion from $15.6 billion in FY 1971, but estimated expenditures would rise only from $15.3 to $15.6 billion. NSF would get new obligational authority of $622 million in FY 1972 but would spend estimated $546 million. (*Science*, 2/5/71, 459–60)

- Astronomers had asked Tucson, Ariz., to modify its outdoor lighting so their view of heavens would not be obstructed by nighttime glare, *Science* reported. Astronomers at Kitt Peak National Observatory, National Radio Astronomy Observatory, Steward Observatory, Lunar and Planetary Laboratory, and Smithsonian Astrophysical Observatory had complained. (Bazell, *Science*, 2/5/71, 561))

February 6: New York Times editorial commented on *Apollo 14* lunar landing: ". . . no one who sat entranced before the picture of yesterday's historic events can have doubted that Antares has scored a major triumph in the effort to advance lunar exploration. Today Shepard and Mitchell will face new problems and grasp new opportunities as they lope their way toward Cone Crater and its geological riches. These two brave and highly skilled men—along with their partner, astronaut Roosa, in lunar-orbiting Kitty Hawk—are writing another brilliant and memorable chapter in man's still-infant probing of earth's nearest neighbor in the vast reaches of space." (*NYT*, 2/6/71, 28)

February 7: RM Radiation/Meteoroid satellite launched pickaback with *Ofo* satellite Nov. 9, 1970, reentered atmosphere and disintegrated. Satellite had demonstrated that advanced radiation system was feasible and accurate and had verified improved instrumentation for measuring meteoroid impact, flux, direction, and speed. (NASA Release 71–40)

February 7–8: U.S.S.R.'s *Lunokhod 1* lunar rover, which landed on moon on board *Luna 17* Nov. 18, 1970, traveled 597 m (653 yds) across moon during first two traverses since end of third lunar night. (*SBD*, 2/10/71, 182)

February 8: Apollo 14 Astronauts Alan B. Shepard, Jr., Edgar D. Mitchell, and Stuart A. Roosa held 25-min press conference, televised from spacecraft during transearth coast. Mitchell called mission "quite a success" in spite of failure to reach Cone crater during second EVA: "I think the majority of the type rocks we would find at the rim were in the boulder field that we were working, although it was a disappointment, just as a matter of challenge, not to get up there. I think we accomplished the scientific objectives that we went for."

Shepard explained major problem was "the undulating terrain where you simply couldn't see more than 100 to 150 yards [90 to 140 meters] away from you. Consequently, you were never quite sure what landmark would appear when you topped the next ridge. We were very surprised when we . . . approached the ridge which we thought to be the rim of Cone Crater, to find there was another one behind it. And that was the beginning of the real problem. As far as we were concerned, our only problem was the amount of time allotted for the excursion. To us it was just a matter of working against the clock. I don't believe that we were disoriented or lost any time. I think if we had wanted to reach the top of the crater and did nothing else, that we could have done that within the time period allotted. But I think the method to which we reverted . . . of collecting rocks from a point not quite near the top of the crater provided a lot more geologically and gave us a better cross-section of the rocks in the area . . . than had we gone to the crater and back and not collected as many rocks." (*NYT*, 2/9/71, 26)

- *Natosat 2* military comsat, launched by NASA for USAF and NATO Feb. 2, was adjudged successful by NASA. Spacecraft was in satisfactory orbit and was functioning normally. (NASA Proj Off)
- Aerobee 170 sounding rocket was launched by NASA from WSMR carrying Univ. of Michigan experiment to observe airglow. Rocket and instruments functioned satisfactorily. (SR list)
- USN P–3C Orion turboprop antisubmarine aircraft, piloted by Cdr. Donald H. Lilienthal (USN), broke four "time-to-climb" records by climbing from brake release to 3000 m (9843 ft) in 2 min 59 sec, to 6000 m (19 685 ft) in 5 min 48 sec; to 9000 m (29 528 ft) in 10 min 31 sec, and to 12 000 m (39 370 ft) in 19 min 53 sec. Aircraft also posted new maximum altitude record for its class of 14 050 m (46 100 ft). (*AF Mag*, 7/71, 15)
- Sen. Clinton P. Anderson (D-N. Mex.), Chairman of Senate Committee on Aeronautical and Space Sciences, said in Washington, D.C., that cutback in funding for NERVA development in NASA FY 1972 budget was gross error reflecting on management capability of NASA. Cutting program funding to $30 million would mean virtual halt in work on NERVA. (AP, Cleveland *Plain Dealer*, 2/9/71)
- NASA announced it had awarded Bellcomm, Inc., 17-mo, $11.5-million extension to contract for systems analysis, study, planning, and tech-

nical support to manned space flight program. Extension, which would continue through May 1972, brought total value of contract since March 1962 to $91 million. (NASA Release 71-17)

- President Nixon sent message to Congress outlining 1971 environmental program. Program included measures to strengthen pollution control, control emerging problems including noise pollution, promote environmental quality in land use decisions, further institutional improvement, expand international cooperation, and establish "World Heritage Trust to preserve parks and areas of unique cultural value throughout the world." (PD, 2/15/71, 187–204)

February 8–9: NASA Aerospace Advisory Panel met at MSFC to discuss safety aspects of LRV, Skylab spacecraft, and proposed space shuttle. (MSFC Release 71-23)

February 9: Following *Apollo 14* splashdown, President Nixon telephoned *Apollo 14* astronauts aboard recovery ship U.S.S. *New Orleans* to express personal congratulations on successful completion of mission. He released statement comparing *Apollo 14* with May 5, 1961, *Freedom 7* mission flown by Astronaut Alan B. Shepard, Jr.: "The difference between that first flight and this latest is a measure of how far we have progressed in space in the short span of 10 years. But two things have not changed: the courage and determination of the men who fly these missions, and the skill and dedication of the thousands here on earth who make their flights possible." (PD, 2/15/71, 205)

- *Cosmos 394* was launched by U.S.S.R. into orbit with 613-km (380.9-mi) apogee, 572-km (355.4-mi) perigee, 96.5-min period, and 65.8° inclination. (GSFC *SSR*, 2/28/71)

- NASA's M2–F3 lifting body vehicle, piloted by Maj. Jerauld R. Gentry (USAF), successfully completed fifth flight after air-launch from B–52 aircraft at FRC. Purpose of flight was to compare flight characteristics of M2–F3 with those of M2–F2. Maj. Gentry was only active pilot who had flown both vehicles. Mission was his last before leaving for duty in Southeast Asia. (NASA Proj Off)

- Rep. George P. Miller (D-Calif.), Chairman of House Committee on Science and Astronautics, introduced H.R. 3981, $3.271-billion FY 1972 NASA authorization bill. Bill authorized appropriations of $2.518 billion for R&D, $56.3 million for construction of facilities, and $697.4 million for research and program management. (CR, 2/9/71, H588; NASA *LAR* X/11)

- *Christian Science Monitor* editorial commented: "Technically, the feats of Apollo 14 . . . are invaluable. The scientific data obtained have opened many new paths to broader knowledge. But far more noteworthy should be the effect which these can have upon man's view of his place and role in the universal scheme. It is already historical fact that America's space program has helped focus sharper attention on challenges at home. This is bound, in the long run, to be one of the program's greatest contributions." (CSM, 2/9/71)

- Former New Mexico Governor David F. Cargo presented *Apollo 11* moon fragments to Museum of New Mexico at ceremony in Santa Fe. Later, AP said incumbent Governor Bruce King had claimed lunar samples for people of New Mexico, while Cargo said they had been presented to him, personally, by President Nixon. AP said White House records

showed plaque presented with moon fragments was inscribed "to the people of the State of New Mexico." (AP, *W Post*, 2/11/71)
- DOD said USA had been directed to continue development of Hardsite Defense, supplementary ABM system that included modified Sprint interceptor missiles and smaller, less expensive radars than those used in Safeguard ABM system. (AP, B *Sun*, 2/10/71, A1)

February 10: Soviet President Nikolay Podgorny sent message to President Nixon: "Please accept . . . our congratulations on the successful conclusion of the flight of the spacecraft Apollo 14 and on the safe return of the American astronauts." Tass said he also asked President Nixon to forward best wishes to "courageous crew." (Reuters, *C Trib*, 2/11/71)
- European interest in U.S. space program had "with notable exceptions" dwindled since *Apollo 11*'s first lunar landing and "trouble-fraught" *Apollo 13*, *New York Times* said. Despite extensive press and TV coverage, Europeans had taken *Apollo 14* for granted. "In some countries local events overshadowed man's third visit to the moon," but in West Germany opinion poll had shown that 64% favored continuation of manned space flights. *Frankfurter Rundschau* editorial had said, "What Shepard and Mitchell carried off couldn't have been done by a Lunokhod."

  Polish newspapers had given equal prominence to *Lunokhod 1* and *Apollo 14*, but stressed technological feat of unmanned vehicle.

  Soviet press had highlighted *Apollo 14* difficulties to suggest superiority of unmanned vehicles. Soviet man in street "seemed bored with space exploration, although some Russians went out of their way to ask Westerners about the United States moonshot." *Komsomolskaya Pravda*, Soviet youth daily, had praised astronauts' "indisputable bravery."

  In U.K. *Apollo 14* had been overshadowed by unrest in Northern Ireland and collapse of Rolls-Royce Ltd. Live TV coverage of *Apollo 14* had been less than of previous Apollo missions.

  Earthquake and civil disorders had crowded *Apollo 14* from front page of Italian newspapers, and Feb. 7 power blackout in New York had received as much space as launch. Italian radio and TV coverage had been thorough but audiences seemed smaller than for previous Apollos. Pope Paul VI had praised *Apollo 14* as advance for all mankind.

  In France interest had waned as mission progressed. Newspapers and TV offered comprehensive coverage but "excitement of two years ago" had vanished.

  Spaniards, too, were losing interest in space exploration and public interest had declined in Hungary and Yugoslavia. Communist student leader in Zagreb had said, "If the Americans and Russians would cooperate in space instead of carrying their rivalries to the moon, maybe there could be some scientific progress on earth and maybe the smaller nations could be helped to contribute their knowledge instead of being shut out of science by the expense." (*NYT*, 2/10/71, 24)
- *Apollo 14* technical problems were discussed by Dr. George M. Low, Acting NASA Administrator, in interview published by *New York Times*: "I feel that what we've seen on Apollo 14 is about the number of problems one must expect on any given flight with such a complicated

- piece of hardware. Yet the spacecraft design is such that it is forgiving of failures. And we hope we'll always be able to work around them and press on with the mission." (Witkin, *NYT*, 2/10/71)
- MSC scientists reported instruments left on moon by Apollo astronauts had apparently survived sudden drop in temperature during lunar eclipse. Instruments registered no unusual seismic events or radiation fluxes during eclipse. Temperature at *Apollo 12* site dropped from 348.9 K (168.3°F) to 170.4 K (−153°F), a 178.5 K (321.3°F) change. Temperature at *Apollo 14* site dropped from 341 K (154.1°F) to 170.4 K (−153°F). (*NYT*, 2/11/71, 40)
- Dr. Charles A. Berry, Director of Medical Research and Operations, announced at MSC that *Apollo 14* Astronaut Alan B. Shepard, Jr., had become first man to gain weight while in space. Other two crew members had lost weight: Edgar D. Mitchell had lost 0.5 kg (1 lb) and Stuart A. Roosa had lost 4.5 kg (10 lbs). Shepard had gained 0.5 kg (1 lb). NASA physicians were making fluid tests on astronauts to determine reason for weight loss during space flight. (Altman, *NYT*, 2/11/71, 40)
- *Milwaukee Journal* editorial on *Apollo 14* commented: "The real return on Apollo lies in the future—beyond the more difficult missions of flights 15, 16 and 17. It is in Skylab, reusable space shuttles and sophisticated space stations to come. It is in the complex monitoring of the earth and its resources from space. It is in the co-operation and internationalization of space programs that appear to be developing—hopefully to replace competition. Here is the payoff. To this Apollo 14 and its crew have made their extraordinary contribution." (*MJ*, 2/10/71)
- NASA announced renaming of five recently reorganized units of Office of Advanced Research and Technology (OART). Aeronautical Research Div. became Aerodynamics and Vehicle Systems Div.; Environmental Systems and Effects Div. became Aeronautical Life Sciences Div.; STOL Program Office became STOL Technology Office; ATET Program Office became Supercritical Technology Office; and Lifting Body Program Office became Entry Technology Office.

    Dr. Leo Fox, Deputy Director of Biotechnology and Human Resources Div., was named Director of Aeronautical Life Sciences Div. (NASA Special Release)
- NASA announced appointment of Harry W. Johnson, Vice President for Engineering in Gates Learjet Corp.'s Aircraft Div., as Director of OART's Aeronautical Propulsion Div. (NASA Release 71–19)
- Sen. Clinton P. Anderson (D-N. Mex.), Chairman of Senate Committee on Aeronautical and Space Sciences, introduced S. 720, $3.271-billion FY 1972 NASA authorization bill, identical to House bill introduced Feb. 9 (*CR*, 2/10/71, S1193)
- Lockheed Aircraft Corp. President Charles S. Wagner announced that Lockheed would lay off up to 6500 employees as result of financial collapse of Rolls-Royce Ltd. British concern was unable to make scheduled deliveries of engines for Lockheed L–1011 TriStar airbus. Lockheed had orders for 178 TriStars from seven airlines and three holding companies at approximately $16 million each. (Witkin, *NYT*, 2/11/71, 67)
- Long Beach, Calif., Harbor Commission refused to renew Howard Hughes'

lease on 3-hectare (7.2-acre) site where eight-engine wooden flying boat *Spruce Goose* had been in hangar for more than 23 yrs. Lease would be terminated on Sept. 1 expiration because property could be more profitably developed as marine terminal. Hughes had designed and built aircraft at estimated $24-million cost. Aircraft, 67 m (219 ft) long with 97.5-m (320-ft) wingspan, had been flown once, by Hughes Nov. 3, 1947. It had not been seen by public since. (UPI, *W Post*, 2/12/71, A8)

- President Nixon submitted nomination to Senate of Herbert F. DeSimone to be Assistant Secretary of Transportation for Environment and Urban Systems. (PD, 2/15/71, 209, 217)

February 11: President Nixon signed Seabed Arms Control Treaty in State Dept. ceremony held simultaneously with ceremonies in Moscow and London. He said: "It has been very properly pointed out that the seabed is man's last frontier on earth, and that frontier can either be a source of peril or promise. By the signing of this treaty, we have pledged to seek its promise and to remove its peril." Treaty was "indication of progress that has been made and continues to be made toward the goal that we all seek: the control of instruments of mass destruction, so that we can reduce the danger of war." (PD, 2/15/71, 211–2)

- Dr. Robert Jastrow, Director of GSFC Institute for Space Studies, was quoted as saying in London that people often forgot intelligence factor in debate over manned versus unmanned space missions. Dr. Jastrow had pointed out that by time unmanned space exploration technology reached level of achievement already exhibited in manned exploration, lunar exploration—or even Martian exploration—might have advanced to stage of extended expeditions or space station. Then human performance would again have jumped ahead of machines. (Cowen, CSM, 2/11/71)

February 11–12: Delegation from 11-nation European Space Conference met at Dept. of State with State Dept., NASC, and NASA officials to continue discussion of possible cooperation in post-Apollo space program, particularly space shuttle and space station. Conference President Theo Lefevre, Belgian Minister for Scientific Policy and Planning, in address raised question of availability to European nations of U.S. launchers if European collaboration in post-Apollo projects failed to materialize, or materialized to smaller degree than U.S. requested. U.S. condition that its launch vehicles be made available only for uses compatible with international agreements raised difficulties for Europeans, especially since Intelsat agreement had not been completed. ESC was proposing that each party supply its partner "with any element of the space system which it is asked to supply, while leaving to its partner the final responsibility for complying with the international agreements it has undertaken. . . . these agreements are identical for the U.S. and the European countries." (Text; Joint Communiqué)

February 12: LRL scientists opened first box of lunar samples returned by *Apollo 14*. Samples included one rock almost pure white with some grey flecks and one with 10 times more potassium, thorium, and uranium than was found in *Apollo 11* and *12* samples. Dr. Paul Gast, Director of Planetary and Earth Sciences at MSC, said preliminary studies of samples were "not inconsistent" with idea that lunar highlands might represent lunar crust. "If this additional evidence that the (highlands) materials on the moon contain abundant rocks with high potassium and high uranium and high thorium . . . it means the highlands are almost as old as the moon itself. What we're saying is that there is a layer of unknown thickness . . . on the surface of the moon which is of very different composition than the average moon." (AP, W *Star*, 2/13/71, A1)

- *Apollo 14* Astronaut Edgar D. Mitchell had tried personal experiment in mental telepathy during flight to moon and back, AP reported. Mitchell, who had life-long interest in psychic phenomena, had concentrated at prearranged times on symbols. Psychic in Chicago had concentrated at same time in attempt to receive signals. Mitchell confirmed that he had conducted experiment, but declined to reveal details. (AP, B *Sun*, 2/13/71, A1)
- Analysis of low-frequency sound from *Apollo 14* spacecraft passing 188 km (117 mi) over Bermuda, obtained from infrasonic sensors in Bermuda, was presented in *Science* by Columbia Univ. geologists. Signals, reminiscent of N-waves from sonic booms, were horizontally coherent, had nearly identical appearances and frequencies, and had supersonic trace velocities across tripartite sensor arrays. They had identical arrival times after spacecraft launch from KSC and were only coherent signals recorded over many hours. Observations "seem to establish that the recorded sound comes from the rockets at high elevation" but values of surface pressure appeared explainable by combination of "kinetic theory approach to shock formation in rarefied atmospheres with established gas-dynamics shock theory." (Cotten, Donn, *Science*, 2/12/71, 565-7).
- Age and composition effects on alteration of lunar optical properties were described in *Science*. Evidence from *Apollo 11* and *Apollo 12* samples and telescopic spectral reflectivity measurements indicated that lunar seas "are similar in mineralogy on a regional scale and that the highlands are consistent with an anorthositic-gabbro composition. Bright craters and rays in both regions expose materials that are relatively crystalline compared with their backgrounds, which are richer in dark glass. With age, bright craters and rays in the maria darken in place by meteorite impact-induced vitrification and mixing with the surrounding material. Highland bright craters and rays may, however, darken primarily through regional contamination by iron and titanium rich mare material." (Adams, McCord, *Science*, 2/12/71, 567-71)
- MSFC announced appointment of Dr. William R. Lucas, Director of MSFC's Program Development Directorate, as MSFC Deputy Director, Technical, effective Feb. 15. Dr. Lucas' position would be assumed by James T. Murphy in acting capacity. (MSFC Release 71-25)
- NASA announced publication of second annual Research and Technology Operating Plan (RTOP) Summary. FY 1971 edition identified and described hundreds of NASA research and technology efforts. (NASA Release 71-18)
- USN announced cause of Dec. 30, 1970, crash of F-14 Tomcat fighter aircraft on second test flight was two pencil-thin hydraulic lines that ruptured after severe vibration. Findings confirmed those of aircraft man-

ufacturer, Grumman Aerospace Corp. USN said problem could be corrected without major redesign or changes. (AP, *NYT*, 2/15/71, 44)

- Philadelphia *Evening Bulletin* editorial commented on signing of seabed treaty: "A quarter century of fear and mistrust has left both the U.S. and U.S.S.R. with vast nuclear power, but little real security. The Seabed Treaty is not going to reverse this, but it is one more slim strand across the chasm of nationalistic rivalry upon which mankind may eventually suspend a bridge." (P *Bull*, 2/12/71)

February 13: President Nixon announced intention to nominate *Apollo 14* Astronaut Edgar D. Mitchell for promotion to grade of captain in USN. Astronaut Stuart A. Roosa had been promoted by USAF to grade of lieutenant colonel. (*PD*, 2/22/71, 266)

- NASA released 9 color photos, 14 black-and-white photos, and segment of 16-mm film taken during *Apollo 14* mission. Pictures showed LM, footprints, tire tracks, Astronauts Shepard and Mitchell during EVA, American flag on surface, and large boulders. (AP, W *Star*, 2/14/71, A3)

February 14: Nike-Tomahawk sounding rocket was launched by NASA from Fairbanks, Alaska, carrying Rice Univ. experiment to conduct auroral studies. Rocket and instruments functioned satisfactorily. (SR list)

- Historical value of *Apollo 14* was appraised by Thomas O'Toole in *Washington Post* article. Mission "may well be remembered as a turning point in the sense that it taught man what his role could be in the exploration of space." *Apollo 11* and *Apollo 12* landings had been "feats of technology that showed the world what its richest nation could do when it set its mind to something. Apollo 14 went a giant step further, to a place where man might appreciate what he had done and what he could do in the future on the surface of the moon or another planet." (*W Post*, 2/14/71, C1)

- JPL announced selection of Howard H. Haglund, former Surveyor project manager, to be project manager for DOT's People Mover transit system at Morgantown, W. Va. Dr. Albert R. Hibbs would be deputy project manager. JPL was initiating research and design of automated, computer-controlled system under $1 353 000 DOT contract. (JPL Release 570)

February 15: Lockheed Aircraft Corp.'s second L–1011 TriStar, piloted by Lockheed test pilot Ralph C. Cokeley, made maiden flight from Lockheed facility at Palmdale, Calif. Aircraft was flown at speeds well under 966-km-per-hr (600-mph) capacity to test control surfaces and hydraulics systems. First TriStar had been flown Nov. 16, 1970. (AP, B *Sun*, 2/16/71, A6)

- *Aviation Week & Space Technology* editorial commented: "Apollo 14 . . . demonstrated that man still has much to learn about the moon. Even with the ability to land in relatively rough areas and to extend useful working time on the moon, the experience of the Apollo 14 duo in the Fra Mauro area indicates that better new equipment is needed to extend the range and accuracy of their explorations. And once the various typical areas of the lunar surface are reached it is obvious that vehicles with payloads much larger than Apollo will be required to haul the equipment needed to establish permanent scientific working stations on the lunar surface. For not until that is accomplished will man really begin to reap the full harvest of knowledge from his lunar capabilities." (Hotz, *Av Wk*, 2/15/71, 9)

- British Aircraft Corp. said in London that Concorde test flights had "met with success unique in the annals of commercial aircraft flight development." BAC chief test pilot Brian Trubshaw told press conference he was optimistic since British and French prototypes had flown total 393 hrs 56 min, including 92 hrs 35 min at supersonic speeds. Aircraft had maintained directional stability at twice speed of sound and on stalling two of its four engines at same speed. Modifications would be made on nacelles before next phase of test flights because of Jan. 22 incident to French prototype 001 at mach 2 over Atlantic. Air pressure had broken off forward intake ramp of No. 4 engine and engine had fallen off into sea. (Berger, B *Sun*, 2/16/71, C6)

February 16: Univ. of Tokyo's Space and Aeronautics Institute successfully launched Japan's second satellite, 63-kg (139-lb) *Tansei* (MST-1), from Uchinoura Space Center with four-stage Mu-4S booster. Orbital parameters: apogee, 1110 km (689.7 mi); perigee, 990 km (615.2 mi); period, 106 min; and inclination, 29.7°. Purpose of mission was to conduct "engineering test for the launching of scientific satellites." *Tansei*, "Light Blue," was named for Univ. of Tokyo colors. First satellite, *Ohsumi*—named for district from which it was launched—had been launched Feb. 11, 1970. (UN General Assembly Release 71-03697; *SBD*, 2/17/71, 211; NASA Off Int Aff; W *Star*, 2/16/71, A7)

- Scientists at MSC, after studying *Apollo 14* photos and questioning crew, reported Astronauts Mitchell and Shepard had turned back within 25–50 m (80–165 ft) of Cone Crater during second EVA period. Crew was apparently misled by slight downhill slope around top of Cone Crater which led them to believe it was not part of crater. Ridge astronauts were struggling to reach appeared to have been southeast of Cone. (AP, B *Sun*, 2/17/71, A1)

- European Space Conference President Theo Lefevre, Belgian Minister for Scientific Policy and Planning, told Brussels press conference that U.S. had refused at Feb. 11–12 conference in Washington, D.C., to sell U.S. launch vehicles for European comsats. Lefevre said State Dept. had told him, however, that if two thirds of Intelsat's 72 members agreed that launch vehicles should be sold U.S. would sell. (*NYT*, 2/17/71, 14; *W Post*, 2/17/71, A13)

- NASA announced publication of *This Island Earth* (NASA SP-250). Book, edited by LaRC Deputy Director Oran W. Nicks, contained color photos of earth taken on U.S. space flights. In foreword Dr. George M. Low, NASA Acting Administrator, said: "Photographs such as this book contains increase our understanding of the relationships between our activities and our environment. It is somewhat paradoxical that man's new ability to voyage in space has provided him with a valuable way to appreciate his Earth." (NASA Special Release, 2/16/71)

February 17: USAF launched four satellites from Vandenberg AFB by one Thor-Burner II booster. Unidentified satellite entered orbit with 832.0-km (517-mi) apogee, 767.7-km (477-mi) perigee, 100.8-min period, and 98.8° inclination. *Calisphere 3, 4,* and *5*—launched to provide targets for radar calibration and evaluate surface material erosion and drag effects vis-a-vis their inert gold or aluminum surfaces—entered orbits with 833.6-km (518-mi) apogee, 762.8- to 772.5-km (474- to 480-mi) perigee, 100.8-min period, and 98.8° inclination. (Pres Rpt 72; *SBD*, 2/18/71, 224)

- Unidentified satellite launched by USAF from Vandenberg AFB by Thor-Agena booster exploded about 40 sec after liftoff. (UPI, W *Star*, 2/18/71)
- U.S.S.R. launched *Cosmos 395* from Plesetsk into orbit with 545-km (338.7-mi) apogee, 529-km (328.7-mi) perigee, 95.3-min period, and 74.0° inclination. (GSFC *SSR*, 2/28/71; *SBD*, 2/19/71, 224)
- Presidential Management Improvement Certificate was presented to NASA Interplanetary Monitoring Platforms (IMP) Management Team in NASA Hq. ceremony. Award was for "efficient program management and design improvements which permitted deferment of a launch for a year while maintaining the achievement of program objectives." (NASA Org & Mgmt Off)
- NAS and NAE released *Jamaica Bay and Kennedy Airport: A Multidisciplinary Environmental Study*. Report, prepared at request of Port of New York Authority, recommended against expansion of John F. Kennedy International Airport. "Any runway construction will damage the natural environment of the Bay and reduce its potential use for conservation, recreation, and housing." (NAS–NAE Release 2/17/71)
- U.S. Dept. of Justice asked Supreme Court to deny atheist Mrs. Madalyn Murray O'Hair's request that U.S. astronauts be prohibited from undertaking religious activities during lunar missions. Solicitor Gen. Erwin Griswold said astronauts carried only personal religious items which improved their morale and contributed "to the success of the flight." (UPI, *H Post*, 2/17/71)
- MSC announced award of $800 000 cost-plus-fixed-fee contract to Lockheed Electronics Co. for first year performance of scientific and technical support services for Earth Resources Laboratory at MTF. (MSC Release 71–07)

February 18: U.S.S.R. launched *Cosmos 396* from Plesetsk into orbit with 268-km (166.5-mi) apogee, 189-km (117.4-mi) perigee, 89.0-min period, and 65.4° inclination. Satellite reentered March 3. (GSFC *SSR*, 2/28/71; 3/31/71; *SRD*, 2/19/71, 224)
- USAF's X–24A lifting-body vehicle, piloted by NASA test pilot John A. Manke, successfully completed 23rd flight at FRC. Objectives of powered flight were to expand flight envelope to mach 1.5, determine lateral-directional derivatives, and determine longitudinal trim and lift-to-drag ratio with 40° upper flap setting and 0° rudder bias. Pilot performed pushovers, pullups, and aileron doublets at a number of angles of attack. (NASA Proj Off)
- NASA Contractor Equal Employment Opportunity Program activities were transferred from Procurement Office to Industrial Relations Office. Joseph M. Hogan would continue to direct group and would report to Robert E. King, Director of Industrial Relations. Personnel and functions concerned with participation of minority-owned business in NASA procurements would remain in Procurement Office. (NASA Special Ann)
- NASA selected Wyle Laboratories to receive cost-plus-award-fee contract for instrument repair, maintenance, and calibration services at LaRC. Contract was for one year with two one-year renewal options at $7.5 million for three years. (NASA Release 71–22)

February 19: President Nixon was expected to name Univ. of Utah President, Dr. James C. Fletcher, as NASA Administrator, *Washington Post* reported. Dr. Fletcher, a physicist, had been an aerospace executive and

February 19

consultant to NACA, DOD, and President's Science Advisory Committee. (O'Toole, *W Post*, 2/19/71, A1)

- Evidence of nature of clouds of Venus derived from comparing polarization observations with theoretical calculations was reported in *Science* by James E. Hansen and Dr. Albert Arking of Goddard Institute for Space Studies. Linear polarization of sunlight multiply scattered by atmosphere and Venus cloud particles had been computed and compared with observations over wavelength range from UV to infrared region. Refractive index of cloud particles was 1.45 ± 0.02 at 0.55μ wavelength. There was indication of slight decrease in value from UV to near-infrared region; mean particle radius was very near 1μ and most of particles were spherical. Cloud layer occurred high in atmosphere where pressure was about 50 millibars or equivalent to approximately 20-km (12.4-mi) altitude on earth. Results for index of refraction eliminated possibility that visible clouds were composed of pure water or ice. Aqueous solution of hydrochloric acid and carbon suboxide "are not absolutely excluded, but the liklihood for either is not high. A new look at the question of the Venus cloud composition seems in order." (*Science*, 2/19/71, 669–72)

- Boosted Arcas I sounding rocket was launched by NASA from Churchill Research Range carrying Univ. of Houston experiment to measure low-altitude x-ray spectrum and electron flux above 30 500 m (100 000 ft). Rocket reached 59.1-km (36.7-mi) altitude and mission was 100% successful. (NASA Rpt SRL)

- ComSatCorp reported net income for 1970 was $17 501 000 ($1.75 per share), up from $7 129 000 (71 cents per share) for 1969. Operating revenues for 1970 reached $69 598 000—increase of $22 564 000 (48%) over $47 034 000 received in 1969. (ComSatCorp Release 71–10)

- Ten Western European nations agreed to cooperate in building 300-bev particle accelerator on frontier between France and Switzerland. Accelerator would fit into tunnel forming 2-km (1¼-mi) ring. Decision to proceed with $260-million project had been awaited since 1964. (*NYT*, 2/20/71, 28)

- *The Decision To Go to the Moon*, book by John M. Logsdon, was reviewed in *Technology Review*: "Logsdon has done a considerable service in pulling together the elements of the histories of official U.S. space policy, of German-American rocket engineering and the obsessive aiming-for-the-stars that went with it, of congressional enthusiasm for extraterrestrial responsibilities, and finally of the months of Kennedy's greatest vulnerability. The factual part of the book leaves one ready to draw one's own conclusions." Review had concluded "that we cannot hope for any useful Apollo-sized program in the absence of any counterpart of the von Braun team, tirelessly bringing to perfection a thing that had not been asked for." (Wheeler, *Tech Rev*, 2/19/71, 22)

February 20: National Urban Coalition released *Counterbudget—A Blueprint for Changing National Priorities, 1971–1976*. Report noted diminishing public support for science and technology and recommended 4% increase in R&D funding to prevent deterioration of Government's role in science and technology in national affairs. Space program should be kept at funding level just over $3 billion, rather than $5.9-billion level reached in 1966. Space shuttle and station programs should be continued but on stretched schedule. Mars mission should be undertaken only when it could be done internationally. Value of space program as whole lay in its scientific yield, practical results, and national leadership. (*CR*, 5/21/71, E4836–98; NUC PIO)

- Award of contracts up to $100 000 each to LTV Aerospace Corp. and Rohr Corp. for preparation of proposals for installation of tracked air cushion vehicle (TACV) system along access road to Dulles Airport in Chantilly, Va., was announced by DOT. Negotiations for similar contract with Grumman Aerospace Corp. were continuing. (DOT Release 3071)

February 21: Aerobee 350 sounding rocket, launched by NASA from Wallops Station at 7:43 pm EST, carried 440.4-kg (971-lb) Columbia Univ. payload to 196.3-km (122-mi) altitude to search for x-ray polarization of Crab Nebula. Rocket and instruments—including two polarimeters sensitive to x-rays at 2.6 kev and 5.25 kev—functioned satisfactorily. Payload recovery would be attempted after daybreak Feb. 22. (WS Release 71–1)

- Text of Dr. Wernher von Braun interview by West German magazine *Der Spiegel* was released to U.S. press. NASA Associate Administrator for Planning was convinced that in 20 yrs "we shall have reached the point where space travel will earn more than it is costing." Earnings from "bread and butter programs"—comsats, weather-monitoring spacecraft, and ERS satellites—could pay for permanent lunar bases and manned exploration of Mars. "Most important product of the seventies" would be application of "all that the space program has taught us." (Wilford, *NYT*, 2/21/71, 75)

- LaRC engineers were using automobile fitted with diagonal braking system to measure slipperiness of wet runways and predict aircraft stopping distance to prevent accidents, NASA announced. Technique could provide data for realistic calculations of crosswind limitations. Stopping distances of car were measured from 97-km-per-hr (60-mph) speed for skidding locked wheels on wet and dry runways. Tests showed ratio of wet-to-dry stopping distances for auto correlated well with those of representative aircraft. (NASA Release 71–21)

February 22: Portable magnetometer carried during *Apollo 14* EVA had detected unexpected bulge in moon's magnetic profile, ARC announced. Data indicated lunar magnetic field at point 297 m (325 yds) northeast of *Apollo 14* landing site was about 100 gammas—three times higher than measurements at *Apollo 12* site. Another reading, 457 m (500 yds) away near edge of Cone crater, measured about 40 gammas. (ARC Release, 71–4)

- First flight model LRV had begun six-week acceptance testing by Boeing Co. preliminary to scheduled delivery to NASA on April 1, NASA announced. Model would transport two astronauts on three extravehicular traverses during Apollo 15 mission in July. Second LRV flight model, being built at Boeing, was scheduled for testing in last week of March and for April delivery. Third model was scheduled for May delivery. (NASA Release 71–25)

- Julian Scheer resigned position as NASA Assistant Administrator for Public Affairs. He had joined NASA as consultant in November 1962 and directed NASA public affairs activities since November 1963. Announcing Scheer's resignation Dr. George M. Low, Acting NASA Administrator, said, "Under Julian Scheer's leadership, the story of our achieve-

February 22

ments in space has been told well, and the public was fully and openly informed." Under Scheer's direction, NASA had handled press requirements for Apollo flights, including worldwide communications network for disseminating live TV photos from moon; developed worldwide exhibit program; begun regular TV and radio network services; started publications program; produced more than 100 major films on NASA programs; and developed education program. (NASA Release 71-26)

- President Nixon accepted resignation of *Apollo 11* Astronaut Michael Collins as Assistant Secretary of State for Public Affairs effective April 11. Collins would become Director of Smithsonian Institution's National Air and Space Museum. (*PD*, 3/1/71, 384)
- Aerobee 350 sounding rocket was launched by NASA from Wallops Station carrying Columbia Astrophysics Laboratory payload to study x-ray polarization. Mission was unsuccessful. (SR list)
- Reuters quoted sources as saying DOD would pay Lockheed Aircraft Corp. $10 million incentive award to 1969 $1-billion contract for development and first year production for superior work on Poseidon missile. Award increased contract price from $70 million to $80 million. Lockheed had bettered contract's specifications for delivery schedule, reliability, and range. (*NYT*, 2/23/71, 41)

February 23: Dr. George M. Low, Acting NASA Administrator, testified before Senate Committee on Aeronautical and Space Sciences and Joint Committee on Atomic Energy on NERVA space nuclear propulsion program: "As we prepared our FY 1972 budget request, we were faced with one of three choices for NERVA"—to continue with full scale development of flight engine; proceed at reduced rate, emphasizing technology and long-lead time development; or cancel NERVA program. "Because there is simply no need to proceed with full scale development now, and in view of fiscal constraints, we decided to . . . continue with the development of long lead items and technology, but to not proceed with the full development effort. This will preserve the technology that has already been gained, and the hard core capability that now exists, without requiring a major expenditure of funds."

NASA and AEC support of NERVA project was "clear example of sustained support over many years of an advanced development even though we have not yet come to the point of proposing the specific missions on which it would be used." While best chemical propulsion systems operated at specific impulse of 450 sec, "with nuclear propulsion we have demonstrated . . . equivalent specific impulse of 825 seconds and have every confidence that the technology we are developing . . . can raise this to at least 900 seconds. Thus, our justification for the large and continuing investment in the NERVA program is that we have a way of breaking one of the principal limiting constraints on space flight by doubling our propulsion efficiencies for many important classes of large space missions."

Dr. Low, in response to questions, testified that NASA plan to stretch out NERVA development program would necessitate immediate employment cutback from 2500 persons to 800. U.S. had invested $1.4 billion in NERVA since 1955. (Transcript; *W Post*, 2/24/71; *B Sun*, 2/24/71, A7)

Milton Klein, Manager, NASA–AEC Space Nuclear Systems Office, testified on NERVA program status: "Eighteen rocket reactors have been tested, each adding importantly to our knowledge. Two experimental engines, which couple the reactor with the other principal components necessary to form a complete nuclear rocket engine, have been operated. The last of these engines, XE, was started up 28 times and tested over a wide variety of conditions. We have accumulated over 14 hours of system operating experience, including more than four hours at or near design power." Specific impulse "of 825 seconds has been demonstrated for extended durations in a flight size reactor. A range of thrust levels, including 75,000 pounds [34,000 kg], has been demonstrated over a wide range of conditions, including the ability to throttle down to about one-half design thrust at full specific impulse. The ability to stop and start at will has been demonstrated. Finally, this system has proved to be quite predictable, an important feature from a development and operational standpoint. All of the technology goals of the program have been met and the potential for performance growth shown." (Transcript)

- Activity at Shuang-ch'eng spaceport—from which Chinese Communists launched *Chicom I* on April 24, 1970—indicated imminent launch of another satellite, according to sources quoted in *Washington Post*. U.S. space specialists had said second Communist Chinese launch would be further proof of steady scientific progress regardless of internal political turmoil in Red China. (Wilson, *W Post*, 2/23/71, A16)
- Citizens Bank and Trust Co. of Maryland would submit requests for space-derived technical information to NASA's Regional Dissemination Centers, NASA announced. NASA centers would advise bank's patrons as to available data and outline charges to cover service cost. (NASA Release 71-23)

February 24: NASA development of short-haul jet transportation system was discussed by Dr. George M. Low, Acting NASA Administrator, at Council for Advancement of Science Writing Seminar on Science and Public Policy in Washington, D.C.: "We estimate that the market for short-haul transportation will approach 40 billion passenger miles by 1980, and may well be between 100 and 300 billion passenger miles by 1995. The problems of developing such a system are exceedingly complex. Involved are aircraft, airways, airports, avionics, ground transportation, and appropriate regulations. To be acceptable, the system must be low in noise, low in pollution, and low in its contribution to the congestion of local transportation. The service must be easily accessible, dependable, and comfortable. And the fare must be reasonable." If development were to take normal course, "it would move ahead very slowly. And in the meantime, a domestic need would go wanting." Foreign competition, with funding by foreign governments, "would step in to fill the gap."

"Here we have a good case where the government should help. The need exists; an integrated approach involving government and industry is required; and the initial development risk is high." DOT was responsible for overall short-haul transportation system; NASA had responsibility for developing experimental research aircraft. "But NASA will not do this alone; we intend to approach this development in partnership—a joint enterprise—with industry. NASA and industry will share the development responsibility . . . and costs. The production airplanes . . . will be built by industry on a purely commercial, competitive basis. But the government will have served the industry in speeding the commercial availability of the aircraft, and will have served the people by speeding the availability of a needed service." (Text)

- NASA launched three sounding rockets. Arcas launched from Barking Sands, Kauai, Hawaii, carried GSFC experiment to 56.7-km (35.2-mi) altitude to obtain ozone measurements in conjunction with *Nimbus IV* overpass. Rocket and instruments functioned satisfactorily and payload was recovered successfully.

   Black Brant VB, launched from Wallops Station at 8:14 pm EST, carried 216.8-kg (478-lb) NRL instrumented telescope to 281.6-km (175-mi) altitude for infrared astronomy studies. Telescope was double-walled with unique helium cooling system that maintained unit's temperature at 5.4 K (−450°F) to prevent telescope's own radiation from interfering with measurements. Rocket and instruments functioned satisfactorily; payload would be recovered after daylight Feb. 25.

   Aerobee 150, launched from WSMR, carried ARC experiment to study micrometeorites. Rocket and instruments functioned satisfactorily. (NASA Rpt SRL; WS Release 71–2; SR list)

- ComSatCorp announced election of John L. Martin, Jr., as Assistant Vice President for Domestic and Aeronautical Satellite Systems. (ComSat Corp Release 71–11)

- Crises facing airlines in early 1960s and 1970s were contrasted by Floyd D. Hall, Chairman and Chief Executive Officer of Eastern Airlines, Inc., in speech before northeast region of Aviation/Space Writers Assn. in New York. Jet aircraft had offered "radically improved seat-mile factor" in 1960s. There were no appreciable unit-cost savings in wide-bodied equipment of 1970's. "In sharp contrast with a decade ago, technology promises little help." Economists were predicting no general economic boom of mid-1960 proportions for at least five years and there was "no traffic surge in sight so great as to mask or counterbalance our other problems." There was "new and ominous militancy" in labor outlook which had "sent costs soaring . . . at rate greater than normal even in our inflation-ridden economy." In 1959, U.S. trunkline aviation industry had been committed to $1.7 billion in new aircraft. "Ten years later, in 1969, we were committed to $4.7 billion—$6 billion when the cost of spares and associated ground equipment is added —or more than three times as much. The value of flight equipment on order at the end of 1969 . . . equals nearly 87 percent of the net book value of existing flight equipment. The total commitment amounts to more than the debt and equity capital of all eleven trunklines combined as of year-end 1969." In 1960s "when airlines sought new capital to finance their new jet fleets," capital had been more readily available "and airlines were able to compete for it against other industries on fairly favorable terms. Today we, a high-risk industry, are competing for it against low-risk industries, and at a time when the availability of capital is limited." (Text)

February 25: U.S.S.R. launched *Cosmos 397* into orbit with 2241-km (1392.5-mi) apogee, 584-km (362.9-mi) perigee, 113.8-min period, and 65.8° inclination. (GSFC *SSR*, 2/28/71)

- President Nixon addressed Nation by radio upon transmitting foreign policy report for 1970s to Congress: "Over the past two years in some fields the Soviet Union and the United States have moved ahead together. We have taken the first step toward cooperation in outer space. We have both ratified the treaty limiting the spread of nuclear weapons. Just 2 weeks ago, we signed a treaty to prohibit nuclear weapons from the seabeds." Most significant result of negotiations between U.S. and U.S.S.R. in past year "could be in the field of arms control." SALT had produced "most searching examination of the nature of strategic competition ever conducted between our nations." If talks continued in cooperative vein there was "reason to hope that specific agreements will be reached to curb the arms race." (*PD*, 3/1/71, 298–304)

- President Nixon transmitted to Congress second annual report on foreign policy, *United States Foreign Policy for the 1970's: Building for Peace*: Space was "clearest example of the necessity for international scientific cooperation and the benefits that accrue from it. The world community has already determined and agreed that space is open to all and can be made the special province of none. Space is the new frontier of man, both a physical and an intellectual frontier. . . . As mutual help and cooperation were essential to life on the American frontier, so it is on the frontier of space. It is with that sense that we approach the sharing of both the burdens and the fruits of our space activity. . . . We have some 250 agreements with 74 countries covering space cooperation." Space was "the only area of which it can be literally said that the potential for cooperation is infinite. . . . We have opened virtually all of our NASA space projects to international participation."

  President said he would submit to Senate shortly ICAO treaty which "recognizes aircraft hijacking as a crime . . . and ensures that hijackers will be subject to prosecution or extradition if apprehended on the territory of contracting states." U.S. intended "to exert every effort to ensure the widest possible international acceptance of this convention."

  President's report said it was "settled U.S. policy to encourage international cooperation in basic science." Closely allied was national policy on technology exchange. U.S. preeminence in both fields posed question as to extent of its sharing scientific and technological knowledge. "There are obvious security implications in many technological developments, for example in the nuclear and space fields." U.S. policy, however, "is to keep those areas as circumscribed as possible, and to take the leadership in encouraging the exchange of scientific and technological information." (*PD*, 3/1/71, 305–77)

- Appointment of Dr. Marshall E. Alper, Manager of JPL's Applied Mechanics Section since 1964, to succeed Howard H. Haglund as Manager of JPL's Civil Systems Project Office was announced by Dr. William H. Pickering, JPL Director. Haglund had been appointed Project Manager of DOT's automated transit system, JPL's principal civil systems project. (JPL Release 571)

February 25–26: Crew station review, during which stowed equipment was fitted to first flight model LRV for first time, was held at Boeing plant in Kent, Wash. Review concentrated on manual tasks of astronauts in setting up and operating LRV. Activities included fitness checks of hardware to be installed on LRV after deployment from LM, walk-through rehearsal of manual loading and preparation of LRV equipment, and reverification of relationships between stowed components and LRV. (MSFC Release 71-33)

February 26: Cosmos 398 was launched by U.S.S.R. into orbit with 232-km (144.2-mi) apogee, 191-km (118.7-mi) perigee, 89.0-min period, and 51.6° inclination. (GSFC *SSR*, 2/28/71)

- NASA's M2–F3 lifting-body vehicle, piloted by test pilot William H. Dana, completed sixth flight after air launch from B-52 aircraft over FRC. Objectives were to obtain M2–F3 and B-52 adapter compatibility by captive flight at 9000-m (30 000-ft) altitude and mach 0.46 to 0.67, expand flight envelope to mach 0.85 by climbing to 13 700 m (45 000 ft) and releasing M2–F3, obtain stability and control data at mach 0.8, investigate lateral phugoid, and obtain performance data at mach 0.8. Alternate flight plan was flown after one of three engine chambers malfunctioned, but vehicle reached mach 0.8 and most objectives were achieved. (NASA Proj Off)
- LeRC announced it was accurately producing and measuring surface contamination to spacecraft by harsh elements in outer space under simulated space vacuum conditions in liquid-helium-cooled space tank linked to computer. Facility could achieve and maintain very high vacuum. LeRC Aerospace Environment Branch also would study effect of exhaust from small reactor control thruster—like those used for Apollo attitude control—on optical properties of nearby surfaces in space, such as Skylab surface. (LeRC Release 71-6)
- UPI released rare photo of sun in which silhouette of jet aircraft created optical illusion that sun was only five times wider than aircraft. Photo, showing sun spots as black patches and large white areas described as possible incipient solar flares, was taken by Pennsylvania State Univ. astronomers using special filters on automatic camera. (UPI Photo Ed)
- NASA announced availability to general-aviation manufacturers of *A Study of NACA and NASA Published Information of Pertinence in the Design of Light Aircraft,* Volumes I, II, and III. Information on structural design, propulsion subsystems, landing-gear loads, flutter, refined performance calculation procedures, and high-horsepower propellers was obtained from 10 000 aeronautical documents that NASA had organized, cataloged, and evaluated. (NASA Release 71-27)
- *The Supersonic Transport,* study by George N. Chatham and Franklin P. Huddle, was released by Science Policy Research Div. of Library of Congress Congressional Research Service. Study found many uncertainties concerning SST: "as to the verity of the competition from foreign SST developments; as to whether the Boeing SST will produce the economic gains claimed for it; as to the extent of engineering risk this vehicle represents; and as to whether it can stand alone, without other supporting elements of a complete system of air transportation." SST's environmental aspects had received bulk of critical attention but "most of postulated effects are found to be non-existent or of a scale making detection difficult." Most uncertainties were in field of economics. "It is likely that these can be resolved only by actual experience with the product in use." (Text)

February 27: President Nixon announced intention to nominate Dr. James C. Fletcher to be NASA Administrator succeeding Dr. Thomas O. Paine, whose resignation was effective Sept. 15, 1970. Dr. Fletcher had been President of Univ. of Utah and College of Eastern Utah since 1964. He had organized Space General Corp., subsidiary of Aerojet-General Corp. in 1960 and served as President 1960–1962 and Chairman of Board 1962–1964. From 1960 to 1962 he was President of Space Electronics Corp., which he founded. Dr. Fletcher had been Associate Director of guided missile laboratory at Ramo-Woolridge Corp. He was member of Air Force Science Advisory Board and Naval Warfare Panel.

Nomination was submitted to Senate March 1. (*PD*, 3/1/71, 383–4; 3/8/71, 437)

- Dr. James C. Fletcher, President Nixon's nominee for NASA Administrator, told Salt Lake City press conference he favored grand tour of solar system's larger planets by unmanned spacecraft. "It would be very exciting for man to go beyond the moon, but I suspect that's beyond the country's budget." Dr. Fletcher did not think it would be necessary to curtail space exploration sharply in coming years despite budget cuts. "We'll have to do better for the same amount of money.... And we may tend to reduce manned flights in favor of unmanned flights." (AP, W *Star*, 2/28/71; UPI, *LA Times*, 2/28/71)
- Special Apollo Awards Ceremony was held at MSC to recognize individuals and support groups for their contributions to *Apollo 13* and *Apollo 14* missions. Dr. George M. Low, Acting NASA Administrator, presented NASA Distinguished Service Medals to Walter J. Kapryan, KSC Launch Operations Director; Eugene F. Kranz, *Apollo 14* Flight Director; Glynn S. Lunney, *Apollo 13* Flight Director; former astronaut James A. McDivitt, Apollo Spacecraft Program Manager; Dale D. Myers, NASA Associate Administrator for Manned Space Flight; and Sigurd A. Sjoberg, *Apollo 13* Flight Operations Director.

  NASA Distinguished Public Service Medal was presented to James J. Gavin, Jr., of Grumman Aerospace Corp.'s LM team and Public Service Medal to Donald E. Eyles of MIT. NASA Exceptional Service Medals were presented to 23 employees of MSC and KSC. Group Achievement Awards were presented to DOD Recovery Forces and Manned Spaceflight Support Network. (NASA Special Ann, 2/26/71; NASA Release 71–28)

February 28: Analysis of discolorations on camera that had landed on moon with *Surveyor 3* April 19, 1967, was reported in *Science* by JPL scientist Dr. Leonard D. Jaffe. Camera had been retrieved from lunar surface by *Apollo 12* astronauts Nov. 20, 1969. Discoloration pattern indicated that camera had been whitened by impact of particles blown from lunar surface by exhaust of LM as it landed and also that many particles had moved at very low angles to the horizontal. (*Science*, 2/28/71, 798–9)

- George H. Stoner, Senior Vice President of Boeing Co., died in Seattle at age 53. Stoner had previously worked on pilotless aircraft and Saturn programs and was to have received NASA's highest industry award for his role in development of Saturn V booster. (AP, *NYT*, 3/2/71, 39)

During February: Opportunity for participating in post-Apollo program had resulted in deep division of opinion among European countries, Dr. Peter Stubbs reported in *Technology Review*: ". . . never has so much furore been generated on the basis of so little hard fact. But politics have taken over the scene to the almost entire exclusion of rational procedures." France and West Germany had not forgiven U.K. for "welshing on them over the—by U.S. standards archaic!— Europa launcher." There was fear that without independent launch vehicle for commercial satellites, Europe could not compete on fair terms with U.S. "Beneath the European tussle of threat and counter-threat . . . run deeper waters: France's desire for her own telecommunications satellite system to propagandize former African colonies; West Germany's urge to become a fully fledged space power; the British Government's intense anxiety to enter the Common Market— even possibly, in part, by blackmailing its way in over the space program." ESRO had built competency "which should not be allowed to go by the board simply because a bunch of politicians can't decide to what proper purpose it should be put." (*Tech Rev*, 2/71, 14)

- Future of U.S. space program was subject of articles in *America Illustrated*, USIA publication distributed in U.S.S.R. Jay Holmes described plans for next two decades in space including planetary exploration and Grand Tours of outer space, practical applications of space technology, and international cooperation in space. Technical advancements directly traceable to U.S. space program were described by James J. Haggerty, who forecast additional space benefits in future. Reusable space shuttle and earth-orbiting stations, continued lunar exploration, and manned voyages to Mars were discussed in article on earth, moon, and Mars. (*Am Ill*, 2/71)

- U.S.S.R. issued 1971–1975 five-year plan with new emphasis on practical benefits from space. Main attention was focused on spaceborne communications, meteorology, earth-resources survey, geographical research, and solution of other economic tasks. (*AF Mag*, 6/71, 54–9)

- NSF issued *Scientific Activities of Independent Nonprofit Institutions 1970* (NSF 71-9). Independent nonprofit institutions employed 23 700 scientists and engineers in January 1970, down from 25 600 employed in 1967. Decrease resulted from shift of several large research institutes from nonprofit sector to other sectors of economy and from slackened rate of increase in R&D activities by nonprofit institutions during 1966–1969.

    Life scientists composed largest occupational group—$33\frac{1}{3}$% of total. Engineers followed with 22%. R&D expenditures amounted to $845 million in 1969, actual increase of 6%, but in terms of constant dollars only 1.3% over 1966 total. Research institutes and nonprofit-administered Federally funded R&D centers together accounted for more than $66\frac{2}{3}$% of scientists and engineers and about 75% of intramural R&D expenditures. (Text)

- NSF released *Federal Funds for Academic Science, Fiscal Year 1969* (NSF 71-7). Ten Federal agencies reported total $2.314 million for academic science activities during FY 1969—same level of funding reported for FY 1968. HEW was primary source, accounting for $1.245 million, or more than 50%, of all obligations from Federal Government. NSF was second with $362 million, DOD was third with $272 million. Dept. of Agriculture accounted for $155 million, NASA for $125 million, and AEC for $121 million. (Text)

March 1971

March 1: President Nixon submitted to Congress nomination of Dr. James C. Fletcher as NASA Administrator [see Feb. 27]. (*PD,* 3/8/71, 437)

- President Nixon presented NASA Distinguished Service Medal to *Apollo 14* Astronauts Alan B. Shepard, Jr., Edgar D. Mitchell, and Stuart A. Roosa at White House dinner. President recalled briefing of President Eisenhower after Soviet launch of *Sputnik 1*: "And the briefing was a very exciting one by one of the scientific advisers to the President. . . . But then I shall never forget that at the conclusion . . . the one who was briefing the President . . . turned to him and said, 'Mr. President, members of the Cabinet, I simply want to say that probably the most important discovery we will make in our exploration of space is not on this chart.' And that was a lesson for me. It was a great lesson for the American people. Exploration . . . involves going into the unknown."

    In space, "we think we know what we want to find and what we may find, but the more we explore we break into new vistas of knowledge. . . . America must continue to be a great nation. We must explore the unknown, not because of what we are going to find or think we are going to find, not because of the uses that we expect to get from space, but because there is something there, something there that we must explore. It's there to find. . . . this Nation with all of its capability and with all of its promise is first in space today and America, as a great nation, must set as its goal remaining first in space, because that's the way to continue to be a great nation. This is a goal that we must set for ourselves.

    "Now, to do this requires thousands of men and women on the ground, devoted and dedicated men in the scientific field, in the engineering field and other areas. It requires very brave men going on these journeys into outer space with all of the uncertainties that we saw on Apollo 13 and very brave women waiting at home with their children, their fathers, and their mothers." (*PD,* 3/8/71, 407–11)

- *Apollo 14* Astronauts Alan B. Shepard, Jr., Edgar D. Mitchell, and Stuart A. Roosa held news briefing in Washington, D.C. Crew showed film and slides and described mission and preliminary results.

    To question on future of manned space flight Shepard said: "The relative comparison of the efficiency of the manned versus the unmanned systems, I think we tried over the years to show where we feel that man has his point, has his place in space . . . and with 14, certainly, there were several instances where basic reflexes, basic judgment, and human instincts made it a successful flight where it would not have been successful otherwise.

    "Obviously, we brought back a lot more rocks than the Lunokhod has brought back. But I don't think that is the point. I think the point is that we have chosen to exercise part of the expenditures of our money along the areas that we have found to be successful; that is, manned flight. I think we have shown on the 14 . . . that manned space

flight can contribute on a scientific basis and contribute efficiently when it comes to assessing the dollar value put on it." (Transcript)
- NASA announced selection of Mississippi Test Facility near New Orleans, La., as site for sea-level testing of space shuttle rocket engines. Testing under simulated altitude conditions would be done at USAF's Arnold Engineering Development Center at Tullahoma, Tenn. Test program would include some 1200 development and acceptance tests from 1973 through 1979, with 45 to 50 sustaining engineering tests per year afterward, and some 100 development tests under simulated altitude conditions from 1974 through 1976. Site Evaluation Board had selected sites after surveying existing Government-owned or controlled properties. Space shuttle engine would be reusable, high-performance, high-chamber-pressure engine using liquid hydrogen and liquid oxygen. First stage would be larger than J–2 engine in Saturn V, with 249 500-kg (550 000-lb) thrust. Orbiter stage engine would have extendable nozzle skirt for use in space. (NASA Release 71–30)
- MSFC issued RFPs to Aerojet-General Liquid Rocket Co., United Aircraft Corp. Pratt & Whitney Div., and NR Rocketdyne Div. on development of main engines for two-stage reusable launch vehicle (space shuttle). Companies had been performing preliminary design and definition studies of shuttle under independent, $6-million, parallel contracts since June 1970. In June 1971 one company would be chosen to develop engine. (MSFC Release 71–37)
- National Civil Service League announced selection of 10 Federal officials to receive annual Career Service Awards. Recipients included NASA Associate Deputy Administrator Willis H. Shapley, cited for "his administrative expertise which enables him to help design major policy directions at NASA and interpret them to the Congress and the public." Honorees would receive $1000, watches, and citations at banquet April 23 in Washington, D.C. (W Star, 3/2/71, A2)
- Aerobee 170 sounding rocket was launched by NASA from WSMR carrying Univ. of Wisconsin experiment to study stellar UV. Rocket and instruments functioned satisfactorily. (SR list)
- Multipurpose system of three large in-orbit satellites and network of 132 earth stations to provide nationwide communication services to various customers was proposed by ComSatCorp President Joseph V. Charyk in letter to FCC Chairman Dean Burch. Proposal would require initial investment of $250 million. (Text)
- AIAA published *The Supersonic Transport: A Factual Basis for Decision*. Report by ad hoc committee concluded that DOT plans for SST prototype program were "reasonable and proper and in the best interests of this nation." Committee found SST performance objectives could be met "with as high a confidence level as on any previous developmental program prior to flight test." Technically, successful production SST could be built following prototype phase, which was estimated to require additional Federal funding of $250 million over two years. Stretchout in prototype funding would cause "extensive immediate cost increases"; resulting delay in production aircraft market penetration would "curtail total revenues, reduce new job opportunities and impair favorable impact on international balance of trade." Abandonment of U.S. SST program could lead to "takeover of U.S.'s traditional civil aircraft sales dominance by foreign interests." SST would meet demands for inter-

continental air travel during 1980s while providing productivity gain. SST would "meet or better all engine-noise specifications" and two prototype SST's would "not produce any equilibrium environmental changes." (Text)

March 2: Apollo 14 astronauts and families visited both houses of Congress. Astronauts reported to Senate and House on mission. Commander Alan B. Shepard, Jr., told Senate they had "a totally successful mission, not only from the areas of scientific and technical endeavors achieved, but also in the areas of international prestige and the tremendous posture of this country." (*CR*, 3/2/71, S2306–7)

- Appointment of Dr. Alfred J. Eggers, Jr., NASA Assistant Administrator for Policy, as Assistant Director for Research Applications, NSF, was announced by Dr. William D. McElroy, NSF Director. Dr. Eggers would head major new organizational unit of NSF designed to support research applied to environmental and societal problems. (NSF Release 71–118)

March 3: Communist China launched her second satellite, from Shuang-ch'eng at 8:15 pm local time. Satellite entered orbit with 1815-km (1127.8-mi) apogee, 266-km (165.3-mi) perigee, 105.9-min period, and 69.9° inclination. Launch was unannounced, but it was monitored by NORAD and accidently observed by USN pilot returning to carrier off Vietnam coast after mission over Southeast Asia. First satellite, *Chicom 1*, had been launched April 24, 1970. (GSFC *SSR*, 3/31/71; Finney, *NYT*, 3/4/71, 1)

- U.S.S.R. launched *Cosmos 399* from Baykonur into orbit with 438-km (272.2-mi) apogee, 196-km (121.8-mi) perigee, 90.8-min period, and 64.9° inclination. Satellite reentered March 17. (GSFC *SSR*, 3/31/71; *SBD*, 3/8/71, 40)

- NASA announced prime and backup crews for Apollo 16 mission, scheduled for launch in March 1972. Prime crewmen were John W. Young (commander), Thomas K. Mattingly II (CM pilot), and Charles M. Duke, Jr. (LM pilot). Backup crewmen were Fred W. Haise, Jr., Stuart A. Roosa, and Edgar D. Mitchell. Mission would last about 12 days, with lunar surface stay time of about 67 hrs—including three EVA periods totaling about 20 hrs. Site for lunar landing had not yet been selected. (NASA Release 71–31)

- Dr. George M. Low, Acting NASA Administrator, presented NASA FY 1972 budget request to House Committee on Science and Astronautics: FY 1972 projects—including Apollo 15 and 16, two Mariner spacecraft, first ERTS satellite, and continuing work toward future flights of Apollo 17, Skylab, earth resources and ATS satellites, and Viking probes —represented "the fulfillment of enterprises of the 1960's, the tailing off to completion of work in progress for many years. By 1974 the number of NASA space launches per year will have declined from 26 in 1966 to eight. After the Skylab missions in 1973, we face at least 4 years in which there will be no United States manned space flight.

"Five years ago there were over 390 000 people in industry employed on NASA work. By the end of fiscal year 1971 that figure will be about 108 000. The decline will continue for a few more months, but we expect it to start increasing by the middle of fiscal year 1972, with the end-of-the-year total being about equal to that at the beginning." U.S. was "running a serious risk of losing too much of the aerospace capability that is an essential ingredient of our long term national strength and security."

There was every indication U.S. would lose leadership in space "unless we move forward now with our space programs for the 1970's. In 1970 the Soviets launched 88 payloads into space compared to . . . 36 for the United States. Their R&D budget is continuing to increase. They have seldom missed a launch window in the exploration of Venus and Mars while we have many gaps in our planetary program. We would expect them to use automated vehicles and sample return spacecraft from the planets, while we have no such program. Finally the Soviets have a systematic continuing manned space flight program and have stated that they expect to fly a space station before we do." U.S. should continue to compete with U.S.S.R. "because accomplishments in space represent a measure of the state of our technology; and in today's world, the level of our technology is of first importance, strategically, economically, and politically." Dr. Low believed U.S. could cooperate with U.S.S.R. "in the areas we have under discussion without prejudicing any of our vital national interests."

NASA was reexamining procurement policies and practices and "beginning a serious critical appraisal of the proper size, composition, and organization of NASA's institutional base." Position reduction of 1500 in FY 1972 budget was in anticipation that "reduction of this magnitude would be found possible without unduly impairing our ability to carry out the program. Our budget justifications show this reduction as a flat percentage cut—about 5.1%—in all of our major activities. The actual adjustments . . . will be refined as our study of the NASA institutional base proceeds."

NASA program was 1.4% of U.S. budget, contrasting with 53% for domestic programs related to human and physical resources, yet U.S. stood to gain "tremendous benefits now and in the future" from exploration and applications of space science and technology. And whole future "as a nation and as mankind" depended on space science. "For in order to survive in the changing environment on our planet Earth, we must understand the processes that govern our universe."

Dr. Wernher von Braun, Deputy Associate Administrator for Planning, said purpose of NASA Meteorology Program was to "develop and demonstrate sensors and spacecraft for making those meteorological measurements that will enable a continued improvement in weather forecasting throughout the world with an ultimate view to a limited degree of weather modification and control of severe weather systems. Projections indicate that by about 1980 it should be possible to reach a global capability to acquire information throughout the atmosphere and to develop comprehensive weather models. Once developed, these models can be used to make 14-day weather forecasts and long-range climate estimates." Ability to develop weather models would permit basic experimenting in weather control.

Space shuttle was key NASA development believed essential in 1970s, Dr. von Braun said, presenting U.S. with its greatest technology challenge for space in this decade. "Preliminary analysis indicates that if the shuttle is used in lieu of current expendable systems, the transportation costs will be reduced by at least a factor of ten. . . . Only by an investment now will we be able to have operational an economical transportation system by the end of the 1970's to support a vigorous and balanced space flight program." (Transcript)

- Findings that linked light flashes experienced by astronauts in periods of otherwise total darkness to high-speed cosmic rays passing through eye retina were reported at NASA and American Nuclear Society Conference in Las Vegas, Nev. Findings were from studies by scientists of Univ. of Washington at Seattle and Univ. of California at Berkeley. Dr. Cornelius A. Tobias and Dr. Thomas F. Budinger of Univ. of California at Berkeley had twice exposed themselves to low doses of cyclotron-accelerated cosmic rays aimed at their eyes in effort to reproduce flashes described by *Apollo 11* Astronaut Edwin E. Aldrin, Jr., and *Apollo 14* Astronaut Edgar D. Mitchell. Scientists said they saw flashes that differed from visual phenomena associated with x-rays or electrical and magnetic exposures. They concluded that flashes came from ionization and excitation of atoms in retina as it was bombarded by cosmic rays. Evidence indicated flashes were detected by retinal rods, structures in eyes that allowed night vision. Radiations were not believed to be hazardous to astronauts on short flights, but long-duration effects had not yet been evaluated. (Wilford, *NYT*, 3/4/71, 18)

- NASA launched two Arcas sounding rockets. Boosted Arcas I, launched from Churchill Research Range, carried Univ. of Houston experiment to 64.9-km (40.3-mi) to measure low-altitude x-ray spectrum and electron flux above 30 500 m (100 000 ft). Rocket and instruments functioned satisfactorily. Data would be correlated with high-altitude data from Black Brant launched 30 sec earlier by Canadian National Research Council.

  Arcas launched from Barking Sands, Kauai, Hawaii, carried GSFC experiment to 58.6-km (36.4-mi) altitude to obtain ozone measurements in conjunction with overpass of *Nimbus IV* satellite. Rocket and instruments functioned satisfactorily. Launch was second in series of two; first had been launched Feb. 24. (NASA Rpts SRL)

- White House announced that Dr. Edward E. David, Jr., President's Science Adviser, had chaired working conference with industry, education, and Government representatives on employment problems of scientists and engineers in aerospace and defense industries. Meeting had been called at President Nixon's request. (*PD*, 3/8/71, 436)

- Rep. Olin E. Teague (D-Tex.) introduced H.R. 5529 "to authorize the coinage of 50-cent pieces to commemorate the Apollo 11 lunar landing and to establish the Apollo Lunar Landing Commemorative Trust Fund." Measure was referred to House Committee on Banking and Currency. (*CR*, 3/3/71, H1254)

March 4: Dale D. Myers, NASA Associate Administrator for Manned Space Flight, testified before House Committee on Science and Astronautics during hearings on NASA FY 1972 authorization bill: "With the Apollo 14 scientific instruments operating properly, we now have begun to develop the networks of instrumentation at different locations on the lunar surface which will allow us to use normal triangulation procedures to establish the details of the moon's structure. We have two passive seismometers and two suprathermal ion detectors in operation, one each at the Apollo 12 and Apollo 14 sites, about 115 miles [185 km] apart. Last week several moonquakes were reported . . .

when the moon reached its perigee. Our instruments pinpointed the location of these events . . . close to the Fra Mauro crater area.

"We also have a laser ranging retroflector at the Apollo 11 Tranquility Base and another at Fra Mauro—750 miles [1200 km] apart. The laser ranging team at the MacDonald Observatory in Texas has reported high-quality returns from the Apollo 14 reflector. As the world's observatories refine their ability to use these instruments, we hope that they will be able to detect random wobbles of the earth's rotational axis. If so, this may contribute to the ability to predict major earthquakes."

During past year, design and essentially all phases of development testing had been completed for Skylab and flight hardware was in fabrication. Definition of space shuttle was nearing completion. To develop limited capability to rescue Skylab astronauts from space, NASA had initiated design action on modification kit to give Skylab CM capacity to carry two men up to orbit and five men back to earth. Stranded astronauts could use Skylab cluster as shelter while modification kit was installed and Apollo-Saturn IB launch vehicle assigned to next revisit was made ready for launch. (Transcript)

- *New York Times* editorial commented on nomination of Dr. James C. Fletcher as NASA Administrator: "Mr. Fletcher's highest priority at NASA will undoubtedly be to try to restore morale, insofar as possible, in a program whose ranks have been decimated by the budget cuts of recent years. NASA's new chief apparently takes the job with few illusions. This is suggested by his recent comments that budgetary problems will probably force a shift from manned to unmanned space exploration, especially beyond the moon." Such realism should "serve Mr. Fletcher well in the difficult days ahead as he tries to lead NASA toward a new equilibrium with a Congress largely disillusioned about the worth of showy and ultra-expensive ventures in space." (*NYT*, 3/4/71)

March 5: Cold-cathode ion gauge left on moon by *Apollo 14* astronauts had discovered first hard evidence of gas escaping from lunar surface, Dr. Gary V. Latham, Apollo program chief seismic investigator, said at MSC. At same time, seismometer left at *Apollo 14* mission landing site had detected moonquake, suggesting quake might have released gas from moon or that sudden venting of gas trapped in cavern beneath lunar surface initiated quake. (O'Toole, *W Post*, 3/6/71, A1)

- Analysis of magnetometer data from *Explorer 35* (IMP-E) (launched July 19, 1967) was described in *Science* by ARC scientists J. D. Minalov and Charles P. Sonett, MIT scientist J. H. Binsack, and Univ. of Manchester, England, scientist M. D. Moutsoulas. Magnetization of selected nonmare areas on moon's far side was inferred from positive disturbances in magnitude of magnetic field exterior to magnetic signature of lunar cavity. (*Science*, 3/5/71, 892–5)

- NASA launched two sounding rockets. Nike-Cajun, launched from Kourou, French Guiana, carried GSFC experiment to 23.1-km (14.4-mi) altitude to verify rocket vehicle and launch facility compatibility and rocket vehicle and radar tracking system compatibility in preparation for future launches. Rocket performed satisfactorily; no instrumentation was carried on flight. Launch was first in series of four.

Arcas, launched from Barking Sands, Kauai, Hawaii, carried GSFC experiment to 58-km (36-mi) altitude to obtain ozone measurements in conjunction with overpass of *Nimbus IV* satellite and to study anomalous absorption detected by Arcas launched Feb. 24. Rocket and instruments functioned satisfactorily and good data were acquired. (NASA Rpts SRL)

- FAA plan to create V/STOL Special Projects Office to formulate and maintain development plan for V/STOL program, provide executive management of FAA V/STOL R&D, and make economic studies and market analyses of V/STOL sales and service potential was announced by John H. Shaffer, FAA Administrator. Jerold M. Chavkin, Special Assistant to Associate Administrator for Engineering and Development, would be acting Director (FAA Release 71-24)
- Election of Dr. Glenn T. Seaborg, AEC Chairman, as honorary member of Soviet Academy of Sciences was reported in *Washington Post*. (*W Post*, 3/5/71, A18)

March 5-6: Apollo 14 Astronauts Alan B. Shepard, Jr., Edgar D. Mitchell, and Stuart A. Roosa spent 26 hrs in Chicago, during which they were honored at City Council meeting, attended civic luncheon, and presided at press conference with 3000 high school students. (*Chicago Today*, 3/7/71)

March 6-8: Apollo 14 Astronauts Alan B. Shepard, Jr., Stuart A. Roosa, and Edgar D. Mitchell, and wives on three-day visit to New York taped TV programs, visited U.N., and rode in motorcade from hotel to City Hall for official reception. (Montgomery, *NYT*, 3/9/71, 1)

March 8: USAF X-24A lifting-body vehicle, piloted by Maj. Cecil Powell (USAF), reached mach 1 and 18 300-m (60 000-ft) altitude during 24th flight from FRC in joint NASA-USAF research program. Purpose of powered flight was to determine lateral-directional derivatives with power on and to determine longitudinal derivatives at mach 0.6 (NASA Proj Off)

- U.S. Supreme Court refused to grant hearing to atheist Mrs. Madalyn Murray O'Hair on appeal of suit charging NASA had knowingly used taxpayers' money for religious activities by permitting astronauts to conduct Bible readings and prayer services while orbiting moon. (Reuters, B *Sun*, 3/9/71, A3)
- Lockheed Aircraft Corp. and Microwave Communications, Inc., would petition FCC for authorization to construct and operate $169.4-million domestic comsat system under name MCI Lockheed Satellite Corp., Washington *Evening Star* reported. Firms said system could be operational by 1975 and produce $70 million per year revenues. (W *Star*, 3/8/71, A12)
- Economic slump that hit Cape Kennedy and Brevard County, Fla., after space program cuts appeared to be bottoming out, *New York Times* reported. Three Chambers of Commerce and county economic development council were spending $260 000 annually, "compared with next to nothing three years ago," to promote KSC tours and area's natural assets. Decline in jobs since 1968 had leveled off and panic sale of homes at bargain prices had ended. (Janson, *NYT*, 3/8/71, 25)

March 9: NASA supercritical wing was flown successfully from FRC during first flight on TF-8A jet aircraft piloted by NASA test pilot Thomas C. McMurtry. Purposes of flight were to conduct functional check of aircraft and stability-augmentation system and evaluate low-speed han-

March 9: *Dr. Richard T. Whitcomb (left), developer of NASA's supercritical wing, congratulated test pilot Thomas C. McMurtry after the successful first flight of the wing on the TF-8A jet aircraft. Program Manager John G. McTigue looked on. The new airfoil shape, undergoing tests at the Flight Research Center, was designed to reduce buffeting at high subsonic speeds and permit increased aircraft performance.*

dling qualities. Aircraft reached 3000-m (10 000-ft) altitude at 220 and 300 knots. (NASA Proj Off)

- U.S.S.R.'s *Lunokhod 1* lunar rover was activated on moon's Sea of Rains after surviving fourth lunar night since landing on moon on board *Luna 17* Nov. 17, 1970. (*SBD*, 3/10/71, 55)

- U.S.S.R. had again tested "hunter" satellites, designed to locate and destroy other satellites, George C. Wilson reported in the *Washington Post*. *Cosmos 394* (launched Feb. 9) had entered circular orbit and had been intercepted by *Cosmos 397* (launched Feb. 25), he said. Satellites' inclination suggested launches were from Plesetsk rather than Tyuratum [Baykonur] and "probably means that the Soviet Union considers its satellite inspection system in the operational rather than experimental category." (*W Post*, 3/9/71, A3)

- Dr. John E. Naugle, NASA Associate Administrator for Space Science and Applications, testified in FY 1972 authorization hearings before House Committee on Science and Astronautics: ". . . true boundaries of our ecological system have not been determined, but they certainly extend to the known dimensions of the universe. The forces and laws which affect the most distant star are directly relevant to the problems of energy generation and control on Earth. Mankind is just beginning to grasp the fact that all natural systems are intimately related and interdependent" and man must learn "to manage wisely what he can, and to live in equilibrium with what is beyond his control. It is to these ends ... that we have structured the nation's continuing space activities in science and applications."

Most significant new astronomy program element was HEAO, "designed to unlock some of the most profound and puzzling secrets of modern physics." There was evidence that in pulsars and quasars "a whole new regime of power generation exists and operates at levels that cannot be explained by modern nuclear physics." Understanding these energy processes might lead to harnessing of powers "heretofore unimaginable."

Both planetary and lunar exploration should provide clues to earth's past and future, to factors that caused long-period environmental changes and how to control them. Earth observations program also was addressed to fundamental problems of environmental and ecological management. Satellites could see, record, and transmit information on causes and effects in global air-ocean-earth interactions.

Trend in U.S. space science and applications had been decreasing budget and launch rates; OSSA had had to "cut back ongoing projects, dispense with backup payloads, stretch launch schedules and mission intervals, and defer promising new projects," but had "preserved a steady momentum and a balanced program thrust wherever possible." High level of productivity in science, exploration, and applications had been possible "because of the background in technology, in engineering, in conceptual thinking that was built up earlier." But many valuable and important missions had been deferred into future, "a future we expect will permit us to revolutionize the way space is explored and used." Key feature of this revolution would be reusable space shuttle "that makes access to space for man and machines an economical, routine excursion rather than the difficult task it presents now." (Transcript)

- NAS released *Priorities for Space Research, 1971–1980*. Report of study group of NAS–NRC Space Science Board recommended priorities for NASA OSSA programs at three funding levels for space missions in 1970s. Study was made at NASA's request. Highest priorities were recommended for probes of Venus and Jupiter beginning in 1975; completion of three remaining Apollo manned lunar missions and unmanned landers and rovers for lunar and planetary studies; series of HEAO satellites for x-ray, gamma-ray, and cosmic-ray exploration; Small Astronomy Satellites (SAS); increased astronomy rockets, balloons, and aircraft; development of mirror for Large Space Telescope (LST); continued study of earth-orbiting gyroscope and ESRO sun-orbiting satellite for studies of gravitational physics in mid-1970's; pair of satellites to study interaction between earth's magnetosphere and solar wind and doubling of funds for solar-terrestrial physics; three satellite programs and increased data analysis for meteorology and earth resources surveys; and ground-based research to support Viking softlander and future exobiology missions. Report recommended next larger level of NASA funding provide for 1.5-m (60-in) space telescope, OSO, and additional Atmospheric Explorers and Earth Resources Satellites. Grand Tour of planets between 1975 and 1980 was recommended only if NASA's budget was increased still further. (NAS–NRC–NAE *News*, 3/9/71)

- William A. Anders, NASC Executive Secretary and *Apollo 8* Astronaut, and Astronauts John L. Swigert, Jr., William R. Pogue, and Jack R. Lousma met with President Nixon at White House to report on their visits to 17 college campuses from October to December 1970. At White House press conference later, Anders said meetings with students had been "very positive." Students had been "very curious about the Space Program." (*PD*, 3/15/71, 475; Transcript)

- Communist China possibly had made limited-range 3200-km (2000-mi) shot of prototype ICBM in late 1970, Secretary of Defense Melvin R. Laird said in posture statement before House Armed Services Committee. Range probably had been limited to keep test within Chinese national borders. "Estimated earliest possible [Communist Chinese] ICBM capability is 1973, with the more likely time being the mid-1970s." (Testimony)

March 10: Senate Committee on Aeronautical and Space Sciences unanimously approved nomination of Dr. James C. Fletcher as NASA Administrator, following one-hour hearing. Dr. Fletcher told Committee U.S. had made "some small steps" toward cooperation with U.S.S.R. in space. "I think we can make these even larger steps." Asked if he thought NASA needed any change in direction Dr. Fletcher said, "No organization is perfect, and a new administration will always look at the program to see if it can make any changes and I intend to do that." In response to question from Sen. Margaret Chase Smith (R-Me.), Dr. Fletcher said it was "not NASA's mission" to act as technological WPA in aiding depressed areas through the channel of space contracts. (*CR*, 3/10/71, D144; O'Toole, *W Post*, 3/11/71)

- NASA launched series of five sounding rockets from Wallops Station to collect data for comparison with data collected during March 7, 1970, solar eclipse. Two Nike-Cajuns each carried 31-grenade payload to 113-km (70.2-mi) altitude. All but one grenade ejected and exploded as planned. Three Nike-Apaches carried pitot-static probe payloads to 179-km (111.2-mi), 178-km (110.6-mi), and 171-km (106.3-mi) altitudes. Rockets and instruments functioned satisfactorily and good data were obtained. (NASA Rpts SRL)

- First flight model LRV was formally delivered to NASA in ceremony at Boeing Co.'s Space Center at Kent, Wash. Dr. Eberhard F. M. Rees, MSFC Director, accepted first of three flight models from Boeing Co. Group Vice President for Aerospace O. C. Boileau. Model would be shipped to arrive at KSC March 15 for final checks and installation aboard Apollo 15 LM. Delivery would be two weeks before April 1 contract delivery date. (NASA Release 71-38; Boeing Release 3/10/71)

- NASA plans to develop first space rescue capability during 1972 Skylab program were described by Dr. Wernher von Braun, Deputy Associate Administrator for Planning, before AAS meeting in Washington, D.C.: Studies indicated NASA could have rescue spacecraft standing by on launch pad 20 days before scheduled Skylab launch in spring 1973. Rescue technique would be facilitated by equipping Apollo spacecraft with two extra seats beneath three regular crew seats. In space emergency, modified spacecraft would be launched with two-man crew rather than three. It would rendezvous with distressed Skylab vehicle to recover three-man Skylab team. Rescue vehicle would have im-

proved hydraulic telescoping landing legs to absorb impact on land if water landing was not practical. (AP, *Huntsville Times*, 3/11/71)
- NASA Associate Administrator for Tracking and Data Acquisition Gerald M. Truszynski testified on FY 1972 budget request during House Committee on Science and Astronautics authorization hearings. During 1970 OTDA supported some 50 ongoing scientific, applications, planetary, and lunar experiment missions as well as some 14 new flight projects launched in year by NASA and DOD. During *Apollo 13* mission, with safe return after inflight explosion, MSFN "demonstrated that it is, in fact, the astronauts' life-line to earth." Valuable clues to origin of explosion were found in analysis of recorded telemetry data. MSFN was still supporting experiments left on moon by *Apollo 12* and *14* missions. After FY 1972's Apollo 15 and 16 missions, network would be continuously monitoring 16 ALSEP experiments. Satellite network provided tracking and data acquisition for average monthly workload of some 40 earth-orbital satellites, including NASA's scientific and applications satellites and projects of other Government agencies and foreign countries. DSN supported *Mariner 6* and *7* probes in extended mission phase and *Apollo 13*, as well as four ongoing Pioneer missions (*6* through *9*). Data acquired by Goldstone antenna on effect of solar gravitation on Mariner spacecraft signals had verified Einstein theory of delay of 200 microseconds in return of signals from gravitational field of sun. Construction continued on second 64-m (210-ft) antenna at Tidbinbilla, Australia, and ground was broken for third near Madrid, Spain. (Transcript)
- ComSatCorp released 1970 annual report. Highlights of 1970 included: declaration of corporation's first dividend; increase in net income from 70 cents per share to $1.75; submission to FCC of proposal for domestic comsat system for lease to AT&T; completion of Bartlett earth station at Talkeetna, Alaska, and nine new foreign earth station antennas, bringing total antennas to 51; beginning of construction of new antenna at Andover, Me.; launch of the last three Intelsat III satellites; and election of Joseph H. McConnell as ComSatCorp Chairman. (Text)
- Informatics Tisco announced it had received one-year, $4.8-million, incentive contract extension from NASA for operation of NASA Scientific and Technical Information Facility at College Park, Md. (Informatics Release)
- Sen. Joseph M. Montoya (D-N. Mex.) introduced S. 1184, "to establish a Department of Science and Technology, and to transfer certain agencies and functions to such Department." Bill provided that NASA, NSF, and AEC be transferred to new department. Sen. Montoya had introduced similar legislation on Oct. 12, 1970. (*CR*, 3/10/71, S2800–12; NASA *LAR*, X/22)
- Sen. Winston L. Prouty (R-Vt.) introduced S. 1168, $619-million NSF FY 1972 authorization bill. (*CR*, 3/10/71, S2783–4)

March 11: Senate confirmed nomination of Dr. James C. Fletcher as NASA Administrator. (*CR*, 3/11/71, S3088)
- Goddard Memorial Trophy for 1971 was presented by Mrs. Esther Goddard to James E. Webb, NASA Administrator from 1961 to 1968, at 14th annual Goddard Memorial Dinner in Washington, D.C. National Space Club award—named for late Dr. Robert H. Goddard, "father of American rocketry"—was given to Webb for "unprece-

dented accomplishments." Citation said his management of Apollo program leading to manned lunar landing "was matched only by the accomplishments in international scientific cooperation, the technology utilization program, the sustaining university program and the greatest industrial team ever assembled."

Astronautics Engineer Award was presented to Dr. Rocco A. Petrone, Apollo Program Director, "for his outstanding personal leadership of the Apollo mission to the Moon."

Wallace H. Deckert of ARC Advanced Aircraft Programs Office was awarded Hugh L. Dryden Memorial Fellowship "in recognition of the need for the continued scientific and intellectual growth of the individual, who is already pursuing the challenges of research, exploration and administration in astronautics, so that he may continue to seek and maintain a pre-eminence in outer space for the United States."

Nelson P. Jackson Aerospace Award was presented to NASA–NOAA Space Meteorology Team "for opening a revolutionary dimension in observing earth's weather, and for significant help in mankind's perpetual battle with the elements." National Space Club Press Award was presented to Kenneth F. Weaver of *National Geographic* magazine "for his illuminating and imaginative editorial contributions to astronautics and his excellence in conveying to the public at large man's expanding concept of his cosmos."

Dinner, which honored "decade and more of international cooperation in space," was attended by *Apollo 14* astronauts, Soviet Ambassador Anatoly F. Dobrynin, Indian Ambassador Lakshmi Kant Jha, comedian Bob Hope, and U.S. Ambassador to U.N. George Bush. (Program; Dean, W *Star*, 3/13/71, C1)

- NASA OART officials testified before House Committee on Science and Astronautics during hearings on FY 1972 NASA authorization. Roy P. Jackson, Associate Administrator for Advanced Research and Technology, said aeronautics was largest part of OART program for first time. In military aviation, "work is under way at the Centers in support of the F–14 for the Navy. . . . Over a dozen wind tunnels and simulators . . . are used in this support. We are helping the USAF with . . . the F–15, their new B–1 bomber and we plan to assist" with the A–X. One ARC facility was on indefinite loan to USA. "In all, we use over a third of our available facility time to support military needs. Research on transonic aerodynamics, the NASA supercritical airfoil, fly-by-wire flight controls, spin stabilization, and advanced structures using beryllium or composites are some of the areas where NASA is making research contributions to meet military requirements in the future."

  In civil aviation effort, OART was providing DOT with assistance on SST. ARC was assisting FAA in certification studies of Concorde and SST. ARC flight simulator for advanced aircraft was "proving to be the most useful facility available in the free world for studying airworthiness standards for supersonic transports." NASA and DOT were completing joint study of "relationships between the benefits that accrue to the Nation from aviation and the level of R&D effort. The study displays the ways in which civil aviation can affect regional development, population distribution, [and] land use, and contribute to many other social and economic goals of the Nation." Problems under study included aircraft noise, air pollution from aircraft engine emissions, and aeronautical operating problems "such as approach and landing safety, delays, congestion, and community noise caused by operation of aircraft." OART Aeronautical Operating Systems Div. had established joint STOL operating problems experiments program with DOT and FAA, and Aeronautical Life Sciences Div. "to study man's role in advanced aeronautical systems . . . as an operator, as a passenger, and as a member of the community exposed to air operations."

Apollo 11 Astronaut Neil A. Armstrong, Deputy Associate Administrator (Aeronautics), described improvements in aircraft materials and structural design: "I have been particularly impressed with the progress of the composite materials. The potential of boron and carbon fibers imbedded in an epoxy resin matrix to provide high strength, low weight structures has been discussed with you for several years. This year the testing of special glass fibers has revealed the possibility that low cost, stiff composites may be fabricated with much greater strengths than previously thought possible. Composites are finding application both as a replacement for metallic components and by augmenting metal strengths using selective reinforcement. In the forthcoming year we propose to construct several airplane components of primary structure for installation and flight demonstration. We believe that these new materials will find widespread use in the industry when the technical risk has been demonstrated to be acceptable."

Milton Klein, Manager of Space Nuclear Systems Office, reported on progress of NERVA program: "Through the systematic engineering design process . . . design concepts to be incorporated in NERVA were formulated, requirements were specified, and a baseline engine design has been established. The design provides a thrust level of 75 000 lbs [334 000 newtons], and a specific impulse of 825 seconds. High reliability and safety in operation are provided through the use of redundancy in critical areas and conservative design approaches. The endurance goal for the reusable system is 10 hours, coupled with the ability to undergo 60 start-and-stop cycles, and the key fuel element technology continued to make good progress toward that goal." (Transcript)

- RCA Corp. announced it had filed application with FCC for construction and operation of $198-million domestic satellite system for 50 states and Puerto Rico. System would consist of three satellites and 13 major earth stations. At Washington, D.C., news conference, Howard R. Hawkins—President of RCA Global Communications, Inc., and RCA Alaska Communications, Inc.—said system could begin first phase in 1974. System would include distribution of motion pictures by closed circuit telecasts instead of film. (AP, *NYT*, 3/12/71, 62)
- NASA announced award of $18 704 300 cost-plus-award-fee contract to Boeing Co. for installation and technical support services at KSC. Contract was for 11 mos with four one-year options. New contract combined work previously performed under three separate contracts. TWA, which had been performing support services, had been denied preliminary injunction to halt contract award, in Federal court March 10. (NASA Release 71-39; *WSJ*, 3/12/71)
- Sen. George McGovern (D–S.D.) introduced, for himself and cosponsors, S. 1191 "to assist workers whose jobs may be terminated by reduction in defense and space spending, to establish a system of benefits to ease

the transition of such workers to civilian occupations, to minimize the hardships encountered by communities which are dependent upon defense industry and employment, to encourage cooperation between the United States and defense contractors in meeting the challenge of economic conversion and diversification brought on by changing defense needs, . . . and to facilitate the transfer of public and private resources to new national priorities." Bill would require that portion of profits from military, space, and atomic energy contracts be deposited in reserve to be administered by National Commission on Peacetime Transition. Funds would be available to benefit workers who lost income as a result of cutbacks. Bill was referred to Senate Committee on Government Operations. (CR, 3/11/71, S2926-35)

- *Kansas City Times* editorial commented on appointment of Dr. James C. Fletcher as NASA Administrator: Dr. Fletcher had "earned respect at Utah as a master of administrative analysis and detail, a man able to wring the last measure of usefulness out of every available dollar and, at the same time, to command intense loyalty from those who work with him. No qualities are likely to prove of more service to a man responsible for leading the American space program out of its nadir of tight money and flagging morale." (*KC Times*, 3/11/71)
- *New York Times* commented on orbiting of second Communist Chinese satellite: It had provided "impressive evidence of the advance of Communist China in war-related technology. It hardly needs underlining that a rocket capable of orbiting a sputnik could also deliver a hydrogen bomb to a distant target on earth." It was evident that "Mao Tse-tung has been trying to insulate a small group of scientists and their activity in weapons development from the general turmoil of the cultural revolution as well as from the inadequate supply of practically everything in normal Chinese economic life. That effort has apparently been successful." (*NYT*, 3/11/71, 38)

March 12: Apollo 14 astronauts were presented Distinguished Service Medals at DOD ceremonies. Secretary of the Navy John H. Chaffee presented USN DSM to Astronauts Alan B. Shepard, Jr., and Edgar D. Mitchell. Secretary of the Air Force, Dr. Robert C. Seamans, Jr., presented USAF DSM and Command Pilot Astronaut Badge to Astronaut Stuart A. Roosa. (*NASA Activities* 4/15/71, 61; DOD Release 218-71; MSC Hist Off)

- New photographic process that produced nickel image by exposing ammonium, lithium, lead, and nickel hypophosphite to electrons or x-radiation had been developed by LeRC scientists Dr. Warren H. Philipp, Dr. Charles E. May, and Stanley J. Marsik, *Lewis News* reported. Advantages were that nickel compound was cheaper and more abundant than silver used in standard photographic process and, because nickel compound was not sensitive to visible light, process eliminated need for darkroom or cassette containers for undeveloped film. (*Lewis News*, 3/12/71, 1)

March 13–17: Explorer 43 (IMP-I) Interplanetary Monitoring Platform was launched by NASA from ETR by three-stage, thrust-augmented Thor-Delta M-6 booster at 11:15 am EST. Spacecraft entered highly elliptical orbit with 206 049-km (128 032-mi) apogee, 241-km (149.8-mi) perigee, 6012-min period (4 days 12 hrs 12 min), and 28.6° inclination. Primary objective was to investigate during period of decreasing solar activity and through several solar rotations, the nature of the interplanetary medium and interplanetary-magnetospheric interaction, including characteristic features of solar wind, interplanetary fields and sector structure, and modulation effects on cosmic rays. By March 17 onboard computer and 9 of 12 science experiments were operating satisfactorily.

Explorer 43 was first of series of second-generation spacecraft designed to study solar-lunar-terrestrial relationships. It was eighth of 10 IMPs and at 288 kg (635 lbs) was largest as well as most advanced spacecraft in Explorer series. Seventh IMP, *Explorer 41*, had been launched June 21, 1969, and was still operating. IMP program was managed by GSFC under OSSA direction. (NASA Proj Off; NASA Release 71–35; *SSR*, 3/31/71)

March 13: U.S.S.R.'s *Lunokhod 1* unmanned lunar rover explored 457-m (500-yd) crater on moon, made chemical analysis of soil, photographed area around crater, and "engaged in Cosmo-physical measurements," Tass announced. (UPI, W *Star*, 3/14/71, A15)

- Nike-Apache sounding rocket was launched by NASA from WSMR carrying Univ. of Colorado airglow experiment. Mission was unsuccessful. (SR list)
- In cable to Sen. William Proxmire (D-Wis.), Jean-Jacques Servan-Schreiber, member of French National Assembly, said Anglo-French Concorde supersonic transport "looks to us...like an industrial Vietnam." Every Concorde cost analysis had proved to be wrong. Cost had "multiplied here, as it will everywhere, four times the initial evaluations." Facts had not been available until recent months. "Now the public eye is on the project and what it sees is bankruptcy." (Kenworthy, *NYT*, 3/14/71, 21)

March 14: FRC and MSC scientists and engineers would begin research in adapting Apollo hardware to "fly-by-wire" electronic control system for future aircraft, NASA announced. System substituted lightweight wires for push-rods and added redundant wire paths at different locations in aircraft, to make overall system less vulnerable to damage. Electronic system could soften bumps, sways, and lurches of aircraft in rough weather and ease pilot's load. (NASA Release 71–36)

- Moscow newspaper *Sotsialisticheskaya Industriya* quoted unidentified "Chief Designer of Spaceships" as saying U.S.S.R. was preparing further manned orbital flight of long duration as prelude to creation of permanent space laboratory. Soyuz spacecraft had "undergone necessary modifications to insure fulfillment of a long and extensive program." Later *New York Times* said designer was believed to be Mikhail K. Yangel, member of Soviet Academy of Sciences. (*NYT*, 3/15/71)

March 15: Detection of "possible spinar"—celestial object which could be link between galaxies and quasars in outer space—was described by NRL team headed by Dr. Herbert Friedman in *Astrophysical Journal*. Spinar, 200 million light years from earth, could be entire galaxy of collapsed and rotating neutron stars or single, massive pulsar. Object had been found by x-ray detector aboard Aerobee sounding rocket.

In interview with Washington *Evening Star* later, Dr. Friedman said new source, classified as a Seyfert Galaxy, emitted "soft x-rays."

Spinar would radiate 1000 times less visible light than quasar and 12 000 times more infrared than visible light energy—approximately what had been found in Seyfert Galaxy. But x-ray emissions perhaps as much as 10 trillion suns indicated it might be powered by same mechanism as pulsar. Dr. Friedman speculated that pulsar such as spinar could be missing link between conventional galaxy like Milky Way and quasars and that spacecraft-mounted x-ray detectors would find "tens of thousands" of such sources. (*Astrophysical Journal*, 3/15/71, L-81-5; Lannan, W *Star*, 3/19/71, B5)

- NASA launched two Nike-Tomahawk sounding rockets—one from Fox Main, Dew Line Station, Alaska, and one from Pin Main—to study electric fields. Rockets and instruments functioned satisfactorily. (SR List)
- House Committee on Appropriations by vote of 26 to 15 approved full $290 million requested by DOT for continued SST development. (*CR*, 3/15/71, D164; AP, *NYT*, 3/16/71, 62)
- FCC deadline for filing proposals to build domestic comsat systems. Eight proposals had been filed, calling for launch of 19 satellites and expenditure of more than $1 billion.

  Fairchild Hiller Corp. and Tele-Communications, Inc., had said they would petition FCC for permission to orbit domestic comsat systems, *Wall Street Journal* reported. Fairchild Hiller proposal called for two satellites in equatorial orbit with third reserve satellite on ground. Tele-Communications envisioned $66-million comsat system of two satellites and six ground stations to start. (*WSJ*, 3/15/71, 13; 3/16/71, 7)
- Feasibility of comsats for helicopter communications had been demonstrated by tests conducted with DOD's *Tacsat 1* comsat over Pacific Ocean, *Aviation Week & Space Technology* reported. Tests at UHF and super-high (X-band) frequencies used different antennas for each, mounted on Army/Bell UH–ID helicopter atop rotor mast to avoid signal modulation by rotating blades. (*Av Wk*, 3/15/71, 39)
- U.S. and U.S.S.R. opened fourth round of Strategic Arms Limitation Talks (SALT) in Vienna. (AP, W *Star*, 3/15/71, A3)
- President Nixon, in Key Biscayne, Fla., sent message to Congress transmitting 10th annual report of U.S. Arms Control and Disarmament Agency: "For the first time a realistic dialogue is taking place between the Soviet Union and ourselves about the management of our strategic relations. The mutuality of interests which brought us to the table encourages our hope that the Strategic Arms Limitation Talks will succeed." (*PD*, 3/22/71, 492)
- *New York Times* editorial commented on Pioneer F: "The mission of Pioneer F—the spacecraft that will be launched next year to visit giant Jupiter and send back photographs and scientific data—opens some of the most exciting vistas yet in the entire space program. If accomplished ... this mission would mark a major advance in man's understanding of a hitherto unpenetrated section of the solar system." (*NYT*, 3/15/71, 38)

March 15–16: NASA launched two Nike–Cajun sounding rockets from Kourou, French Guiana, carrying GSFC experiments to measure ozone distribution at low latitudes near equinox from 70- to 15-km (44- to 9-mi) altitude. Rockets reached 75.5- and 72.5-km (46.9- and 45.1-mi) altitudes with rockets and instruments functioning satisfactorily. Launches were second and third in four-launch series; first had been conducted March 5. (NASA Rpts SRL)

March 15–18: AIAA Space Shuttle Development, Testing, and Operations Conference/NASA Space Shuttle Technology Conference was held in Phoenix, Ariz. Charles J. Donlan, Deputy Associate Administrator (Technical), OMSF, said NASA plans called for unmanned testing of reusable space shuttle beginning in 1976 and manned tests in 1977. Fully operational vehicle would be ready in 1979.

Col. John G. Albert, USAF Director of Space Operations, said DOD was "putting its faith in the shuttle and as a result we are not developing any other space rocket beyond the present Titan 3. We intend to use the shuttle for all military space operations." Use would include communications, navigation, weather-watching, and surveillance. Shuttle would place military satellites in orbit and perform specialized missions in orbit for periods up to seven days. Col. Albert told news conference later that DOD would "certainly make a sizable funding contribution at the proper time" to assist NASA in shuttle development. To be useful to the military, shuttle must be able to operate like jet transport with quick turnaround on runway after conventional landing. (AP, B *Sun*, 3/16/71)

Jean Bernard LaGarde, liaison officer for European Space Conference and French representative to ESC, said Europe would contribute $1 billion to U.S. manned space programs in future if U.S. lifted ban on launching comsats for European nations. LaGarde said U.S. would have to give positive answer by late April or early May. (*LA Times*, 3/17/71)

AIAA President Martin Goland at banquet March 16 presented Haley Astronautical Award of $500, medal, and certificate, to *Apollo 13* Astronauts James A. Lovell, Jr., Fred W. Haise, Jr., and John L. Swigert for "exceptional manner in which the Apollo 13 crew conducted themselves and their spacecraft under extraordinary circumstances of extreme stress."

Octave Chanute Award of $500 and certificate was presented to William M. Magruder, Director of SST Development for DOT, for "outstanding contributions by a pilot to the development of safer, more efficient air transportation."

AIAA History Manuscript Award for "best historical manuscript dealing with the science, technology, and impact of aeronautics and astronautics on society" went to Constance McLaughlin Green and Milton Lomask for "A History of Project Vanguard" [published in NASA Historical Series in March as *Vanguard—A History*, SP-4202]. (AIAA Release; *AIAA Bull*, 3/71; NASA Hist Off)

March 16: Peking Radio announced that Communist China's second satellite, launched March 3, was transmitting scientific data in "whistles, pips, bleeps, and blurrs." Announcement broke two-week silence on launching of satellite. New China News Agency said satellite weighed 220.5 kg (486 lbs) and was in elliptical orbit. (Reuters, *W Post*, 3/17/71, A12)

• NASA Associate Administrator for Organization and Management Richard C. McCurdy testified on research and program management (R&PM) budget request during House Committee on Science and Astronautics hearings on NASA FY 1972 authorization: FY 1972 request was $21 million less than in FY 1971, primarily because of continuing reduction in permanent positions. Request of $697.4 million was based on further reduction of Government personnel of 1500 positions, lowering employment to 28 350—total reduction of 6971 positions from 1967 peak and lowest level since 1963. Cut of "some 20 percent in the 1967–72 period is by a considerable margin the greatest, proportionately, to be made by any Government agency in the period." Of total NASA force, scientists and engineers made up 44.8%, technicians 26.7%, administrative personnel 13.9%, and clerical 14.6%. Hiring restrictions had reduced number of new college graduates added to work force and raised average age to 41.3, with rate of 8/10 a year each year, endangering health and vigor of NASA.

"In the past decade, the aerospace industry, overall as well as that portion serving NASA, has passed through a substantial perturbation. Contractor employment climbed to a high peak, then dropped drastically. The current estimate for industry employment on NASA work at the end of this year and next is about 108,000—less than one-third of what it was at the peak of the space program. But with the major reductions in the space program over the last several years, this is how it must be."

Daniel J. Harnett, Assistant Administrator for Industry Affairs and Technology Utilization, testified on NASA management of its new technology as "a repository of national resources": Some 20 000 technical contributions had been disclosed by NASA contractors, 3500 in past year. Another 4000 had been reported by NASA employees from inhouse projects. Contributions reported were evaluated, defined, and published for use by others. Allocations for FY 1972 were increased over previous year as NASA began to work with innovative events of early stages of 1970s R&D programs. (Transcript)

- President Nixon transmitted NASA's 22nd Semiannual Report to Congress covering July 1 to Dec. 31, 1969.

  In letter accompanying report, Dr. George M. Low, Acting NASA Administrator, wrote, "The two successful Moon missions [*Apollo 11* and *12*] fulfilled the national goal of a manned lunar landing and safe return within the decade of the sixties and convincingly demonstrated the technological competence of the Apollo program." Flights had shown "value of the space program as a unifying force in international relations, for interest in the Moon landings and in the astronauts transcended national boundaries. This Nation, in turn, took the view that the achievement was a triumph for all mankind even though the deed itself was performed by American astronauts." (Text; *CR*, 3/17/71, H1631)

- President Nixon transmitted to Congress NSF annual report for FY 1970. He said: "Science has entered an era of unprecedented fruitfulness. The investment we have made in the last three decades offers us an array of scientific talent and a heritage of scientific achievement unprecedented in human history." (*PD*, 3/22/71, 495–6)

- NASA launched Nike-Tomahawk sounding rocket from Fairbanks, Alaska, carrying Univ. of Alaska experiment to conduct auroral studies. Mission was unsuccessful. (SR list)

- Sen. Edward M. Kennedy (D-Mass.) introduced S. 1261, to authorize

NSF "to undertake a loan guarantee and interest assistance program to aid unemployed scientists and engineers in the conversion from defense-related to civilian, socially-oriented research, development, and engineering activities." (*CR*, S3225–31)

- Senate passed S.R. 17, establishing Joint Committee on the Environment. (*CR*, 3/16/71, S3317–25)

March 17: NASA supercritical wing, flown on TF–8A jet aircraft piloted by test pilot Thomas C. McMurtry, successfully completed second flight from FRC. Aircraft reached 7600-m (25 000-ft) altitude and 300 knots (mach 0.75). Purpose of flight was to continue evaluation of aircraft's handling qualities at higher altitudes and speeds. (NASA Proj Off)

- Dr. George M. Low, NASA Deputy Administrator, testified before Senate Committee on Aeronautical and Space Sciences during hearing on U.S.–U.S.S.R. space cooperation: Recent agreements had been directed toward "achieving common objectives," not "exchange of technology." Review of relations between NASA and U.S.S.R. "suggests that we should exercise considerable caution in assessing the prospects for prompt and effective implementation of these agreements." But U.S. negotiators of October and January agreements had observed "definite improvement" over past decade. Negotiations had been "more straight forward, less political. . . . A closer kind of relationship seems to be intended by the Soviet side. The top level of the Soviet Academy for the first time participated directly. The negotiations were attended by the first instances of meaningful access to Soviet facilities."

  Dr. Low said he had "spent a half-hour or more inside the Soviet simulator with a young Russian engineer who explained in detail the working of that spacecraft and answered every question that I had. Later on, I was in the docking and rendezvous trainer with one of their cosmonauts and was allowed to attempt to fly a docking of one Soyuz against another. I did not succeed in bringing the two together. In addition, in January our delegation was invited to the principal Soviet tracking station in the Crimea. Unfortunately, the time available . . . did not permit us to make this visit, but we have every intention of doing so in the future."

  Advocating "air of cautious optimism," Dr. Low said U.S.S.R. had, "in effect, met the schedules laid down in the first agreement with a variance of only a few days. In any event, we have laid out and agreed to specific milestones for an even-handed exchange, so that neither side will gain more than the other."

  In response to question, Dr. Low said military components of U.S. and Soviet space programs had not been discussed at NASA–Soviet Academy of Sciences meetings. (Transcript)

- Nike-Tomahawk sounding rocket was launched by NASA from Fairbanks, Alaska, carrying Univ. of Alaska experiment to conduct auroral studies. Rocket and instruments functioned satisfactorily. (SR list)

- Doyle G. Berry, owner of Berry Aviation Co., said in New Orleans that he had reached exclusive agreement with U.S.S.R. to sell Tu-144 supersonic transport in Western Hemisphere. Contract was being drawn up subject to State Department approval. (*NYT*, 3/19/71)

- Two Sprint missiles successfully intercepted ICBM nosecone over Pacific in latest test of Safeguard defense system. It was first time two missiles were used in Safeguard test. Target was launched from Vandenberg AFB, Calif., 6700-km (4200-mi) distance. (Reuters, *C Trib*, 2/23/71, 8)

March 18: House voted 215 to 204 to end Federal funding for SST development. It was first time in seven years that House had voted against measure though it had passed House in 1970 by only 20 votes. *Washington Post* said turnaround was due to growing ecological and economic lobby against SST program, to new House members who voted 33 to 18 against it, and to spotlight on record vote. It was first time House had cast recorded vote on SST issue. *Post* said supporters of SST program predicted House vote would kill program, at least for present; private industry could not raise funds needed to finance it. (*CR*, 3/18/71, H1717–49; Lyons, *W Post*, 3/19/71, A1)

- NASA launched two sounding rockets. Nike-Cajun, launched from Kourou, French Guiana, carried GSFC experiment to 125-km (77.7-mi) altitude to measure upper atmospheric pressures, temperatures, densities, and winds near equatorial latitudes and to verify operation of ground instrumentation in preparation for future grenade missions. Rocket and instruments functioned satisfactorily. All 18 grenades detonated and were recorded on ground. Launch was last of four in Kourou series. Previous launches had been conducted March 5 and March 15–16.

  Nike-Tomahawk, launched from Fox Main, Dew Line Station, Alaska, carried experiment to study electric fields. Rocket and instruments functioned satisfactorily. (NASA Rpt SRL; SR list)

- Dr. Marvin J. Kelly, former President and Board Chairman of Bell Telephone Laboratories, died at age 77. After retiring from Bell in 1959 Dr. Kelly had been named adviser to NASA Administrator and consultant to IBM, Ingersoll-Rand Co., and Kennecott Copper Corp. (*NYT*, 3/20/71, 32)

March 19: U.S.S.R. launched *Cosmos 400* into orbit with 1005-km (624.5-mi) apogee, 983-km (610.8-mi) perigee, 104.9-min period, and 65.8° inclination. (GSFC *SSR*, 3/31/71; UPI, *W Star*, 3/19/71, A6)

- NASA launched three-Nike-Tomahawk sounding rockets from Alaska: one from Fox Main, Dew Line Station, carrying GSFC experiment to study electric fields; one from Fairbanks, carrying Univ. of Alaska experiment to conduct auroral studies; and one from Fox Main carrying GSFC experiment to study magnetic fields. Rockets and instruments functioned satisfactorily.

- Senate passed unanimously S. 1117, to regulate public exposure to sonic booms by prohibiting supersonic flights of civil aircraft over continental U.S. (*CR*, 3/17/71, S3466–76)

- GAO study had disclosed that cost overruns on 61 weapon systems had reached $33.4 billion, of which some $9.5 billion had taken place before production began, *Washington Post* reported. Report blamed "deliberate underestimating," unanticipated development difficulties, faulty planning, poor management, bad estimating, inflation, and desire to exploit new technology. It said newer DOD programs "are characterized by a slower development pace and more conservative procurement practices." (Nossiter, *W Post*, 3/19/71, A1)

- Astronaut Edwin E. Aldrin, Jr., received Honorary Knight of St. Patrick Award from College of Engineering, Univ. of Missouri at Columbia, Mo. (*NASA Activities*, 4/15/71, 61)

- Boeing Co. had presented to its management results of diversification study completed by panel of Boeing executives, *Wall Street Journal* reported. Principal conclusion had been that best chances for new nonaerospace work lay in fields of surface transportation and community development. Study envisioned broadening Boeing's business base so that in 10 years' time one-third of Boeing's output would be outside its traditional product lines. (*WSJ*, 3/19/71)
- *Washington Post* editorial commented on House vote against SST funding: "... we do not see the need to hurry—either across the oceans or into the building of an aircraft which will convenience only a tiny fraction of the population at a time when great numbers of our people have more urgent needs. Some would say that we can have the SST and everything else at the same time, and perhaps we can. But we aren't doing the other things, which is precisely why the SST has become a very meaningful symbol of misplaced values and disordered priorities." (*W Post*, 3/19/71)

March 20: NASA launched two sounding rockets. Nike-Tomahawk launched from Fox Main, Dew Line Station, Alaska, carried GSFC experiment to study electric fields.

Aerobee 170 launched from WSMR carried Univ. of Wisconsin experiment to study soft x-rays, but mission was unsuccessful. (SR list)

- Tass reported on meeting of Soviet Transport Commission of Council of Mutual Economic Assistance in Moscow. Soviet bloc's economic alliance had discussed "coming" introduction of Soviet supersonic transport Tu-144 and future operation of subsonic jumbo jets on Soviet bloc routes. Report had mentioned no date for introduction of Tu-144 into commercial service. (Shabad, *NYT*, 3/21/71)

March 21: USAF launched unidentified satellite from Vandenberg AFB by Titan IIIB-Agena booster into orbit with 39 266.4-km (24 399-mi) apogee, 328.3-km (204-mi) perigee, 701.8-min period, and 63.2° inclination. (Pres Rpt 72; *Av Wk*, 4/26/71, 19)

- NASA launched two Nike-Tomahawk sounding rockets from Alaska. Rocket launched from Fox Main, Dew Line Station, carried GSFC experiment to study electric fields. Rocket launched from Fairbanks carried Univ. of Alaska experiment to conduct auroral studies. Rockets and instruments functioned satisfactorily. (SR list)

March 22: Nike-Tomahawk sounding rocket was launched by NASA from Fox Main, Dew Line Station, Alaska, carrying GSFC experiment to study magnetic fields. Rocket and instruments functioned satisfactorily. (SR list)

- New York attorney Paul Sawyer was sworn in as consultant to NASA Administrator on NASA Public Affairs Program. Sawyer was member of law firm of Phillips, Nizer, Benjamin, Krim, and Ballou. (NASA Release 71-46)
- Approval of full appropriations for National Air and Space Museum and increases in funding for National Museum of Natural History were urged by Sen. Barry M. Goldwater (R–Ariz.) in testimony before Senate Appropriations Committee's Subcommittee on Appropriations for Dept. of Interior and Related Agencies. Top priority in Air and Space museum budget was $1.9 million for planning and redesign of new building to be opened July 4, 1976. "The building project has been on dead center for five years now, though Congress provided

authority for it in 1966. Land is set aside on the Mall and a complete set of blueprints are in hand, ready for use; but a combination of resistance by the budget bureau and a recommendation from the Senate Committee on Rules and Administration have held the building back." Museum was needed to "tell the story of America's monumental achievements in flight. It will offer a mixture of regular and shifting exhibits of original aircraft and spacecraft from which the average citizen can see for himself the significant breakthroughs in flight and what lies beyond." Sen. Goldwater supported request for redesign of original plans for Museum as "surest and swiftest route toward completion of the project." Since building was authorized in 1966, estimated expense of $40 million had climbed to $70 million. (CR, 3/23/71, S3605–7)

- FAA announced start of major program to criss-cross U.S. with area navigation airways. Four transcontinental routes would link New York City with San Francisco and Los Angeles, effective April 29. FAA would announce additional 115 area navigation routes within next two months. Shorter interim area navigation routes had already been established to test and develop air traffic control and piloting techniques. Area navigation used airborne computers that permitted pilots to extrapolate signals from existing ground navaids, allowing greater choice and flexibility in route selection. (FAA Release 71–33)
- DOD was developing three aircraft for "what some influential lawmakers view as essentially the same mission," close air support, *Washington Post* reported. To buy USA's Cheyenne helicopter, USAF's AX, and USMC's British-built Harrier—about 1000 aircraft all told—would cost taxpayers $4 billion to $5 billion over next few years. (Getler, *W Post*, 3/22/71, A1)

March 23: Dr. George M. Low, Acting NASA Administrator, testified before House Committee on Appropriations' Subcommittee on HUD-Space-Science in support of NASA FY 1972 budget request: "We have recently settled on a single set of performance characteristics for the preliminary design of the two-stage fully reusable [space] shuttle. Alternate approaches are also still under study. We expect to be in a position to make decisions this summer on how we will proceed with airframe design or development in FY 1972. The $100 million recommended in the FY 1972 budget will support, in addition to the engine development, either continued intensive design studies and related efforts or the initiation of airframe development."

Dr. Low noted "tremendous interest in exploration during Apollo 11, when an estimated 107 million people in the U.S. watched our first manned lunar landing mission; and we saw it again last month, when 104 million watched Apollo 14." (Text)

- NASA announced award of $5-million, cost-plus-award-fee contract to Mason-Rust for support services at Michoud Assembly Facility. Contract was for one year with two one-year options. (NASA Release 71–47)

March 24: USAF launched unidentified satellite from Vandenberg AFB by Thorad-Agena booster. Orbital parameters: 235-km (146-mi) apogee, 172.2-km (107-mi) perigee, 88.5-min period, and 81.5° inclination. Satellite reentered April 12. (Pres Rpt 72)

- Senate voted 51 to 46 to cut off all Federal funding for SST, apparently ending controversial project. House had voted to stop Federal funding

March 18. Specific proposal before Senate had been to provide $134 million more for continued work on two SST prototypes for next three months. Sen. William Proxmire (D-Wis.) said Nixon Administration had requested $235 million for SST development in FY 1972 budget, but he predicted that if White House continued to seek appropriations it would be defeated, *Wall Street Journal* reported. (*CR*, 3/24/71, S3818–69; *WSJ*, 3/25/71)

- President Nixon issued statement following Senate disapproval of funds for SST, calling action "a reversal of America's tradition of staying in the vanguard of scientific and technological advance." He said: "It has always been America's pride, and the source of much of our strength, that we have constantly reached out toward new horizons in the search for knowledge—not from a chauvinistic desire to be number one, but from the conviction that we must continue to develop the countless new benefits that flow from exploration of the unknown. Development of the SST has been a part of that proud, creative, and deeply humanistic tradition. Though the Congress has declined to continue helping fund this development, I shall strive to ensure that the tradition is maintained." (*PD*, 3/29/71, 541)
- Boeing Co. Vice President for Industrial Relations Lowell P. Mickelwait said company would lay off about 7000 workers and disband SST program as result of Senate's rejection of SST financing. Plans were in anticipation of notice of SST program termination from DOT. Of workers to be laid off within seven weeks of notice, 4500 were force working directly on SST production. Others were clerical and office workers. Mickelwait said Seattle-area work force could drop to 26 500 or remain as high as 31 000 after SST program cancellation. Estimate of 29 000 workers was "reasonable." (AP, B *Sun*, 3/25/71, A1)
- NASA announced selection of RCA Corp. to receive $12-million, cost-plus-award-fee contract for Atmosphere Explorer (AE) satellites C, D, and E. Spacecraft would be launched in 1973, 1974, and 1975 to study upper atmosphere between 80 and 240 km (50 to 150 mi). NASA Release 71-48)
- NASA announced availability of computer information packages NASA/STIMS (Scientific and Technical Information Modular System) and NASA/RECON (Remote Console) from Computer Software Management and Information Center (COSMIC), operated by Univ. of Georgia under NASA contract. Packages would enable users to set up on-line information retrieval systems similar to NASA's. (NASA Release 71-44)

March 25: ARC researchers had used sensors on twin-engine Cessna 401 light aircraft to detect and measure oil spills in Pacific Ocean off California, NASA announced. Sensors had detected slicks from heavy and light crude oils and light diesel oil. ARC researchers John P. Millard and John C. Arveson had said approach should be applicable to satellite-borne sensor system. (NASA Release 71-42)

- West Germany's Bochum Observatory reported Communist China's second satellite (launched March 3) had stopped transmitting signals on March 23. Scientists did not know whether satellite had failed or had been turned off. (AP, W *Star*, 3/25/71, A1)
- President Nixon sent message to Congress proposing establishment of Dept. of Natural Resources, Dept. of Community Development, Dept. of Human Resources, and Dept. of Economic Affairs. Dept. of Natural

Resources would absorb Dept. of Interior and would include functions of DOT, NOAA, and civilian power functions of AEC. While DOT was relatively new entity it was "organized around methods and not around purposes." Much of DOT would be moved into new Dept. of Economic Affairs, "but those functions which particularly support community development would be placed in the Department which is designed to meet that goal."

Dept. of Economic Affairs would include National Transportation Safety Board, Transportation Systems Center, FAA, Science Information Exchange program from Smithsonian Institution, and Office of Technology Utilization from NASA (PD, 3/25/71, 545–60)

- President Nixon spoke by telephone with Boeing Co. employees in Seattle, Wash., and Wichita, Kans., to express thanks for work on SST project: "The reason I fought so hard to keep the SST project alive was that I believe deeply that America must remain in the vanguard of scientific and technological progress—the kind of progress your team represents." (PD, 3/25/71, 560–1)

- Report that Japan planned to build supersonic transport aircraft since U.S. had dropped SST program was denied by Shoichi Akazawa, official of Japanese International Trade and Industry Ministry in Tokyo. (Reuters, *W Post*, 3/25/71, A23)

- Economic impact of end of SST subsidy was described by Frank C. Porter in *Washington Post*: While move could idle 13 000 workers across U.S., "it is unlikely immediately to cause more than a ripple in the $1 trillion American economy." Some experts had suggested that continued SST development could cost more jobs ultimately than it would create. Meanwhile, it would cost U.S. Government extra $97 million to scrap program: contract termination costs of $52 million to Boeing Co., $33 million to General Electric Co., $10 million in miscellaneous termination costs, and $2 million to close SST project office at DOT. Government also would have to return $22.4 million in downpayments made by airlines. (*W Post*, 3/25/71, A23)

- Tass announced construction by Soviet scientists of "world's first operating installation" to convert atomic energy directly into electricity. Installation had produced "several kilowatts" for "new major achievement of Soviet atomic science and technology." AEC said later that, while U.S. had done same thing on smaller scale, U.S.S.R. might be closer to building unit large enough to power giant, multichannel comsats, long-range space missions, or compact power station under sea or in remote places. (Cohn, *W Post*, 3/26/71, A1)

- Sen. Edmund S. Muskie (D-Me.) introduced, for himself and cosponsors, S. 1382, "to authorize the Secretary of Transportation to carry out a special program of transportation research and development utilizing the unique experience and manpower of the airframe and defense industries." (CR, 3/25/71, S3900)

- French government approved plans for construction of world's first commercial air-cushion railroad to relieve auto traffic crisis. (Hess, *NYT*, 3/26/71, 65)

March 26: Intelsat-IV F–2, launched by NASA for ComSatCorp Jan. 25, began commercial service, transmitting to 15 earth stations in Western Europe, Latin America, and North America. Satellite was initially carrying 830 circuits and transmission was excellent. (INTELSAT Release 71-20; ComSatCorp PIO)

- NASA announced plans for first launchings of NASA sounding rockets from French Guiana Space Center at Kourou in cooperative France-U.S. upper-atmosphere research project. Project would launch 20 French Centaure and NASA Nike-Cajun rockets to obtain data on structure and circulation of equatorial atmosphere between 30- and 95-km (18- and 60-mi) altitudes. Under agreement between French National Center for Space Studies (CNES) and NASA, CNES would provide Centaure rockets with sodium vapor payloads, F-1 launcher, launch team, and tracking and ground support. NASA would be responsible for Nike-Cajuns with grenade and ozone payloads, one Nike with an inert 2nd stage, and ground support for payloads. Each agency would bear cost of its agreed responsibilities. (NASA Release 71-53)

- LeRC tests had shown turbofan built by General Electric Co.'s Aircraft Engine Group was quieter than any in use on commercial jets, NASA announced. Noise data from January tests of three GE turbofans had been analyzed. Francis J. Montegani of LeRC Quiet Engine Project had said if four quiet turbofans were installed in long-range subsonic transport, noise level would be within FAA standards. (NASA Release 71-81)

- Despite basic differences, lunar samples were not completely different from terrestrial rocks, Northern Illinois Univ. geologist S. S. Goldich wrote in *Science*. Basalt hornfels from Keweenawan Duluth complex in Minnesota, containing 7% by weight of titanium, were similar in many respects to *Apollo 11* lunar samples. Hornfels' texture, as well as primary textures in lunar rocks, resembled those in Keweenawan rocks. (*Science*, 3/26/71, 1245-6)

- Preliminary results of studies of nuclear tracks in glass lens cover of TV camera landed on moon by *Surveyor 3* April 19, 1967, and retrieved by *Apollo 12* mission were reported in *Science* by Washington Univ. physicists G. Crozaz and R. M. Walker. Studies gave information about low-energy nuclear particles from sun and provided basic calibration for nuclear tracks in lunar rock surfaces—erosion rate of nearly 10 millionths cm per year. Results also suggested "small-scale erosion process in lunar rocks."

    Use of *Surveyor 3* TV camera glass filter had enabled General Electric Co. scientists R. L. Fleischer, H. R. Hart, Jr., and G. M. Comstock to determine, for first time, energy spectrum of solar cosmic-ray particles of iron group over energy range from 1 to 100 million ev per nucleon. Difference between observed spectrum and limiting spectrum derived previously from lunar rock tracks gave erosion rate of 0 to 2 Å per year. "High-energy fission of lead, induced by galactic cosmic ray protons and alpha particles, has also been observed." (*Science*, 3/26/71, 1237-41)

- Element variation and possible material source of *Apollo 12* crystalline rocks was discussed in *Science* by Univ. of Tokyo geologists Ikuo Kushiro and Hiroshi Haramura. Chemical analysis of nine rocks had shown five rocks contained normative quartz; others had normative olivine and hypersthene. Rocks showed wide range in ratio of iron to magnesium. Findings led investigators to suggest rocks, with one exception, represented different parts of differentiated magmatic body, in which magmatic differentiation by crystallization and settling of olivine was most effective. Source material of original magma might be peridotite with or without plagioclase or spinel or garnet, depending on depth of magma generation. (*Science*, 3/26/71, 1235-7)

- Dr. Robert R. Gilruth, MSC Director, had been selected to receive 1971 James Watt International Medal from Institution of Mechanical Engineers in London, MSC *Roundup* announced. Dr. Gilruth had been nominated by American Society of Mechanical Engineers and was cited for his "distinguished services to Aeronautical and Space Research and for . . . engineering leadership which inspired and directed the Manned Space Flights and successful landings on the Moon." (MSC *Roundup*, 3/26/71, 1)
- NASA announced appointment of B. Porter Brown, Director of Operations Support Requirements in OMSF, as Special Assistant for Missions, Skylab Program, OMSF. (NASA Release 71-45)
- New York attorney Paul Sawyer, who had joined NASA to evaluate agency's public relations efforts on March 22, had taken 30-day leave from law firm in which he represented astronauts in negotiating exclusive rights to their stories for magazines, *Washington Post* reported. Sawyer had said he would not be representing astronauts during NASA assignment. (*W Post*, 3/26/71, A3)

March 27: U.S.S.R. launched *Cosmos 401* into orbit with 291-km (180.8-mi) apogee, 186-km (115.6-mi) perigee, 89.2-min period, and 72.8° inclination. Satellite reentered April 9. (GSFC *SSR*, 3/31/71; 4/30/71)

- *Washington Post* editorial commented on NASA's public information policy: "There is something strange about the decision . . . to bring in a New York lawyer, Paul Sawyer, as a consultant on its public affairs policies. NASA has had, for a decade now, what seems to us and to most journalists to be the most effective and most honest public information program in government. It is hard to see why public monies should be spent to study it, particularly when the man chosen to make the study has represented the astronauts in recent months in negotiating private contracts for the stories of their activities." (*W Post*, 3/27/71, A18)
- Pattern of recent ICBM construction in U.S.S.R. might presage deployment of new Soviet offensive weapon system, according to U.S. officials quoted in *New York Times*. Construction detected Dec. 20, 1970, had shown 20 holes large enough to accommodate Soviet SS-9 ICBM or even larger weapons. Holes were distributed in five clusters along wide arc forming Soviet offensive missile system from Polish border to Chinese frontier. (Szulc, *NYT*, 3/21/71, 1)
- *New Republic* published interview with Communist China Premier Chou En-lai by Edgar Snow. Premier had said in Peking: ". . . our nuclear tests are still in the experimental stage, and every test carried out is limited and made only when necessary. The aim of our nuclear tests is to break the nuclear monopoly and nuclear blackmail and prevent a nuclear war. Therefore, each time when we conclude a test, we declare that at no time and in no circumstances shall China be the first to use nuclear weapons." (*New Republic*, 3/27/71, 20-3)

March 28: Foreign sources interested in buying SST development package had contacted DOT, William M. Magruder, Director of DOT's SST program, told press in Washington, D.C. Inquiries had come from Japanese trading firm Ataka, from Middle East source, and from West

German source. Magruder said congressional leaders "have made no attempt to discourage" plan to seek package to recoup termination costs of SST program, estimated at between $97 million and $157 million, depending on whether airlines got back SST purchase deposits. (AP, B *Sun*, 3/29/71, A3)

- JPL announced that Dr. Gary A. Flandro, former JPL scientist who formulated planetary Grand Tour concept, had been named winner of British Interplanetary Society's Golovine Award for "outstanding contributions to astronautics" through basic mathematical work he did at JPL from 1965 to 1968. Dr. Flandro, now associate professor at Univ. of Utah, would receive award in June. (JPL Release 574)
- Eugene C. Draley, LaRC Director for Center Development, died at age 55. Draley had joined LaRC research staff in 1938 and had held increasingly responsible positions, from Head of 8-Foot High-Speed Tunnel to LaRC Assistant Director. (*Langley Researcher*, 4/2/71, 1)
- Sherman M. Fairchild, founder and board chairman of Fairchild Hiller Corp., died at age 74. Fairchild had invented first aerial camera and had been honored in 1970 by Smithsonian Institution for 50 yrs' service in aviation industry. (Fowle, *NYT*, 3/29/71, 32)

March 29: X–24A lifting body reached its maximum speed of 1687 km (1048 mph) and 21 300-m (70 000-ft) altitude during 25th flight from FRC. USAF aircraft, launched from B–52 aircraft, was piloted by NASA test pilot John A. Manke on powered flight to expand flight envelope, determine lateral-directional derivative with rudder bias at 2° toe out, and check lateral-acceleration feedback to control system. (NASA Proj Off; Pres Rpt 72)

- U.S.S.R. announced that Soviet supersonic transport Tu-144 had reached designed 2400-km-per-hr (1500-mph) speed at 16 700-m (55 000-ft) altitude during test flight Nov. 12, 1970. (*NYT*, 3/29/71, 53)
- NASA announced award of seven-month study contracts totaling $1 387 576 to nine universities and one NASA center to define requirements for seven astronomy experiments for proposed High Energy Astronomy Observatory (HEAO). (MSFC Release 71–52)
- FAA and DOT announced awards totaling $14 246 542 in amendments to contracts with IBM Federal Systems Div. for additional equipment and services to automate air traffic control system. (FAA Release 71–35)
- Immediate one-year freeze on deployment by U.S. and U.S.S.R. of land-based missiles was urged by Sen. Henry M. Jackson (D-Wash.) in Senate speech. Both countries would retain freedom to ensure survivability of strategic land-based forces as long as they did not add to offensive potential. (*CR*, 3/29/71, S4035–6)
- Editorial in *Aviation Week & Space Technology* commented on rejection by Government of further SST funding: "The fate of the SST should serve notice . . . that the aerospace industry can no longer survive with its political naivete and aloofness from the fray." It was "sad day when a deliberate national decision is made to abandon a major frontier of technological advance. But it will be even sadder when the victors in this significant debate gird for their next assault on the space shuttle program." (Hotz, *Av Wk*, 3/29/71, 9)

March 30: Dr. George M. Low, Acting NASA Administrator, testified on aerospace unemployment before Senate Committee on Aeronautical and Space Sciences during FY 1972 authorization hearings: From number of studies of unemployment of scientists and engineers, estimated 30 000 to 40 000 were out of work at end of 1970. "The majority of these came from aerospace-related industries. Projections indicate that this number could more than double by June 1972, which means that as many as 30 percent of the Nation's aerospace scientists and engineers will be unemployed. The effect of this national trend—in terms of an individual scientist or engineer being reemployed—is magnified by the concentration of this unemployment, geographically and by training and experience. . . . This not only causes great personal hardship but . . . represents the loss of a tremendous national resource." NASA-sponsored survey by Battelle Memorial Institute of persons laid off by group of NASA contractors from June 1968 to September 1970 had made preliminary findings: only 31% had found permanent jobs, 32% had found temporary jobs, 32% were still unemployed, and 5% had left work-force. Average person was unemployed 31 weeks. Of those who had found permanent jobs, only 18% returned to aerospace. Of those permanently employed outside of aerospace, only 12% reported having positions highly related to aerospace skills and 50% reported new positions unrelated to aerospace. Unemployment rate of those over 50 was 48%. Of those who had found permanent work outside, 31% said they would not return to aerospace if given opportunity; 16% said they would. Preliminary results of survey indicated "we are losing much of the aerospace capability that is an indispensable ingredient of our long term economic strength and national security."

In closing, Dr. Low made point that there was "no either/or choice between technological advances and social advances. Without technology, we cannot maintain the capability to deal effectively with any national problem."

Dr. Low was questioned on his testimony that U.S.S.R. was spending more on R&D than U.S. was spending in total science and technology area. He answered that Soviet space program, with apparent Soviet policy of steadily increasing R&D investment, would soon match and then surpass U.S. program in size and accomplishment. "For this comparison, I am dealing with the combined civil and military space programs of both nations; to try to differentiate between civil and military activities of the U.S.S.R. would be speculative at best. The most difficult comparison to try to make between the U.S. and U.S.S.R. economies or programs is in terms of absolute costs: each national economy is geared and motivated to produce a different, nonparallel mix of goods and services." (CR, 7/31/71, E8611-2)

Dale D. Myers, NASA Associate Administrator for Manned Space Flight, discussed future of Apollo program: "We can now move with confidence to the final phase. . . . In these last missions, we will expand the capability of the system very dramatically. Design changes in the spacecraft and supporting equipment will allow the two astronauts in the lunar module to remain on the moon up to 67 hours, as compared with 34 hours on Apollo 14. The scientific payload landed on the moon will be doubled to above 1200 lbs [544 kg]. Changes in the command and service modules will permit up to 10 days total flight time and major increases in orbital scientific activities. During the Apollo 15 and 16 missions small subsatellites will be placed in lunar orbit and will continue to make scientific measurements long after the astronauts return to earth."

In Skylab program, "three separate three-man Skylab flight crews will be selected during the coming year. Scientist astronauts will be included. . . . They plan to perform about 50 experiments in various disciplines. Twenty of these are in the life sciences, to determine how human beings adjust and perform under the conditions of prolonged space flight, up to 2 months' duration.

"They will also operate the Skylab Earth Resources Experiment Package in the second space flight phase of NASA's earth resources program. These observations will be in conjunction with [and] complementary to those [of] the automated earth resources test technology ERTS, to be launched in 1972."

FY 1972 request for space shuttle of $100 million would provide for broad spectrum of studies, including analysis of new alloys and promising external materials for thermal protection; test demonstrations of design concepts for structures; wind-tunnel tests to define aerodynamic heating, launch aerodynamics, etc.; design of long-leadtime hardware, subsystems, and test devices; initiation of scaled model for flight tests; dynamics testing; electronic-data-bus-system demonstration testing; and integration of upgraded fuel-cell electrical power subsystem with other components for proof-of-concept testing. (Transcript)

Dr. Robert C. Seamans, Jr., Secretary of the Air Force, testified on DOD applications of space transportation system and DOD–NASA teamwork on space shuttle: "Present DOD satellite communications program consists of two major systems: The long distance point to point or strategic system and the local-area multiple-user or tactical system. . . . We are pursuing a new system called Defense Satellite Communications System, Phase II, which can provide many voice channels to selected limited areas of the earth. These present and proposed systems give us the flexibility for voice and teletype communications from our forces overseas and allow rapid transmission of photographic images from operational theaters to the United States."

In conjunction with NASA studies, USAF was analyzing defense applications of space shuttle to provide better insight into utility of concept for DOD. USAF was providing NASA with data to help ensure that configuration was of maximum utility to DOD.

NASA and USAF had worked jointly on engine development. Air Force "has supported advanced development efforts on liquid rocket engines in the past, and more recently concluded a hydrogen/oxygen high performance engine demonstration program." Engine, XLR–129, had "provided the basis for selection of the reusable high pressure rocket engine concept for the shuttle. It also provided design data for a 250 000-pound [1100-kilonewton] thrust engine which supports the current shuttle engine design concepts."

In operational testing area, "USAF and NASA recently conducted unpowered approach and landing demonstrations with F–111s and B–52s. . . . These demonstrations investigated terminal area energy management for maneuvering reentry vehicles and have a relationship to the Space Shuttle operational concepts." (Transcript)

- DOT request to Congress for $97.3 million to terminate SST Development program was prepared by James M. Beggs, Under Secretary of Transportation. Sum did not include repayment of $58.5 million risk capital contributed by nine U.S. airlines and KLM Royal Dutch Airlines to support SST R&D. (AP, *NYT*, 3/31/71, 3:27)
- U.K. government and Lockheed Aircraft Corp. agreed on terms for perpetuating Rolls-Royce engine developed for Lockheed's L–1011 TriStar jet transport after two-day meetings in Washington, D.C. Terms would be submitted to airlines that had ordered aircraft and to banks financing sales. (Witkin, *NYT*, 3/31/71, 1)
- ABC, CBS, and NBC TV networks had filed statement with FCC saying they planned use of other domestic comsats for program transmission rather than building separate satellite system of their own, *Wall Street Journal* reported. Networks were negotiating with several companies on use of domestic comsats but would like to keep their own option open in case negotiations failed to lead to "satisfactory" arrangements. (*WSJ*, 30/30/71, 6)
- U.S.S.R. at Geneva disarmament conference introduced draft treaty that would bind nations "not to develop, produce, stockpile or otherwise acquire" microbiological or other biological agents or toxins "not designed for the prevention of disease or for other peaceful purposes." (Lusinchi, *NYT*, 3/31/71, 1)

March 31: Canadian *Isis 2* (ISIS–B) International Satellite for Ionospheric Studies was launched by NASA from WTR 6:57 pm PST by thrust-augmented three-stage Thor-Delta booster. Satellite entered orbit with 1423-km (884.2-mi) apogee, 1355-km (842.0-mi) perigee, 113.5-min period, and 88.2° inclination. Primary NASA objectives were to place *Isis 2* into circular earth orbit that would permit study of topside of ionosphere above electron peak of F region and to extend cooperative Canadian-U.S. program of ionospheric studies initiated by *Alouette 1* (launched Sept. 28, 1962) by combining sounder data with correlative direct measurements for time sufficient to cover latitudinal and diurnal variations during high solar activity. Spacecraft was operating satisfactorily and was expected to be fully operational by April 17.

Canadian-built *Isis 2* was eight-sided spheroid that weighed 264 kg (582 lbs), was 127 cm (50 in) in diameter with two sounder antennas, and carried 12 ionospheric investigation experiments. Satellite was fourth in series of five satellites to improve understanding of ionospheric physics. First launch in series (Nov. 28, 1965) had orbited Canada's *Alouette 2* and U.S. *Explorer 31. Isis 1* had been launched Jan. 30, 1969. ISIS program was joint undertaking of NASA and Canadian Defence Research Board (DRB) under December 1963 Memorandum of Understanding. DRB was responsible for spacecraft design, fabrication, electrical testing, experiment integration, and satellite control. NASA provided launch vehicle and launch facilities. (NASA Proj Off; NASA Release 71–41)

- NASA announced confirmation of text of agreement with Soviet Academy of Sciences on exchange of lunar samples and on procedures toward recommendations for joint consideration of objectives and results of space research [see Jan. 18–21]. Agreement had been confirmed by exchange of letters between Dr. George M. Low, Acting NASA Administrator, and Academician Mstislav V. Keldysh, President of Soviet Academy. (NASA Release 71-57)

- Data from *Explorer 42* (*Uhuru,* launched Dec. 12, 1970) was reported at American Astronomical Society meeting in Baton Rouge, La., by Dr. Riccardo Giaconni of American Science and Engineering, Inc., principal investigator for satellite. *Explorer 42* had detected new x-ray pulsar Cygnus X-1 generating x-ray pulses at 15 per sec in Constellation Cygnus, 13 new x-ray objects in Milky Way and remote galaxies, and confirmed that quasar 3C273 in Seyfert Galaxy was x-ray emitter. Quasar was billion light years away—most distant object from earth known to emit x-rays. (NASA Release 71-50)
- Nike-Tomahawk sounding rocket was launched by NASA from Churchill Research Range carrying TRW Inc., experiment to conduct auroral studies. (SR list)
- MSFC announced award of $2 249 700, one-year contract to Air Products and Chemicals, Inc., to supply and deliver liquid hydrogen for all Government aerospace use in eastern U.S. (MSFC Release 71-55)

During March: Prestige of U.S. science was being undermined by assaults from many sides—which might leave U.S. "a second-rate power and a third-rate place to live"—Lawrence P. Lessing wrote in *Fortune* article.

Funds had been "damagingly cut in such basic areas as the life and medical sciences . . . while a large amount of advanced but abstract technology, which could begin to solve the problems of overcrowding, energy, pollution, transportation, and waste, . . . are going neglected." New hostility to science and technology had arisen among many "ordinary Americans," even from within science itself, and among young activists. But if man's store of knowledge were torn down, man would "backslide not a few centuries but two hundred thousand years."

Severe cutback in research and development because of Vietnam war, inflation, and contracted economy—with Federal R&D expenditures declining in real dollars by more than 20%—possibly endangered whole structure of science. Decline followed post-World War II growth that had produced antibiotics, atomic power, cryogenics, computers, jet planes, rocket vehicles, radar, transistors, masers, and lasers and had spurred U.S. to do its own basic research. "From . . . only a sparse dozen Nobel Prizes in the forty years up to 1940, U.S. scientists went on in the next thirty years to win forty-five."

"While men and talent are . . . going to waste the things that need doing . . . keep piling up." U.S. was beginning to lag behind other nations in high-energy physics, radioastronomy, plasma physics, conventional energy research, space science, transportation, and life sciences. "With two of the later, more scientifically oriented Apollo moon flights canceled, though equipment is bought and paid for, work on NASA's forward-looking space station, nuclear-powered rocket, and unmanned planetary exploration vehicles has been so cut back that after 1975 almost nothing will be scheduled. And research funds in life sciences were still lagging 20% behind research capacity—in field that held promises for much safer biological insecticides, genetic repair of congenital defects, and basic attacks on cancer and heart disease. (*Fortune,* 3/71, 88-9, 153-5)

- Production of formaldehyde, acetaldehyde, and glycolic acid in UV tests under simulated Martian conditions indicated organic material was probably being produced by sunlight on Mars' surface, JPL scientists Dr. Norman H. Horowitz, Dr. Jerry S. Hubbard, and Dr. James P. Hardy reported in *Proceedings of National Academy of Science of USA*. Same organic compounds were believed to have been precursors to biological molecules on primitive earth. Dr. Horowitz said findings, obtained from irradiation of fine soil and powdered glass samples by high-pressure xenon lamp and low-pressure mercury lamp, provided "most favorable indication for a possible Martian biological evolution that we have had in the last five years." There were "still many uncertainties, however, which won't be resolved until we land on the planet." (*Proceedings of NAS*, 3/71, 574–8)
- Promise of zero g in manufacturing in space was described by A. R. Sorells in *America Illustrated*, USIA publication distributed in U.S.S.R. Technicians anticipated future world of factories in space, where machines were grown molecule-by-molecule, as nature grew plants. First space workshops were scheduled to fly during decade as pioneers of later manufacturing islands circling above earth. Even dream of perpetual motion machines might materialize. (*Am Ill*, 3-71, 34–37)
- NASA published *Manned Space Program Accident/Incident Summaries*. Result of NASA Safety Office survey of accidents and incidents that occurred during Apollo Program development years could be used in NASA training programs. Of some 10 000 case documents reviewed, 508 mishaps had been studied closely. Of these, 47% had occurred during operational test and checkout. Procedural deficiencies were largest contributor to accidents—46%—and 74% had human error as contributing factor. (NASA *Awareness*)
- Scientific impact of SST was described by James J. Haggerty in *Air Line Pilot* magazine: "SST . . . represents one of the most sweeping aeronautical advances since Kitty Hawk, one that, brought to fruition, will have follow-on impact comparable to the introduction of jet propulsion or the first conquest of the sound barrier. It has particular importance as a research vehicle, because there is no other airplane flying or in development with its particular combination of performance characteristics. It is equally important as an impetus to technology, since the SST contains thousands of components that have never been built before." SST was "atmospheric Apollo, a prober of new regions of flight and . . . a thrusting force pushing aeronautical technology to a new plateau." (*Air Line Pilot*, 3/71)
- *The Making of an Ex-astronaut* by former astronaut Dr. Brian T. O'Leary was reviewed by Edwin E. Salpeter in *Bulletin of the Atomic Scientists*: "The most apparent human drama in the book is the clash between two alien cultures, or rather the struggle of a minority culture, the scientist surrounded by the dominant culture of the technologist-testpilot-administrator combination." It was interesting psychological sidelight that Dr. O'Leary had not worried about "large but inevitable dangers inherent in spaceflight itself, but bitterly resented the additional danger of flying jets (about a 10 per cent probability of death in a jet accident over a 10-year service in the program)." While jet flights had been emphasized, opportunities for scientist-astronauts to participate in space missions had receded and opportunities for keeping up scientific

career in space program "were severely limited." Dr. O'Leary had become, in 1968, first scientist-astronaut to resign. (*Bull of Atomic Scientists*, 3/71, 44)

- Second edition of one-volume Soviet space encyclopedia *Cosmonautics* was published in Moscow. Book listed as editor Valentin Petrovich Glushko, reputed chief designer of Soviet rockets and missiles, and featured biographical article on him describing his work. *Pravda*, in post-publication article on encyclopedia, had identified chief editor as "Professor G. Petrovich." (*NYT*, 3/19/71, 3)

- Freeze on missile testing was advocated by Stanford Univ. geneticist Dr. Joshua Lederberg in *Bulletin of Atomic Scientists:* "From a technical standpoint, the most amenable place for controls is testing; a comprehensive freeze on all missile tests would be most easily verified and would provide the utmost assurance against the perpetuation of a costly technology race. It would complicate some peaceful applications of space technology. However, none of these require precise reentry after a brief, high velocity flight. Furthermore, nothing would be lost in requiring a definite pattern of international participation in space missions to assure that these were a net benefit to the whole earth from which they have embarked." (*Bull of Atomic Scientists*, 3/71, 4–6, 43)

April 1971

April 1: Discovery that comets were made of water-ice and discovery of new concept of dynamics of interstellar hydrogen gas from data obtained by *Ogo 5* (launched March 4, 1968) and *Oao 2* (launched Dec. 7, 1968) were announced. Dr. Charles Lillie and Dr. Gary Thomas of Univ. of Colorado and Prof. Jacques E. Blamont of Univ. of Paris said at NASA Hq. press conference that OAO observations indicated comets were water with traces of methane, ammonia, and cyanogen. OGO observations showed gas surrounding comets to be composed almost totally of hydrogen and that interstellar wind previously believed to be coming from constellation Hercules actually seemed to be coming from Sagittarius and Scorpius. Scientists now surmised that there was no stellar wind, but only apparent wind caused by passage of solar system through thin and nearly stationary interstellar medium of hydrogen. (Lannan, W *Star*, 4/2/71, A9)

- U.S.S.R. launched *Cosmos 402* from Baykonur into orbit with 1035-km (643.2-mi) apogee, 948-km (589.1-mi) perigee, 104.9-min period, and 64.9° inclination. (GSFC *SSR*, 4/30/71; *SBD*, 4/5/71, 181)

- Dr. George M. Low, Acting NASA Administrator, summarized reasons NASA proposed space shuttle development as top space priority for 1970s, during Senate Committee on Aeronautical and Space Sciences NASA FY 1972 authorization hearings: U.S. "should and will continue an active space program from now on" and new approach was needed to make space "practical and economical" for all users—NASA, DOD, other Government agencies, and commercial enterprises. Reusable shuttle, "entirely new concept in space operation," would deliver and retrieve manned and unmanned payloads; "we will be able to repair, modify, or update payloads in orbit and . . . reduce drastically the costs of payloads by taking advantage of the much larger weight and volume that will be available. . . . When future space programs using the shuttle are compared with those using conventional launch vehicles, the shuttle offers a very real economic advantage."

    Shuttle development should proceed in FY 1972 because necessary technology was available, delay in development would widen existing four-year gap in U.S. manned space flight schedule, and shuttle was "keystone to the total plan for the U.S. space program."

    NASA Associate Administrator for Manned Space Flight Dale D. Myers discussed candidate concepts for space shuttle: "The principle objective in examining various systems is to determine which systems have the greatest potential for low operational costs together with low development costs. The most attractive system . . . is the fully reusable system where nothing is discarded in the course of the flight and the orbiter and booster are 'turned around' with minimum maintenance for reuse. This concept has been the subject of intense 'phase B' definition studies over . . . the past 9 months."

    Dr. John E. Naugle, Associate Administrator for Space Science and Applications, reviewed OSSA programs, goals, and motivations: "When we speak of 'space' science and 'space' applications, we are talking about using a new technology . . . to enhance scientific progress and its translation into benefits. In this role, NASA is a service agency; we provide the technology, the engineering, the management, and the leadership that make possible the exploration of space and the practical use of space flight. The scientific community, which represents the focus of man's unceasing drive for understanding, identifies the problems and the unknowns, and designs the experiments that will answer the key questions. Society levies demands upon us for useful services, valuable products, and improvements in the quality of life." There was no difficulty in identifying needs or new missions; acute problem was to select those offering greatest reward as range of highest national priorities expanded but resources remained limited. (Transcript)

- Rep. George P. Miller (D-Calif.), Chairman of House Committee on Science and Astronautics, introduced H.R. 7109, $3.434-billion FY 1972 NASA authorization bill. Bill added $163 million to Administration's requested $3.271 billion, including additional $45 million for Skylab and $25 million for space shuttle, $39.9 million for NERVA engine, and $25 million for aeronautical research and technology. Bill provided $2.668 billion for R&D, increase of $150 million over requested $2.518 billion; $58.6 million for construction of facilities, up $2.3 million from $56.3-million request; and $706.9 million for research and program management, increase of $9.5 million over requested $697.4 million. (H Rpt 92–143)
- Soviet Academy of Sciences President Mstislav V. Keldysh told session of Soviet Party Congress that "we must to a large extent apply [space activities] to the solution of practical problems. *Armed Forces Journal* later observed that previously, U.S.S.R. had spoken mostly of furthering "scientific knowledge" as basis of its space program. (*AFJ*, 6/71, 54–9)
- Leaders of aerospace industry, professional societies, and academic community met with President Nixon; Dr. Edward E. David, Jr., Science Adviser to President; and James D. Hodgson, Secretary of Labor. Meeting discussed Technology Mobilization and Reemployment Program to alleviate unemployment in aerospace industry. (*PD*, 4/5/71, 592)
- Tenth anniversary of Air Force Systems Command. AFSC—in charge of planning, developing, and acquiring all USAF weapon and space systems and supporting technologies—administered annual contracts worth about $48 billion. Annual expenditures accounted for more than one fourth of total USAF budget. AFSC was staffed by 9600 officers, 18 500 airmen, and 31 000 civilians. (AFSC Release 94.71; *AFM*, 4/71, 43)

April 2: U.S.S.R. launched *Cosmos 403* from Plesetsk into orbit with 213-km (132.4-mi) apogee, 205-km (127.4-mi) perigee, 88.6-min period, and 81.3° inclination. Satellite reentered April 14. (GSFC *SSR*, 4/30/71; *SF*, 10/71, 386)

- NASA's supercritical wing, flown on TF-8A jet aircraft piloted by test pilot Thomas C. McMurtry, successfully completed third flight from FRC. Objectives of flight—to expand flight envelope to higher altitudes and higher speeds, evaluate augmented stability and control, and determine loads and buffet boundaries—were achieved. Aircraft reached 10 700-m (35 000-ft) altitude and mach 0.8. (NASA Proj Off)

- NASA announced it was expanding its airborne research program by acquiring two Lockheed U-2 high-altitude aircraft under loan agreement with USAF and concentrating flights over four ecological test areas in U.S. Program objectives were to simulate data output of ERTS scheduled for 1972 launch, collect data simultaneously over various test sites with passes of ERTS and Skylab after they were orbited, support earth resources survey programs of other agencies, and make observations in astronomy, atmospheric physics, and geophysics for NASA's physics and astronomy programs. (NASA Release 71-60)
- Dr. John S. Foster, Jr., Director of Defense Research and Engineering, testified on DOD space activities during authorization hearings before Senate Committee on Aeronautical and Space Sciences: DOD space funding program for FY 1972 called for $1552.6 million, slightly under the $1553.1 million of FY 1971. "The design and fabrication of a new set of satellites intended to replace the interim Defense Satellite Communications System (DSCS) will be completed in 1971. These satellites, to be launched next month, have significantly higher power and greater bandwidth than their predecessors, and each is equipped with a versatile set of antennas capable of serving fixed and transportable users dispersed over large areas of the earth. Among these are steerable antennas which can be oriented to satisfy contingency needs in changing global situations. To exploit the potential of this new system, we will need better surface terminal equipment. Accordingly, we intend to continue the intensive development of new reliable terminals and modulation and antijamming equipment."

    In FY 1972, DOD expected "to start developing a satellite communications system that operates in the military ultra-high-frequency (UHF) band and is tailored to the characteristics of the mobile users. When deployed, this system should allow us to retire much of our HF radio equipment."

    Use of navigational satellites "has the potential of attaining a continuous, highly accurate worldwide position-fixing system. That system could support many ground, sea, airborne and space applications." In FY 1972, DOD planned "to undertake system definition and design studies of the next-generation Defense Navigation Satellite System (DNSS)."

    Under Doppler Geodetic Satellite Program coordinated by Defense Intelligence Agency, geodetic Doppler receivers would observe USN navigation and other Doppler-equipped satellites. Program would provide worldwide geodetic data.

    DOD proposed "to continue a small effort in FY 1972 to develop propulsion technology for high-energy orbit-to-orbit stages that will be compatible with the shuttle system, with a view to future space maneuvering missions projected for the late 1970's and early 1980's." (Transcript)
- USAF planned to measure static electrification on Titan IIIC missile from preignition to orbit at KSC during second quarter of 1971, AFSC announced. Data would be gathered from two field meters—one on missile and one on ground—and would help define electrostatic requirements for future missile systems. (AFSC Release 82.71)
- AEC announced it was developing plutonium-fueled water recovery and waste management system for crew on extended space flights. Device

converted liquid wastes into sterile drinking water by high-temperature processing and disposed of nonmetallic cabin wastes by incineration. System would meet needs of four astronauts for up to 180 days. (AEC Release 0–47)

- LeRC announced award of $4 264 734, fixed-price, supplemental agreement to United Aircraft Corp. Pratt & Whitney Div. for 12 RL–10 rocket engines for Centaur boosters. (LeRC Release 71–11)

April 4: U.S.S.R. launched *Cosmos 404* from Baykonur into orbit with 1009-km (627-mi) apogee, 817-km (507.7-mi) perigee, 103.1-min period, and 65.7° inclination. Satellite reentered same day. (GSFC SSR, 4/30/71; SF, 10/71, 386)

- NASA manned space flight officials proposed four-launch program to place astronauts in earth orbit to map U.S., study earth resources, and study pollution. NASA Associate Administrator for Manned Space Flight Dale D. Myers said in AP-reported interview that program was in preliminary stages and would require additional funding from Congress. MSC Deputy Director Christopher C. Kraft, Jr., said project would keep U.S. in manned space flight business during period when U.S.S.R. would probably be very active in the field. Project would fill four-year gap between Skylab and manned space shuttle tests. It would also directly benefit mankind, Kraft explained: "The two-man crews will fly at a higher inclination than we've ever flown before in earth orbit. Using photographic and other sensing equipment developed in the space program, they could accurately map the entire United States, including Alaska. The astronauts also could make earth resources and environmental measurements." (AP, W *Star*, 4/4/71, A16)

- Team of 108 scientists from U.S. and six foreign countries had been chosen from some 500 scientists who submitted proposals to participate in definition phase of proposed outer planets missions in late 1970s, NASA announced. Teams from 13 investigating areas would be represented by team leader on Outer Planets Mission Steering Group responsible for integrating overall science. Scientists represented 26 U.S. institutions and institutions in Canada, Denmark, West Germany, France, Sweden, and U.K. (NASA Release 71–56)

April 5: First simulated altitude test of manned Apollo 15 CM in KSC vacuum chamber was postponed after oil was found leaking from light fixture. Problem did not affect spacecraft. (*SBD*, 4/6/71, 183)

- Senate Committee on Aeronautical and Space Sciences continued hearings on NASA FY 1972 authorization. Roy P. Jackson, NASA Associate Administrator for Advanced Research and Technology, testified that LeRC quiet engine program was proceeding on schedule: "Tests of full-scale fans for the engine have revealed much new information on the interrelationships among noise, aerodynamic performance, and structural integrity. The complete experimental engine will be operated in about 6 months. Results . . . indicate that we may expect to better the original noise abatement goals. Program results are given to industry on a continuing basis. . . . We can expect some of our improvements to be adopted by industry in new engine designs before completion of our program."

NASA–AEC Space Nuclear Systems Office Manager Milton Klein said: "Reactor power sources will be required to provide power in amounts greater than a few kilowatts for many of the missions of the future.

For unmanned military satellites, radiation hardening—mandatory for certain military missions—and low recurring costs for multiple missions are important advantages of small reactor systems. For unmanned NASA earth orbit missions, reactors become cost effective above a few kilowatts, a power range needed for such applications missions as communications satellites. In manned missions . . . overriding advantage of reactor systems is their ability to accommodate unforeseen large increases in power demand." Advanced reactor systems "are needed for electric propulsion systems capable of high energy missions . . . to the far planets."

Associate Administrator for Tracking and Data Acquisition Gerald M. Truszynski reported that satellite network included "ten electronic ground stations . . . operated by NASA, which are supplemented by an optical tracking network operated by the Smithsonian Astrophysical Observatory. The electronic stations provide a worldwide system which . . . tracks each satellite, . . . determines the status of onboard systems, . . . commands the satellite's functions, and . . . acquires stored or real-time data from the satellite. The optical stations, which include laser capabilities at selected sites, provide specialized services in precision orbital tracking." (Transcript)

- LeRC was studying blown-flap STOL aircraft using series of fans mounted atop each flap semispan, *Aviation Week & Space Technology* reported. Concept was similar to fan/flap arrangements studied in wind tunnel tests. Bleed air from main engines was ducted to periphery of each fan nacelle and drove fans through tip turbines. Flow from fans then passed over flaps, which could be deflected up to 60° for landing. (*Av Wk*, 4/5/71, 44)

April 5–7: Conference on Remote Sensing of Chesapeake Bay was held at Wallops Station to identify primary environmental problems of area and determine extent to which remote sensing from aircraft or satellites could contribute to solution. States of Maryland and Delaware, Commonwealth of Virginia, NASA, and Smithsonian Institution participated. Chesapeake Bay area was one of four regional test sites under consideration for evaluation of remote sensing techniques. Experiments would be conducted in area using data to be acquired by first Earth Resources Technology Satellite (ERTS–A) scheduled for launch by NASA in spring 1972. (NASA Release 71-55; WS PAO)

- National Endowment for Humanities sponsored Thomas Harriot Symposium at Univ. of Delaware, honoring Elizabethan mathematician, astronomer, geographer, navigator, and philosopher sent to New World in 1585 by his patron, Sir Walter Raleigh. Harriot had written first English account of wealth of America, *A Briefe and True Report of the New Found Land of Virginia*. After 350 yrs of obscurity, Harriot was being revived by scholars who were mentioning his name with Kepler and Galileo, *New York Times* reported. Harriot had independently begun telescopic observation of heavens in July of 1609, at about time Galileo did in Italy, and had drawn what was believed to be earliest map of moon in which vast seas and craters were recognizable. He had observed and counted sun spots through telescope using London fog as filter and had calculated period of solar rotation and traced moons of Jupiter. In 1601 he had discovered law governing refraction of light. (Reinhold, *NYT*, 4/8/71, 39)

April 6: Intelsat 1 (Early Bird), in orbit over Pacific, was reactivated to mark its sixth anniversary and relay first message ever transmitted directly from Hawaii to U.S. East Coast via commercial satellite. Satellite—first commercial comsat—had been launched from KSC April 6, 1965. (ComSatCorp Release 71–23)

- Distribution of about 6.8 kg (15 lbs) of the 42.6 kg (94 lbs) of samples returned from moon's Fra Mauro area by *Apollo 14* astronauts had begun, NASA announced. To date, 187 scientific teams in U.S. and 14 foreign countries had been scheduled to receive samples. Additional proposals for consideration under *Apollo 14* through Apollo 17 sample program were under review. More than 700 scientists would analyze samples, believed to include material formed at same time as original lunar crust. Investigating teams in U.S. were at 65 different institutions in 30 states and Virgin Islands. The 56 foreign teams were at 41 institutions. (NASA Release 71–62)
- Third attempt to test manned Apollo 15 LM in simulated environment was completed at KSC. Two attempts during previous week had been unsuccessful because technician failed to open oxygen valve and because two oxygen regulators failed. (*SBD*, 4/7/71, 192)
- LeRC announced award of $7 939 400 follow-on contract to General Dynamics Corp. Convair Div. for Centaur launch and test support services at ETR. Cost-plus-award-fee contract would cover one year, during which six Centaurs would be launched. (LeRC Release 71–13)
- Dr. F. A. Speer, head of MSFC's Mission Operations Office in Program Management, became manager of newly augmented HEAO task team at MSFC. Team would continue feasibility and definition work for proposed 1975 mission. (MSFC Release 71–75)
- AP quoted Fairchild Hiller Corp. spokesman as saying firm was serving as clearinghouse for proposals for SST funding. "Sketchy and tentative plans" existed, AP said, to finance SST through public stock issue backed up by $1 billion in bank loans. (AP, *B Sun*, 4/7/71, A6)

April 6–7: Space Shuttle Technology conference at MSFC discussed space vehicle propulsion systems with operational life requirements of 100 flight missions. (MSFC Release 71–59; MSFC PAO)

April 7: U.S.S.R. launched *Cosmos 405* from Plesetsk into orbit with 681-km (423.2-mi) apogee, 674-km (418.8-mi) perigee, 98.2-min period, and 81.2° inclination. (GSFC *SSR*, 4/30/71; *SBD*, 4/8/71, 197)

- U.S.S.R.'s *Lunokhod 1* lunar rover began sixth lunar day on moon's Sea of Rains, traveling 301 m (988 ft) across lunar surface and completing two communications sessions with ground. Vehicle had landed on moon on board *Luna 17* Nov. 17, 1970. (*SBD*, 4/9/71, 205)
- NASA and Dept. of Agriculture announced advance plans were under way for ground and air study of possible southern corn leaf blight during 1971 growing season. Remote aerial sensing techniques would be used in coordination with special ground observations. NASA high-altitude aircraft would photograph cornbelt area using special infrared and natural color film. Data would be sent to photo-interpretation team at Purdue Univ. for analysis. (NASA Release 71–64)
- Computer method being developed by NASA and Stanford Univ. team would provide simple means of viewing heart in action, NASA announced. System projected three-dimension animated cartoon-like image

of any desired chamber of patient's heart in lines of light on computer screen. Display, which would enable doctors to detect cardiac malfunctioning more simply than by conventional diagnostic methods, was derived from two-dimensional "x-ray movies" made by injecting x-ray contrast dye into desired heart chamber. (NASA Release 71-58)
- President Nixon sent message to Congress proposing improvements for District of Columbia: "A number of construction projects included in my budget for fiscal year 1972 . . . point to an attractive new look for Federal Washington by 1976. These include the Smithsonian Institution's plans to build a new National Air and Space Museum on the Mall and a new display area for cultural and technological advances of the past two centuries in the National Museum of History and Technology." President asked Congress to appropriate necessary funds for these and other projects. (PD, 4/12/71, 605–10)
- President Nixon transmitted *Marine Science Affairs: Annual Report of the President to the Congress on Marine Resources and Engineering Development*: U.S. marine science program in 1970 had been marked by sound accomplishments and "new policies and programs that fulfill the objectives of the Marine Resources and Engineering Development Act." FY 1972 budget request provided $609.1 million for marine science, technology, and services, "an increase of more than $70 million over my request of a year ago. These funds will permit NOAA to undertake priority programs of fundamental importance to the Nation's marine science interests; they will permit us to continue the accomplishments of the Sea Grant program; to further our participation in the International Decade of Ocean Exploration; to insure that necessary marine research and development is conducted for national security purposes; and to make certain that marine research and development, generally, continue to make productive contributions to our growing use of the sea" (PD, 4/12/71, 611)

April 8: Miniature analytical clinical laboratory to monitor astronauts' health on spacecraft was being developed at Oak Ridge National Laboratory, AEC announced. "Gravity-Zero" system would permit astronauts to perform quickly and automatically 16 parallel chemical tests on samples of plasma or serum, based on colorimetric determinations. Test results would be automatically radioed to ground control, which would recommend action to astronauts. (AEC Release 0-51)
- MSFC announced award of $29 136 622 contract modification to Chrysler Corp. for additional work in Saturn IB program. Chrysler, prime contractor for Saturn IB first stage (S-IB), would maintain nine boosters in storage and make prelaunch checkout of three of the boosters for Skylab program. (MSFC Release 71-60)
- Fritz von Opel—engineer, pilot, and industrialist—died in St. Moritz, Switzerland, at age 71. Von Opel had flown in rocket-propelled glider in 1929 and had developed rocket car which he drove at more than 3 km (2 mi) per minute in 1928. (UPI, *NYT*, 4/12/71, 38)

April 9: Cosmic ray tracks found in helmets of Apollo astronauts had indicated that exposure during space mission was sufficient to damage certain nonreplaceable cells on long flights, General Electric R&D Center team reported in *Science*. On two-year mission, fraction of cells killed could rise to 0.12% in cerebral cortex, 0.05% in retina, and more than 1.5% for giant cells. Proportions might be "highly worrisome, since additional safe shielding would impose important weight considerations on spacecraft design." (Comstock, Fleischer, et al., *Science*, 4/9/71, 154–6)

April 12: Tenth anniversary of U.S.S.R.'s *Vostok 1*, first manned space flight, piloted by Cosmonaut Yuri A. Gagarin. Anniversary was celebrated in Moscow with 6000 guests attending memorial meeting in Kremlin's Palace of Congresses and with Soviet newspapers filled with articles commemorating flight.

At Moscow ceremony Mstislav V. Keldysh, President of Soviet Academy of Sciences, said U.S.S.R. was looking forward to time "when man will make interplanetary flights. Tremendous potentialities in the study of the earth's natural satellite were uncovered by the flights of American astronauts to the moon and the development in our own country of new types of automatic lunar stations." Successful operation of U.S.S.R.'s *Lunokhod 1* lunar rover represented start of new era envisioned by Soviet scientists, Keldysh said. U.S.S.R. would emphasize unmanned flights with particular stress on long-term orbiting space stations.

Cosmonaut Andrian G. Nikolayev told meeting U.S.S.R.'s 1971–1975 economic plans called for "fundamental development of scientific research, with a significant role for lunar and solar investigation."

April 12: *Tenth anniversary of the first manned space flight, made by Cosmonaut Yuri A. Gagarin in the* Vostok 1. *In the April 14, 1961, photo Gagarin and U.S.S.R. Premier Nikita Khrushchev, standing atop the Lenin-Stalin Mausoleum, responded to a welcome by crowds during a ceremony in Moscow's Red Square. (Photo by courtesy of the National Air and Space Museum, Smithsonian Institution.)*

In letter to *Pravda*, group of Soviet cosmonauts said "it is our opinion that businesslike cooperation of space researchers of different countries, including the U.S.S.R. and the United States, should develop and grow stronger in the interests of peace and friendship among the peoples of our planet."

Soviet Academician Boris N. Petrov said in article reprinted in a number of Soviet journals that "main trend of the next development of cosmonautics and space exploration . . . will probably be as follows: Further investigation of near-Earth space, studying the Earth from space for purposes of space meteorology, geology, agriculture, oceanology, and marine and air navigation. The task of constructing space exploration facilities and the usage of outer space for peaceful purposes is the favorable basis for fruitful international cooperation wherein both big and small countries can effectively participate." U.S.S.R. had "always directed its efforts into an arena of peace and international cooperation. . . . The time is not behind the mountains when the remarkable idea about construction of long-duration, piloted orbital stations will be implemented."

Soviet Military Review editorial said Soviet space program had demonstrated "the benefiting the whole of mankind and the contribution to worldwide progress. The Gagarin flight, as all subsequent achievements of Soviet space exploration, demonstrated the organizing and guiding role of the Communist Party of the Soviet Union, and the inexhaustible potentialities of the socialist social system and of the Soviet economy, science, and technology. Consistency and purposefulness are a feature of the Soviet program of exploration in outer space, on the Moon and on the planets of the solar system." (Yuenger, *C Trib*, 4/13/71; Shabad, *NYT*, 4/13/71; UPI, *W News*, 4/12/71; UPI, *LA Her-Exam*, 4/12/71; *SBD*, 4/12/71, 263)

- *Apollo 11* Astronaut Michael Collins assumed duties as Director of Smithsonian Institution's National Air and Space Museum. He had served as Assistant Secretary of State for Public Affairs after leaving NASA. (Smithsonian Ann; Smithsonian PIO)
- NASA selected Boeing Co. Commercial Airplane Group, General Dynamics Corp. Convair Div., and Lockheed-Georgia Co. to conduct nine-month studies to determine benefits of applying advanced aeronautical technologies to 1975–1985 transport aircraft. (NASA Release 71–66)
- NASA announced award of $2-million, one-year, fixed-price contract to General Dynamics Corp. Convair Div. for definition and preliminary design of Research and Applications Module (RAM) that could be attached to space shuttle. (NASA Release 71–67)

April 13: NASA's supercritical wing, flown on TF–8A jet aircraft piloted by test pilot Thomas C. McMurtry, successfully completed fourth flight from FRC. Purposes of one-hour flight—to expand flight envelope to higher altitudes and higher speeds, evaluate augmented stability and control, determine handling qualities with stability augmentation system on and off, determine structural loads and buffet boundaries, and make preliminary evaluation of flutter characteristics—were achieved. Aircraft reached 10 700-m (35 000-ft) altitude and mach 0.9. (NASA Proj Off)

- NAE announced election of 29 U.S. engineers to NAE, bringing total membership to 356. New members included L/G Samuel C. Phillips (USAF),

Commander of SAMSO; Joseph F. Shea, Senior Vice President and General Manager of Raytheon Co. Equipment Div.; and Ronald Smelt, Vice President and Chief Scientist, Lockheed Aircraft Corp. (NAE Release)

April 14: U.S.S.R. launched *Cosmos 406* from Plesetsk into orbit with 222-km (137.9-mi) apogee, 199-km (123.7-mi) perigee, 88.6-min period, and 81.3° inclination. Satellite reentered April 24. (GSFC *SSR*, 4/30/71)

- Representatives of 79 INTELSAT member nations opened negotiating and organizing meeting at Dept. of State. Twelve observers, including representatives from Communist countries, attended. Purpose of meeting was to discuss proposed agreement to base member nation's voting power on its use of INTELSAT and limit U.S. power to 40%. (AP, *NYT*, 4/15/71, 14)

- Communist Party General Secretary Leonid I. Brezhnev had been guiding Soviet space research and rocket engineering since 1963, *Wall Street Journal* reported. Information had been contained in article circulated to Moscow correspondents by a Soviet news agency. (*WSJ*, 4/14/71, 1)

- Discovery of "first incontrovertible evidence" of existence of natural plutonium 244 by team of Univ. of California at Berkeley scientists was described by Dr. Calvin Alexander in Washington, D.C., press interview. Dr. Alexander and team had baked 15 mg of Pu 244—two thirds of world's supply—in vacuum system at 2144.3 K (3400°F). Pu 244 had been made expressly for experiment by Oak Ridge National Laboratory. It survived experiment almost intact, having lost only "a hundred milliom atoms of mass" in form of rare gas, xenon. Those atoms had provided "fingerprint" for Pu 244 similar to fingerprint found in certain meteorites. Scientists had suspected meteorite fingerprint had been remnant of completely decayed Pu 244, but until Berkeley experiment they had had no "file" against which to check it. (Lannan, *W Star*, 4/15/71, A6)

- AAAS held symposium in Boston on use of long-baseline interferometry—antennas thousands of miles apart, working together, to determine angular width and precise direction of distant radio sources. Three teams of astronomers who had worked independently reported observation in distant space of what appeared to be two objects flying apart at 10 times speed of light, but were unable to agree on plausible explanation for phenomenon that defied laws of physics that said such velocity was impossible. Discovery had been made originally by nine-man team of scientists from MIT, GSFC, JPL, and Univ. of Maryland. Team, under direction of Dr. Irwin I. Shapiro of MIT, using MIT Haystack antenna and JPL's Goldstone antenna, had observed two components of quasar 3C-279 accidentally during experiments conducted in October 1970 to observe passage of 3C-279 around sun. Discovery also was described by Dr. David L. Chauncey of Cornell Univ. and team from Cal Tech and National Radio Astronomy Observatory. Dr. Alan T. Moffet of Cal Tech reported on joint observations of phenomenon with Australian astronomers.

Following symposium, Rumford Premium, oldest scientific prize given in U.S., was presented to 21 scientists from U.S. and Canadian observatories who had pioneered in development of baseline interferometry. (Sullivan, *NYT*, 4/15/71, 33)

- MSC announced signing of $11 577 561 supplemental agreement with Grumman Aerospace Corp. for changes in Apollo LM contract. Agreement brought total value of contract since January 1963 to $1 617 510 000 (MSC Release 71-18)
- Dr. Armand N. Spitz, astronomer and author, died at age 66 after heart attack. Dr. Spitz had founded Spitz Laboratories, organized NASA Moonwatch program of using volunteer astronomers to track satellites, and created Spitz planetarium—small, inexpensive instrument used in schools and small museums. He was member of AAS, fellow of AAAS, consultant to NSF, and author of numerous publications. (W *Star*, 4/16/71, B5)

April 15: France launched 90-kg (198-lb) D2–A satellite *Tournesol* (Sunflower) from Kourou, French Guiana, by Diamant-B booster. Satellite, which would measure radiation for six months, entered orbit with 696-km (432.5-mi) apogee, 457-km (284-mi) perigee, 96.2-min period, and 46.3° inclination. (GSFC *SSR*, 4/30/71; *SBD*, 4/16/71, 239; *SF*, 6/71, 197; *Spacewarn*, 3/23/71, 6)

- President Nixon transmitted to Congress *World Weather Program, Plan for Fiscal Year 1972*. Report described significant activities of program and planned participation of Federal agencies. In message of transmission President said: "Satellite technology is being used with increasing effectiveness to gather global information for earlier, more accurate predictions and warnings of hazardous weather. New stations are being established for long-term measurement of atmospheric change. Computers have been programmed to determine the effect of pollution upon the atmosphere. A major international experiment in the Atlantic Ocean is being prepared under the Global Atmospheric Research Program. During the past year many nations, including the United States, have indicated their support of this tropical experiment and have made tentative commitments to provide ships, aircraft, satellites, and other observing facilities. Linked with an increased computer capability to assess and integrate results, this experiment should be an important step toward attaining a true understanding of the global atmosphere." (*PD*, 4/19/71, 630)
- President Nixon transmitted to Senate *Convention for the Suppression of Unlawful Seizure of Aircraft* and requested advice and consent to ratification. Convention "to ensure that all hijackers, wherever found, would be subject to severe punishment for an act which endangers the safety and lives of passengers and crew aboard" had been signed at Diplomatic Conference at The Hague Dec. 16, 1970. President said: "Because of the worldwide threat of hijacking, the Convention provides that all States may become parties. I hope that it will be applied universally." (*PD*, 4/19/71, 630)
- New U.S.–U.S.S.R. missile gap was described by Sen. Henry M. Jackson (D-Wash.) in speech before American Society of Newspaper Editors: "Today the United States bomber force is less than 600, our sub-launched missiles have numbered 656 for four years, and our land-based missiles have totaled 1,054 for four years. During this time the Soviet bomber force has remained almost constant, but their submarine missile force has grown to almost 400 launchers and is expected to overtake ours in two to three years. And the Soviet land-based intercontinental force has risen to over 1,440 operational launchers—with

more on the way. Thus, while the Soviets are gaining and will soon exceed us in missile weaponry at sea, we have already fallen distinctly behind them on land.

"Over the same period, the actual destructive power of the U.S. forces has continually decreased, while the Soviet's destructive power has greatly increased, until the total megatonnage that the Soviets could deliver against U.S. targets is several times our own.

"The only strategic area in which we are staying ahead is in the number of individual warheads, and we are doing that only by going to small-yield multiples. Even here, the Soviets could in time overtake us, since their missile forces have considerably greater payloads than ours." (Text)

April 16: NASA announced development of new underwater camera system designed to photograph algae on bottom of Lake Erie. Developed by LaRC for Environmental Protection Agency, 35-mm Nikon-F camera with Braun Lite 515 flash unit, timing electronics, and water-tight plastic cases could automatically click off one frame per hour for more than 10 days underwater and unattended. Summer 1970 joint study with Canadian Centre of Inland Waters—Project Hypo—had investigated causes of eutrophication (process leading to overabundant algae growth, robbing lake of oxygen). The 400 color slides taken traced developments in seasonal surge of algae growth and revealed previously unobserved effects of bottom currents on sediment resuspension. (NASA Release 71–65)

- Government of Jamaica had asked U.S. to assist in surveying Jamaica's natural resources with aircraft specially equipped for purpose, NASA announced. U.N.'s Food and Agricultural Organization had invited NASA to provide instrumented C–130 aircraft to gather experimental, thermal imagery data by cameras for high- and low-altitude infrared photography and sensors during flights at altitudes from 900 to 7600 m (3000 to 25 000 ft) over Jamaica and selected coastal waters. Jamaican Geological Survey and U.S. Geological Survey would provide personnel and equipment for gathering ground data. (NASA Release 71–70)

- U.S. and Canadian scientists were planning to build $12- to $15-million upper-atmosphere observatory to study relationship of space energy to weather, radio communications, and other phenomena on earth, *Science* reported. Transmitter and four receivers would be erected near Great Lakes, "where the earth's magnetic field causes ionized layers of the upper atmosphere to form a low-density 'trough.'" Preliminary engineering studies for observatory, which was scheduled for completion in 1974, were being conducted under $99 950 NSF grant. (*Science,* 4/16/71, 244)

- ComSatCorp reported net income for first quarter of 1971 had increased to $6 691 000 (67 cents per share) from $3 345 000 (33 cents per share) for first quarter of 1970. Operating revenues for first quarter of 1971 totaled $21 934 000, increase of $6 499 000 (42%) over $15 435 000 received in first quarter of 1970. (ComSatCorp Release 71–25)

- *Science* editorial commented on underemployment of scientists and engineers: "Some companies and congressmen would like to believe that high technology can be effectively directed to solving problems of society such as pollution. This is only wishful thinking. A limited number of individual scientists and engineers are being hired by municipalities, states, and industries, but no mass hiring is likely." It would be "mistake to place too much dependence on government." Retraining without specific job openings was "waste of money, time, and emotion." Experience had shown scientists and engineers quickly developed needed skills on the job. "What is needed is a strenuous effort to identify possible openings that match to some degree the potentialities of individuals." (Abelson, *Science*, 4/16/71, 221)

April 17: U.S.S.R. launched *Meteor 8* meteorological satellite from Plesetsk into orbit with 633-km (393.3-mi) apogee, 609-km (378.4-mi) perigee, 97.1-min period, and 81.2° inclination. (GSFC *SSR*, 4/30/71; *SBD*, 4/21/71, 265)

April 17–18: Wallops Station held open house for public in joint Federal activities in area that included Assateague National Seashore Park, Chincoteague Coast Guard Station, and Chincoteague National Wildlife Refuge. (WS Release 71–5; WS PIO)

April 18: FAA issued multiyear aviation forecast: passengers carried by scheduled airlines would increase from 173.2 million expected in FY 1971 to 186.9 million in FY 1972. Increase of 8% would contrast with growth of about 2% in FY 1970 and 1% increase expected in FY 1971. Annual growth rate of 10% was projected through remainder of decade, with total passengers reaching 513.5 million in FY 1982. Over 12 yrs covered by forecast, number of hours flown in general aviation was expected to double, reaching 49.6 million in FY 1982. Scheduled revenue passenger-miles flown by U.S. certificated airlines was expected to grow from 133.7 billion in FY 1971 to 147.3 in FY 1972, 10.2% increase. By FY 1982, figure would reach 485 billion. (FAA Release 11–49)

April 19: U.S.S.R. launched unmanned *Salyut 1* orbital scientific station from Baykonur into orbit with 269-km (167.2-mi) apogee, 256-km (159.1-mi) perigee, 89.7-min period, and 51.5° inclination. Tass said experiment was designed for "perfecting the elements of the design and the on-board systems and for conducting scientific research and experiments in space." Western speculation, later confirmed [see April 23–25], was that *Salyut 1* would dock with manned spacecraft in attempt to establish first manned orbital workshop. (GSFC *SSR*, 4/30/71; *SBD*, 4/20/71, 254; Mills, B *Sun*, 4/20/71, A1)

- MSC announced issuance of RFPs for development of human waste collection and storage system for space shuttle orbiter stage. Contractor would receive $90 000, cost-plus-fixed-fee contract under which firm would fabricate for testing one fully functional unit that could be used individually during week-long test by team of three men and one woman. (MSC Release 71–23)

- Nike-Apache sounding rocket was launched by NASA from TERLS carrying Indian experiment to study luminescent clouds. Rocket and instruments functioned satisfactorily. (SR List)

- Space Program Advisory Council, established by NASA to advise on space program goals and objectives and aid in reviewing plans for and work in progress on space programs and accomplishments, held first meeting. Council of prominent scientists would report to NASA Deputy Administrator on findings of four committees—Applications, Physical Sciences, Life Sciences, and Space Systems. (NASA Release 71–71)

April 20: Dr. George M. Low, Acting NASA Administrator, and Alan Gotlieb, Deputy Minister of Canadian Dept. of Communications, signed agreement at NASA Hq. for first cooperative international project for experimental communications technology satellite. Cooperative Applications Satellite C (CAS–C) would be launched into geostationary orbit by NASA in 1974 to conduct communications experiments with ground terminals operating at extremely high frequencies, test one-kilowatt solar-power-cell arrays that would unfurl in space like window shades, and test electric propulsion and stabilization systems. Canadian DOC would be responsible for design, construction, integration, and subsystems testing of spacecraft and for tracking, data acquisition, command, and control of spacecraft after it reached orbit. NASA would provide Thor-Delta booster, launch services, superefficiency power tubes and associated equipment, and environmental testing of integrated spacecraft. There would be no exchange of funds between DOC and NASA. Project was fifth in series of cooperative satellites which began with launch of *Alouette 1* Sept. 28, 1962. (NASA Release 71–72)

- National Space Club luncheon honored Dr. Hugh L. Dryden, late NASA Deputy Administrator, physicist, and Director of NACA from 1947 to 1958. Dr. Robert C. Seamans, Jr., Secretary of the Air Force, said in speech: "Dr. Dryden held that the decentralization of authority was a desirable goal and dictatorial control of projects, funds, and facilities should be avoided. He knew that ideas and skills would be widely dispersed. Even though the goals of applied research and development projects were concrete and specific, the approach to these goals had to be diversified. Freedom of the agencies and scientists engaged in the work was a necessity here, as it was in fundamental research, if the best use was to be made of their abilities. These concepts have lived and grown, and I believe, still represent the fundamental policy for the management of programs at NASA today."

  During luncheon, bust of Dr. Dryden, commissioned by National Space Club and executed by sculptress Una Hanbury, was officially presented to NAS President, Dr. Philip Handler. (Text; NSC *News Letter*)

- MSFC announced award of $1 081 343 contract to Research, Inc., to design, fabricate, install, and check out electrical heating devices for space shuttle prototype heat protection systems. Device would be installed in an existing structural test facility at MSFC and would heat up to 1600 K (2500°F) on lower surface and 1300 K (1800°F) on top surface of test articles. Work on device would be funded in increments, beginning with $100 000 for preliminary design phase. (MSFC Release 71–67)

- Astronaut Alan B. Shepard, Jr., received *Golf Magazine's* All-America Award for making first golf shot on moon during *Apollo 14* mission (Jan. 31–Feb. 9). (*NYT*, 4/21/71, 57)

April 21: LeRC had turned over to St. Vincent Charity Hospital in Cleveland, Ohio, small analog computer designed by LeRC scientists Vernon D. Gebben and John A. Webb, Jr., at cost of $1500, NASA announced. Computer measured increase or decrease of stroke volume of heart, or changes in length of time heart valves remained open, to detect certain abnormalities early in development. Hospital planned to use computer

experimentally before application to human patients recovering from open heart surgery. (NASA Release 71-69)
- MSC announced signing of $10 699 404 contract modification with Martin Marietta Corp. for equipment in support of Skylab earth resources experiment package (EREP). Modification brought total value of contract since February 1970 to $60 million. (MSC Release 71-25)
- Soviet ICBM construction program was increasing, Secretary of Defense Melvin R. Laird said in speech before American Newspaper Publishers Assn. in New York: "In December and January it began to look as if the Soviet Union was slowing down its rapid rate of ICBM deployments after having reached a level of land-based ICBM's that gave them approximately 400 more than the 1054 possessed by the United States. The situation began to change in February and March. . . . More recent evidence confirms the sobering fact that the Soviet Union is involved in a new—and apparently extensive—ICBM construction program." (Text)
- FAA–DOT program to lower base of positive control airspace to 5486 m (18 000 ft) over entire U.S. by end of 1971 was announced by FAA Administrator John H. Shaffer. Purpose was to reduce potential for midair collisions. Current positive control base was between 5486 and 18 288 m (18 000 and 60 000 ft) over northeastern and north central U.S. and between 7315 and 18 288 m (24 000 and 60 000 ft) over remainder of conterminous 48 states. (FAA Release 71-50)
- Secretary of Transportation John A. Volpe announced appointment of John E. Hirten, former President of Institute of Planners of California, as Deputy Assistant Secretary of Transportation for Environment and Urban Systems. Hirten would succeed Michael Cafferty, who had returned to private law practice. (DOT Release R-24)

April 22: USAF launched unidentified satellite on Titan IIIB-Agena booster from Vandenberg AFB into orbit with 400.7-km (249-mi) apogee, 130.4-km (81-mi) perigee, 89.8-min period, and 110.9° inclination. Satellite reentered May 13. (GSFC *SSR*, 4/30/71; 5/31/71; Pres Rpt 72)
- House Committee on Science and Astronautics favorably reported with amendment H.R. 7109, $3.434-billion NASA authorization bill [see April 1]. (Text; NASA *LAR* X/35)
- Nike-Apache sounding rocket launched by NASA from WSMR carried Dudley Observatory payload to 123.5-km (76.7-mi) altitude to measure particulate matter in upper atmosphere by photometer, microphones, plasma detector, and particle collection. Rocket and instruments functioned satisfactorily and payload was recovered as planned. (NASA Rpt SRL)
- Largest influx of immigrant scientists, engineers, and physicians over past 20 yrs was reported by NSF for FY 1970. After sharp decline of 21% between 1968 and 1969, rise of 30% in 1970 had brought number of entries to over 13 000 level. (NSF *Highlights*, 4/22/71, 1)

April 23–25: U.S.S.R.'s *Soyuz 10*—carrying Cosmonauts Vladimir A. Shatalov, Aleksey S. Yeliseyev, and Nikolay N. Rukavishnikov—was launched from Baykonur at 4:54 am local time April 23 (6:54 pm EST April 22). Orbital parameters: apogee, 224 km (139.2 mi); perigee, 200 km (124.3 mi); period, 88.2 min; and inclination, 51.3°. Tass said spacecraft would conduct joint experiments with *Salyut 1* (launched April 19), make comprehensive check of onboard systems, test manual and automatic control systems, check out spacecraft orientation and stabilization in different flight conditions, and conduct medical-biological research on influence of space flight factors on human organism.

On April 24 *Soyuz 10* linked with *Salyut 1* for 5½ hrs, undocked, and pulled away. Taped pictures released later on Moscow TV showed *Soyuz 10* moving away from *Salyut 1* after docking experiment while commentator said: "The principles of rendezvous and docking with the use of new equipment were tested and checked out in the course of the joint experiment. A rigid mechanical link-up was achieved, followed by undocking of the manned ship from the orbital station. After the end of this experiment in maneuvering, both vehicles continued to fulfill their program." According to information released later, *Soyuz 10* caught up with *Salyut 1* in two stages. First stage automatically brought *Soyuz 10* within 180 m (590 ft) of target vehicle; in second, approach and docking were effected manually. After separation *Soyuz 10* circled *Salyut 1* for about one hour, photographing and filming it from various angles.

Tass announced *Soyuz 10* reentered and landed near Karaganda, Kazakhstan, at 4:40 am Baykonur time April 25 (6:40 pm EDT April 24) after "fulfilling the program of studies. The studies carried out during this flight are a stage in the general program of work with the orbital scientific station." (GSFC *SSR*, 4/30/71; *NYT*, 4/23–26/71; *W Post*, 4/25/71, A1)

April 23: U.S.S.R. launched *Cosmos 407* from Plesetsk into orbit with 818-km (508.3-mi) apogee, 791-km (491.5-mi) perigee, 100.9-min period, and 74.0° inclination. (GSFC *SSR*, 4/30/71; *SBD*, 4/26/71, 290)

- NASA announced tentative schedule for 12-day Apollo 15 manned lunar landing mission. Spacecraft, carrying three-man crew, would be launched from KSC at 9:34 am EDT July 26. Apollo 15 LM would land near moon's Hadley Rille at about 6:15 pm EDT July 30. Two astronauts would spend 67 hrs on moon making three trips across surface on LRV. Spacecraft and crew would splash down in Pacific at 4:46 EDT pm Aug. 7. (NASA Release 71–73)

- USAF F-111 on test flight from Edward AFB, Calif., crashed on gunnery range near Barstow, Calif., killing pilot and copilot, who had ejected. (UPI, *W Post*, 4/25/71, A2)

- NASA launched two Nike-Apache sounding rockets from WSMR. One carried Univ. of Colorado and GSFC experiment to study airglow, and one carried Dudley Observatory and GSFC experiment to collect micrometeoroids. Rockets and instruments functioned satisfactorily. (SR list)

- Perspective in science funding was drawn by Dr. Homer E. Newell, NASA Associate Administrator, in *Science* editorial: "The funding proposed for 1972 for basic science ($2.4 billion), or even that for the entire space exploration program ($3 billion), is a very small fraction of the funding that is proposed for efforts to ameliorate societal problems today ($90 billion). The real need is not so much for additional dollar attention as it is for attention of a different kind. Ideas, new approaches, and new insights into the wise management and utilization of our human and natural resources are what is required. Many

experimentally before application to human patients recovering from open heart surgery. (NASA Release 71-69)
- MSC announced signing of $10 699 404 contract modification with Martin Marietta Corp. for equipment in support of Skylab earth resources experiment package (EREP). Modification brought total value of contract since February 1970 to $60 million. (MSC Release 71-25)
- Soviet ICBM construction program was increasing, Secretary of Defense Melvin R. Laird said in speech before American Newspaper Publishers Assn. in New York: "In December and January it began to look as if the Soviet Union was slowing down its rapid rate of ICBM deployments after having reached a level of land-based ICBM's that gave them approximately 400 more than the 1054 possessed by the United States. The situation began to change in February and March. . . . More recent evidence confirms the sobering fact that the Soviet Union is involved in a new—and apparently extensive—ICBM construction program." (Text)
- FAA-DOT program to lower base of positive control airspace to 5486 m (18 000 ft) over entire U.S. by end of 1971 was announced by FAA Administrator John H. Shaffer. Purpose was to reduce potential for midair collisions. Current positive control base was between 5486 and 18 288 m (18 000 and 60 000 ft) over northeastern and north central U.S. and between 7315 and 18 288 m (24 000 and 60 000 ft) over remainder of conterminous 48 states. (FAA Release 71-50)
- Secretary of Transportation John A. Volpe announced appointment of John E. Hirten, former President of Institute of Planners of California, as Deputy Assistant Secretary of Transportation for Environment and Urban Systems. Hirten would succeed Michael Cafferty, who had returned to private law practice. (DOT Release R-24)

April 22: USAF launched unidentified satellite on Titan IIIB-Agena booster from Vandenberg AFB into orbit with 400.7-km (249-mi) apogee, 130.4-km (81-mi) perigee, 89.8-min period, and 110.9° inclination. Satellite reentered May 13. (GSFC *SSR*, 4/30/71; 5/31/71; Pres Rpt 72)
- House Committee on Science and Astronautics favorably reported with amendment H.R. 7109, $3.434-billion NASA authorization bill [see April 1]. (Text; NASA *LAR* X/35)
- Nike-Apache sounding rocket launched by NASA from WSMR carried Dudley Observatory payload to 123.5-km (76.7-mi) altitude to measure particulate matter in upper atmosphere by photometer, microphones, plasma detector, and particle collection. Rocket and instruments functioned satisfactorily and payload was recovered as planned. (NASA Rpt SRL)
- Largest influx of immigrant scientists, engineers, and physicians over past 20 yrs was reported by NSF for FY 1970. After sharp decline of 21% between 1968 and 1969, rise of 30% in 1970 had brought number of entries to over 13 000 level. (NSF *Highlights*, 4/22/71, 1)

April 23–25: U.S.S.R.'s *Soyuz 10*—carrying Cosmonauts Vladimir A. Shatalov, Aleksey S. Yeliseyev, and Nikolay N. Rukavishnikov—was launched from Baykonur at 4:54 am local time April 23 (6:54 pm EST April 22). Orbital parameters: apogee, 224 km (139.2 mi); perigee, 200 km (124.3 mi); period, 88.2 min; and inclination, 51.3°. Tass said spacecraft would conduct joint experiments with *Salyut 1* building spacecraft and conducting launch operations. NASA provided Scout booster, two experiments, technical consultations, launch crew training, and spacecraft tracking and data acquisition. (NASA Release 71-63; NASA Proj Off; GSFC SSR, 4/30/71)

- U.S.S.R. launched *Cosmos 408* from Plesetsk into orbit with 1510-km (938.3-mi) apogee, 200-km (124.3-mi) perigee, 102-min period, and 81.8° inclination. Satellite reentered Dec. 29. (GSFC SSR, 4/30/71; 12/31/71; SF, 10/71, 386)

- *New York Times* editorial commented on Soyuz-Salyut mission: "The latest Soviet space spectacular follows a long series of indications by Kremlin leaders that their present priority in manned space flight is establishment of a long-lasting station in space. This latest experiment seems an important—as well as expensive—step on the road toward that objective." But there was no guarantee that space stations would be restricted to benign purposes. "Logically, there should be no national space stations in space. All such facilities should be operated by the United Nations employing crews from many nations and they should all be open to appropriate international inspection." Principle had been recognized in Antarctic "where the presence of Soviet scientists in American bases and American scientists in Soviet bases has a long, useful and honorable history. With the latest Soyuz and Salyut vehicles in orbit, the sensitive issues raised by space stations must be faced in the nearest future." (*NYT*, 4/24/71)

- Soviet bloc geologists meeting in four-day conference at Riga, Latvia, had decided to establish International Coordinating Center of Marine Exploration in U.S.S.R., *New York Times* reported. Center, to ensure "rational use of mineral resources of the oceans," would be open to members of Council of Mutual Economic Assistance (Comecon), economic alliance of U.S.S.R. and Eastern Europe. (Shabad, *NYT*, 4/24/71, 9)

April 25: First flight-model Lunar Roving Vehicle (LRV) was installed on board Apollo 15 LM on test stand at KSC. LM would be moved to Vehicle Assembly Building May 8 to be mated to Saturn V booster and CSM. Complete Apollo 15 vehicle would be rolled out to launch pad May 11. (MSFC Release 71-71; *Marshall Star*, 4/28/71, 1)

- At press conference following *Soyuz 10* landing Cosmonaut Aleksey S. Yeliseyev described *Salyut 1* as "overwhelming. It was a little like a train entering a railroad terminal. That's how we felt as our rather big Soyuz eased up to the station.

"The rendezvous maneuver began from a long distance away. At first we could not see the orbital station and we corrected our orbit to get closer. We first saw the Salyut from a distance of about 15 kilometers [9 mi]. At that distance we were able to see the station through a special optical device. At first it looked very tiny, just a black spot... of small size. The station flashed special light beacons for easier recognition. From that moment on we kept it constantly in view as we proceeded with the rendezvous maneuver.

"As we approached the station, we kept looking at the station through instruments. It was quite an imposing sight. Then the docking took place. All the time we kept the required components of the station in the field of vision of our television and optical devices. Then we separated and circled the station, looking at it from all sides and shooting movie film of it.

"It was an overwhelming sight: There was this structure in flight, with a tremendous amount of apparatus and all sorts of antennas." (Shabad, *NYT*, 4/26/71, 1; 4/27/71, 30)

Cosmonaut Konstantin P. Feoktistov later said that in future, "it will become necessary to learn to dock a relatively small transport spaceship with a high flying multipurpose laboratory. . . . it is already necessary to look for more national technical solutions, in particular of docking units, as much depends on their design and perfection. A new version of docking units was tested during the flight."

Cosmonaut Boris Yegorov said "operations of docking and rendezvous lead to a considerable emotional burden on the cosmonauts. Medical men and engineers are looking for and finding the most rational combination of man and automatic equipment in such a complex maneuver in orbit. . . . quite a few problems of engineering and cosmic psychology must be solved to achieve the best possible coordination of man and automatic equipment."

Soviet comments on its space station program had shifted in emphasis to possible future docking experiments between Salyut satellites and manned spacecraft and had "left open the question of whether . . . Soyuz 10 . . . succeeded in doing all it set out to do," Anthony Astrachan later suggested in the *Washington Post*. Comments by Feoktistov strongly suggested that the Soyuz-Salyut experiment was only prelude to creation of permanent orbital lab. (*W Post*, 4/26/71, A1)

- Texas A&M Univ. scientists had irradiated five grams of lunar soil to determine whether its elements, apparently soluble in water, were absorbed in growing lettuce plants, AP reported. Laboratory plants growing in moon soil mixture had grown better than those in ordinary soil. Dr. Paul Baur, Jr., botanist working at MSC, had said if plants took up activated elements from lunar soil they could be detected on x-ray plates. "This may give us new clues to the fertilizer requirements of plants." (W *Star*, 4/25/71, A16)

- NASA launched two Nike-Apache sounding rockets from TERLS. One carried India-Japan experiment to study ionosphere. Mission was partial success. Second rocket carried Indian experiment to study luminescent clouds. Mission was unsuccessful. (SR list)

- Interview with Chien Wei-chang, professor of engineering at Chinghua Univ. in Peking who had worked at JPL from 1940 to 1946, was published in *Washington Post*. Asked if he had had anything to do with China's earth satellite program, Chien had replied, "What do you think?" (Roderick, AP, *W Post*, 4/25/71, A26)

- U.S.S.R. was "clearly a big jump ahead" of U.S. in "post-moon race operations," *Detroit News* editorial said. "Credit must be given the Russians for having the shrewdness to go all out for orbital space stations when they realized, as they must have done before Armstrong and Aldrin landed on the moon . . . that they were sure to lose the moon race. We shall have to watch for a few years because the decision has been made to slash the space budget." (*Detroit News*, 4/25/71, 4M)

- KSC tourist facilities were described in *New York Times* travel section: "In these days of popular denigration of space exploration, the tourist is likely to approach this vast space-age outdoor science museum in a defensively skeptical frame of mind. Three or more hours later he is likely to leave the center convinced that it was a valuable and exciting experience." KSC was "one of the least commercialized major tourist attractions of the world." NASA was "selling . . . on tapes and in guide lectures, but not offensively and not at the expense of information you want." There was little evidence of security "and the guides emphasize that you may photograph anything you see." (Friedlander, *NYT*, 4/25/71)

April 26: NASA launched series of four Nike-Cajun sounding rockets from Arenosillo, Spain, carrying Spanish experiments to study luminescent clouds. Rockets and instruments functioned satisfactorily. (SR list)

- "The short and uneventful flight of Soyuz 10 suggested . . . that the Soviet Union ran into technical difficulties in its first attempt at establishing a manned workshop in earth orbit," Thomas O'Toole said in *Washington Post.* "Just what the difficulties might have been is unclear, but the landing of Soyuz 10 after only two days in space, the brief time Soyuz spent docked with the Salute workshop and the apparent failure to transfer crew members . . . all added up in American minds to a bungled mission." (*W Post*, 4/26/71, A21)

- Possibility of using electrically controlled optical sensors and signal processes—techniques used to guide military missiles to targets—was being examined as substitute for sightless eyes by Huntington Institute of Applied Medical Research team led by Dr. Robert H. Pudenz, *New York Times* reported. Electrical impulses, as coded messages, would stimulate visual cortex of central nervous system and communicate pulses of information to brain by way of optic nerve. Artificial system would bypass eyes and optic nerve and radio signals directly to brain. (Holles, *NYT*, 4/26/71, 23)

- House Committee on Armed Services' subcommittee investigating Dec. 30, 1970, crash of USN F-14 Tomcat fighter aircraft during test flight at Grumman facility in Calverton, N.Y., received communication from identified source containing 13 allegations of defects and deficiencies in F-14A design, manufacture, testing, and administrative controls. Rep. F. Edward Hebert (D-La.), as Chairman of investigation subcommittee, ordered full inquiry into charges. (Subcom Rpt, 12/20/71)

- Sonic booms from USAF jets had smashed prehistoric ruins and caused disintegration of cliff faces at Yellowstone and Teton National Parks, Director of National Park Service George B. Hartzog said in testimony before House Committee on Appropriations' Subcommittee on Department of the Interior and Related Agencies. He believed jets originated in California but had received no meaningful response from correspondence with USAF. (Testimony)

April 26–28: NAS held annual meeting for first time in $4.25-million Washington, D.C., facility that contained 684-seat Dryden Auditorium. Auditorium had been funded in part by Hugh L. Dryden Memorial Fund as tribute to former NACA director and NASA Deputy Administrator who died in December 1965. Dr. Dryden had been NAS Home Secretary for 10 yrs.

Dr. Frederick Seitz, former NAS President, paid tribute to Dr. Dryden at opening session: "I am particularly pleased that these remarks can be made . . . following several successful lunar manned missions. In a sense, Dr. Dryden's professional life was directed as by fate to the scientific and technological watershed represented by a successful Apollo program. Today we stand beyond that watershed and are privileged to inspect new vistas."

Dr. Dryden had become "perhaps the central figure" in U.S. transition to space age following 1957 launch by U.S.S.R. of first successful Sputnik. He had helped prepare legislation to establish NASA. "In this process, he . . . drew upon his creative genius in helping to devise an administrative and operational instrument which would make flexible use of the existing and potential resources of our country and at the same time, focus, in a relentlessly systematic way, upon well-defined missions. Many individuals have contributed to the success of our space program, but none have been more crucial than Hugh Dryden working at his broadest as the scientist, the administrator, and the visionary."

Dr. Seitz estimated Dryden Memorial Fund contribution to auditorium's construction represented nearly 40% of total cost.

Following Dr. Seitz' remarks, bust of Dr. Dryden presented to NAS by National Space Club was unveiled.

At business session, NAS announced decision to increase membership to include leaders in medical, social, and behavioral research. Limit of 50 new members elected annually would be raised to 75 in 1972 and 100 in 1973. Thereafter limit would be reduced over several years to 60. (NAS–NRC–NAE *News Rpt*, 3/71, 1; Text; NASA Special Ann; Program; Schmeck, *NYT*, 4/28/71, 3:32)

April 26–29: American Physical Society held spring meeting in Washington, D.C.

Dr. Gary V. Latham, chief Apollo program seismologist, said most quakes shaking two Apollo seismometers left on moon appeared to be coming from one place that could be 644 km (400 mi) below lunar surface. Dr. Latham guessed quakes were result of molten rock being moved about under moon by tidal forces on earth. Larger quakes were too deep to be caused by escaping gas, which led to his theory that sublunar volcano caused quakes. He was unable to pinpoint quake source, except that they emanated from somewhere along 1290-km (800-mi) line between seismometers left on moon by *Apollo 12* and *Apollo 14* astronauts. (O'Toole, *W Post*, 4/27/71, A3)

Discovery by *Uhuru* (*Explorer 42*) Small Astronomy Satellite (launched Dec. 12, 1970) of slowdown in spin rate of x-ray star Centaurus X–3 in Milky Way Galaxy was reported by Dr. Riccardo Giacconi and Dr. Wallace Tucker of American Science and Engineering, Inc. Change, deduced April 12 from star's pulsation's, was caused, scientists thought, by some violent event, such as vast in-fall of material onto star. (Sullivan, *NYT*, 4/29/71, 52)

Possible first interception of solar neutrinos—"ghost" particles that sped earthward from sun's core—was described by Dr. Raymond Davis, Jr., of Brookhaven's National Laboratory. Using tank containing cleaning fluid 1.6 km (1 mi) underground in gold mine in South Dakota, Dr. Davis was detecting average of one solar neutrino every third day. Detection was difficult because neutrinos had no electric charge and floated freely through atom unless they hit its nucleus squarely. If Dr. Davis' observations were confirmed, they would contribute to understanding of how sun produced energy. Neutrinos were thought to be largely produced by reactions that began with fusion of two hydrogen nuclei (protons). (Sullivan, *NYT*, 4/30/71, 3:23)

Dr. V. Paul Kenney of Notre Dame Univ. told press briefing that "every 20 years or so we get a 1000 fold increase in the energy of our accelerators and each time we enter a new era of physics." (Lannan, W *Star*, 4/30/71, A16)

- Forty-second annual scientific meeting of Aerospace Medical Assn. was held in Houston, Tex. Approximately 5000 aerospace doctors attended. Dr. Charles A. Berry, MSC Director of Medical Research and Operations, was out-going president. (MSC Release 71-22)

April 27: Dr. James C. Fletcher was sworn in as NASA Administrator by D.C. Superior Court Judge James A. Belson in White House ceremony attended by President Nixon. (NASA Off Admin; *W Post*, 4/28/71, A2)

- USN TF–8A jet aircraft with NASA's supercritical wing, piloted by test pilot Gary Krier, successfully completed fifth flight from FRC. Flight achieved planned purposes: to extend flight envelope to higher dynamic pressures; evaluate augmented stability and control; determine handling qualities, structural loads, and buffet boundaries; and make preliminary evaluation of flutter characteristics at higher dynamic pressures. Aircraft reached 7600-m (25 000-ft) altitude and mach 0.9. (NASA Proj Off)

- Paul F. Bikle, Director of FRC since 1959, would retire May 31, NASA announced. To ensure smooth transition until appointment of new

April 27: Dr. James C. Fletcher was sworn in as NASA *Administrator by District of Columbia Superior Court Judge James A. Belson as President Nixon, Mrs. Fletcher, and Dr. George M. Low, Deputy Administrator (behind), watched in the President's office.*

director De E. Beeler, FRC Deputy Director, was named Acting Director with Bikle acting as an adviser. (NASA Release 71–77)

April 28: NASA announced decision to discontinue quarantine of astronauts, spacecraft, and lunar samples for remaining three Apollo flights. Dr. George M. Low, NASA Deputy Administrator, said NASA had concluded from analysis of *Apollo 11, 12,* and *14* quarantine information "that there is no hazard to man, animal, or plants in the lunar material. These results have been reviewed by the Interagency Committee on Back Contamination, and that committee has recommended that further lunar missions need not be subject to quarantine." (NASA Release 71–78)

- U.S.S.R. launched *Cosmos 409* from Plesetsk into orbit with 1228-km (763-mi) apogee, 1172-km (728.3-mi) perigee, 109.4-min period, and 74.0° inclination. (GEFC SSR, 4/30/71; *SF,* 10/71, 386)
- Dr. Clarence L. Johnson, Senior Vice President of Lockheed Aircraft Corp., received NAE's sixth Founders Medal for designing advanced aircraft and for experimental and theoretical investigations in aerospace sciences. Dr. Johnson was best known for his work in designing Hudson bomber, Constellation and Superconstellation transports, P-38, T-33 trainer, F-90, JetStar, U-2, Warning Star, YF-12A, and SR-71. (NAE Release, 7/13/71; NAE *Bridge,* 7/71, 1)

April 29: USN TF-8A jet aircraft with NASA's supercritical wing, piloted by test pilot Thomas C. McMurtry, successfully completed sixth flight from FRC. Purposes of flight—to explore flutter boundary and evaluate stability and control characteristics—were achieved. Aircraft reached 12 000-m (40 000-ft) altitude and mach 0.95. (NASA Proj Off)

- Aerobee 170 sounding rocket was launched by NASA from WSMR carrying Lockheed Aircraft Corp. and GSFC experiment to study solar x-ray emissions. Rocket and instruments functioned satisfactorily. (SR list)
- NASA announced selection of Boeing Co. to build Mariner Venus-Mercury '73 spacecraft—first to explore two planets on single mission. Boeing, one of four companies that had submitted proposals, submitted estimated cost of $47 million, with award fee based on performance. (NASA Release 71–81)
- Major Soviet space station project included "powerful new launch vehicle in the class of America's Saturn 5," Kenneth Gatland said in *New Scientist and Science Journal* article. Prototype had been destroyed by fire in 1969 while undergoing static test on launch pad at Baykonur, he reported. (*New Sci & Sci J,* 4/29/71, 256–7)
- Change in name of Fairchild Hiller Corp. to Fairchild Industries, Inc., was approved by stockholders at annual meeting in Germantown, Md. (*W Post,* 4/29/71, B11)

April 29–30: Geology field trip and lunar EVA practice session using lunar roving vehicle simulator was conducted at China Lake, Calif., by Apollo 15 prime and backup commanders and LM pilots. Astronauts would use LRV to explore moon's Sea of Rains during July lunar landing mission. (NASA Special Release; W *Star,* 4/30/71, A9)

April 30: Nike-Apache sounding rocket was launched by NASA from TERLS carrying India-GSFC experiment to study ionosphere. Rocket and instruments functioned satisfactorily. (SR list)

- Recent Soviet launches "indicate continued progress toward a system for inspecting and possibly destroying American satellites," George C.

Wilson reported in the *Washington Post*. *Cosmos 400*, target satellite launched March 19, was hunted by *Cosmos 404*, launched April 4, which "went through a series of maneuvers under propulsion supplied by the giant SS-9 rocket. . . . The Soviet Union has conducted such space marksmanship tests before . . . But this time the test showed more sophistication . . . as the hunter satellite stayed in phase with the target for a longer period of time than on previous shots." (W *Post*, 4/30/71, A1)

- USAF placed its 257 operational F-111s on suspended status after investigation of April 23 fatal crash showed cause to be failure of ejection capsule to function. (Schmidt, *NYT*, 5/2/71, 66)
- Marshall Space Flight Center announced retirement of Director of Program Management Lee B. James, effective May 31. James, who would return to academic community, would be replaced in acting capacity by James T. Shepherd, Deputy Director (Technical), Program Management. (MSFC Release 71-77)

During April: NASA issued *Funds for Research, Development, R&D Plant and Scientific and Technical Information, Fiscal Years 1970-1972: Annual Report to The National Science Foundation.* In FY 1972 budget, NASA was seeking authorization of $3.271 billion. Effect on programs would be: Funds for basic research would remain at FY 1970 level in FY 1971 and rise in FY 1972, reflecting increased funding for Viking and initiation of outer planets Grand Tour project. In applied research, estimates for FY 1971 reflected 10% increase over 1970 and slight increase in FY 1972, primarily for earth resources surveys and space station. Budget authority for development activities would decline in 1971 and 1972, primarily as result of completion of Apollo program and reduction in NASA-AEC NERVA program effort. R&D funds were requested in FY 1972 for space shuttle technology and engine development facilities. (Text)

- AIA Aerospace Research Center published *National Technology Support: A Study of Research and Development Trends and Their Implications:* New patterns were emerging in magnitude and direction of R&D in U.S. "Recent R&D trends reflect diminishing governmental leadership in R&D which could lead to an erosion of the national research effort. The impact of continued inflation, the higher costs and longer leadtimes associated with increasingly sophisticated projects, plus accelerated efforts to meet specific national goals, suggest an even greater degradation of the total R&D effort than an examination of expenditures would indicate."

  Recent growth of industrial financing and leveling of Federal funding indicated shift in direction of national R&D activities. "Whereas industrial research and development has focused primarily upon product improvement and product development, most of the nation's basic research and the high risk, high cost activity has been financed by the Federal Government. Although their respective R&D programs frequently are complementary, certain technological projects traditionally initiated or sponsored by the Federal Government are beyond the financial scope of private enterprise. Consequently, a reduction in Federal R&D activity could have a negative impact on the level of sophisticated effort nationally."

Investments in higher education, level and nature of Federal R&D support, and utilization of existing manpower "provide some indication of the long-term scientific and technical capability of the nation. Previous Federal support of certain programs has had a feedback effect on the desirability of pursuing certain careers. Thus the failure to consider the long-term relationship among various fields of science implies future imbalances." Establishment of longer range R&D priorities and well defined national technological strategy was needed. (Text)

- Challenge of educational satellite telecommunications was discussed in *Bulletin of the Atomic Scientists* by Delbert E. Smith: While advances in comsat technology promised to revolutionize global communication, "it is not certain that educational uses of such a satellite system will develop concomitantly. They may, in fact, be lost in the speed with which our society utilizes the more spectacular and commercially viable facets of the medium." Educational Satellite Center at Univ. of Wisconsin had been established to work toward "fuller understanding of the social, cultural, political and legal consequences of the inevitable massive transformations in global communication patterns which are imminent." Objectives of EDSAT Center were: to provide focus for multidisciplinary research and training in educational and social applications and impact of comsats, to develop working models for application of satellite telecommunications systems to educational and social problems, to develop and maintain satellite transmission and reception capability for integration of hardware and software research, and to disseminate information on educational and social applications of space telecommunications. (*Bull of Atomic Scientists*, 4/71, 14–8)

- Changes and chances in U.S.–U.S.S.R. foreign affairs were described by Dr. Hans J. Morgenthau of Univ. of Chicago and City Univ. of New York in *Foreign Affairs*: ". . . future of American-Soviet relations is shrouded in uncertainty. Neither amity nor enmity is foreordained. Those who proclaim the inevitability of conflict on ideological grounds are as wrong as those who assert the inevitability of peace, or even friendship, because the United States and the Soviet Union have become more restrained in words and deeds in dealing with each other. The future depends first of all upon how the two governments conceive of their respective interests and how they will go about defending and protecting them. If they conceive of them in compatible terms and pursue them with appropriate concern for each other's sensibilities, the future might well witness the realization of Roosevelt's dream, Stalin's grand design, and Mao's nightmare: the cooperation of the United States and the Soviet Union in establishing and maintaining a modicum of order in the world. Otherwise the world will continue to hover on the brink of self-destruction." (*Foreign Affairs*, 4/71, 428–41)

- Study completed by TRW Inc. for DOD and CIA estimated that multiple warheads flight-tested to date with Soviet SS–9 ICBMs were not accurate enough to knock out U.S. Minuteman ICBMs in surprise attack. Study said warhead accuracy probably could not be improved enough with current techniques to achieve first-strike capability. (Getler, *W Post*, 6/17/71, A1)

- NSF published *Research and Development in Industry, 1969: Funds, 1969; Scientists & Engineers, January 1970* (NSF 71–18). In 1969 industry spent $18.5 billion for R&D—6% above 1968 level of $17.5 billion and 5 times amount spent on R&D in 1953. Increase in 1968–1969 was due to increase in companies' own funds. Between January 1969 and January 1970 full-time equivalent number of R&D scientists and engineers employed in industry dropped from 387 100 to 380 600. Decline —first recorded by NSF in 13-yr series of reports, was primarily caused by cutbacks in personnel working on Federal R&D programs. Most of R&D unemployment among scientists and engineers occurred in aircraft and missiles industry. Number of R&D professionals in DOD programs remained level between 1968 and 1969. Federal Government was source of 47% of R&D dollars spent by industry during 1969, down from 59% in 1959. DOD and NASA furnished 89% of Federal R&D funds to industry in 1969 and supported 89% of industrial scientists and engineers working on Federal programs, at average annual cost of $56 200 each. In 1969 industry financed 53% of its R&D with its own funds, increase from 41% in 1959. Industry allocated 3% of its 1969 R&D funds to basic research with more than 50% of this amount spent on physical sciences. (Text)

- *Government Executive* editorial noted that many experts feared "nation's wealth of technological brainpower has . . . been clobbered by budget cuts close to the point of no recovery. And history records so consistently it's become a politico-economic truism that the civilization which does not face up to these technological challenges sooner or later becomes captive of the civilization which does." (*Govt Exec*, 4/71, 9)

May 1971

May 1: NASA launched two sounding rockets from WSMR. Aerobee 150 carried GSFC experiment to study stellar UV, and Aerobee 170 carried MIT–GSFC experiment to study stellar x-ray sources. Rockets and instruments functioned satisfactorily. (SR list)

May 2: FY 1972 NASA budget placed agency "in good shape" for next year, Dr. Edward E. David, Jr., Presidential Science Adviser, said in AP interview published in *Chicago Tribune*. "I think there will have to be some major decisions in the next 18 months about the future of the space program." He also said that there was "activity" in Communist Chinese missile program, "but at the moment . . . I would say their activities are more threatening to the Soviet Union than they are to us." (AP, *C Trib*, 5/2/71, 8)

- Previously unidentified Soviet "specialist in space engineering" had been identified as Boris V. Raushenbakh, corresponding member of Soviet Academy of Sciences, *New York Times* said. Raushenbakh, jet combustion engineer, had joined Soviet Cosmonauts Konstantin P. Feoktistov and Boris B. Yegorov in giving interviews to Soviet reporters on *Salyut 1–Soyuz 10* mission April 23–25 at mission control center for Salyut-Soyuz. *Times* said it was believed to be first time that Academy member had been publicly associated with operation of space experiment. Normally, only former cosmonauts were identified with mission control on ground. (Shabad, *NYT*, 5/2/71)

May 3: U.S.S.R. was "showing signs of an accelerated effort" in space "stretching far beyond the implications of last month's start on an orbiting space station," *Aviation Week & Space Technology* said. "Latest analysis of orbital changes achieved by Cosmos 382, launched by a Proton booster . . . last fall . . . shows that the large unmanned satellite was moved once at an acceleration corresponding precisely to that required for lunar orbit insertion and once at exactly the velocity change needed for trans-earth injection." (*Av Wk*, 5/3/71, 13)

- Dept. of Labor's Manpower Administration began Technology Mobilization and Reemployment Program to help find jobs for 30 000 of 100 000 unemployed scientists, engineers, and technicians in 14 target areas hit by layoffs in aerospace and defense industries. (Labor Dept PIO)

- Defeat of SST program and "widespread public antipathy toward technology" had "encouraged the political liberals to intensify their attack on defense spending and press for a larger welfare budget," Robert Hotz said in *Aviation Week & Space Technology* editorial. "The Air Force B–1 supersonic bomber program will inherit all the supersonic transport's onus since it has basically the same flight characteristics." (*Av Wk*, 5/3/71, 9)

May 3–14: International Workshop on Earth Resources Survey Systems was held at Univ. of Michigan to inform representatives of 51 countries and international organizations of latest techniques for interpreting earth resources data acquired by aircraft and satellite remote sensing systems. Workshop was sponsored by NASA, Dept. of Agriculture, NOAA, U.S. Geological Survey, USN Naval Oceanographic Office, AID, and Dept. of State.

Ambassador George H. W. Bush, U.S. Representative to U.N., in welcoming address called space technology a global tool and said, "for a global tool to be used productively and efficiently, there must be a climate of international cooperation." Workshop and U.N. could "provide policy makers and program managers in developing countries with the basic information to enable them to consider seriously how this new technology may help them meet their needs."

Dr. William T. Pecora, Under Secretary of the Interior, said in keynote address that earth-surveying satellites were "must" for world's welfare. "Whether or not our great population can avoid intolerable social problems of a crowded world and maintain the hoped-for living standards is a matter of conjecture, but our ability to survive will depend in large part on critical assessment of all the earth's resources."

Leonard Jaffe, NASA Deputy Associate Administrator for Applications, described NASA Earth Resources Aircraft Program (ERAP): "Instead of waiting for data acquired from spacecraft, users are presently obtaining multispectral data of Earth phenomena from sensors carried by NASA aircraft . . . to evaluate the sensors and to develop a solid foundation for observational and interpretive techniques for earth-resources space missions." ERAP included RB–57F, P–3A, and C–130B aircraft. "We shall shortly add two more high altitude (U–2) aircraft." Lockheed C–130B was flying optical and infrared laboratory with 24-channel multispectral scanner to provide information on signatures used to separate, classify, and identify specific earth resources. "This is an expansion of the 12-channel scanner of the University of Michigan C–47 aircraft which we also use for the same purposes. The P–3A or Electra is our radio-wave instrument development laboratory." Radio antennas on underside of aircraft were used heavily for oceanographic and hydrological studies. Since remotely sensed data had to be verified with actual ground data, "complementary ground efforts are necessary during the research phase of the program."

In spacecraft program to begin in 1972, "ERTS satellites will be placed into a 920-kilometer [570-mi], sun-synchronous (near polar) orbit. This will permit the satellite, with its narrow-angle sensors, to observe the same spot on Earth once every 18 days. It also ensures that observations are made with a nearly constant solar-illumination angle required for developing identification signatures. The altitude permits narrow-angle sensors to be used to obtain virtually undistorted images." (NASA Release 71–74; GE *Challenge*, Summer 1971; Transcript)

May 4: Apollo 15 Astronauts David R. Scott and James B. Irwin demonstrated LRV Rover for press at KSC. Newsmen were permitted to take turns driving vehicle at maximum 16-km-per-hr (10-mph) speed. Scott told press, "We expect this little buggy to work just fine on the moon. It's really a very straightforward little vehicle, just like driving your own automobile." On first LRV excursion, Scott and Irwin would drive about one kilometer (six tenths mile) from Apollo 15 landing site at Hadley Rille to west edge of foothill for Hadley Delta at average speed of eight kilometers per hour (five miles per hour). On second excursion next day, they would drive to location as far as eight kilometers (five miles) from landing site and seek place to climb front and survey valley. "We'll be looking for bedrock or some unique feature that might have sprung up from the moon with the front," Irwin told press. "If we see it and the trail upward looks traversable, we might climb as high as 600 ft [180 m] to get to it." Third and last trip on third day on moon would take astronauts to edge of Hadley Rille and north along ravine to sample crater cluster that appeared to scientists to be volcanic in origin. On return, astronauts would park LRV 90 m (300 ft) east of LM with TV camera pointed directly at LM. Irwin told press if all went well worldwide TV audience would get first live picture of spacecraft rocketing off moon on first leg of return journey to earth. (O'Toole, *W Post*, 5/5/71)

- MSFC announced plans to launch 914-mm (36-in) Stratoscope II balloon-borne astronomical telescope from MSFC in August. Objective of mission was to study Galaxies M31 and M32, Planetary Nebula (NGC 7662), Orion Nebula, and planet Saturn. Stratoscope II project management, previously directed by Princeton Univ., had been recently assigned to MSFC as part of project reorganization. Seven previous launches had been from National Center for Atmospheric Research's Scientific Balloon Flight Station at Palestine, Tex. (MSFC Release 71-78)

- *Apollo 11* Astronaut Neil A. Armstrong received 14th annual Sylvanus Thayer Award from U.S. Military Academy Assn. of Graduates in ceremony at West Point, N.Y. Citation read: "For his selfless devotion to this nation's aviation and space efforts reflecting the ideals symbolized in the West Point motto—Duty, Honor, Country. His outstanding accomplishments, spanning more than two decades of aeronautical history, have made him an acknowledged leader among American space pioneers."

  Earlier he received sabre from 1800-member cadet corps and told cadets, "As I stood on the Sea of Tranquility and looked up at the Earth, my impression was of the importance of the small, fragile, remote blue planet." Asked about future of U.S. space programs, Armstrong told press, "I am certainly optimistic, but we're doing considerably less than we'd like to. But space is here to stay for all humanity." (NASA PAO; Everly, AP, *NY Post*, 5/5/71, 17)

- U.S. space program was "beyond the days of reacting to each major Soviet space event," Dr. George M. Low, NASA Deputy Administrator, said in speech before Aviation/Space Writers Assn. meeting in Wichita, Kans. NASA's FY 1972 request was typical of those to come for several years; "we have made no commitment to any new program beyond the 1972 budget." In response to questions, Dr. Low said Soviet *Salyut 1–Soyuz 10* mission was "not clearly successful or unsuccessful." It was "inconceivable" that total planned mission was 48-hr flight to rendezvous and dock briefly. But U.S.S.R. had "strong program. They are concentrating very hard on a major space effort." (*Aero Daily*, 5/10/71)

- NASA planned to study "dial-a-plane" system in which computer would accept telephone requests, determine best aircraft itinerary to minimize trip lengths and passenger waiting, and provide effective air transportation system for smaller cities and less densely populated areas, NASA announced. If studies proved concept feasible, proposal for demon-

May 4

stration project would be made to DOT and FAA, with NASA supplying route-scheduling computer and software. (NASA Release 71–79)

- Spokesman for Army Corps of Engineers announced personnel at KSC District Office would be cut from 30 to 15, effective June 30. Change was "result of a greatly diminished workload for NASA and the Air Force in the Cape Kennedy area." (Cocoa, Fla, *Today*, 5/5/71, B1)
- U.S.S.R. had been running more than seven years behind U.S. in development of third-generation computers using integrated circuits, Mikhail Y. Rakovsky, Deputy Chairman of Soviet State Planning Committee, said at Moscow news conference. Integrated circuit computers had appeared in U.S. in 1964 but were scheduled for production in U.S.S.R. under five-year plan that went into effect this year. U.S.S.R. ranked fifth among world's computer users, behind U.S., West Germany, U.K., and Japan. (Shabad, *NYT*, 5/5/71, 13)
- Bipartisan 115-member congressional group—Members of Congress for Peace Through Law—issued report criticizing proposed B–1 bomber aircraft as unnecessary, ineffective, and obsolete in nuclear age. Report was prepared by Sen. George S. McGovern (D–S.D.) and Rep. John F. Seiberling (D–Ohio). (*CR*, 5/5/71, H3558–63)

May 4–6: Conference on aircraft safety and operating problems was held at LaRC. NASA speakers reported on supercritical wing work, effects of aircraft noise, general aviation aircraft experience, STOL operations, collision hazard warning, steep instrument approaches, ditching, trailing vortex, exhaust pollutants, and sonic booms. (NASA Release 71–76; LaRC PIO)

May 5: USAF launched unidentified satellite from AFETR on Titan IIIC booster into eccentric orbit with 35 787-km (22 237-mi) apogee, 295-km (183.3-mi) perigee, 631-min period, and 26.4° inclination. According to press reports satellite was reconnaissance mission to monitor Soviet and Communist Chinese missile tests and provide 30-min warning of long-range rocket attack. The 800-kg (1800-lb) satellite would be transferred to synchronous orbit and stationed over Asia at 36 000-km (22 300-mi) altitude. (GSFC *SSR*, 5/31/71; World Data Center A; *SBD*, 5/6/71, 28; AP, B *Sun*, 5/6/71, A8)

- Tenth anniversary of first U.S. manned space flight. *Freedom 7*—Mercury spacecraft launched May 5, 1961—had achieved objective of putting man into suborbital flight when Astronaut Alan B. Shepard, Jr., went to 185-km (115-mi) altitude during 15-min flight. Anniversary was observed at KSC with exhibition of Mercury spacecraft and Redstone launch vehicle in main auditorium, motion pictures on *Freedom 7* and *Apollo 14*, and guided tours. Following commemorative ceremony, Shepard visited Freedom 7 Elementary School in Cocoa Beach, Fla., where he told students he had made his last venture into space and would step aside to give younger astronauts a chance. (MSC Release 71–28; MSC PAO; W *Star*, 5/6/71)
- U.S.S.R.'s *Salyut I* orbital scientific station launched April 19 had completed 425 orbits of earth by 1:00 pm Moscow time (6:00 am EDT). All systems were functioning normally. (FBIS–Sov–71–95, 5/10/71, L11)
- NASA and DOT released information given in *Joint DOT–NASA Civil Aviation Research and Development Policy Study Report* dated March 1971. Study had been recommended by Congress and made by NASA,

May 5: *Dedication ceremonies for the historic launch site at Kennedy Space Center marked the 10th anniversary of the first U.S. manned space flight, flown on the Freedom 7 Mercury spacecraft. The first American to go into space, Astronaut Alan B. Shepard, Jr.—who also was commander of the Apollo 14 lunar landing mission Jan. 31–Feb. 9 this year—stood by the plaque marking the site at Launch Complex 5/6 during ceremonies attended by 3000 guests. The site had been restored to its original condition.*

DOT, and FAA, with assistance from DOD, CAB, and eight other Federal agencies. Report concluded that aircraft noise abatement deserved highest priority because of widespread concern for environment and because noise abatement program's success would affect solution to other aviation problems. Study recommended reductions of at least 10 decibels every 10 yrs until aircraft noise was suppressed into community background noise.

Continued Federal support at high level of aeronautical R&D was necessary to ensure strong technical base. Continued R&D was essential if current and future problems of civil aviation were to be solved and if civil aviation was to contribute to such future areas as regional development. Study recommended Government evaluate its regulatory role to be certain that policies were not inhibiting innovations by industry and that NASC develop permanent mechanism to review and recommend policies affecting civil aviation that embraced several agencies. Study called for exchange of middle management personnel be-

tween aviation-related agencies like NASA, DOT, DOD, and CAB. (NASA Release 71–82; Text; *Av Wk*, 5/10/71, 20–1)
- Eastern Airlines signed conditional agreement with Lockheed Aircraft Corp. to purchase 50 L–1011 TriStar jet aircraft which it had previously ordered. Eastern thus became first of nine companies with L–1011 orders to reaffirm purchase agreement since February bankruptcy of Rolls-Royce, Ltd., manufacturer of aircraft's RB–211 engines. (Bedingfield, *NYT*, 5/6/71, 63)
- Soviet pilots in Egypt were testing "superjets," probably MIG–23s, according to diplomatic sources quoted by UPI in *Boston Globe*. Aircraft were understood to be capable of reaching altitudes to 24 000 m (80 000 ft) and speeds up to three times speed of sound. They outmatched U.S. F–4 Phantoms used by Israeli air force. (*B Globe*, 5/5/71, 6)
- USN was concerned that F–14 fighter aircraft might go into potentially dangerous flat, fast spin, Capt. Lionel E. Ames, Jr. (USN), F–14 Project Manager, said in testimony before Senate Committee on Armed Services. (Transcript)

May 6: U.S.S.R. launched *Cosmos 410* from Baykonur into orbit with 288-km (179-mi) apogee, 203-km (126-mi) perigee, 89.2-min period, and 65° inclination. Satellite reentered May 18. (*SBD*, 5/7/71, 39; GSFC *SSR*, 5/31/71)
- Sen. Edward W. Brooke (R-Mass.) introduced S. 1805, "to provide relocation assistance, training assistance, and interest supplements to adversely affected workers separated from their employment because of the termination of defense and space contracts." (*CR*, 5/6/71, S6341)
- After meeting with President Nixon at White House, Secretary of the Treasury John B. Connally, Jr., announced President's intention to send to Congress request for legislation providing $250 million in loan guarantees for Lockheed Aircraft Corp. (*PD*, 5/10/71, 738)
- Administration officials had been "playing politics with some of the most sensitive secrets that come into the Pentagon—namely, the intelligence gleaned from our space photography," Jack Anderson said in *Washington Post*. Satellite reconnaissance intelligence, known by code name "Tango-Kilo" was so tightly guarded that some DOD intelligence analysts could not get clearance. "Yet Defense Secretary Mel Laird and CIA Chief Richard Helms have been giving out selective T–K intelligence to favorite senators to win support for the defense budget." (*W Post*, 5/6/71, F7)
- Federation of American Scientists issued report *Is There an R&D Gap?* Report examined DOD charges that U.S.S.R. R&D expenditures exceeded those of U.S. and would result in Soviet assumption of technological superiority. Conclusion reached was: "*This entire episode has been a classical numbers game featuring selective disclosure, questionable assumptions, exaggeratedly precise estimates, misleading language, and alarmist nonsequiter conclusions.*" There was "no claim, much less any evidence" that U.S.S.R. was spending more than U.S. on military technological advances as measured by "military technology base—research, exploratory development and a fraction of advanced development." No one had claimed to be able to measure Soviet expenditures in this category. "These expenditures would amount to a few billion at most

and be most difficult to estimate. *Necessary expenditures become progressively larger as one moves from research on basic technological discoveries to development of weapons. This shows the extreme difficulty in making meaningful comparisons on a financial basis of efforts to protect against technological surprise.* Neither the funding nor numbers of personnel involved are a sensible measure of original technological advance. Of far greater importance is the organization and application of available intellectual and other resources." (Senate Com on Armed Services, Hearings on FY 1972 Auth; *Science*, 8/29/71, 707–9)

May 7: U.S.S.R. launched eight Cosmos satellites from Plesetsk on one booster. Each satellite was about 0.8 m (2.5 ft) in diameter, 0.9 m (3 ft) long, and weighed 36–45 kg (80–100 lbs). *Cosmos 411* entered orbit with 1493-km (927.7-mi) apogee, 1317-km (818.3-mi) perigee, 113.8-min period, and 74° inclination. Others entered following orbits:

Cosmos 412, 1536-km (954.4-mi) apogee, 1482-km (920.9-mi) perigee, 116.1-min period, and 74° inclination;

Cosmos 413, 1508-km (937-mi) apogee, 1476-km (917.2-mi) perigee, 115.7-min period, and 74° inclination;

Cosmos 414, 1495-km (928.9-mi) apogee, 1428-km (887.3-mi) perigee, 115.1-min period, and 74° inclination;

Cosmos 415, 1501-km (932.7-mi) apogee, 1453-km (902.9-mi) perigee, 115.4-min period, and 74° inclination;

Cosmos 416, 1493-km (927.7-mi) apogee, 1373-km (853.1-mi) perigee, 114.4-min period, and 74° inclination;

Cosmos 417, 1494-km (928.3-mi) apogee, 1345-km (835.7-mi) perigee, 114.1-min period, and 74° inclination;

Cosmos 418, 1494-km (928.3-mi) apogee, 1401-km (870.5-mi) perigee, 114.8-min period, and 74° inclination. (GSFC *SSR*, 5/31/71; *SBD*, 5/11/71, 53)

- Nonexplosive Sprint missile guided by Safeguard radar system successfully intercepted Polaris missile warhead over Pacific Ocean during DOD test. DOD spokesman later said Sprint, launched from Kwajalein Missile Range in Pacific, had passed close enough to target fired from USN missile-firing ship U.S.S. *Observation Island* to have destroyed it if it had been armed with operational nuclear warhead. (UPI, *W Post*, 5/13/71, A4)
- French President Georges Pompidou was passenger aboard Concorde 001, French prototype of Anglo-French supersonic airliner, in 75-min flight from Paris to Toulouse. Aircraft reached twice speed of sound. Later, Concorde 002, British prototype, was forced to turn back after taking off from Toulouse because of faulty landing gear. Aircraft circled for half hour at 600 m (2000 ft) but made perfect landing 38 min after takeoff. (Reuters, B *Sun*, 5/8/71)
- AIA President Karl G. Harr, Jr., told press in Washington, D.C., that aerospace industry employment would decline almost 12% during 1971—much less than the 17½% loss in 1970. "This lends support to our expectations that 1972 will see a reversal of the declining employment trends experienced during the past three years." AIA forecast industry's payroll would shrink by year's end to 943 000 persons, first drop below one million since record keeping began in 1959. From 1969 to 1970, industry had been forced to release more than one

third of its labor force. AIA survey indicated aircraft industry employment would decline by 67 000 to 506 000; missiles and space by 39 000, to 291 000; and commercial transport aircraft by 14 000, to 76 000. (AP, W *Star*, 5/7/71, A3)

- Wallops Station announced selection of Lockheed Aircraft Service Co. to receive $300 000, cost-plus-award-fee contract for aircraft maintenance and operations at Wallops Station. Contract would cover one year, beginning in July, with two one-year options. (WS Release 71-7)
- Library of Congress Congressional Research Service released report, *The Soviet SST*. Report by John D. Holmfeld summarized progress of Tu-144 from press reports and U.S. and U.K. technical journals. Tu-144 was "clearly an aircraft which is competitive with the Concorde." It had 27.1-m (89-ft) wing span and Concorde's was 25.6 m (84 ft); it was 56 m (184 ft) long and Concorde was 62.1 m (204 ft). Tu-144 weighed 149 700 kg (330 000 lbs); Concorde weighed 174 600 kg (385 000 lbs). Tu-144 would cruise at 2490 km per hr (1550 mph) at 19 800-m (65 000-ft) altitude; Concorde cruised at 2170 km per hr (1350 mph) at altitude of 18 000 m (60 000 ft). On New York-London route, Concorde flight time was estimated at 3 hrs and 30 min; Tu-144 probably would require 3 hrs and 10 min. Tu-144 maximum range was estimated to be 6400 km (4000 mi). Tu-144 was powered by four Kuznetsov NK-144 turbofan engines with 127 480-newton (28 660-lb) thrust. Concorde's four Rolls-Royce Olympus engines had 146 000-newton (32 825-lb) thrust. (Text)

May 7–8: Working Group on Philosophy, Science, and Technology met at MIT. NASA Associate Administrator, Dr. Homer E. Newell, reviewed U.S. space program: "It is improper, I believe, to claim the ability to assess at this time the impact of space on human history. Our involvement in space has not been an isolated event, nor even a short term episode, in man's growing awareness of his universe and his built-in drive to dominate it. Space is simply another dimension of human endeavor, of success and failure, of progress and achievement. The reasons that the Space Age began when and how it did could be discussed at length; the fact of the matter is that we stand now well within it, and cannot have the perspective to grasp its full implications. But one can discern numerous impacts, many of which contribute in a variety of ways to our country and to the world. Among these contributions have already been: An uplifting of the human spirit, with an undoubtedly major impact on the human horizon, the scale of man's thought, and his estimate of his own capabilities. The opening of a new frontier. Major advances in technology and in our expectation of technology. The development of practical applications, with the special long range impact of a quantum jump in communications. Substantial contributions to the advancement of science and human knowledge. The exploration of our solar system.

"These benefits are worth a substantial investment of the Nation's resources, not to the neglect of other needs of our Nation and humanity, but as part of our investment in the continuing future and strength of our Nation." (Text)

May 8: Mariner 8 (Mariner-H) Mars probe, launched by NASA from ETR, Launch Complex 36, Pad A, at 9:11 pm EDT, failed to enter orbit when Centaur stage of Atlas-Centaur booster malfunctioned after normal countdown and liftoff. Anomalies began to appear with Centaur main engine start. Centaur stage oscillated in pitch in diverging manner and subsequently tumbled out of control, engines shut down from starvation caused by tumbling, and Centaur and spacecraft separated and reentered earth's atmosphere approximately 1500 km (900 mi) down range and 400 km (250 mi) north of Puerto Rico.

Failure investigation team composed of LeRC, General Dynamics Convair Div., KSC, and JPL personnel had been established to determine cause of failure and recommend corrective action required before launching Mariner-I. (NASA Proj Off)

- Interview with Tu-144 designer, Academician Aleksey N. Tupolev, was published in *Pravda*. Asked to comment on report Tu-144 would go into commercial service before Concorde Anglo-French supersonic airliner, Tupolev said: "In accordance with world standards, a new plane must fly several million kilometers before it is granted a ladder for the boarding of the first passengers. Many stages of development and testing are already behind the TU-144. And things here are good absolutely according to the program." (FBIS-Sov-71-97, 5/17/71, L13)

May 9: Ofo Orbiting Frog Otolith satellite, launched by NASA Nov. 9, 1970, reentered atmosphere. Mission had been adjudged successful Dec. 11, 1970. All objectives—including maintenance of two bullfrogs in space to obtain information on functioning and adaptability in weightlessness of vestibule, portion of inner ear which controlled balance—had been achieved. (GSFC *SSR*, 5/31/71)

- First detailed Soviet analysis of U.S. plan for Grand Tour of outer planets was reported by *New York Times*. In *Vestnik*, principal publication of Soviet Academy of Sciences, Soviet space expert Timur M. Eneyev had suggested that four missions covering two planets each might be more successful than the two missions to three planets contemplated by NASA. He indicated Soviet scientists favored approach of delivering automatic space stations to outer planets and suspending them by balloons in planetary atmosphere. He said technique, "for all its exotic aspects and difficulties of realization, is probably the most promising because it would yield far more data about the nature of the giant planets than could be obtained from fly-by trajectories." (Shabad, *NYT*, 5/9/71)

May 10: Dr. James C. Fletcher held first press conference as NASA Administrator. In response to questions Dr. Fletcher said strength of U.S. depended critically on development of new technology. "And this means new technology of all kinds—electronics, computers, materials, as well as military technology. And I think that's . . . why NASA . . . programs should be vigorously pursued." NASA had studied possibility of second Skylab and "many other manned missions in space, most of them somewhat cheaper than the second Skylab; and we intend to look at these very carefully to see what programs, if any, made sense after . . . the first Skylab is flown." Dr. Fletcher thought NASA STOL experimental aircraft program was "long overdue."

Asked his personal plans for NASA, Dr. Fletcher said, "I can't say that I want to go in a new direction from the way NASA has been going, except maybe up instead of down." He was interested in international cooperation in space and "whole business of commercial use of the things that NASA has been developing," including "strides in the applications satellites area." He was "great supporter of the shuttle," and had been "for many years before coming to NASA." With development of shuttle, or "a cheap transfer system to orbit . . . it'll open up all kinds of new things you can do in space. Since the costs will be much less—the costs per launch will be so much less. That is, all kinds of new applications, all kinds of new science programs, new manned, near-earth programs, almost anything you can think of can be done much cheaper once you have the shuttle developed." With sufficient funds, NASA would probably pursue space station. "But in my judgment, if you have to decide between the shuttle and the space station, you pick the shuttle first because you have to have that for the second." (Transcript)

- U.S.S.R. launched *Cosmos 419* from Baykonur into orbit with 340-km (211.3-mi) apogee, 203-km (126.1-mi) perigee, 87.5-min period, and 51.5° inclination. Satellite reentered May 12. (*SBD*, 5/18/71, 88; GSFC *SSR*, 5/31/71)

- NASA announced signing of contract with Telesat Canada under which NASA would launch two Telesat satellites for Canadian Domestic Communications System. Telesat's Anik (Eskimo for "brother") satellites would each provide 12 RF channels—10 for commercial use and two in reserve. Each RF channel would be capable of carrying one color TV channel or up to 960 voice channels. System would be first operational comsat system established entirely for domestic communications. Two satellites would be launched from KSC, one in late 1972 or early 1973 and one six months later. Telesat would reimburse NASA for Thor-Delta booster and launch services and would have options for additional launches as required. (NASA Release 71-85)

- LRV qualification test unit, replica of flight LRVs, was delivered to MSFC by Boeing Co., prime LRV contractor. Second flight model LRV, scheduled for use on January 1972 Apollo 16 mission, would be turned over to NASA May 12 and stored at Boeing's Kent, Washington, Space Center until after Apollo 15 mission in July 1971. (MSFC Release 71-80)

- MSFC announced new projects in research program on application of remote sensing techniques. Land-use survey of five north Alabama counties would photograph area from air with remote sensing camera to study how land was being used. Results would be used by Top of Alabama Regional Council of Governments. In "ground truth survey" with Alabama A&M Univ., MSFC would supplement aerial photography; university personnel would make sample ground study to correlate remote sensing information. (MSFC Release 71-79)

- Experiments by airlines to fill "growing gap in transportation for small isolated communities" caused by rising operating costs of larger aircraft, loss of train service, and steady population drop were described in *New York Times*. Frontier Airlines was using 15-passenger Beechcraft 99s to serve seven communities in rural Nebraska, Wyoming, and South Dakota. Trans-Texas International Airlines was using Beechcraft on some routes and planned to evaluate other small aircraft as replacement for full sized airliners, and Allegheny Airlines had turned over its service in 16 cities to nine small air taxi lines, some of which used Beechcraft 99s.

More than 50 small towns had lost regularly scheduled air service in past five years. Growing number of independent air taxi commuter airlines using 10- to 20-passenger aircraft were gaining economic stability, improving safety records, and carrying increasing number of passengers over short routes that regular airlines did not want. (Lindsey, *NYT*, 5/10/71, 29)

May 10–14: International Solar Energy Society held conference at GSFC. In keynote speech Dr. Manfred Altman, Univ. of Pennsylvania nuclear engineer, said there was "great need" for U.S. and other countries to develop energy sources "for which they will not have to compete." There was "total absence of air and water pollution" in use of solar energy. "We are facing an uncertain future as far as energy resources and energy converters are concerned. One thing which we have learned . . . in the aerospace industry is that one cannot make choices among alternatives unless they are all roughly at the same stage of development. Let us make sure that solar energy is brought up to the level of the other contenders—lest we make poor choices later on. Solar energy is not just a dream—if we will it—it can play a major, beneficial role in our lifetime." (*CR*, 6/4/71, S8262)

May 11: NASA announced availability of new publications: *Space Shuttle Technologies, FY 1971 Programs,* compilation of 110 projects making up NASA shuttle technology program, described projects and listed FY 1971 funding for each.

Significant NASA Inventions Available for Licensing in Foreign Countries listed abstracts of inventions in which NASA owned principal or exclusive rights and which were available for patent licensing in various countries. (NASA Special Release; NASA Release 71-83)

• General Dynamics Corp. demonstrated CL-215 Air Tanker over Anacostia River in Washington, D.C. Amphibious aircraft scooped up 5.5 cu m (1440 gal) of water from river, took off, and released water from two 2.7-cu-m (720-gal) tanks to show what it could do to forest fire. General Dynamics was interested in selling three CL-215s at $1.4 million each to U.S. Forest Service, *Washington Post* reported. (Barnes, *W Post*, 5/12/71, B1)

• Visit to U.S. by Soviet Americanologist Georgy Arbatov reflected continuing tension between U.S. and U.S.S.R., Joseph Kraft said in Baltimore *Sun*. While Arbatov had been received cordially "in very high places," difficulties arose "whenever conversations came round to specifics." Subject of strategic arms had been particularly thorny, "for the Russians appear to be forcing the pace in the arms race with the new holes apparently set out for deployment of an improved SS-9 or some other monster weapon." (*B Sun*, 5/11/71, 17)

• *New York Times* editorial advocated reimbursement by Government of airlines' investment in SST: "The bill which the SST lobby seeks to amend is a Supplemental Appropriations bill that contains money for the Government's closing costs arising from cancellation of the contract for the plane. The bill does provide an opportunity for the Government to consider its debt of honor to the airlines. They were strongly advised by the Department of Transportation in 1967 to advance money to the Boeing Company, to be credited against the future purchase of SST's. The airlines realistically recognized that this was risk capital if the plane should prove a technological failure, but in view of the Federal Government's insistent if unwise pressure in behalf of the plane, they had good reason to rely upon continued Government backing for the project. Although the $58.5-million involved is not a legal obligation, Congress should honor this commitment." (*NYT*, 5/11/71)

May 12: USAF X–24A lifting-body vehicle, piloted by Maj. Cecil A. Powell (USAF) in joint NASA–USAF research program, reached 21 000-m (70 000-ft) altitude and mach 1.35 during 26th flight after air launch from B-52 aircraft from FRC. Objectives were to determine lateral-directional derivative with rudder bias at 2° toe out, check out lateral acceleration feedback to control system, and check out automatic mach schedule for upper flaps. (NASA Proj Off)

- Nike-Cajun sounding rocket was launched by NASA from WSMR carrying MSC experiment to study airglow. Rocket and instruments functioned satisfactorily. (SR list)

- NASA's work on quiet engines was described by Roy P. Jackson, Associate Administrator, OART, at 1971 Society of Automotive Engineers Meeting in Atlanta, Ga.: "The first objective of our work on design principle for quiet engines is an experimental CTOL [conventional-takeoff-and-landing] engine and nacelle with 15 to 20 PNdb [perceived noise in decibels] less noise than the comparable JT3D class of engines in service today. Program began in 1969 and would be completed by end of 1972. General Electric Co. and NASA's LeRC were working together on program. "It is on schedule and should meet or surpass the noise reduction objectives. We believe that production of quiet engines using this technology could begin to be available for retrofit and new aircraft beginning in 1976. . . . Such quiet engine installation in aircraft and the size of the 707 and DC–8 could permit this class of aircraft to operate fully 10 db lower than the present FAR 36 standard for new aircraft." NASA expected "to continue technology that will show how to drive the noise level down another significant increment like 6 db within a few years." Aircraft "the size of the 707 should be able to operate at a noise level of 92 EPNdb [effective perceived noise in decibels]. Our objective is to develop the necessary technology for this by 1977. Production engines meeting such noise standards could then be developed by industry through normal development and certification cycles and be available for next-generation aircraft of all types beginning in the late 1970's or early 1980's." (Text)

- House, in Committee of the Whole, agreed to amendment to H.R. 8190, supplemental FY 1972 appropriations bill, to appropriate $85.3 million for continued SST development. Record teller vote was 201 to 195, with 2 voting "present." After committee rose on request, for separate vote, SST amendment was agreed to by vote of 201 to 197. (*CR*, 5/12/71. H3837-62)

- President Nixon congratulated House for reversing earlier position on Federal funding for SST: "Congress has today taken an important first step on behalf of thousands of workers across the country who have been engaged in the SST program—and whose vital skills and experience might otherwise be lost to the Nation." (*PD*, 5/17/71, 755)

- LM similar to those used on *Apollo 11, 12,* and *14* was placed on permanent display at Smithsonian Institution's National Air and Space Museum. (*W Post*, 5/7/71, B1; Smithsonian PIO)

- NASA released *Apollo 11 Mission Report* (NASA SP–238), summarizing events and results of July 16–24, 1969, *Apollo 11* manned lunar land-

ing mission. Report was for sale by National Technical Information Service in Springfield, Va. (Text)
- New York State Court of Appeals approved sale of Cornell Aeronautical Laboratory to EDP Technology, Inc., affirming lower court ruling that sale of research facility to private firm was legal. (*W Post*, 5/13/71, K11)

May 13: *Uhuru* (*Explorer 42*) Small Astronomy Satellite—launched Dec. 12, 1970—was adjudged successful by NASA. Satellite had completed full systematic scanning of galactic plane and substantially complete scan (95%) of celestial sphere. Experiment instrumentation was functioning in outstanding manner, exceeding many design objectives. Spacecraft control section performance—except for tape recorder and slight decrease in telemetry modulation, which had not affected data quality—was excellent. Acquired scientific data had touched on every aspect of observational x-ray astronomy and significant unexpected phenomena had been observed. Discovery of three new x-ray pulsars—Cygnus X–1, Centaurus X–3, and Lupus X–1—revealed class of pulsating x-ray source that differed in many respects from previously known x-ray pulsar in Crab Nebula. (NASA Proj Off)
- Nike-Cajun sounding rocket was launched by NASA from WSMR carrying MSC experiment to study airglow. Rocket and instruments functioned satisfactorily. (SR list)
- Senate Committee on Appropriations favorably reported H.R. 8190, supplemental FY 1972 appropriations bill that contained $85.3-million for further SST development.

 In *New York Times* Richard Witkin said bipartisan House effort to resurrect SST program, led by minority leader Rep. Gerald R. Ford (R-Mich.), had been "kept at low key so as not to alert the opposition." Knowing issue could be raised when H.R. 8190 came up, "rescue team" had "sounded out colleagues to see if the job layoffs and other effects of the SST's defeat had raised doubts in the minds of some who had voted to cancel the program." Help had been enlisted from Administration and from AFL President George Meany. (*CR*, 5/13/71, D412; *NYT*, 5/13/71, 26)
- Federal court in Orlando, Fla., had dismissed $5-million damage suit filed by widow of Astronaut Virgil I. Grissom, killed in Jan. 27, 1967, Apollo spacecraft fire, on grounds it was filed after statute of limitations expired, *Washington Post* reported. (*W Post*, 5/13/71, A9)
- MSC announced selection of Dynalectron Corp. Land-Air Div. to receive $1 805 000, one-year, cost-plus-award-fee contract for support services at White Sands Test Facility, N. Mex. (MSC Release 71–33)
- William M. Allen, Boeing Co. Chairman, received 1970 Robert J. Collier Trophy during Washington, D.C., ceremonies. Citation was to "the Boeing Company as leader of the Industry-Airline-Government team which successfully introduced the 747 into commercial service with particular recognition to Pratt & Whitney division of the United Aircraft Corporation and to Pan American World Airways." (NAA Release)
- USAF cleared 40 F–111s to fly after main parachute compartments had been fitted with new ejection explosives. Entire F–111 fleet had been restricted April 29 following April 23 crash near Edwards AFB, Calif., that killed both crewmen. (UPI, *NYT*, 5/15/71, 9)

May 13

- President Nixon designated Robert C. Tyson Chairman of U.S. delegation to World Administrative Radio Conference for Space Telecommunications in Geneva, June 7 through July 17. (*PD*, 5/17/71, 759)
- Sen. Clinton P. Anderson (D-N. Mex.), Chairman of Senate Committee on Aeronautical and Space Sciences, submitted to Senate resolutions authorizing printing of Committee reports *International Cooperation in Outer Space: A Symposium* and *Soviet Space Programs, 1966–70* as Senate documents. (*CR*, 5/13/71, S6844–5)

May 14: Three men would spend up to 56 days in MSC altitude chamber in early 1972 to obtain medical data and evaluate medical experiment equipment for Skylab program, NASA announced. Test would closely simulate Skylab conditions so that differences observed during actual flight might be attributed to causes such as weightlessness which could not be simulated in chamber tests. Test would evaluate 16 medical experiments and selected items of experiment equipment and aid in training ground-based medical operations team for their participation during actual space flight. Chamber was 6 m (20 ft) in diameter and was being modified to resemble Skylab Workshop crew quarters. Atmosphere would be 70% oxygen and 30% nitrogen at 34.5 kilonewtons per sq m (5 psi) pressure with 45% to 60% humidity. (NASA Release 71–86)

- Evidence obtained by two Aerobee sounding rockets launched from WSMR March 14, 1969, supported "closed universe theory" that expanding universe would slow down and then reverse itself until cosmos compressed again into fireball, team of NRL and Johns Hopkins Univ. scientists said in *Nature*. X-ray telescopes aboard rockets had observed x-ray flux in Coma and Virgo galaxies 20 times as great as would have been expected from individual galaxies. This proved gas in clusters was different from intergalactic gas and supported theory that gases were spread around universe in sufficient density to bring on collapse of universe. (Meekins, Fritz, *et al.*, *Nature*, 5/14/71, 107–8)
- Lockheed Aircraft Corp. issued statement saying it would be forced into bankruptcy if it could not find sufficient financing for L–1011 TriStar airbus. Bankruptcy could "create great confusion, greatly increased costs to the government and displacements and hazards for Lockheed employes and shareholders." (Text)
- MSC announced issuance of RFPs on $400 000 design study of space shuttle auxiliary propulsion system. Study would define oxygen-hydrogen system compatible for use in both booster and orbital vehicles. (MSC Release 71–35)

May 16: MSC announced issuance of RFPs for technological development of new surface materials that could stand environmental extremes expected to be experienced by space shuttle. MSC would award fixed-price contract to company that designed and developed best ceramic insulator. (MSC Release 71–36)

May 17: Completion of first phase of high-altitude photography for joint NASA, Dept. of Agriculture, and cornbelt states 1971 Corn Blight Watch Experiment [see April 7] was announced by NASA. RB–57F aircraft based at Scott AFB, Ill., had obtained black and white photos of areas of cornbelt region to orient personnel using remote aerial sensors and ground observations to study possible southern corn leaf blight during 1971 growing season. Second phase, to obtain baseline color infrared photos, was in progress for completion later in month. Infrared photos would be used for analysis of soil conditions before emergence of corn crop. Third phase, from June to September, would consist of high-altitude infrared photography in test areas every two weeks. (NASA Release 71–88)

- U.S.S.R.'s *Lunokhod 1* lunar rover completed six months of lunar exploration and was still functioning satisfactorily. *Pravda* reported vehicle was being activated on a month-to-month basis by Crimea tracking station. According to *Pravda, Lunokhod 1*'s systems "began to give out" at beginning of fifth lunar day in early April, but recuperated. "Now at the height of the sixth lunar day the vehicle, heeding command from earth, continues to live, work and move about as if it [were] only yesteday, and not Nov. 17 [1970], that it descended from the landing stage that had brought it to the Sea of Rains. The viability of the Lunokhod's design had turned out to be far greater than its creators expected." (*NYT*, 5/18/71)

- D. J. Fink, Vice President of General Electric Co. and General Manager of Space Div., told Ninth International Symposium on Space Technology and Science in Tokyo that U.S. did not have monopoly on earth resources programs. "All of us need the ingenuity of the entire world on how to utilize this technology, so I encourage your expanding participation and cooperation." (GE Reprint)

- NASA's 13 scientist-astronauts were "fuming," Thomas O'Toole said in *Washington Post*. They had been selected to be members of Apollo crews landing on moon, primarily as geologists, and members of Skylab crews, as physicians, physicists, and chemists. But "no geologist has landed on the moon yet, and only one has a chance of making it," O'Toole said. Among names of nine prime Skylab crew members and six backups recently circulated at MSC, three prime and two backup members were scientists and "this enraged the scientists who were hoping for two-thirds representation on Skylab." Three meetings had taken place between scientist-astronauts and senior NASA officials—Dr. John E. Naugle, Associate Administrator for Space Science and Applications; Dr. Homer E. Newell, Associate Administrator; and Dale D. Myers, Associate Administrator for Manned Space Flight. Scientist-astronauts had asked for more freedom in developing their scientific careers. O'Toole said NASA officials had been said to be sympathetic, "but the gripe about getting their own experiments aboard Skylab fell on deaf ears." (*W Post*, 5/17/71, A1)

- Eastern Airlines announced it was introducing advanced magnetometer at departure gates of major airports in U.S., Puerto Rico, and Virgin Islands to detect potential hijackers carrying weapons. Device, with existing equipment, could provide Eastern with 100% screening capability, according to Eastern Manager of Operational Safety Jack E. Shields. (*NYT*, 5/19/71, 45)

May 18: U.S.S.R. launched *Cosmos 420* from Baykonur into orbit with 248-km (154.1-mi) apogee, 200-km (124.3-mi) perigee, 88.9-min period, and 51.7° inclination. Satellite reentered May 29. (GSFC *SSR*, 5/31/71)

- Benefits of U.S.–U.S.S.R. cooperation in space under agreements signed by NASA and Soviet Academy of Sciences in October 1970 and January 1971 were described by Arnold W. Frutkin, NASA Assistant Administrator for International Affairs, during testimony before House Committee on Science and Astronautics' Subcommittee on International Cooperation in Science and Space: "If a cautious optimism proves justified, we can expect definite benefits. Each side will gain access to lunar materials from sites additional to those it had explored on its own. We will nearly double the coverage of our operational meteorological satellites. We shall have created a global network for sounding rocket observations of basic importance to world meteorological research. We shall have demonstrated a mutual participation in earth resources survey work. We can greatly expand our knowledge of the behavior of man in flight through data on Soviet missions in addition to our own. We shall make it possible to avoid any undesirable duplication of scientific work in space, to plan complementary missions, and to verify anomalous results—all helping us to develop opportunities for more rational and economic conduct of space science and exploration." Successful performance of rendezvous and docking agreement "will give us greater flexibility in space emergencies and for joint activities in space which could go a long way toward reducing international tensions and demonstrating common human interest." (Testimony)

- Sen. Robert P. Griffin (R-Mich.) introduced for Sen. Edward J. Gurney (R-Fla.) S.R. 101, "to authorize and request the President to issue a proclamation designating July 20, 1971, as 'National Moon Walk Day.'" (CR, 5/18/71, S7165)
- FAA released annual statistical report on aircraft landing facilities in U.S. and its territories: Airports, heliports, and seaplane bases totaled 11 261 as of Dec. 31, 1970—net increase of 211 over end of 1969. Included were 20 facilities in Puerto Rico, 4 in Virgin Islands, and 11 in South Pacific Island territories. During 1970, 204 airports, 35 heliports, and 11 seaplane bases had been reported abandoned. Of national total, 4260 landing facilities were classified as publicly owned and 7001 as privately owned and operated. (FAA Release 71–68)

May 19: U.S.S.R. launched *Mars 2* probe toward Mars from Baykonur at 9:23 pm local time (12:23 pm EDT). At 10:59 pm Baykonur time (1:59 pm EDT) spacecraft was injected into flight trajectory for Mars from "artificial earth satellite" in terrestrial orbit, according to Tass. Instruments were said to be working normally and flight was said to be close to calculated course.

Primary objective of six-month, 470-million-km (292-million-mi) mission, Tass said, was to conduct complex research of Mars and its atmosphere and study characteristics of solar plasma, cosmic rays, and radiation along Mars route. Satellite was scheduled to reach Mars vicinity in November, when planet was 129 million km (80 million mi) from earth. *Mars 2* was said to weigh 4650 kg (10 250 lbs)—almost five times weight of 1000-kg (2200-lb) Mariner-I scheduled for launch by NASA toward end of May. Size of *Mars 2* had led U.S. space experts to predict that it might attempt Mars orbit.

Mars 1, launched Nov. 1, 1962, had failed when radio contact was lost 106 million km (66 million mi) from earth. *Zond 2*, launched Nov. 30, 1964, had also ended in communication failure. (GSFC *SSR*, 5/31/71; *SBD*, 5/21/71, 112; Shabad, *NYT*, 5/20/71, 1; FBIS–Sov–98–71, 5/20/71, L1; *A&A, 1962; 1963; 1964*)

- U.S.S.R. launched *Cosmos 421* from Plesetsk into orbit with 464-km (288.3-mi) apogee, 273-km (169.6-mi) perigee, 91.9-min period, and 70.9° inclination. Satellite reentered Nov. 8. (GSFC *SSR*, 5/31/71; 11/30/71; *SBD*, 5/21/71, 119)

- Vice President Spiro T. Agnew presented Harmon International Aviator's Trophy for 1970 to *Apollo 11* Astronauts Neil A. Armstrong, Edwin E. Aldrin, Jr., and Michael Collins in Washington, D.C., ceremony. Other recipients were RAF Squadron Leaders Graham Williams and Leslie Lecky-Thompson, first to cross Atlantic nonstop in VTOL aircraft, and Turi Wideroe of Norway, first female airline pilot. (*W Post*, 5/20/71, A28)

- Aerobee 150 sounding rocket was launched by NASA from WSMR carrying MSC experiment to study UV spectra. Rocket and instruments functioned satisfactorily. (SR list)

- Vanguard satellite whose launch had been attempted Dec. 6, 1957, was presented to National Air and Space Museum, Smithsonian Institution, by Dr. John P. Hagen, former director of U.S. IGY satellite program. Spherical payload of Test Vehicle 3 (TV–3), first complete Vanguard launch vehicle with three live stages, had been thrown clear when vehicle exploded on pad. (NASA Hist Off; *Vanguard—A History*, SP-4202, 1970)

- Senate, by unanimous vote of 94, passed H.R. 8190, supplemental FY 1972 appropriations bill that authorized use of $10 million provided in Independent Offices and HUD Appropriations Act, 1971, for improvements to MTF/Slidell Computer Complex. Before passage Senate voted 58 to 37 to strike from bill $85.3 million for continued SST development and, by vote of 92 to 3, adopted amendment appropriating $155.8 million for expenses to terminate SST program. Expenses included $58.5 million refund to airlines; $85.3 million refund to Boeing Co., prime airframe contractor, and to General Electric Co., prime engine contractor; and $12 million to cover administrative costs. Earlier, White House Press Secretary Ronald L. Ziegler had told press Administration had decided after May 18 meeting with Boeing officials that it would cost more to continue with SST development than to terminate program. Boeing had added condition to completion of contract; condition called for development, at $350-million cost, of quieter engine for prototypes. (*CR*, 5/19/71, S7343–63, S7442–56; Lyons, *W Post*, 5/20/71, A1; *WSJ*, 5/21/71, 1)

- NASA announced award of $61.6-million, cost-plus-award-fee contract to General Electric Co. for hardware development of Earth Resources Technology Satellite (ERTS) system. Under provisions of award—which definitized July 27, 1970, letter contract—GE would develop two flight spacecraft (ERTS–A and ERTS–B), provide equipment and services needed for GSFC ground data-handling system, and develop spacecraft receiver and six ground platforms for remote-site data-collection-system experiment. (NASA Release 71–89)

- Administration officials expected U.S.S.R. to test-fire SS–9 ICBMs within few months, *New York Times* reported. Two silos were being rebuilt at Baykonur test center in Kazakhstan to same dimensions as larger silos that had been appearing all over western Russia since December. (Beecher, *NYT*, 5/19/71, 1)

May 20: NASA held Hq. news conference on Mariner status. Dr. George M. Low, NASA Deputy Administrator, reported on failure of May 8 Mariner-H launch. Review by JPL, LeRC, and General Dynamics Convair Div. personnel had isolated failure to integrated circuit in pitch channel of rate gyro preamplifier in Centaur stage autopilot. When Atlas-Centaur ignited, "amplifier output was . . . only 20 to 40 percent of what it should be." Second anomaly had been noted 25 sec after main engines started; "engines did not gimbal all the way to the stops. . . . At 1.2 degrees they stopped and hung up. So the signal for the first 20 or 25 seconds indicated that the engines oscillated with increasing amplitude in response to autopilot signals but the signals were at a lower gain than they should have been. . . . With the engine then thrusting at an angle, the vehicle wound up and started tumbling." Amplifier had functioned properly during checkout 40 min before liftoff, "so sometime between T-30 minutes and the Atlas sustainer engine shutdown ignition of the Centaur stage, this component failed." Investigators had assumed "that we had in the circuit a failed diode or a diode with a loose connection in it, that a voltage transient during powered flight . . . then got into the integrated circuit and caused all of the other events."

To avoid repetition of anomaly, NASA was "subjecting the autopilot cans to additional temperature cycling and vibration tests" to ascertain they were operating within limits "under some off-nominal conditions." New test was being conducted on autopilots for Mariner-I Atlas-Centaur 23 "to make sure that the diodes are indeed there and properly functioning and properly protecting the integrated circuit . . . before we fly again."

Failure of three Atlas-Centaurs in 15 flights had led to appraisal of "general quality of the vehicle," Dr. Low said. Conclusion "was that, if anything, quality has improved. There is no trend away from improving quality. . . ."

Vincent L. Johnson, Deputy Associate Administrator, OSSA, described checkout problem with propellant utilization in Atlas-Centaur 23. Short had been discovered in system that measured amount of hydrogen and oxygen in tanks and controlled flow of each propellant to engine. "Therefore it became necessary to open the tank and get inside to locate the fault, find it and fix it." Reassembly was expected to be completed by weekend and five- to seven-day checkout on Mariner-I begun. Launch date could occur from May 28 until June 6 with good launch windows available. (Transcript)

- Dr. James C. Fletcher, NASA Administrator, reviewed NASA program before 25th annual conference of AIAA Board of Governors in Williamsburg, Va.: "This country needs space programs that move out vigorously to create new technology. A space effort that limps along on yesterday's knowhow will not do much for the country. A slow program may cost less, but be worth nothing in terms of meeting the country's needs. And a slow program—or no program—is all we are going to have without the strong support of the public. . . . We should invent and fly more self-supporting satellites. We are seeking and getting valuable scientific knowledge from space we could not get in any other way . . . about the Earth and its atmosphere, the sun and the planets, and the universe. And about man himself." National security was at stake.

"It would not be safe for the United States, with its great responsibilities for world peace, to lag behind any other country in space technology. This is an axiom we did not quibble about in the '60s and should not quibble about in the '70s." Space program had proved "excellent hotbed for forcing new technology, which in turn raises our national productivity and prosperity and increases our ability to solve pressing social problems of today's urban society." Finally, space exploration was needed as inspiration for modern man. "I think we are inspired, and our children are inspired, to be living in an age when man first moved out into space and began the exploration of our solar system. I think we should be ashamed of ourselves, as a society, if we withdrew from space exploration now after such an auspicious beginning." (Text)

- Astronaut James A. Lovell, Jr., was named Deputy Director of Science and Applications at MSC. Veteran of four space flights totaling nearly 30 days in space, Lovell would assist in overall management of Science and Applications Directorate. (NASA Special Release; MSC Release 71-34)

- NASA-inspired Earth Awareness Foundation—nonprofit, educational and advisory body originated by NASA officials and Apollo astronauts—marked first anniversary. Group was chartered to develop community solutions to environmental problems using NASA-developed knowledge to maximum extent practicable. During first year group had established community chapters called "Earth I Care" clubs, mostly on school campuses; started "Summer Scientist" program to study pollution; established newsletter; presented awards for concern with environmental quality; and produced slides and films on environment from NASA photos taken in orbital flight. (MSC PAO)

- President Nixon announced agreement between U.S. and U.S.S.R. "to concentrate this year on working out agreement for the limitation of the deployment of antiballistic missile systems (ABMs)" and to "agree on certain measures with respect to the limitation of offensive strategic weapons." Agreement had been made "in the conviction that it will create more favorable conditions for further negotiations to limit all strategic arms." President called agreement "major step in breaking the stalemate on nuclear arms talks." Agreement was announced simultaneously in Moscow. (PD, 5/24/71, 783)

- House, by rollcall vote of 116 to 157, rejected entire sum of $155.8 million that had been approved by Senate May 19 for termination of SST program. Senate and House conferees then met in executive session to resolve differences between Senate and House versions of H.R. 8109, supplemental FY 1972 appropriations bill.

    At evening session, House agreed by voice vote to include in bill $97.3 million for SST termination costs, but rejected inclusion of $58.5 million for refunds to airlines. (CR, 5/20/71, D449; 5/21/71, S7594)

- Plans for offshore airport in Atlantic Ocean off Long Beach, N.Y., were described by Lawrence Lerner of Litton Industries Environetics Div. at national convention of NOISE—organization to represent persons bothered by aircraft noise. Called "wetport," facility designed to handle jet aircraft would consist of 14-km (9-mi) platform 8 km (5 mi) offshore, constructed on reinforced concrete. Facility could be prefabricated on land and floated to its location. With nine runways in operation, wetport could serve 50% more air traffic than New York's airports combined. "Fingers" of structure would contain basins and harbors for 80 to 120 ships. (Carper, *W Post*, 5/21/71)

May 21: Draft treaty to establish new organization to operate worldwide comsat system was approved by 73 of 79 nations holding membership in INTELSAT. Key provisions of treaty included creation of new policy-making board of governors on which U.S. control would shrink from 52% to 40%, appointment of new director general to assume control of INTELSAT system currently operated by ComSatCorp, new six-year contract with ComSatCorp for continued operation and technical management of INTELSAT system, and limitation of board of governors to 12 nations that used INTELSAT most. Board actions on important matters were to require two-thirds vote. No single nation nor any three-nation group would be able to block decision under board's consideration. Agreement could be amended either by two thirds of members holding two thirds of investment or by 85% of members holding 45% of investment. Signatories would meet every year to determine membership shares on which seats on board would be based.

Following approval of agreement, President Nixon said: ". . . this kind of breakthrough, through which it will be possible to have instant communication around the world, will reduce the ignorance. It will increase the information. It will reduce those areas of difference which exist because of ignorance and lack of information to a minimum." (W *Star*, 5/21/71; *PD*, 5/24/71, 786–7; INTELSAT PIO)

- President Nixon presented National Medal of Science to nine scientists in White House ceremony. Recipients, announced Jan. 27, included Dr. George E. Mueller, Senior Vice President of General Dynamics Corp. and former NASA Associate Administrator for Manned Space Flight. (*PD*, 5/24/71, 784)

- Dr. Donald L. Holmquest, scientist-astronaut selected in August 1967, had been granted one-year leave from NASA for teaching and medical research at Baylor Univ. School of Medicine in Houston, NASA announced. Dr. Holmquest would pursue "sub-specialty studies in the field of nuclear medicine in addition to general medical research and research related to NASA programs, such as a more convenient method of measuring changes in body calcium." He would return to astronaut program May 15, 1972. (MSC Release 71-73)

- Senate voted 27 to 25 to adopt conference report on H.R. 8190, supplemental FY 1972 appropriations bill. Report deleted $58.5 million for refunds to airlines from SST termination funds. (*CR*, 5/21/71, S7593–7697, S7610–20)

- William M. Magruder, Director of SST Development for DOT, told press in Washington, D.C., that Ataka and Co. of Japan had suggested it could obtain financing of $500 million or more from major Japanese industrial concerns to resurrect SST program. In return Ataka would become 50% partner in enterprise. On May 23, however, Ataru Takizawa, Vice President of Ataka and Co., said in Tokyo there was no truth to report that his firm made exploratory proposal to revive U.S. SST program. (Aug, *W Star*, 5/22/71; AP, *W Post*, 5/23/71)

- U.S.-sponsored resources survey of Jamaica, in which NASA provided C-130 aircraft [see April 16], had been described as among most complete and efficient jobs accomplished in air and ground data gathering, MSC *Roundup* reported. Allen H. Watkins, MSC Manager of Earth Observations Aircraft Program, had said 95% of data acquisition objectives had been achieved in "most complete ground truthing that the aircraft program has ever experienced." (MSC *Roundup*, 5/21/71, 1)

- Boeing Co. Chairman William M. Allen received National Transportation Award of silver bowl from Vice President Spiro T. Agnew during Washington, D.C., ceremonies. Award was for Boeing's construction of 747 jumbo jet aircraft. (AP, B *Sun*, 5/22/71)
- FAA said air traffic congestion had eased sufficiently at New York's Kennedy International Airport to permit suspension of hourly limits on landings and takeoffs that it imposed in 1969. Flight quotas would probably be removed after June 15. (Lindsey, *NYT*, 5/22/71, 54:3)
- *New York Times* editorial commented on NASA under new Administrator, Dr. James C. Fletcher: "The task of reversing NASA's downhill course will not be easy, but there are grounds for believing that this period of doldrums can be overcome. It will help . . . if the agency's efforts in respect to Mariner 9 and Apollo 15 succeed as planned." Two forces at work suggested "NASA may yet recapture some of the public support and interest which evaporated after its first spectacular Apollo flights to the moon's surface. One is the capability of satellites to survey the earth with an efficiency never before known. . . . Beyond this there is the promising possibility of use of space research as a means of encouraging international cooperation. . . . over the next decade NASA may make its most important contribution as an instrument for demonstrating that men and women from many nations and many ideologies can work together for the benefit of all the earth." (*NYT*, 5/21/71, 23)
- *New York Times* editorial praised end of SST: "This action expresses a sound sense of national priority and of concern with ecological values. It also exposes last week's surprise vote in the House to revive the SST as a political power play that reflects no credit on the leadership of either party. To reopen a major issue without advance warning to opponents was highhanded; to do so without getting up-to-date cost estimates from the Boeing Company was frivolous." (*NYT*, 5/21/71)

May 21–22: Application of laminar air-flow techniques in medical surgery was discussed during symposium on Clean Room Technology in Surgery Suites at KSC. (NASA Special Release 5/14/71; KSC PAO)

May 22: U.S.S.R. launched *Cosmos 422* into orbit from Plesetsk with 1011-km (628.2-mi) apogee, 986-km (612.7-mi) perigee, 105-min period, and 74° inclination. (GSFC SSR, 5/31/71; SF, 11/71, 428)

- U.S.S.R.'s *Lunokhod 1* lunar rover would begin "a new mode of operation requiring only minimum movements," Tass announced. Vehicle would remain in semihibernation, photographing surrounding area and transmitting data to earth until its instruments wore out. (UPI, *NYT*, 5/23/71; *SBD*, 5/25/71, 129)
- Reuters quoted MSC spokesman as saying scheduled technical talks between U.S. and U.S.S.R. on space cooperation had been postponed because Soviet specialists were too busy to attend. (*NYT*, 5/23/71)
- Future space shuttle missions were described by Dale D. Myers, NASA Associate Administrator for Manned Space Flight, in speech before

May 22

South Carolina Lions Clubs in Greenville, S.C.: "These payloads cover the areas of space physics, astronomy, space applications, life sciences, unmanned planetary exploration, and earth resources, military and commercial." Payloads ranged from 113-kg (250-lb) Explorer and Space Physics Satellites to 9100-kg (20 000-lb) astronomy mission "requiring annual revisits by the shuttle for servicing maintenance and updating of instrumentation." Sortie flights conducted solely by shuttle could be made within its seven-day orbit capability. Scientist could "conduct his own experiment in the orbiting shuttle and return with the data as well as his own observations. When the space station becomes operational, crew rescue missions could also be flown." (Text)

- Activities of Soviet aircraft industry were described in *Izvestia* by U.S.S.R. Aviation Industry Minister Pavel Dementyev: "Recently the IL-62, TU-134 and YAK-40 aircraft, which are of a high modern technical standard, have been developed and are being produced and used. This year the major route TU-154 aircraft comes into service. These liners are the basic aircraft of our civil fleet and are to replace their predecessors—the TU-104, AN-10, IL-18 and IL-14." (FBIS–Sov–71–103, 5/27/71, K5)

- Soviet progress in thermonuclear synthesis was described in *Pravda* article by Mstislav V. Keldysh, President of Soviet Academy of Sciences: "Over recent years we have successfully developed work on high-temperature plasma in connection with the search for ways of solving the problem of thermonuclear synthesis. In the 5-year period that has begun the creation of a Tokamak-type installation with optimum parameters is planned in order to increase two or three times, the density of the particles and the retention-time of the plasma, which is an essential condition for the creation of a thermonuclear reactor. Evidently the time has already come for starting research into the transformation of the energy emitted in the process of thermonuclear synthesis into electrical energy." (FBIS–Sov–71–103, 5/27/71, L1)

- Washington *Evening Star* editorial said "sad little commentary on the times" had arisen from NASA report of space age spinoff, a miniaturized ultrasonic alarm system that had been used to aid accident and illness victims. "But the purpose which NASA foresees is what's depressing." NASA OART saw invention's main value "in preventing minor challenges to teachers or fights between students from developing into major disturbances or even riots. That's the view from space of this poor earth beneath, but what a pity. Whatever happened to that luminous blue ball, ashine in the limitless, starry night?" (W *Star*, 5/22/71)

May 23: Chinese-built version of Soviet MIG-19 fighter produced at Mukden, Manchuria, at rate of 200 per year had been "backbone" of Communist Chinese Air Force for number of years, *New York Times* said. So many had been produced from blueprints supplied by U.S.S.R. before Sino-Soviet split in early 1960s that Communist China had sold some to neighboring Pakistan. Western sources quoted by *Times* had recently reported that 2800 combat aircraft of Communist Chinese Air Force were being supplemented by new jet fighter of Chinese design being produced in China at rate of 10 per month. Aircraft, called F-9 in West, flew at twice speed of sound, had combat range of 480 to 800 km (300 to 500 mi) and could operate above 15 000 m (50 000 ft). (*NYT*, 5/23/71)

May 24: NASA supercritical wing, flown on TF-8A jet aircraft piloted by NASA test pilot Thomas C. McMurtry, successfully completed seventh flight from FRC. Aircraft reached 12 900-m (42 500-ft) altitude and mach 0.98. Objectives were to expand flight envelope to higher altitudes and explore flutter and buffet boundaries, evaluate stability and control characteristics, and evaluate control system. (NASA Proj Off)

- At Apollo 15 press briefing in Washington, D.C., Astronaut Alfred M. Worden, CM pilot, described planned activities in lunar orbit, including first "working walk" in space. At 320 000 km (200 000 mi) from earth on return journey from moon, he would move outside CM, handwalk on 7.6-m (25-ft) tether along rail, and open bay in SM to retrieve two canisters of film used to photograph moon while spacecraft was in lunar orbit. It was necessary to retrieve film before SM was jettisoned 160 km (100 mi) from earth before CM's reentry. Worden said one hour had been allotted to space walk but underwater practice had indicated retrievals could be completed in 20 min. Two roundtrips from CM to SM were necessary because only one canister could be carried at a time.

  First canister would weigh 36.3 kg (80 lbs) and hold 2000 m (6700 ft) of film, including more than 1600 exposures of lunar surface taken from 113-km (70-mi) lunar orbit. Photos would be enlarged on earth to compose blowup 15.2 m (50 ft) wide and 160 km (100 mi) long showing details on lunar surface as small as 1.5 m (5 ft) in diameter. NASA hoped these would include photo of LRV. Second canister would weight 3.6 kg (8 lbs) and hold 1100 m (3600 ft) of film used in mapping camera.

  While Worden retrieved cannisters, CM's hatch would be opened. Astronaut James B. Irwin would guide Worden along tether and Astronaut David R. Scott would stow canisters when Worden passed them through hatch. Worden told press, "We're going to be doing something out there, not just testing out an engineering concept or working out a procedure to do something." (Transcript)

- Grumman Aerospace Corp. pilots Robert K. Smyth and William H. Miller conducted successful 58-min flight of USN F-14 from Grumman facility at Calverton, N.Y. Same pilots had flown first USN F-14 on second flight, on Dec. 30, 1970, when it crashed into woods near Grumman facility seconds after they successfully ejected. Aircraft was kept at speeds below 467 km per hr (290 mph) during maneuvers to familiarize pilots with handling characteristics. (Witkin, *NYT*, 5/25/71)

- Rep. Glenn M. Anderson (D-Calif.) introduced Emergency Conversion Loan Act to aid victims of aerospace industry depression. Legislation would immediately qualify jobless engineers, scientists, and technicians for conversion loans from banks in amounts up to 60% of prior salary, but not to exceed $12 000. (*CR*, 5/24/71, E4915)

- Dept. of Labor released results to date of Technology Mobilization and Reemployment Program to relocate unemployed aerospace and defense industry personnel. While 10 284 professionals had registered with state unemployment agencies since program's May 3 inception, about 300 jobs had come into registry. These did not include aerospace and defense jobs already listed but not yet tallied. Labor Dept. spokesman said program hoped to provide jobs for about 30 000 out of about

100 000 scientists, engineers, and technicians it would serve over next two years. (Shabecoff, *NYT*, 5/25/71)

May 24–25: Technical conference sponsored by MSFC and Huntsville, Ala., section of Optical Society of America at MSFC reviewed research in holography and optical filtering being conducted by NASA, industry, and universities. Holography was technique of using lasers to obtain three dimensional image of object or scene without use of camera or lenses. Method was employed in display systems and testing. (MSFC Release 71–85; MSFC PAO)

May 24–26: Eleventh European Space Symposium was held in West Berlin, Germany, to discuss applications satellites. In welcoming address Dr. Manfred Bodenschatz of German Society for Aviation and Astronautics (DGLR) said: "We have passed through the infancy period of space flight and realized the untold possibilities of utilization. . . . It is only to be regretted that the general public has not apparently fully realized this and one does . . . begin to note a certain indifference to space flight accompanied by the feeling that the major space powers will be doing it for us anyway." Purpose of symposium was "to demonstrate clearly that a whole range of . . . fields of application already exists and must be enlarged upon, and that European space technology, science, and industry can and must render its own important contribution here if it does not want to be left hopelessly behind." (*SF*, 10/71, 393–4)

May 25: USAF's X–24A lifting body, piloted by NASA test pilot John A. Manke, successfully completed 27th flight after air launch from B–52 aircraft over FRC in joint NASA–USAF program. Objectives of powered flight were to determine lateral-directional derivatives with rudder bias at 2° toe out, evaluate handling qualities with y accelerometer feedback and power on, determine transonic pitch damping, and obtain upper-flap pressure with upper flaps at 13°. Although only three of four engine chambers ignited, most objectives were achieved, with vehicle reaching 19 500-m (64 000-ft) altitude and mach 1.1. (NASA Proj Off)

• Tenth anniversary of President Kennedy's message to Congress urging that U.S. "commit itself to achieving the goal, before this decade is out, of landing a man on the moon and returning him safely to earth." Rep. Olin E. Teague (D-Tex.) said on House floor: "As we attempt to plan for a significant and aggressive national space program for the 1970's, it is well to remember that on this date in 1961 the late President John F. Kennedy made an equally difficult and challenging decision. It is time again that we seize the initiative and make a positive decision for a strong national space program." (*CR*, 5/25/71, S7698, E5023)

• Apollo 15 lunar surface experiments—passive seismic, lunar surface magnetometer, solar wind spectrometer, suprathermal ion detector, cold cathode gauge, and heat flow—were described by experimenters during NASA Hq. press briefing on Apollo 15 science objectives. Mission would return up to 68 kg (150 lbs) of lunar samples rather than maximum 44.5 kg (98 lbs) for previous Apollo missions; man-hours spent on moon would increase from 18 to 40; and area covered could extend to 56 km (35 mi) by use of LRV.

Dr. Gary V. Latham, Apollo Program Chief Seismologist, said pri-

May 25: *Tenth anniversary of President Kennedy's message to a joint session of Congress urging that the United States "commit itself to achieving the goal, before this decade is out, of landing a man on the moon and returning him safely to earth." The goal was achieved when Astronaut Neil A. Armstrong, on the* Apollo 11 *mission, became the first man to set a foot on the lunar surface, at 10:56 pm* EDT, *July 20, 1969.*

mary aim of passive seismic experiment was to discover where moonquakes occurred. "So far we have identified 11 zones in which moonquakes occur, but one of these is particularly active, accounting for about 80 percent of the energy we record." Source might be "an isolated pocket of magma under pressure at depths between 30 and 400 miles [50 and 600 km]." Third seismometer, to be placed on moon by Apollo 15, was expected to provide data as to whether all 11 zones corresponded to same source mechanism; i.e., small magma reservoirs. Active zone fell on line that bisected *Apollo 12* and *14* seismometer stations, "so that makes 15 an exceedingly vital station to us. If we just get one perigee passage and a couple of these moonquakes before station 12 or 14 ceases to function, we'll have taken quite a step forward." Apollo 15 would be at edge of prominent mascon basins, Serenatatis and Imbrium. "We expect . . . to impact an S–IVB smack in the middle of one of them, probably Serenatatis. So this will be the first time we get seismic waves propagating through one of these mascon features, and, hopefully, we will be able to say something about how its structures differ from the other mare we have been able to record seismic waves through."

Edward M. Davin, Program Scientist, Lunar Surface Experiments, said *Apollo 12* magnetometer had been "very successfully recording a surprising magnetic field. . . . we now have a general model of the moon based upon three layers . . . a crust, an inner layer, and a core." Apollo 15 lunar surface magnetometer could determine "if there is any asymmetry in the magnetic field" and provide additional information on electrical conductivity that indicated thermal distribution of lunar interior.

JPL scientist Dr. Conway W. Snyder said solar wind spectrometer experiment would be "measuring the weather in space. The solar wind which is a plasma—this is an ionized gas—is quite analogous to weather on earth. . . . it is the phenomenon that determines what magnetic fields are in space, affects cosmic rays, and generally is the fundamental entity in space once you stop considering the planets themselves." With addition of second spectrometer on moon "we hope to be able to start making some definite interpretations of the very complicated kind of plasma effects that go on the moon."

Suprathermal ion detector experiment was described by Rice Univ. scientist Dr. Kent Hills. *Apollo 12* and *14* ion detectors had revealed ion clouds that were not lunar-oriented but had been formed elsewhere "from an expanding neutral gas cloud." With Apollo 15 experiment, third station north of other two would give "extended distance with which to investigate these things and . . . a different direction."

Cold cathode gauge experiment would indicate amount of gas on moon, Dr. Francis S. Johnson of Univ. of Texas at Dallas said. Ion detector already on moon had detected gas clouds in which concentration sometimes increased 100 times. With second gauge "we hope that there will be the prospect of seeing the same event on two gauges, and then we will be able to tell quite a lot more about where these gas sources would be and also learn something about how they propagate across the lunar surface."

Heat flow experiment, with 3-m (10-ft) drill, was to measure amount of heat coming from interior of moon, Columbia Univ. astronomer Dr. Marcus E. Langseth said. Measurement would be made by monitoring heat budget in near-surface material (lunar regolith) over one year with thermometers on probes about 2 m (6 ft) below lunar surface. Measurements could be useful in resolving contradictory data from magnetometer experiments on *Apollo 12* and *14*.

Experiments outside Apollo 15 ALSEP package—laser ranging retroflector, lunar field geology, and lunar soil mechanics—were discussed by Wesleyan Univ. astronomer Dr. James E. Faller; Donald A. Beattie, Program Manager, Lunar Surface Experiments; and Dr. James K. Mitchell of Univ. of California at Berkeley. Laser array would provide third leg of triangle formed by laser retroflectors on *Apollos 12* and *14*, improve ranging reliability from current stations, and "provide for greater efficiencies in telescope use time." Geology experiment objectives were to sample Apennine Mountain front, study Hadley Rille, examine mare, sample volcanic complex to north of landing site, and study cluster of craters southeast of landing site. Soil mechanics experiment would measure density, strength, compressibility and adhesive and cohesive characteristics of material on lunar surface. "The thing that is going to be of considerable interest is the lunar roving vehicle because if we can calibrate this vehicle . . . a knowledge of vehicle tracks and power consumptions should provide . . . continuous record of the soil conditions as the crew moved from point to point." (Transcript)

- Mstislav V. Keldysh was reelected President of Soviet Academy of Sciences at general assembly meeting in Moscow. Following reelection, Tass said: "The successes of the U.S.S.R. in the exploration of space have been associated with Keldysh's work. They included the first Sputniks, the flights to the moon and around it, flights to the planets of the solar system, the first manned space flight, the first experimental orbital station, the return of Luna 16 with lunar soil to the earth and the first Soviet moon rover." (FBIS-Sov-71-102, 5/26/71, L1; Shabad, *NYT*, 5/31/71, 32)
- Sen. Lawton Chiles (D-Fla.) inserted House Memorial 780 from Florida Legislature in *Congressional Record* "to request that Kennedy Space Center be the site for construction of the space shuttle." (*CR*, 5/25/71, S7680)
- Gen. Charles P. Cabell (USAF, Ret.), former CIA Deputy Director and NASA consultant, died at age 68 after heart attack. (Honsa, *W Post*, 5/27/71, B17)

May 26: NASA supercritical wing, flown on TF-8A jet aircraft piloted by NASA test pilot Thomas C. McMurtry, successfully completed eighth flight from FRC. Objectives of flight—to explore flutter boundary and evaluate stability and control characteristics near speed of sound—were achieved. Aircraft reached 14 000-m (46 000-ft) altitude and mach 0.975 with military rated power and 10 600-m (35 000-ft) altitude and mach 1.1 with afterburner. Flight was last in exploratory flight program; aircraft would be checked out and cleaned before beginning performance-data-gathering program in late July or early August. (NASA Proj Off)

- Cooperative U.S. and West Germany barium cloud experiment originally scheduled for launch during April 20-26 launch window, had been postponed until September because of unfavorable weather conditions and minor technical difficulties, NASA announced. Experiment was to study behavior of barium ion cloud at high altitudes to measure magnetic and electrical fields in space and to study plasma physics. (NASA Release)
- RAND Corp. report *The Space Shuttle as an Element in the National Space Program* (RM 6244-1PR) completed in October 1970—study that questioned economic feasibility of NASA's reusable space shuttle program—was entered in *Congressional Record* by Sen. Walter F. Mondale (D-Minn.) Study, undertaken for USAF, concluded "total space funding requirements over the next 20 years are not significantly different for plans that use the shuttle and those that accomplish the same missions without the shuttle." Results indicated "that criteria other than cost should be used to evaluate the desirability of the space transport system." (*CR*, 5/26/71, S7823-9)
- Sen. William Proxmire (D-Wis.) on Senate floor questioned need for space shuttle and space station: "I have written to NASA on a number of occasions to ask whether . . . there is a need for the space shuttle-space station. All NASA has told me in reply is that first, the shuttle would enable us to continue to have an active space program, and

May 26

second, it would reduce the costs of the space program. But why do we actually need it? What would it help us to accomplish that we could not otherwise accomplish? NASA seemingly has no answer to these questions." (*CR*, 5/26/71, S7811-2)

- DOD spokesman Jerry W. Friedheim told press new large missile silos being built in U.S.S.R. appeared to be designed for two different kinds of ICBMs but more than half of new holes might be for relatively small SS-11 ICBM. (Getler, *W Post*, 5/27/71)
- NSF published *Recent Trends in Enrollment and Manpower Resources in Graduate Science Education, 1969–70*. Graduate enrollment in sciences and engineering declined slightly between 1969 and 1970. Decrease and declining growth rate 1967–1969 contrasted with 9% annual rate of increase in enrollment for advanced degrees in science and engineering that characterized seven-year period from 1960 to 1967. Number of faculty and post-doctoral students had increased during 1969–1970, but annual rates of growth were "far below" comparable figures for 1967–1969. (NSF *Highlights*, 5/26/71)

May 26–28: NASA's second Government-Industry System Safety Conference was held at GSFC. In keynote address Rep. Jerry L. Pettis (R-Calif.) said: "The traumatic and inspiring experience of Apollo 13 now can be given profound symbolic meaning. The life on board became vitally important to millions of fellow passengers on Spaceship Earth. For a few moments in history we glimpsed the highest priority. The support crew focused on solving the most urgent problem—and succeeded like seasoned professionals."

Agenda covered results of application of NASA system safety techniques to automotive design, oil drilling and exploration, consumer products, surface ships, helicopters and urban transit, and aeronautics and space. NASA's first System Safety Conference had been held in 1968. (NASA Release 71-52; *CR*, 6/7/71, E5523-4; GSFC PAO)

May 27: Dr. James C. Fletcher, NASA Administrator, announced in released statement he had approved launch of Mariner-I Mars probe: "I have reviewed the results of the Mariner H failure analysis, and of the Mariner I Launch Readiness Review. I am satisfied that a complete and thorough job has been done, that the failure has been identified, and that proper corrective action has been taken. At the same time, I fully recognize that the Atlas-Centaur is an extremely complex vehicle, that there are literally thousands of parts and components that must function perfectly, and that a finite probability of failure must exist on each launch." Failure of Mariner-H May 8 had been traced to integrated circuit in Centaur booster. (NASA Release 71-92)

- U.S.S.R. launched *Cosmos 423* from Plesetsk into orbit with 489-km (303.9-mi) apogee, 279-km (173.4-mi) perigee, 91.9-min period, and 71° inclination. Satellite reentered Nov. 26. (GSFC *SSR*, 5/31/71; 11/30/71; *SBD*, 6/2/71, 169)
- Dr. James C. Fletcher, NASA Administrator, met with MSC officials during his first official visit to Center. Later, at press conference, he said NASA's image could not rest on "absolutely fantastic achievement" of first lunar landing by man. NASA needed to convince taxpayers of basic technology that had been developed on way to moon. Each part of NASA program "has to be defended as a separate piece." (Maloney, *H Post*, 5/28/71)

- Growing climate in U.S. of "irrational hostility" toward science and technology was scored by Dr. Wernher von Braun, NASA Deputy Associate Administrator for Planning, in speech before Washington, D.C., chapter of Aviation/Space Writers Assn.: "It is irrational precisely because those most vocal . . . are the very ones professing the greatest concern about poverty, poor housing, hunger, and the quality of the environment. All of these problems of society depend in varying degree upon our technological capabilities, and certainly on increased productivity for their solutions. Some . . . require advanced research into the nature of the environment and ecological interactions." Anti-science and anti-technology voices were doing U.S. "great disservice. The problems they are rightly anxious and concerned about cannot be solved by a return-to-nature cult." Resources of knowledge that awaited man in space had "great significance to man, to Earth environment, and to the ecology. We are learning of the relationships between Earth and Sun and their effects on our lives which could be learned in no other way save by means of the rocket and spacecraft." (Text)
- LeRC announced NASA had awarded Air Research Manufacturing Co. of Garrett Corp. $4.5-million, 40-mo, cost-plus-fixed-fee contract to build four turbine-alternator compressors (TAC) for experimental nuclear power generation system. System was to produce up to 100 kw as part of LeRC Brayton cycle power conversion system. (LeRC Release 71–23; NASA Release 71–93)
- Soviet scientist Anatoly Fedoseyev, head of Soviet delegation to Paris Air Show, defected to West, according to French sources later quoted by UPI. (UPI, *W Post*, 6/20/71, A1)
- Lockheed Aircraft Corp. had said settlements with U.S. Government on military contracts and other problems had resulted in net loss of $86.3 million in 1970 and $96.6 million in fourth quarter of 1970, *Wall Street Journal* reported. (*WSJ*, 5/27/71, 3)

May 27–June 6: Twenty-ninth Paris International Air Show was held at Le Bourget Airport. More than one million visitors to world's largest air show viewed Soviet Tu-144 supersonic transport in its first appearance in West; Anglo-French supersonic airliners, Concorde 001 and 002; Soviet Mi-12, world's largest helicopter; Lockheed C–5A Galaxy, world's largest aircraft; Lockheed 1011 TriStar; and more than 200 private aircraft in total exhibition of 600 aircraft. U.S. Pavilion exhibited Apollo moon rocks. Canadian Pavilion contained mockup of integrated STOL system envisaging 10-min air hops between suburban shopping areas and city centers. (B *Sun*, 5/26/71, 5/31/71; Hess, *NYT*, 5/26/71)

May 28: Mars 3 probe was launched by U.S.S.R. from Baykonur at 8:26 pm local time (11:26 am EDT) into parking orbit from which it was injected into heliocentric trajectory toward Mars. Payload would supplement that of *Mars 2*, launched May 19. All instruments functioned normally. By 10 pm Moscow time (3 pm EDT), *Mars 2* was 2 495 000 km (1 550 000 mi) from earth and *Mars 3* was 44 000 km (27 000 mi), Tass reported. Each satellite weighed 4650 kg (10 250 lbs) and had as prime mission objective complex research of Mars and its atmosphere. Tass said that *Mars 3* carried instrumentation for exploration of Mars and neighboring space, for investigating solar radio emission in one-meter wave band in experiment developed and manufactured by

May 28

French specialists in French-Soviet cooperative program, and for measuring characteristics of solar plasma and cosmic rays. (FBIS–Sov–71–104, 6/1/71, L1; SBD, 6/2/71, 69; Shabad, NYT, 5/29/71, 1)

- U.S.S.R. launched *Cosmos 424* from Plesetsk into orbit with 287-km (178.3-mi) apogee, 198-km (123-mi) perigee, 88.9-min period, and 65.3° inclination. Satellite reentered June 10. (GSFC SSR, 5/31/71; 6/30/71; SBD, 6/2/71, 169)
- New mission plan had been adopted for Mariner-I (Mariner 9) Mars probe scheduled for launch no earlier than May 29, NASA announced. New plan—which would provide high-quality data return for all of original experiments but less data than planned for original two-spacecraft mission—assigned 65° inclination, 12-hr period, and 1200-km (750-mi) periapsis (low point in orbit around Mars) to Mariner-I, with Nov. 14, 1971, arrival date. New inclination would yield map covering almost 70% of Mars and period would ensure two-tape-load-per-day data return. (NASA Release 71–94)
- Soviet Aviation Industry Minister Pavel Dementyev told Paris press conference Soviet supersonic transport Tu-144 was expected to go into commercial service in late 1973 or early 1974, about same time as Anglo-French Concorde. Only one Tu-144 was operational; two other prototypes under construction were expected to fly at year's end. Tu-144 had logged several hundred flying hours and flown at 2458 km per hr (1528 mph) at 6100 m (20 000 ft). Aleksey N. Tupolev, designer of Tu-144, said at press conference that Pan American World Airways President Najeeb E. Halaby was interested in aircraft, "but it is too soon to speak about markets or negotiations with international airlines until the Tu-144 has completed its test program." (Reuters, W Post, 5/29/71)
- December 1970 *Mariner 6* and *Mariner 7* experiments to simulate UV spectrum that would be viewed by UV spectrometer observing Jupiter or other outer plants were described in *Science* by Univ. of Colorado scientists J. B. Pearce, K. K. Kelly, and C. A. Barth, and JPL scientist A. L. Lane. Mixture of ammonia, nitrogen, and hydrogen had been released from spacecraft 240 million km (149 million mi) from earth and 120 million km (74.5 million mi) from sun. Spectra were recorded while gases were illuminated by sunlight. (*Science*, 5/28/71, 941–3)
- Fourth round of Strategic Arms Limitation Talks ended in Vienna. Talks would resume in Helsinki July 8. (*W Post*, 5/29/71)

May 29: Countdown for launch of NASA's Mariner-I spacecraft toward Mars orbit was interrupted by hold at 4:59 pm EDT 72 min from launch. Hold was caused by difficulty in interpreting data from Centaur stage guidance package. Difficulty was not resolved by 6 pm and launch was postponed to next day. (NASA Proj Off)

- U.S.S.R. launched *Cosmos 425* from Plesetsk into orbit with 550-km (341.8-mi) apogee, 508-km (315.7-mi) perigee, 95.1 min period, and 74° inclination. (GSFC SSR, 5/31/71; SBD, 6/2/71, 169)

May 30: NASA successfully launched *Mariner 9* (Mariner-I) from ETR at 6:23 pm EDT on Atlas-Centaur 23 booster into 398-million-km (247-million-mi) direct-ascent trajectory toward Mars. No problem had been found in Centaur guidance package when May 29 data was fully understood. Launch was second attempt to send spacecraft to orbit Mars

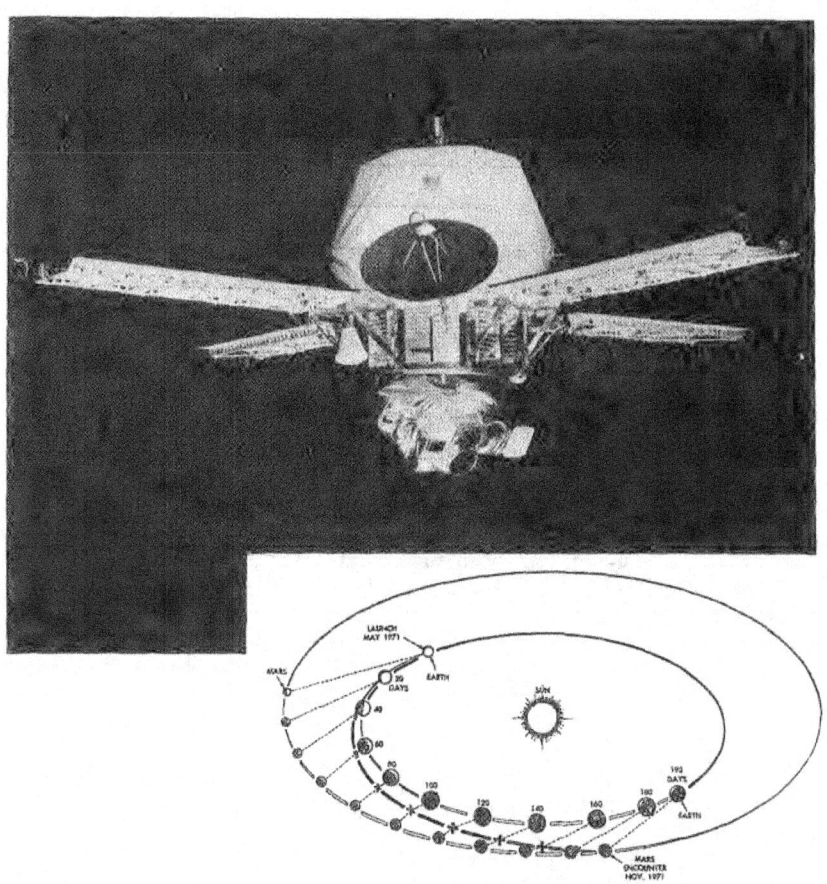

May 30: Mariner 9, *launched from the Eastern Test Range on a 398-million-kilometer (247-million-mile) direct-ascent trajectory, headed toward the planet Mars. Encountering Mars Nov. 13, the spacecraft would be the first to orbit a planet other than earth. Mariner 9—weighing 1000 kilograms (2200 pounds) and spanning almost 7 meters (22⅔ feet) with its solar panels extended—would orbit Mars 90 days to study the atmosphere and surface in detail and map 70 percent of the surface.*

during 1971 opportunity; Mariner 8 (Mariner-H) launch attempt May 8 had failed because of malfunction in Centaur stage of booster. Scheduled for arrival at Mars on Nov. 14, *Mariner 9* would make detailed study of Martian surface and atmosphere and map 70% of surface during planned 90-day orbit of planet.

Launch vehicle performance and spacecraft injection were nominal. Spacecraft separated from Centaur, deployed four solar panels at 6:40 pm, and locked sensors on sun at 7:16 pm, soon after leaving earth's shadow. About four hours into flight Canopus sensor was energized and spacecraft rolled 55° and locked onto star *Achernar*. Canopus acquisition was achieved at 10:26 pm and spacecraft would remain in sun-Canopus orientation for mission except during maneuvers. All spacecraft systems were operating normally.

Mariner 9 weighed 1000 kg (2200 lb) at launch and carried six scientific experiments: infrared radiometer to measure surface temperature; UV spectrometer to investigate composition and structure of atmosphere; infrared interferometer spectrometer to measure surface and atmospheric radiation; S-band occultation experiment to study pressure and structure of atmosphere; TV cameras to transmit low- and high-resolution photographs of surface; and experiment to investigate Martian gravity field. Orbit of spacecraft was designed to guarantee it would not hit Mars for at least 17 yrs, to avoid contamination of planet before studies on surface were made by landing spacecraft.

OSSA assigned project responsibility to JPL, including mission operations and tracking and data acquisition. Launch vehicle was responsibility of LeRC, with contractor General Dynamics Corp. Convair Division. (NASA Proj Off; NASA Release 71–75)

May 30–June 6: Apollo 14 Astronauts Alan B. Shepard, Jr., Edgar D. Mitchell, and Stuart A. Roosa during visit to France attended Paris Air Show; inspected aerospace facilities at Bretigny, Toulouse, and Salon-de-Provence, and attended reception by Mayor of Dijon. (NASA Release 71–84; MSC PIO)

May 31: Since Tu-144 had been exposed to Western inspection at Paris Air Show, Soviet press had been "slowly coming to grips with problems of supersonic flight long passed over in silence," *New York Times* reported. *Sovetskaya Rossiya* had quoted Western comments that Tu-144's engines were quieter and produced less pollution than those of Concorde. "Extensive discussion" of possible environmental and health problems caused by supersonic transport had not been reported in U.S.S.R. previously. (Shabad, *NYT*, 5/31/71)

- Communist China was preparing to test-fire its first liquid-fuel ICBM beyond its borders, probably into Indian Ocean, *New York Times* quoted U.S. China experts as saying. Preparations for ICBM launch included outfitting of tracking ship and extension of missile range tracking stations in Sinkiang Province, but political consideration might cause deferment of testing for several months. It was estimated it would take at least three years after initial testing for China to move its first group of operational ICBMs. (Beecher, *NYT*, 5/31/71)

- Igor I. Sikorsky, founder of Sikorsky Aircraft Div. of United Aircraft Corp., had been named 1971 recipient of USAF Academy's Thomas D. White National Defense Award, presented annually for contributions "to the national defense and security of the United States," *Aviation Week & Space Technology* reported. (*Av Wk*, 5/31/71, 23)

June 1971

June 1: U.S.–Canadian agreement on joint program to use satellites and aircraft in surveys of natural environment was announced by NASA. Objective was to advance remote sensing technology for monitoring air, water, land, forest, and crop conditions and for mapping ice movements, ocean currents, and geologic, hydrologic, vegetation, and soil phenomena. Canadian Dept. of Energy, Mines and Resources (EMR) would receive data at ground station in Prince Albert, Saskatchewan, and data-handling facility at Ottawa directly from NASA's ERTS satellites, scheduled for 1972 and 1973 launch. Canadian and U.S. test areas would be designated. Ground and aircraft data acquired and exchanged would be available to international community as soon as practicable. No funds would be exchanged between EMR and NASA. (NASA Release 71-95)

- British aviatrix Sheila Scott left London Airport in twin-engine Piper Aztec for 55 000-km (34 000-mi) flight around world one and one half times, during which her physical reactions would be relayed via *Nimbus 4* satellite to ground station at Fairbanks, Alaska, and then to GSFC computer. NASA and U.K.'s Institute of Aviation Medicine had installed equipment in aircraft to monitor pulse, breathing, voice, brainwaves, mental alertness, and other functions while Miss Scott spent 18 hrs in air under cramped conditions. Instrumentation also would monitor air pollution at normal flight level of 3000 m (10 000 ft). (Cerutti, *C Trib*, 6/2/71; NASA OSSA)

- Storage lifetime of seals, O-rings, and gaskets in rocket engines and Saturn IB and Saturn V booster stages had been extended from 8 to 10 yrs in program of rocket test firing at MSFC, MSFC announced. (MSFC Release 71-91)

- Comments on U.S. *Mariner 9* and U.S.S.R. *Mars 2* and *3* missions were made in *New York Times* editorial: Three spacecraft—"one American and two Soviet—now speeding toward Mars represent the best type of international competition. Their rivalry centers about the pursuit of knowledge, with the honors going to that nation whose instruments send back the most important data on the red planet that is one of the earth's nearest space neighbors." U.S. led Mars race to date, "because of the brilliant feats by Mariners 4, 6, and 7. In the 1960's the pictures and other data acquired by these pioneering vehicles completely revolutionized scientists' views of the Martian terrain and environment." (*NYT*, 6/1/71)

- NASA issued Management Instruction continuing NASA Historical Advisory Committee to advise and assist in implementation of NASA historical program and designating as members Dr. Louis Morton, Chairman of Dartmouth College Dept. of History, Committee Chairman; Dr. A. Hunter Dupree, Dept. of History, Brown Univ.; Dr. Melvin Kranzberg, Dept. of Humanities, Case Western Reserve Univ.; Dr. Rodman W. Paul, Dept. of History, Cal Tech; and Dr. John B. Rae, Dept. of History, Harvey Mudd College. Executive Secretary for Committee was James P. Nolan, Jr., of NASA Office of Management Development. (NASA NMI 1156.2C)

- President Nixon submitted to Senate nomination of Frederic G. Donner for reappointment as member of ComSatCorp Board of Directors. (*PD*, 6/7/71, 867)

June 2: DOD spokesman Jerry W. Friedheim told press in Washington, D.C., that cost increase for USN F-14 fighter aircraft was about $2 billion. If tentative plans to produce more than 700 aircraft materialized, cost would increase $2.8 million per aircraft, from $11.5 million to $14.3 million. Earlier, manufacturer Grumman Aerospace Corp. had blamed cost increase on inflation and other factors contributing to poor business conditions. (Witkin, *NYT*, 6/3/71, 15)

- Physicist Dr. Edward Teller criticized DOD secrecy in speech at PMR Hq. at Point Mugu, Calif. He said Soviet development of SS-9 ICBM and expanded ABM system posed threat to U.S. deterrent capability but DOD secrecy made it impossible to reveal details of threat. As result, many in Congress and Nation doubted threat's reality. (PMR *Missile*, 6/3/71, 1)

June 3: House, by vote of 302 to 64, passed H.R. 7109, $3.433 billion FY 1972 NASA authorization bill, after agreeing by voice vote to amendment deleting $500 000 for feasibility studies of offshore airports. House rejected amendment that would have eliminated $125 million for space shuttle system. (*CR*, 6/3/71, H4588-628)

- At Paris Air Show *Apollo 14* Astronauts Alan B. Shepard, Jr., Edgar D. Mitchell, and Stuart A. Roosa met in Soviet Pavilion with *Vostok 4* Cosmonaut Pavel R. Popovich and *Soyuz 9* Cosmonauts Vitaly I. Sevastyanov and Andrian G. Nikolayev. Astronauts saw models of Baykonur launch complex, *Vostok 1*, and Soyuz spacecraft; examined display devoted to 10th anniversary of first space flight by Yuri A. Gagarin on April 12, 1961; ascended to Soyuz CM; and admired *Lunokhod 1* lunar vehicle and *Luna 16* unmanned spacecraft that had returned lunar samples to earth. Later astronauts and cosmonauts went to U.S. Pavilion where cosmonauts were shown *Apollo 12* spacecraft and other exhibits. Cosmonauts presented astronauts with model of Soyuz spacecraft; astronauts reciprocated with small Soviet flag they had carried to moon aboard *Apollo 14* and NASA astronaut badges. (FBIS-Sov-71-108, 6/4/71, F1; *NYT*, 6/3/71, 41)

- Ground test version of Skylab workshop arried at MSFC aboard NASA barge *Orion* from MSC. Called "dynamic test article," workshop would be modified for extensive structural testing in MSFC dynamic test tower during two-month program to start about Nov. 1. (MSFC Release 71-93)

- Langley Visitor Information Center was dedicated in formal ceremonies at LaRC. Dr. James C. Fletcher attended, in first official visit of new NASA Administrator to any NASA center. Rep. Thomas N. Downing (D-Va.) said in dedication speech: "Space flight has opened up new vistas of flight and of the mind to understand the universe and our place in it. Our technology has become the world's best and has brought us prosperity in a highly competitive environment."

Dr. Fletcher said center would give public broad understanding of what was happening in aerospace technology. Science was "part of our everyday life." Technological society would continue "for the next 100 or perhaps 1,000 years."

Center displayed model aircraft, model spacecraft, lunar sample, and suit used by Astronaut Alan B. Shepard, Jr., while training for *Apollo 14* mission. (*Langley Researcher*, 5/28/71, 1; 6/11/71, 1: Text)

- NASA Performance Evaluation Board had given NR Space Div. incentive award fees of $234 600 and $307 350 for Apollo CSM and Saturn V 2nd stage (S–II), NASA announced. Awards covered management performance under contracts during 1970. (NASA Release 71–97)
- "USAF in Space" exhibit opened at Smithsonian Institution. Exhibit included two Vela nuclear detection satellites and DOD comsat, scale models of wind tunnels and space chambers, no-torque impact wrench to counter effect of weightlessness, astronaut equipment, prototype space tool for tests simulating weightlessness under water, and sextant used on Dec. 4–18, 1965, *Gemini 7* mission. Also exhibited were scale models of AFCRL solar telescope and Titan boosters and integrating maneuvering life support system—NASA–USAF spacesuit for EVA outside orbiting laboratories. (AFSC *Newsreview*, 9/71, 8–9)
- House agreed to H.R. 461 electing Rep. Mendel J. Davis (D-S.C.) to House Committee on Science and Astronautics. (*CR*, 6/3/71, H4628)

June 4: *Mariner 9*, en route to Mars on orbiting mission, successfully completed first midcourse maneuver with 5.11-sec burn that produced 6.7-m-per-sec (15.1-mph) velocity change. Maneuver placed *Mariner 9* on trajectory that would approach to within 1600 km (1000 mi) of Mars; second maneuver would be conducted Oct. 24 if necessary to correct trajectory further. Spacecraft had traveled 21 779 000 km (13 533 000 mi) since launch May 30 and was 2 076 000 km (1 290 000 mi) from earth, traveling at 10 985 km per hr (6826 mph). It was scheduled to reach Mars vicinity Nov. 13. (NASA Releases 71–96, 71–100)

- U.S.S.R. launched *Cosmos 426* from Plesetsk into orbit with 1996-km (1240.3-mi) apogee, 389-km (241.7-mi) perigee, 109.2-min period, and 74° inclination. (GSFC *SSR*, 6/30/71; *SBD*, 6/14/71, 231)
- USAF X–24A lifting-body vehicle, piloted by NASA test pilot John A. Manke, completed 28th flight from FRC in joint NASA–USAF test program. Objectives were to obtain lateral directional derivatives with rudder bias at 2° toe out, evaluate handling qualities with power and yaw accelerometer feedback system on, and determine supersonic lift-to-drag ratio data and longitudinal trim. Flight was intended to reach maximum mach number, but only two of four chambers ignited. Pilot immediately switched to two-chamber profile and mission plan. (NASA Proj Off)
- Draft treaty on lunar law was submitted to U.N. Secretary General U Thant in letter from Soviet Foreign Affairs Minister Andrey Gromyko with request it be included in agenda of 26th U.N. General Assembly meeting: "In all stages of the exploration of outer space, the Soviet Union has invariably advocated a progressive development of international space law in the interests of all people. The building of a firm international foundation for the activity of states in outer space promotes the cause of peace and helps to strengthen mutual understanding and cooperation between states." Soviet government felt "steps should be taken now with the object of further elaborating

and giving concrete expression to the standards of international law, regulating the activity of states on the moon."

Treaty specified that future exploration and use of moon be based upon consideration for well being of present and future generations; barred use of moon for "hostile actions"; reaffirmed ban on deployment of nuclear and other weapons of mass annihilation on moon and on use of moon for military purposes; ensured against unfavorable change or pollution of lunar environment; specified that surface and interior of moon could not become property of any state, states, persons, or organizations; and urged that signatory states ensure that all steps be taken to preserve man's life and health on moon. (Tass, FBIS–Sov–71–110, 6/8/71, A1)

- President Nixon sent message to Congress on energy sources. Of solar energy, President said: "The sun offers an almost unlimited supply of energy if we can learn to use it economically." NASA and NSF were "reexamining their efforts in this area and we expect to give greater attention to solar energy in the future." President proposed that all important Federal energy resource development programs be consolidated within new Dept. of Natural Resources. "We believe that the planning and funding of civilian nuclear energy activities should now be consolidated with other energy efforts in an agency charged with the mission of ensuring that the total energy resources of the Nation are effectively utilized. The Atomic Energy Commission would still remain intact, in order to execute the nuclear programs and any related energy research which may be appropriate as part of the overall energy program of the Department of Natural Resources." (PD, 6/7/71, 855–66)
- John C. Lindsay Memorial Award for 1971 was presented to Harry E. Peters of GSFC Laser Data Systems Branch during GSFC Scientific Colloquium. Award, given annually to GSFC employee who had made outstanding contribution to science or technology, was presented for individual contributions to the advancement of hydrogen maser frequency standards resulting in most stable frequency standards known and making possible dramatic improvement in space tracking and system precision. (Goddard News, 6/14/71, 1)

June 6–30: Soyuz 11—carrying Cosmonauts Georgy T. Dobrovolsky, Vladislav N. Volkov, and Viktor I. Patsayev—was launched by U.S.S.R. from Baykonur at 9:55 am local time (12:55 am EDT) to join *Salyut 1* "orbital scientific station" (launched April 19). After manned flight record of almost 24 days in space, mission ended in tragedy June 30 when the three cosmonauts were found dead in returned spacecraft.

Orbital parameters for *Soyuz 11* were apogee, 237 km (147.3 mi); perigee, 163 km (101.3 mi); period, 88.4 min; and inclination, 51.5°. Spacecraft docked with *Salyut 1* at 10:45 am Moscow time (3:45 am EDT) June 7. Tass announcement said: "In accordance with the Soviet programme of creating long-term orbital stations, the 'Salyut' orbital scientific station has started to function as the first piloted scientific station. . . . After successfully docking the Soyuz-11 transport spaceship with the scientific station . . . the crew of Soyuz-11 entered the scientific station. Solved for the first time was the engineering and technical task of delivering a crew to an orbiting scientific station by a transport ship." Tass said Salyut-Soyuz space system was 20 m

(66 ft) long, weighed more than 25 tons, and consisted of compartments containing equipment for scientific and technical experiments, for spacecraft control, and for crew work and rest. Cosmonauts would check out spacecraft subsystems; test orientation and navigation equipment; study geological-geographical objects on earth's surface, atmospheric formations, and snow and ice cover; study physical characteristics, processes, and phenomena in atmosphere and outer space; and study medico-biological influence of space flight factors on human organism.

Cosmonauts checked out systems, conducted experiments, and sent messages of thanks to scientists, engineers, and workers who took part

June 6–30: Soyuz 11 *carried Cosmonauts Georgy T. Dobrovolsky, Vladislav N. Volkov, and Viktor I. Patsayev into space for transfer to* Salyut 1, *an "orbital scientific station" launched April 19. After a record-breaking 24-day flight the crew, returning in the* Soyuz 11, *died before landing because an imperfectly sealed hatch permitted air to escape. The interior view of the forward section of the Salyut spacecraft's main working area shows the command and control panel and two seats.*

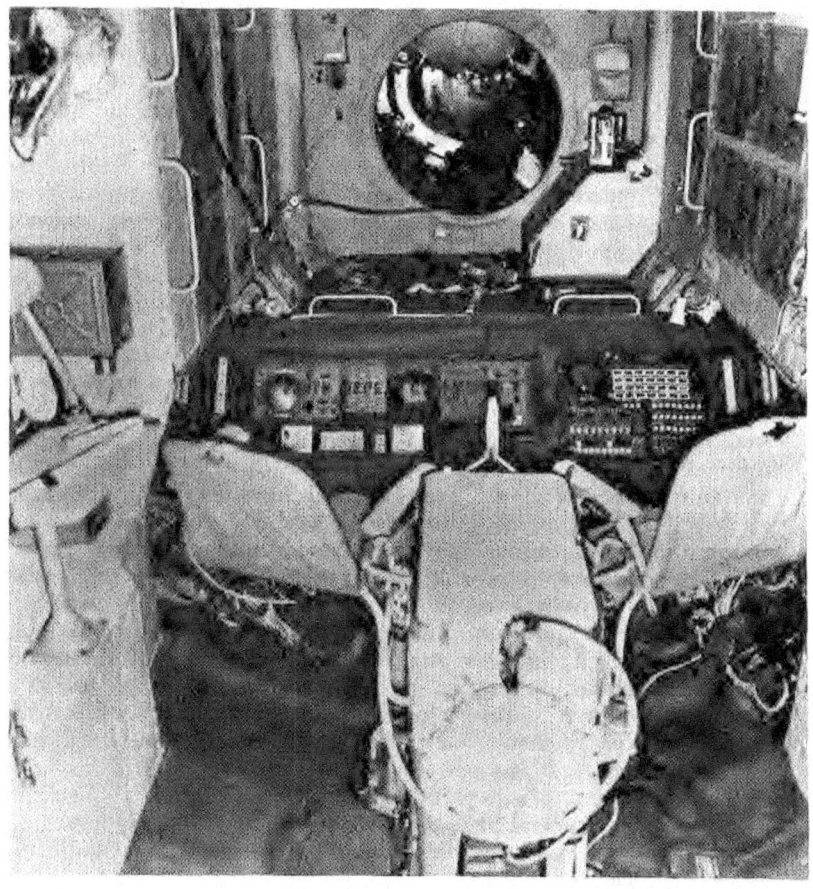

in creation of space station system. TV transmissions showed crew performing duties in spacecraft. After one week in orbit, crew was in good health and had completed first phase of experiments with all systems functioning normally. Crew was allowed one-day rest before continuing experiments. By 1:00 pm Moscow time (6:00 am EDT) June 13 Salyut-Soyuz station had completed 100 orbits of earth.

Cosmonauts continued to follow preplanned program of work, rest, and two hours of exercise per day and reported they were in good health. Experiments included communications with ground via orbiting Molniya I comsat, astronomical observations, medical experiments, observation of cyclone in Pacific near Japan, and experiments with plants in hothouse on board. They reared tadpoles into frogs and grew cabbages and onions. By 12:00 noon Moscow time (5:00 am EDT) June 22 station had completed 245 orbits of earth. Station completed 18th day in orbit June 24, surpassing 17-day 17-hr endurance record set by *Soyuz 9* June 2–19, 1970.

On June 29 cosmonauts completed flight program and prepared for reentry. Crew transferred equipment to *Soyuz 11* and undocked from *Salyut 1* at 9:28 pm Moscow time (2:28 pm EDT). Crew reported successful undocking and said all systems were functioning normally. At 1:35 am Moscow time June 30 (6:35 pm EDT June 29) spacecraft's braking engine was fired for prescheduled period. At end of engine firing communication with *Soyuz 11* crew ceased.

Tass said: "According to the program, after aerodynamic braking in the atmosphere the parachute system was put into action and before landing—the soft-landing engines were fired. The flight of the descending apparatus ended in a smooth landing in the pre-set area. Landing simultaneously with the ship, a helicopter-borne recovery group, upon opening the hatch, found the crew . . . in their seats, without any signs of life. The causes of the crew's death are being investigated." (GSFC *SSR*, 6/30/71; FBIS–Sov–71–109–126, 6/7–30/71; *Newsweek*, 7/12/71)

June 7: Preparations for Apollo 15 mission were proceeding on schedule with liftoff planned for July 26, Capt. Chester M. Lee (USN, Ret.), Apollo Mission Director, told press in Washington, D.C. Astronauts would be able to remain on lunar surface for long EVA periods because of new spacesuit that would give astronauts greater mobility and facilitate experiments. New suit had neck and waist convolutes to permit bending at neck and waist, improved zipper for pressure control, improved thermal garment, more abrasive material protection, and increased water and oxygen supply. Suits also had pocket for fruit bar so astronauts could take bites during mission. (Transcript)

- Dr. James C. Fletcher, NASA Administrator, said in Washington that *Salyut 1–Soyuz 11* mission was "clear demonstration of the Soviet Union's continuing interest in manned space flight and represents a further experiment to determine the value of manned earth orbital operations." He wished Soviet cosmonauts "every success in their mission." (Reuters, B *Sun*, 6/8/71)
- Washington *Evening Star* editorial commented on *Salyut 1–Soyuz 11* mission: "The Russians deserve congratulations for the achievement, which puts them some two years ahead of the United States in the space platform game." News from Moscow implied "that another

kind of breakthrough may be in the making; a development of far greater significance to the future of man in space than the momentary triumph of an orbital linkup or a moon landing. There is evidence that the traditional Soviet passion for secrecy in space is beginning to cool." (W *Star*, 6/7/71)

- Tu-144 Soviet supersonic transport was expected to go on international market at $25 million less than Anglo-French Concorde, *Armed Forces Journal* said. Informed sources had said U.S.S.R. would offer airlines up to 10 yrs credit to obtain sales in West. (*AFJ*, 6/7/71, 13)
- House passed H.R. 7960, $622-million FY 1972 NSF appropriations bill by vote of 319 to 8. FY 1971 appropriation was $511 million. (*CR*, 6/7/71, H4763–85)
- Retired LaRC scientist William J. O'Sullivan, who had conceived and developed *Echo I* comsat (launched Aug. 12, 1960), died at Newport News, Va., at age 55. He had received NASA Exceptional Scientific Achievement Award in October 1961 and National Rocket Club award for conceiving use of lightweight inflatable satellites for communications relay and air density studies. (*W Post*, 6/8/71, C8)

June 8: USAF launched *Sesp* satellite on Thor-Burner II booster from Vandenberg AFB into orbit with 579.4-km (360-mi) apogee, 544-km (338-mi) perigee, 95.8-min period, and 90.2° inclination. Satellite would test infrared celestial mapping sensor system. (GSFC SSR, 6/30/71; Pres Rpt 72; *SBD*, 6/11/71, 221)

- Senate Committee on Aeronautical and Space Sciences favorably reported H.R. 7109, $3.281-billion NASA FY 1972 authorization bill—$152 million less than $3.433 billion approved by House June 3 and almost $10 million above NASA request of $3.271 billion. Committee deleted $45 million increase voted by House for Skylab and additional $25 million House had voted for space shuttle.

  Committee version of bill would provide $2.543 billion for R&D, $55.3 million for construction of facilities, and $681.3 million for research and program management. Total of $3.281 billion was lowest total recommended by Senate Committee since FY 1962 and was $35.1 million less than total Committee recommended in FY 1971. (S Rpt 92–146)

- U.S.S.R.'s *Mars 2* and *Mars 3* spacecraft were functioning normally en route to Mars, Tass announced. At 6:00 am Moscow time June 8 (11:00 pm EDT June 7) *Mars 2*, launched May 19, had completed 14 communications sessions and was 4 897 000 km (3 042 900 mi) from earth. *Mars 3*, launched May 28, had completed 19 communications sessions and was 2 880 000 km (1 789 000 mi) from earth. (*Sov Rpt*, 6/25/71, 7)
- FAA released statistics on aircraft hijacking: Since 1961, 28 persons had been convicted for hijacking U.S. aircraft and had received sentences of up to 50 yrs or life; 12 cases were awaiting trial or outcome of mental examinations; and in 8 others charges had been dismissed when offenders were committed to mental institutions. FAA said 146 persons had participated in 113 successful and unsuccessful hijackings of U.S. aircraft. Of these, 92 persons were listed as fugitives, one of whom was believed to have committed suicide. Fifteen hijackers who fled to Cuba had returned and had been tried. (FAA Release 71–82)

- NR Rocketdyne Div. had been granted $26 228 158 contract modification for continued support on Saturn launch vehicle rocket engines, MSFC announced. Contract—effective July 1, 1971, through Dec. 31, 1972—allowed analysis of F-1, J-2, and H-1 engine performance, field engineering, logistics, and retention of Rocketdyne problem-solving group for remainder of Saturn program. (MSFC Release 71-97)
- *Washington Daily News* editorial commented on *Salyut 1–Soyuz 11* mission: "What is significant . . . is not the docking, in itself, but the evidence of the Soviet Union's zealous determination to exceed everybody in space—for whatever purpose. There is nothing in all this that the U.S. space agency cannot, when it chooses, accomplish—given the funds and priorities. It is not a question of firsts—the U.S. has had its share of those, and more. The difference is that to Russia the priorities and judgments are made in one place, the Kremlin. In the United States, not only the space agency but the President, Congress and the prevailing public mood share the decisions." (*W News*, 6/8/71)

June 9: Newspaper editorials commented on *Salyut 1–Soyuz 11* missions:
New York Times: "It must be recognized that Salyut and its successors raise important new problems for international law and international relations. Back in 1960, during the celebrated case of the American U-2 plane shot down over the Urals, Soviet leaders became very indignant about a manned American vehicle flying over their country. There have been no similar protests about the manned and unmanned satellites of the space era, but uneasiness must grow as political leaders in many countries contemplate the potential military uses of large semipermanent manned space stations such as Salyut." (*NYT*, 6/9/71)

Baltimore *Sun*: "One point the achievement demonstrates is the intensity of Russian concentration on a definite program. As repeatedly defined, and demonstrated in practice, it is a program which sees orbital endeavor as the logical next step, with deeper explorations limited for the time to automatic systems." (*B Sun*, 6/9/71)

- MSFC announced it was evaluating proposals to provide technical systems to develop and calibrate instruments for marine and atmospheric research from 6 firms out of 80 who had received RFPs. Contract winner would install equipment in MTF's Central Instrument Facility. (MSFC Release 71-99)
- Deputy Secretary of Defense David M. Packard, in testimony before Senate Banking and Currency Committee investigating financial status of Lockheed Aircraft Corp., warned against danger of setting precedent in passing legislation to guarantee up to $250 million in private loans to Lockheed: "It's very desirable not to establish a precedent that the government will (aid) any company that gets into trouble." (Samuelson, *NYT*, 6/10/71, 1)
- NASA selected eight firms to negotiate contracts for MSFC engineering support services. One-year contracts to be negotiated were: $8 million to Brown Engineering Co. for services at Astronautics Laboratory, $1.5 million to Hayes International Corp. for services at Product Engineering and Process Technology Laboratory, $4 million to Federal Electric Corp. for services at Quality and Reliability Assurance Laboratory, $3.5 million to Northrop Services Inc. for services at Aero-Astrodynamics Laboratory, $1 million to Planning Research Corp. for

June 10: *Director Lee R. Scherer of NASA's Apollo Lunar Exploration Office received Soviet lunar samples from Vice President Aleksander P. Vinogradov of the Soviet Academy of Sciences in exchange for Apollo 11 and 12 samples. Participating in the ceremony at the Academy in Moscow were (left to right) Dr. Paul W. Gast, Chief of the Manned Spacecraft Center's Planetary and Earth Sciences Division; Scherer; Prof. Vinogradov; interpreter Igor Pochitalin; and Soviet Academician Boris Petrov.*

services at Central Systems Engineering Office, and $10 million to Sperry Rand Corp. for services at Astrionics Laboratory. (NASA Release 71-102)

June 10: U.S. and U.S.S.R. exchanged lunar samples in Moscow ceremony to mark expansion of cooperation in space. Lee R. Scherer, Director of Apollo Lunar Exploration Office, presented three grams (0.10 oz) of *Apollo 11* and three grams of *Apollo 12* lunar material to Academician Aleksander P. Vinogradov, Vice President of Soviet Academy of Sciences, and received three grams of Soviet *Luna 16* samples in return. Scherer said, "Although the quantity of material is small, its significance is very great." U.S. Embassy in Moscow released statement saying exchange provided opportunity for scientists of both countries to benefit from study of material from lunar areas to which they would not otherwise have access. "The exchange of lunar samples exemplifies the desire of NASA, as stated in the agreement of January 21, to expand cooperation in space research and exploration. The U.S. Government looks forward to the broadening and deepening of this cooperation." (NASA Release 71-103; Shabad, *NYT*, 6/11/71, 10)

- Aerobee 350 sounding rocket was launched by NASA from Wallops Station carrying Johns Hopkins Univ. experiment to study airglow. Rocket and instruments functioned satisfactorily. (SR list)
- Appointment of L/G Charles H. Terhune, Jr., (USAF, Ret.), as Deputy Director of JPL was announced by Dr. William H. Pickering, JPL Director. Gen. Terhune would assume duties July 9, succeeding Adm. John E. Clark (USN), who had announced his retirement. (JPL Release 571)

- Management plans for OMSF space shuttle program were announced by NASA: OMSF Hq. in Washington, D.C., would be responsible for overall management, including assignment of responsibilities, basic performance requirements, control of major milestones, and funding allocations to NASA centers. Dale D. Myers, Associate Administrator for Manned Space Flight, had assigned responsibility for program control, overall systems engineering and systems integration, and overall responsibility for definition of elements of total system that interacted with other elements to MSC, which also would be responsible for shuttle's orbiter stage. MSFC would be responsible for booster stage and shuttle's main engines. KSC would be responsible for design of launch and recovery facilities. All three centers would station personnel at MSC for systems engineering and integration activity. (NASA Release 71–104)

June 11: Cosmos 427 was launched by U.S.S.R. from Plesetsk. Satellite entered orbit with 301-km (187-mi) apogee, 207-km (128.6-mi) perigee, 89.7-min period, and 72.9° inclination and reentered June 23. (GSFC *SSR*, 6/30/71; *SF*, 11/71, 456)

- Dr. James C. Fletcher, NASA Administrator, delivered commencement address at Cal Tech: "One of the paths we can take—the obvious one—is to help develop some new area of high technology that will serve to strengthen our Nation's economic base, and thus indirectly, our whole culture. We don't necessarily have to perform this task with productivity foremost in our minds . . . even though that's really the way it works.

  "A second path—and the demand for this path will be growing rapidly, I believe—is to attack directly the social, economic, and other serious problems of our time that have a definite technological component. Problems of ecology . . . have technological solutions. Urban and interurban transportation have technological solutions. As soon as we determine the appropriate political system which will spend the time and money, we already know a good deal about how to solve problems like these, and what remaining technology needs to be developed will undoubtedly appear. Other problems such as rural and ghetto health, housing, hunger, and poverty, have only a partly technological solution; still others, with strong human and sociological components, are even harder to deal with, and it may be a long time before problem-solvers learn to deal with 'people' problems. . . . while it wasn't easy to land men on the Moon—it was nevertheless a straightforward, engineering kind of problem. And as someone has said, there weren't any people between here and there."

  Third path—and he thought few persons "yet realize what a broad and important path it can be in years ahead"—was "to pursue long-range study and research on applications of problem-solving techniques to some of the more purely human problems." (Text)

- NASA Wallops Station participation in Cornell Univ. experiment to obtain data on importance of celestial, magnetic, topographic, and meteorological cues in guidance systems of nocturnally migrating birds was announced by NASA. Station would provide support in tracking birds with 18-m (60-ft) dish on space range antenna (SPANDAR). Birds would be captured in nets, examined, and small magnets would be attached. They would then be released by weather balloon to continue flight under varying cloud conditions. Each bird would be tracked by radar

as it made orientational decision and departed in migration. (NASA Release 71-98)
- Space officials feared budget pressures could prompt either President Nixon or Congress to cancel final two Apollo missions in 1972, *Wall Street Journal* said. In event of cancellation, NASA would propose to use Apollo equipment for "lengthy orbital flights due to start in 1973" in the Skylab program. (*WSJ*, 6/11/71, 1)
- NSF released *Federal Support to Universities and Colleges, Fiscal Year 1970*. Between 1969 and 1970 Federal funds had declined by $227 million, or nearly 7%, to level of $3.227 billion—lowest funding level since 1966 and first drop in actual dollars in Federal support since 1963. Decline might be attributed to shift in Government policy from direct Federal grants for facilities construction to subsidizing interest charges on loans from non-Government sources. (NSF *Highlights*, 6/11/71, 1)
- Lockheed Aircraft Corp. Chairman Daniel J. Haughton told Senate Committee on Banking and Currency he would resign if change in Lockheed management was condition for firms receiving $250-million Government loan guarantee. (Samuelson, *W Post*, 6/12/71, A1)

June 13: Appointment of Laurence T. Hogarth as Deputy Director, Office of Long-Range Plans, in NASA Central Planning Office became effective. (*NASA Activities*, 7/15/71, 119)

June 14: Status of space shuttle technology was described by MSFC propulsion engineer Gerald M. Thompson during AIAA/SAE Seventh Propulsion Joint Specialist Conference at Salt Lake City, Utah. Because "shuttle propulsion technology has been vigorously pursued in a totally organized approach for several years," no major technology work remained to be done before shuttle main engine development phase began. Designs with strong technical bases were ready to proceed. However, if hydrogen propulsion was selected for air-breathing engines that would return shuttle stages to landing fields on mission completion, additional technology development would add about one year to schedule. Main engine development was to begin in 1971, with flight certification targeted for March 1978. (MSFC Release 71-103)
- Plans for joint experiments in which ATS-F (scheduled for May 1973 launch) would be used from September 1973 to May 1974 to explore technical, economic, and educational practicality of regular TV transmissions to low-cost ground receivers in remote areas of U.S. were announced by NASA, HEW, and Corp. for Public Broadcasting. Experiment would begin with transmission of educational and health programs to ground receivers in Rocky Mountain region and Alaska. NASA was designing 15-w transmitters to operate in 2500- to 2690-mhz band and would provide access to ATS ground station. Other ground transmission and receiving facilities would be provided by HEW and CPB. (NASA Release 71-105)
- Aerobee 150 sounding rocket was launched by NASA from WSMR carrying Univ. of Hawaii solar astronomy experiment. Rocket and instruments functioned satisfactorily. (SR list)
- Japanese Space Development Committee's First Technical Div., which had been investigating performance of Mu-4 carrier rocket that orbited *Tansei* satellite Feb. 16, had recommended that plans for launching first "full-fledged" scientific satellite during summer 1971 be approved,

June 14

Space Business Daily reported. Recommendation followed assessment of malfunctions during launch and orbit which caused mission to fall short of its programmed objectives. Satellite, which had orbited earth 96 times to date, had malfunctions of its solar battery performance measuring device, control command device, and stabilization device. (*SBD*, 6/14/71, 228)

- Sen. Hubert H. Humphrey (D-Minn.)—who had been Vice President under President Johnson—addressed Space Seminar of Hugh O'Brian Youth Foundation at KSC: "One of the reasons that I have been active in the Space Program is because I believe that this program did something for all of the things I thought were important in life." Space program was "pioneer in beginning to make the discovery in environmental control." It had proved men could live in pure environment and "that there's a relationship between the living space that we have and the number of people that can be there . . . [and] that you can have clean water and clean air. And it's proven to us that you can work together."

  Space program had helped produce computer and had developed satellite monitoring system. Possibly "one of the greatest efforts for world peace has come right out of the science and technology of space research."

  Sen. Humphrey recalled his role as negotiator of Nuclear Test Ban Treaty: "I went to Moscow when it was signed. I have the pen that President Kennedy used to sign the treaty . . . he said, 'I give you this pen, Hubert, because it's your treaty.'"

  He had been in Moscow when *Apollo 11* Astronaut Neil A. Armstrong made first moon walk. "I was in the offices of Pravda and Izvestia . . . and a dispatch came through from Reuters . . . noting that the launch had been successful, and I had said to the editors of Pravda, 'Why don't you run this as a headline?' There was nothing in their paper that indicated that our launch had taken place. And when I went to Izvestia the same afternoon I said 'Look, the launch has taken place and I see nothing in your headlines in your papers.' I am happy to tell you that they were somewhat embarrassed and the next morning, at the National Hotel, under my door was a copy of Pravda and Izvestia with front page stories saying that there had been a successful launch." When Neil Armstrong "touched down that Sunday night . . . the only information that I was able to get was out of the Voice of America through the United States Embassy because the Russians had blacked out." (*CR*, 6/18/71, H5484–7)

- Representatives of Bankers Trust Co. and Bank of America told Senate Banking and Currency Committee investigating financial condition of Lockheed Aircraft Corp. that they would not make further loans to Lockheed without Government guarantee. (Samuelson, *W Post*, 6/14/71, A2)

- Role of radioelectronics in space was described in *Pravda* by Soviet Academician V. Andreyanov: "Radioelectronics might be called the nervous system of cosmonautics. It links our planet by living threads with the spacecraft and apparatuses when they fly to Mars, work on the moon, or study the earth and circumterrestrial space, like the Salyut station. In less than 15 years, research into space and the nearest planets has not only stimulated the development of many fields of science and technology but has also been accompanied by the emergence of new fields of knowledge and technical methods. A 'spacification' of the sciences had been noted. Concepts which were earlier unusual have also appeared in radioelectronics: space radio communications, space telemetry and television, space radionavigation, and radio methods for studying celestial bodies and interplanetary space." (FBIS–Sov–71–123, 6/23/71, L5)

June 15: USAF launched unidentified satellite at 11:41 PDT from Vandenberg AFB on new Titan IIID booster. Satellite entered orbit with 289.7-km (180-mi) apogee, 178.6-km (111-mi) perigee, 89.1-min period, and 96.3° inclination. Booster used five-segment, 3-m (10-ft) solid-fuel strap-on boosters and was reportedly capable of placing 6800 kg (15 000 lbs) in low earth orbit. Satellite reentered Aug. 6. (GSFC *SSR,* 6/30/71; 8/31/71; *SBD,* 6/18/71, 258; Pres Rpt 72)

- NASA announced completion of supercritical wing shakedown flights aboard modified USN F–8 jet fighter from FRC. Last flight in eight-flight series to allow test pilots to "get feel" of aircraft and investigate effects on wing at various speeds and altitudes, had taken place May 26. FRC officials reported finding no "surprises" during test series and said air foil appeared to confirm wind-tunnel tests. Research aircraft would be off flight status for two months while small irregularities on wing surface were corrected and network of about 250 pressure sensors was added to wing's top surface to locate and measure any shock wave created in air flow. Effectiveness of supercritical wing depended on moving shock wave to near back of wing for increase in total wing efficiency. (NASA Release 71–101)

- Rep. F. Edward Hebert, Chairman of House Armed Services Committee, announced that he would move to block $806-million procurement funds for USN F–14 jet fighter aircraft until USN solved overcost problems. Funds were included in $21.9-billion military weapons authorization bill before House. (AP, *W Post,* 6/16/71, A2)

 Grumman Aerospace Corp. released statement saying it would be "very appropriate" for Congress to stop funding USN's F–14 jet fighter aircraft until cost-overrun problems were settled. (*W Star,* 6/16/71, A4)

- Dr. Vasily V. Parin—Soviet physiologist, former President of Soviet Academy of Medical Sciences, and member of Soviet Academy of Sciences—died at age 68 of liver ailment. Dr. Parin had been known primarily for his work during 1960s on human physiological problems during space flight and had authored studies on effect of weightlessness on functioning of heart and circulatory system. He had received Order of Lenin, U.S.S.R's highest award, in 1963 for his work in developing Soviet medical science. (AP, *NYT,* 6/17/71, 42)

June 15–17: House Committee on Science and Astronautics' Subcommittee on NASA Oversight held hearings on failure of Atlas-Centaur to launch *Ats 4* into planned orbit Aug. 10, 1968, and failure to orbit OAB–B Nov. 30, 1970, and Mariner 8 May 8, 1971.

 Dr. George M. Low, NASA Deputy Administrator, testified June 15 on NASA policy on backup spacecraft and Centaur reliability: "In the manned program, all possible steps are taken to assure the safety of the astronauts. Single point failures are reduced to an absolute minimum. Missions are planned to build upon the previous missions so as to reduce the number of unknowns." Within missions, "plateaus are established so that all systems can be checked out in real time prior to committing to the next plateau. Vigorous testing procedures coupled with redundancy and in many cases dual redundancy are adopted before a system is 'man-rated.'"

In Skylab "we are building a back-up workshop which will be brought right up to the point of actual checkout. This will permit us to proceed with the program even in the event that the first workshop would not be successfully launched. Similarly, we will have a single back-up Command and Service Module-Saturn IB capability for the three manned visits to the workshop." Capability was built into program and was "key to the plan for a rescue capability in Skylab which we recently adopted."

In earth-orbital automated programs "we generally plan a series of related missions, using identical or similar hardware. If we have a mission failure, decisions will then be made, based on all available factors, whether we will (1) prepare and fly a specific back-up to the mission which failed, (2) modify succeeding flights . . . to achieve the objectives, or (3) defer achieving the mission objectives to future years."

In planetary programs, with launch opportunities at only discrete intervals, NASA usually had provided for two identical launches. Each launch in dual mission was planned to complement its twin, while at same time serving as backup if twin were lost. This year *Mariner 9* was backup, flying hybrid mission.

Atlas-Centaur launch vehicle had history of 3 failures out of 16 operational vehicles, without apparent relationship between failures, giving probability of success of 80% if based solely on past experience. Launch vehicles for automated missions—such as Scout, Delta, and Atlas-Centaur—were "complex systems without much redundancy, and cannot be expected to achieve reliabilities much greater than about 90% without very major and costly changes. They certainly cannot be considered 'man-rated' vehicles." Addition of redundant systems would be investigated again but was not complete solution. Redundancy added weight, reduced payload capability, and was expensive. For launch vehicles of size of Titan-Centaur, which would carry Viking and Grand Tour payloads, additional redundancy might be practicable and desirable and would be actively studied. Major increase in reliability of achieving orbit would be one of important attributes of space shuttle.

Dr. John E. Naugle, Associate Administrator for Space Science and Applications, on June 16 continued testimony on backup philosophy, expanding on application in OSSA: "We do not schedule backup missions for observatory spacecraft. We always buy, and intend to fly, at least two flight units of a new series of observatory-class spacecraft. When we have a failure or a malfunction in a spacecraft or a launch vehicle we will always do everything possible to thoroughly understand and correct all other future missions which could conceivably be subject to a similar malfunction. When we have a failure we will always review the effect of the loss on the program, determine the alternative recovery plans by which we can attain the objectives of the lost mission, assess the cost and any other impact of those alternatives on the program and then decide which alternative to choose."

Vincent L. Johnson, Deputy Administrator for Space Science and Applications, testified June 17: "In the basic considerations which establish the philosophical approach to reliability of unmanned space vehicles, the over-riding necessity for protection of the lives of human passengers is absent. The result is an approach to reliability in which cost and simplicity play a much larger part. We have not imposed a requirement for total elimination of single point failure modes, but have concentrated upon reducing the probability of failure by careful design, parts application, major quality and reliability efforts, and, most importantly, tender loving care. Our launch vehicles . . . are characterized as simplex systems, implying lack of the complexity characteristic of redundancy." (Testimony; Subcom Rpt, 10/71)

- Representatives of nine European Space Conference member countries and five Atlantic and Pacific area countries met at FAA Hq. to explore possibilities for unified preoperational program in aeronautical satellites over Pacific and Atlantic Ocean areas. At final meeting, group decided to establish International Collaboration Ad Hoc Group to make specific proposals for unified program to Second Aeronautical Satellite Meeting in Madrid Aug. 3. (FAA Releases 71-91, 71-93)

June 16: NASA interest in "phased approach" to development of reusable space shuttle system was announced by Dr. James C. Fletcher, NASA Administrator. Contractor studies near completion indicated "preferred configuration" was "two-stage delta-wing reusable system in which the orbiter has external propellant tanks that can be jettisoned." While studies to date had favored "concurrent approach" in which development and testing of orbiter and booster stages would proceed simultaneously, NASA also had studied feasibility of sequencing development, test, and verification of system's features. "We now believe that a 'phased approach' is feasible and may offer significant advantages." Additional studies in progress, plus those nearing completion, would enable NASA "to make a decision this fall on the technical and programmatic approach to be followed in the space shuttle program." (NASA Release 71-107)

- NASA and USAF signed agreement to conduct joint Transonic Aircraft Technology (TACT) Program to explore application of supercritical wing technology to highly maneuverable advanced aircraft. F-111 was selected as test bed aircraft. NASA's LaRC, USAF, and General Dynamics Corp. had completed more than 1600 hrs of wind-tunnel tests to develop supercritical airfoil for TACT program. Final configuration development and wind-tunnel tests for low-speed high-lift devices and transonic and supersonic conditions would be conducted at ARC. (AFSC Release 188.71; NASA Release 71-124)

- House considering H.R. 8687, FY 1972 military procurement authorization bill agreed to amendment cutting $806 million for procurement of F-14A jet fighter aircraft. (CR, 6/16/71, H5266-303)

- Thirtieth anniversary of opening of Washington National Airport. Number of passengers using facility annually had grown from fewer than 400 000 in 1941 to nearly 10 million in 1970. At commemorative ceremonies FAA award for extraordinary service was presented to Edward V. Rickenbacker, veteran pilot and retired Eastern Airlines Chairman. Airport received commemorative plaque from Arlington

County (Va.) Board. (FAA Release 71-89; Bernstein, W *News*, 6/16/71, 35; Eisen, *W Post*, 6/17/71, D3)

June 17: NASA announced selection of Descartes as lunar landing site for Apollo 16, scheduled for launch in March 1972. Descartes, in central lunar highlands, provided opportunity to sample volcanic-like highlands basin fill and uplands volcanic terrain. (NASA Release 71-106)

- Impact of NASA installations on local communities was described by NASA Deputy Assistant Administrator for Industry Affairs George J. Vecchietti in testimony before House Select Committee on Small Business's Subcommittee on Small Business Problems in Smaller Towns and Urban Areas. During hearings on impact of Government installations on small business, Vecchietti testified: "The rapid growth of facilities and activities at NASA installations over the past few years have unquestionably generated development of a wide range of commercial facilities. . . . Unfortunately . . . most of the impacts are of an intangible nature and difficult to quantify. Economic impacts, for the most part, have been favorable. But . . . those communities bordering NASA installations have been adversely affected by the wind-down in space activity." (Testimony)

- *Apollo 7* Astronaut R. Walter Cunningham announced his resignation from NASA, effective on or about Aug. 1, to become Vice President of Operations for Century Development Corp., developers of Greenway Plaza in Houston, Tex. (NASA *Release* 71-110)

- New civilian program to "absorb the nation's technical talent, respond to our unmet social needs, and enlist widespread enthusiasm and support" was proposed by Sen. Edward M. Kennedy (D-Mass.) in speech before National Society of Engineers meeting in Washington, D.C. "The development of new cities may be the important challenge we face, because it dramatizes the need for new research and engineering in the way the challenge of the Soviet Sputnik spurred the American space program in the late fifties and early sixties." (Text)

- Launching of eight Soviet Cosmos satellites, *Mars 2*, and *Mariner 9* had raised number of man-made objects in space to 2378, Reuters reported. Of total, little more than one fifth were satellites or payloads; remainder was space junk. Spokesman for USAF Air Defense Command had said USAF had recorded 2840 objects, 760 payloads, and 2080 bits of debris that had decayed since launch of *Sputnik 1* Oct. 4, 1957. (LA *Times*, 6/17/71)

- President Nixon issued Proclamation 4060 commemorating World Law Day. He said: "We can see many heartening evidences that law is becoming stronger and more just around the world under the pressures which reason and necessity exert. Within the nations, human rights and ecological wisdom continue to gain stature in the law. Among the nations, security and cooperation—on every front from space to the seabeds—are being enhanced through negotiations, treaties, and conventions. The United Nations is entering its second quarter of a century, and many other international organizations are working effectively through and for world law." (PD, 6/21/71, 942)

June 17–July 2: Fourteenth Plenary Meeting of Committee on Space Research (COSPAR) was held in Seattle, Wash. International meeting of scientists hosted by NAS included symposia and working groups on

1970 total solar eclipse, astronomical observations from space, use of stratospheric balloons in space research, and lunar lasers.

Grumman Aerospace Corp. scientist Dr. Henry C. Courten, professor of astronomy at Dowling College, said mysterious tracks on photographic plates he made during 1966 and 1970 solar eclipses appeared to be of small planet or asteroid belt inside Mercury's orbit.

Cal Tech scientist and lunar expert Dr. Gerald J. Wasserburg told COSPAR meeting four rocks gathered by *Apollo 14* astronauts had ages ranging from 3.85 billion to 3.95 billion yrs. Ages were surprising "since it was anticipated that Fra Mauro would yield rocks of greater antiquity." Discovery suggested that volcanism was active on moon far longer than had been thought and was still forming lunar crust more than 600 million yrs after moon came into being. It also suggested that Sea of Rains had been carved by collision 600 million yrs after moon's formation, or 100 million to 150 million yrs later than lunar experts had originally believed.

Soviet scientists Yuri G. Nefyuduv, L. I. Kakurin, and A. D. Yegorov said *Soyuz 9* Cosmonauts Andrian G. Nikolayev and Vitaly I. Sevastyanov "found it difficult to keep the upright posture and had to be assisted while walking" during first three hours after return from June 2–19, 1970, mission. "Next day . . . their walk was still inadequate and erect positions were maintained with great efforts." When cosmonauts returned to physical activity, few days after flight, effects of prolonged weightlessness had proved temporary.

Astronaut John W. Young represented U.S. astronaut corps at ceremony to honor 10th anniversary of first space flight by Soviet Cosmonaut Yuri A. Gagarin. He asked, "What better tribute to the memory of Yuri Gagarin can we offer than the hope that the growing unity of efforts in space—which this COSPAR meeting so well exemplifies—may one day extend to all the Earth? (NAS Release, 4/21/71; AP, *W Post*, 6/13/71, A7; O'Toole, *W Post*, 6/27/71, A1; UPI, *W Post*, 6/28/71, A7; *NASA Activities*, 7/15/71, 107)

NAS and NRC submitted to COSPAR *United States Space Science Program*, comprehensive summary of scientific research in space science in U.S. during 1970. Report included summaries of observations from spacecraft, sounding rockets, high-altitude balloons, and aircraft made or reported during 1970 and discussed flights planned for 1971. (Text)

June 18: Discoveries of identical, complex patterns of amino acids and pyrimidines in two meteorites might indicate basic phase of chemical evolution leading to origin of life, ARC scientist Dr. Cyril A. Ponnamperuma told meeting of New York Academy of Sciences. Team of ARC scientists had found that Murray meteorite that fell in Kentucky in 1950 had same 18 amino acids, plus two pyrimidines, as were discovered in December 1970 in Murchison meteorite (which fell in Victoria, Australia, in September 1969). "Starting with these 18 amino acids, it would be theoretically possible to build up a living organism." Discovery appeared to be first conclusive proof of extraterrestrial chemical evolution, strengthening case for chemical evolution theory and increasing likelihood of life elsewhere in universe. (NASA Release 71–709)

• David S. Gabriel, Deputy Manager of AEC–NASA Space Nuclear Systems Office, was appointed by Bruce T. Lundin, LeRC Director, to chair

Centaur Quality and Workmanship Review Board. Board was to "carefully examine and reevaluate the philosophy, practices and procedures employed in the Centaur quality assurance and workmanship program" and to report findings and recommendations by mid-August 1971. (*Review of Recent Launch Failures*, House Sci & Astro Com, NASA Oversight Subcom, 10/71; OSSA)

- U.S.S.R.'s *Lunokhod 1* lunar rover, on moon's Sea of Rains, completed experiments for eighth lunar day and was put in stationary position for lunar night. Experiments had included examinations of lunar rocks, stereoscopic panoramic photography, and measurements of chemical composition and physico-mechanical properties of crater 150 m (164 yds) in diameter. Vehicle had traveled more than 10 km (6 mi) since landing on moon Nov. 17, 1970, and was still functioning normally. (*Sov Rpt*, 7/8/71, 12)
- Second phase of joint NASA, Dept. of Agriculture, and corn belt states 1971 Corn Blight Watch Experiment [see May 17] had been completed and participating RB–57F aircraft was photographing seven-state area of about 11.7 million hectares (45 000 sq mi) in third phase of experiment, NASA announced. (NASA Release 71–112)
- *Pravda* described French bubble hydrogen-filled chamber "Mirabelle," scheduled to enter operation soon for particle research. Active volume of chamber was cylinder 4.5 m (14.8 ft) long and 16 m (52.5 ft) high with 1.6-m (5.3-ft) diameter. In opinion of Soviet Academy of Sciences member Anatoly Logunov, "the considerable volume of 'Mirabelle' will make it possible . . . to trace the mechanism of birth of new particles during the collision of the known ones. It is also planned to carry out a research into the symmetry of elementary particles. The chamber which has almost the largest volume in the world today in the aggregate with the most powerful in the world Serphukov Accelerator is a qualitatively new means for research into . . . elementary particles." (FBIS–Sov–71–120, 6/22/71, L3)

June 19: Soviet space official Anatoly Fedoseyev, head of Soviet delegation to Paris Air Show, had defected to West May 27, UPI quoted French sources as saying. Reportedly Fedoseyev had been refused asylum at Le Bourget Airport in Paris by French officials, who later relented and allowed him to proceed to NATO Hq. in Brussels aboard USAF helicopter. Fedoseyev was believed to be one of highest ranking Russians to defect to West—vice minister in charge of space research for Soyuz and Luna spacecraft programs. Sources had further reported that 80 members of Soviet delegation to Air Show had been recalled to Moscow May 27 and Soviet officials had refused to sit down with French Defense Minister Michel Debré at official luncheon in protest of French action. UPI said both NATO and U.S. officials later had denied knowledge of Fedoseyev and his mission. (UPI, *W Post*, 6/20/71, A19)

June 20: PAET Planetary Atmosphere Experiments Test was launched by NASA from Wallops Station at 3:31 pm EDT on four-stage Scout booster. Primary objective was to investigate means of determining structure and composition of unknown planetary atmosphere. Secondary objectives were to determine profile of water vapor in earth's atmosphere, with altitude, from absorption measurements of reflected sunlight in infrared; investigate effect of radio signal attenuation on relay communications when transmitting from aft hemisphere of a blunt entry

June 18: NASA *announced completion of the second phase of the 1971 Corn Blight Watch Experiment by* NASA, *the Department of Agriculture, and the corn belt states. President Nixon and Secretary of Agriculture Clifford M. Hardin (center left to right) were briefed in Indianapolis June 24 on the RB–57 aircraft's role in the experiment, by Air Force Maj. Francis X. McCabe and Dr. Arch B. Park, Chief of* NASA's *Earth Resources Survey Program. In the second phase, color infrared photos were taken of the test area for analysis of soil conditions before emergence of the corn crop. In the third phase, June to November, high-altitude* IR *photos would be made every two weeks to watch for signs of the southern leaf blight's occurrence in the crop.*

vehicle; determine if nearly circular spiral pitching motion could be achieved with blunt entry vehicle; obtain flight-performance data on low-density ablator; and measure vehicle and spacecraft performance parameters.

Spacecraft achieved planned trajectory within nominal limits, reached 6500-m-per-sec (21 327-fps) entry vehicle velocity at time of separation, and impacted within expected area 965 km (600 mi) downrange about 158 km (85 nm) northeast of Bermuda. PAET entry vehicle instrumentation was functioning during reentry. Real-time and delayed-time playback telemetry records covering reentry period and radar track to near impact were obtained.

PAET entry spacecraft was 64 cm (25.2 in) long and 91.4 cm (36 in) in diameter. Forebody was blunt cone with spherical segment nose and afterbody was hemispherical. PAET program was directed by OART. ARC was responsible for spacecraft design, fabrication, and testing. Wallops Station was responsible for launch, tracking, and data acquisition. (NASA Proj Off; WS Release 71–10)

- ARC would join California Statewide Air Pollution Research Center in smog research program to start in late June, NASA announced. Program would include 12 Cessna 401 aircraft flights over San Francisco Bay area yearly for three years, to trace photochemical production of pollutants and their dispersion in atmosphere. Aircraft would carry some 230 kg (500 lbs) of research equipment for tests at altitudes to 5200 m (17 000 ft). Data collected would be available to research collaborators and data users in Air Pollution Control Office of U.S. Environmental Protection Agency, California State Air Resources Board, Los Angeles Air Pollution Control District, Assn. of Bay Area Governments, and other NASA centers. (NASA Release 71-108)

June 21: Extrasensory perception tests aboard *Apollo 14*, launched Jan. 31, were described by *Apollo 14* Astronaut Edgar D. Mitchell during press conference at Foundation for Research on the Nature of Man at Durham, N.C. In unauthorized experiment, Mitchell had tried to send messages to four "receivers" on earth while en route to moon. He had carried cards bearing ESP test symbols and concentrated on random order of cards while receivers on earth wrote down order in which they believed he was "transmitting." While plan had called for "transmission" once daily for six days, Mitchell had been able to perform experiment for only four days. Two of his receivers had reported "receptions" for all six days, one for one day only, and another for two days. Mitchell said preponderance of wrong answers was significant because it was "so far below chance that the odds are approximately 3000 to one against it being a chance result." If it was not chance, Mitchell said, results indicated "extrasensory exchange." Mitchell declined to identify his receivers, but Olof Johnsson of Chicago had said he was one psychic who had participated. (AP, *NYT*, 6/22/71, 22; UPI, W *Star*, 6/22/71, A1)

- Senate passed S.J.R. 101, to "authorize and request the President of the United States to issue a proclamation designating July 20, 1971, as 'National Moon Walk Day.'" (*CR*, 6/21/71, S9476)

- NAE announced reelection of Clarence H. Lindner as full-time President for three-year term, and election of new members to NAE Council. New council members were Dr. Raymond L. Bisplinghoff, NSF; Francis H. Clauser, Cal Tech; James Hillier, RCA Corp.; Frederic A. L. Holloway, Standard Oil Co. of New Jersey; J. Ross Macdonald, Texas Instruments, Inc.; and Kenneth G. McKay, AT&T. Robert W. Cairns, Hercules, Inc., was reelected to Council. (NAE Release)

- *St. Louis Post-Dispatch* editorial described plight of McDonnell Douglas Corp. in "aerospace depression": Total employment was off almost a third from high of 43 500 in mid-1967 and layoffs were continuing though at declining rate. "The loss of these 14,100 jobs is equivalent to a loss of income for all the families in a city of 35,000 persons. . . ." In one 30-day period, 1600 McDonnell Douglas workers had been laid off. "That is equal to 20 per cent of the workforce from a city the size of Berkeley, where the corporation has its headquarters." Hope for McDonnell Douglas "rests in large measure with the company's plans to produce the F-15 fighter-bomber. Already Congress is looking at this program with a wary eye, and . . . there are indications that production may be slowed as part of an increasingly cautious congressional approach to military spending." (*St Louis P-D*, 6/21/71)

June 21–25: U.S.–U.S.S.R. negotiations on compatible docking systems for spacecraft resumed as three joint working groups, including some 20 Soviet scientists and engineers, met with NASA scientists and engineers at MSC. Purpose of meeting was to devise single set of technical requirements for docking systems. Soviet delegation was headed by Boris N. Petrov, Chairman of Intercosmos Council of Soviet Academy of Sciences. Dr. Robert R. Gilruth, MSC Director, headed NASA group participating in meetings provided for in U.S.–U.S.S.R. agreement reached in Moscow Oct. 28, 1970.

Delegates discussed possibility of joint test flights to test requirements being developed for docking systems. Experiments "might be conducted between spacecraft of the Apollo and Salyut station types and between spacecraft of the Soyuz and Skylab programs," NASA statement said.

Soviet visitors were escorted on tour of MSC by Astronauts Fred W. Haise, Jr., Thomas K. Mattingly II, and John W. Young. Several "flew" computerized simulators used to train Apollo astronauts to dock CM and LM for lunar landing. (NASA Release 71–111; Transcript)

- Jamaican and U.N. officials visited MSC to review data acquired by NASA, Jamaican Geological Survey, and U.S. Geological Survey during earth resources survey of Jamaica made in April at request of Jamaican government and U.N. Food and Agricultural Organization. NASA's instrumented C–130 aircraft had gathered data on Jamaica water supply. Jamaica had average 508-cm (200-in) annual rainfall but much mountain river water never reached reservoirs. It was believed that submarine springs carried portion of water offshore into Caribbean Sea. Allen H. Watkins, Manager of MSC Earth Observations Aircraft program, said excellent data gathered on flights had been supplemented by most nearly complete "ground truth" operation yet attempted by MSC aircraft program. During MSC visit, Jamaican group was trained by NASA to use equipment needed for later analysis of data in Jamaica. (MSC Release 71–42)

June 22: NASA received design patents No. 220 980 through 220 985 for early conceptual designs of space shuttle. Inventors were employees of Lockheed Aircraft Corp. and North American Rockwell Corp. Space Div. (Jones, *NYT*, 6/26/71; Patent Off PIO)

- MSC announced selection of GE Houston Operations to receive $576 730, cost-plus-fixed-fee contract for Skylab flight garments and crew provisions. (MSC Release 71–41)

- Aerobee 150 sounding rocket was launched by NASA from WSMR carrying MSC experiment to study atmospheric composition. Rocket and instruments functioned satisfactorily. (SR list)

- Soviet scientist Anatoly Fedoseyev, said to have defected to West on May 27 [see June 19], had been given permission to stay in England, London *Times* reported. (London *Times*, 6/22/71; British Embassy PIO)

- Sen. Jacob K. Javits (R-N.Y.) introduced for himself and cosponsors amendment to H.R. 7109, FY 1972 NASA authorization bill that would require "reasonable and economic" percentage of recycled scrap materials be used in NASA construction and procurement programs. (*CR*, 6/22/71, S9615–20)

June 23: Dr. James C. Fletcher, NASA Administrator, testified before Senate Committee on Appropriations' Subcommittee on HUD, Space, and Science during hearings on NASA FY 1972 appropriations bill. He urged that "strong program in space and aeronautics is essential to our long-term national well-being. Our capabilities as a Nation to deal with domestic problems of high priority—poverty, cities, the environment . . . and indeed, our pursuit of happiness and cultural advance—ultimately depend on whether we can maintain a strong economy with continually increasing productivity. Increasing productivity, in turn, depends primarily on technological advance, which, in turn, results primarily from the large scale focussed technological efforts like our major space programs."

In addition were both long-term and near-term direct benefits, many already realized. "When 50,000 lives were saved as a result of satellite advance warning at the time of hurricane Camille, the space program probably paid for itself. . . .

"I believe that the space program is a value in itself to the nation and to the whole world. With our everyday preoccupation, necessarily, with the many serious and often depressing problems we face . . . there is, I believe, a human hunger for positive elevating goals to work for at the same time as we do what needs to be done in all these other difficult areas. I believe the strong attraction that the landing of the first man on the moon had on people everywhere, and much of the underlying basis of support in the country for the space program, stems from the fact that the space program provides such goals. The space program gives us goals seemingly beyond human capability, goals which transcend earthly boundaries."

Space shuttle had tremendous impact for future because it not only reduced cost by being reusable itself, but it also gave options of reusing payloads. "You do not have to throw them away once you put them up there. . . . There are many more application satellites we can use with this kind of an arrangement, because we will be able to reuse them."

Skylab, "first and only NASA experimental space station," due to fly in 1973, would be "larger but 2 years later than the Soviet Salyut now in orbit." U.S.S.R. probably would be able to "do things in space, laboratories, 2 years earlier than we . . . to know what the earth is like, to know what the clouds are like, to understand more about the earth's resources. . . . Another aspect, which I am uncertain about, is what they plan to do in a military way. . . . they will have the potential of doing military things on this same vehicle. We do not plan to do this on the Skylab. We also must consider the scientific value of Salyut, which I think is important, but not as important to us as the applications and the military aspects of it."

Associate Administrator for Organization and Management Richard C. McCurdy testified that "models" of anticipated further reduction of 1500 civil service positions in fall "indicate that we have reached a point in which really good people of considerable experience will be going out the door. Certain parts of our operations will be closed. This means that we have come to a place where reductions in force are eating into our capability to do the job." (Transcript)

- House Committee on Appropriations favorably reported H.R. 9382, FY 1972 HUD-space-science appropriations bill which contained $3.272 billion NASA appropriation. Committee recommended reduction in Apollo program funding from $612.2 million requested to $610.2 million; increase in Skylab funding from $535.4 million to $550.4 million to provide rescue capability; additional $5 million over estimate for STOL development, to total $115 million; deferment of $20 million proposed for space shuttle facilities; and reduction of $6.6 million in $726.6 million requested for research and program management. (H Rpt 92-305)
- U.S. and U.S.S.R. reached compromise agreement at Geneva conference on draft treaty to provide compensation for damages caused by objects from outer space and submitted draft articles of proposed treaty to legal subcommittee of U.N. Committee on Outer Space. Key articles provided for arbitration with final decision not binding on signatory states, although there would be moral and political obligation to pay compensation high enough to restore injured person or damaged property to original condition, if possible. (UPI, W Star, 6/23/71, A12)
- NASA held pre-proposal conference for study of tracking and data relay satellite system (TDRSS) to provide design and cost information on tracking and data relay satellites and interfacing elements of overall system. TDRSS was envisioned in RFP as providing extended coverage and service to comsats in near earth orbit. NASA intended to award two separate and concurrent contracts for definition studies of system to include spacecraft, user telecommunications system, ground station, and network scheduling and operations control center. About 90 RFPs had been mailed in advance of conference. (NASA Release 71-113; NASA PAO)
- Rep. J. Edward Roush (D-Ind.) introduced H.R. 9379, "bill to establish an Office of Technology Transfer." He said: "As a nation we have enjoyed many valuable side-effects, spin-offs or 'fallouts' from [Federal] research. Many of the examples of the miracles of modern science, of technology transfer, the path from 'science to sales' come from the annals of NASA, largely because NASA has made a concerted effort to accomplish the transfer of their space technology to everyday usage and have publicized that effort especially well." (CR, 6/23/71, H5825-6)
- NASA announced receipt of more than 600 proposals from potential domestic and international users of data expected from ERTS and earth resources experiment package to be carried on Skylab. It was greatest number of proposals for experiments ever received by NASA in response to announcement of opportunity for analysis of space-derived data. ERTS-A would be launched in spring 1972 and ERTS-B in 1973. First Skylab carrying EREP would be launched in 1973. Proposals in response to February invitation had come from more than 550 domestic and 80 international sources. They were being evaluated by nine panels of more than 100 scientific experts in various disciplines. (NASA Release 71-114)
- U.K. government confirmed to Congress that financing of Rolls-Royce engines for Lockheed Aircraft Corp.'s L-1011 TriStar jet aircraft would "not become effective" unless Congress acted by Aug. 8 to authorize U.S. Government loan guarantee to Lockheed as requested

by Nixon Administration. Confirmation was in letter to Senate Banking and Currency Committee. (*NYT*, 6/24/71, 56)
- NASA selected General Electric Co. and RCA Service Co. for further competitive negotiations leading to award of contract for technical support services at MTF. (MSFC Release 71–109)

June 24: U.S.S.R. launched *Cosmos 428* from Baykonur into orbit with 248-km (154.1-mi) apogee, 204-km (126.8-mi) perigee, 89-min period, and 51.7° inclination. Satellite reentered July 6. (GSFC *SSR*, 6/30/71, 7/31/71; *SBD*, 6/25/71, 299)
- NASA launched three Nike-Cajun sounding rockets carrying GSFC experiments to obtain temperature, pressure, density, and wind data between 35 and 95 km (22 and 59 mi) by detonating grenades and recording their sound arrivals on ground. Rocket launched from Point Barrow, Alaska, reached 110.8-km (68.9-mi) altitude, with all 19 grenades exploding as planned. Rocket launched from Churchill Research Range carried 19 grenades to 124.4-km (77.3-mi) altitude; 17 soundings were recorded. Rocket launched from Wallops Station carried 31 grenades to 114.3-km (71-mi) altitude; all 31 detonated and were recorded on ground. (NASA Rpts SRL)
- Aerobee 350 sounding rocket was launched by NASA from WSMR carrying Lockheed Aircraft Corp. x-ray astronomy experiment. Rocket and instruments functioned satisfactorily. (SR list)
- USAF YF–12 aircraft, one of two in joint NASA–USAF aeronautical research program to advance technology of supersonic military and civil aircraft, crashed on routine flight from Edwards AFB, Calif. Both pilots parachuted to safety.

  Coincidentally, NASA announced addition of another USAF YF–12 to program. New aircraft was specially instrumented for inflight propulsion studies of engine internal flow dynamics, propulsion and airframe interreactions, air inlet dynamics, atmospheric turbulence and temperature variations, rapid airflow changes, and factors affecting inflight stopping and restarting of engines. (AFSC PIO; FRC Release 16–71; *Aero Daily*, 6/28/71, 316)
- Lowell Observatory astronomers had made first accurate mapping of weather of Mars as result of worldwide cooperation of observatories in NASA-funded ground-based planetary patrol program centered at Lowell, *Christian Science Monitor* said. Principal investigator Dr. William A. Baum had said object of program, begun in 1969, was to "maintain nearly continuous photographic surveillance of Mars, Jupiter, and Venus during favorable periods." Mars weather seemed analogous to weather in arid regions of earth. (*CSM*, 6/24/71)
- TRW Inc. would develop conceptual designs for manned research laboratory to support communications and navigation experiments under 10-mo, $288 000 contract, MSFC announced. Study assumed laboratory would be launched in 1980s and could be flown as research and applications module or as segment of orbiting space station. (MSFC Release 71–108)
- Information systems on board manned orbital stations could collect data on geological formations and potential mineral deposits, G. Katys, head of Soviet institute for problems of control, said in interview released by Tass. TV systems, panoramic cameras, radiometers, and spectrometers in infrared range would find application on orbital sta-

tions for earth resources research. There would be need to install heat-sensitive devices to locate volcanoes and sources of geothermal energy and TV cameras to clarify topographic maps. Identification of photos had already produced tangible results in hydrologic and oceanologic research. Photos of rocks, geological structure, and tectonic disturbances had been interpreted. Cosmonauts on orbital stations could conduct research in various sectors of earth's surface, choosing most interesting zones. (FBIS–Sov–71–126, 6/30/71, 16–17)

- MSFC announced selection of NR Rocketdyne Div. for 16-mo, $1.1-million contract to develop turbopump assemblies for hydrogen and oxygen space shuttle auxiliary propulsion system (APS) technology program. Six-phase contract included exploratory study; design for tubopump components; fabrication of turbopump system; testing, refurbishing, acceptance testing, and delivery to NASA; and engineering support. (MSFC Release 71–110)

- MSFC announced it had modified existing contract with Martin Marietta Corp. for work in Skylab Program to incorporate earth resources experiment package (EREP) into flight version of multiple docking adapter, device to permit Apollo spacecraft to dock with Skylab cluster. Total cost of modification was $6 070 500. EREP would contain photographic experiments to study earth from orbit. (MSFC Release 71–111)

- JPL announced selection of Motorola, Inc., Government Electronics Div. for negotiations on $2 179 000 contract with award fee to design, fabricate, and test flight-data subsystem for NASA Mariner Venus-Mercury spacecraft for 1973 mission. JPL also had selected Litton Industries Guidance and Control Systems Div. for negotiations on $10.4-million contract for development, fabrication, and testing of gas chromatograph/mass spectrometer (to be carried on Viking lander) to analyze Mars surface. (JPL Releases 581, 582)

- House passed by voice vote H.J.R. 742 making continuing appropriations for FY 1972, including provisions under H.R. 9382 for NASA, NASC, and NSF [see June 23]. (CR, 6/24/71, H5848–50)

- NSF released *Federal R&D Expenditures Related to Budget Functions, 1960–72*. Between 1960 and 1972 chief areas of Federal R&D funding had been defense, space research and technology, and health—totaling 86% of estimated R&D expenditures of $15.7 billion for 1972. In 1960 same three categories had totaled 95% of $7.3-billion R&D funding. Throughout 1960–1972 period, these three functions had led in R&D expenditures in same order. Ratio of R&D expenditures to total Federal budget outlays rose from 8.5% in 1960 to 12.4% in 1965. Ratio had decreased each year since 1965 and 7.4% estimated for 1972 was lowest in 1960–1972 period. Decline reflected more the absolute increase in total Federal outlays since 1965—from $111 billion to estimated $211 billion in 1972, excluding interest—than any substantial absolute decline in R&D expenditures. These expenditures rose from $13.8 billion in 1965 to $16.3 billion in 1968, but dropped back to $15- to $16-billion range after that.

Defense consistently accounted for bulk of R&D expenditures but its share fell from 87% of Federal R&D total in 1960 to 53% in 1966, when space expenditures peaked. Later defense share increased but leveled off from 1968 at 57% to 58% of total Federal R&D. Space research and technology rose from 5% share of Federal R&D expenditures in 1960 to 36% in 1966. "This function experienced a dramatic rise in funds during that period followed by sharp cutbacks." Space had been only function with smaller R&D expenditure in 1970, 1971, or 1972 than in 1966. Its expected 1972 share was 20%. (NSF *Highlights*, 6/24/71, 1)

- Dr. Addison M. Rothrock, retired NASA Director of Plans and Programs, died in Alexandria, Va., at age 68. During 45-yr career as propulsion research scientist he had joined NASA's predecessor, NACA, in 1926 and had become Assistant Director for Research in 1947. When NASA was established in 1958 he was appointed to similar post. Dr. Rothrock had retired from NASA in 1963 to serve as professor of applied science at George Washington Univ. (*W Post*, 6/27/71, C6)

June 25: Four-stage Javelin rocket, launched by NASA from Wallops Station at 4:44 am EDT, carried NOAA experiment to 662-km (411-mi) altitude. Objective of mission was to investigate region of ionosphere above maximum electron density, where it could be treated as electron gas. Data was to be compared with data from GSFC Langmuir probe in payload and with measurements taken simultaneously by NBS Ionosphere Sounding Station at Wallops Island, Va. Mission was unsuccessful. (WS Release 71–12; SR List)

- NASA announced assignments within MSC Flight Crew Operations Directorate: *Apollo 10* Astronaut Thomas P. Stafford had been named Deputy Director and *Apollo 14* Commander Alan B. Shepard, Jr., would resume duties as Chief of Astronaut Office. Stafford also would assist Donald K. Slayton, Director of Flight Crew Operations, in technical matters related to flight crew activities. Post of Deputy Director had previously been vacant. (NASA Release 71–116)

- ComSatCorp filed with FCC for first-step rate reduction of 25% for Atlantic area satellite services, effective July 1, and proposed second-step reduction for leased voice-grade channels affecting Atlantic and Pacific area in early 1972, provided satellite traffic met forecast increases. (ComSatCorp Release 71–36)

- Tass reported death of Soviet rocket designer Aleksey Isayev and identified him for first time as designer of engines for Vostok and Voskhod rockets and Soyuz manned spacecraft. Tass said Soviet unmanned "interplanetary stations" also had been created under Isayev's direction. *Izvestia* said in biographical sketch that Isayev was one of main creators of aircraft that had made world's first jet powered flight on May 5, 1942. (AP, W *Star*, 6/27/71, A11)

June 26: Anatoly Fedoseyev, who had been described as Soviet space expert, had sought asylum from U.S.S.R. in France on May 27 [see June 19] and was in England "reportedly talking to British and American interrogators," *Chicago Tribune* editorial said. "He could, if he desired, supply information of great value in the launching of our own space laboratory, Skylab, in 1973. More important, he could tell us whether Salyut is being used as a superspy in space to track our Polaris submarines." Some scientists believed Polaris might be tracked by heat produced in its nuclear reactors. "And they suspect that the cosmonauts aboard Salyut are busy testing this very concept with delicate electronic sensors." (*C Trib*, 6/26/71)

- Soviet *Salyut 1–Soyuz 11* cosmonauts were described by Dr. Vladimir Dupik in *Sovietskaya Rossiya* article as best-fed crew in history of space flights. Length of flight had necessitated diet of four hot meals daily including "first breakfast" of veal, bread, candy, and coffee; "second breakfast" of tongue, sausage, and curds; lunch of dried, salted fish, borscht, meat, bread, cake and black currant juice. Fourth meal was not described. Oven had been installed in spacecraft for first time in Soviet space program, spices had been added to meats to improve flavor, and 14 varieties of meat carried had "higher moisture content which makes them juicier and gives off a nice aroma." (UPI, W *Star*, 6/27/71, A11)

- NASA announced resignation of Ronald J. Philips, Director of Office of Technology Utilization, effective July 3. He would become Senior Vice President of technology application program sponsored jointly by International City Management Assn., National League of Cities, U.S. Conference of Mayors, and Ford Foundation. (NASA Release 71-117)

- *New York Times* reported comments on U.S.S.R. acceptance at Geneva conference of idea that claims commission of outside parties could fix amount of compensation for damages caused by objects launched into space. Observers saw acceptance as "major breakthrough in seven-year United Nations effort to complete space treaty covering payment for such damages." While U.S. and smaller powers had urged course, U.S.S.R. had heretofore refused. (Teltsch, *NYT*, 6/26/71, 5)

- AAAS Board agreed to accept invitation from AAAS Soviet counterpart Znaniye (Knowledge) for annual exchange of four scientist-lecturers. (*AAAS Bulletin*, 2/72; AAAS PIO)

- IATA Secretary General Knut Hammarskjold expressed IATA's opposition to proposed use of UHF band in satellite communications along air routes over Pacific Ocean. He said its use would make system twice as costly as one functioning on VHF band already employed in airway communications. Opposition was expressed in letter to U.S. and member governments of ESRO, which planned similar system for Atlantic Ocean. (*NYT*, 6/26/71)

June 28: NASA press conference at MSC summarized June 21–25 U.S.–U.S.S.R. discussions on joint space docking. Dr. Robert R. Gilruth, MSC Director and head of NASA delegation to discussions, said talks had been friendly throughout. "It was a period of hard work covering very difficult technical areas." Meetings had been successful "in reaching conclusions, in principle, on all these subjects that we discussed, and in some detail in several of the subjects." Need had been felt by both sides for "possibility of flight studies to test the technical requirements for docking and rendezvous, . . . one that would use the actual kind of compatible . . . equipment" that would make possible space rescue from spacecraft of either side that might be in trouble. U.S. and Soviet CMs and space stations with mixed crews could orbit earth within four years. Soviet crews and U.S. crews "would no doubt, if they docked, . . . want to open the hatches and visit with one another." Dr. Gilruth said flight test might take place in mid-70s. (Transcript)

- Sen. Walter F. Mondale (D-Minn.), on behalf of himself, Sen. Clifford P. Case (R–N.J.), Sen. Jacob K. Javits (R–N.Y.), and Sen. William Proxmire (D-Wis.), called up his amendment to H.R. 7109, FY 1972

June 28

NASA authorization bill, which would delete from bill $138 million for space shuttle development. (*CR*, 6/28/71, S10061-80; 10084-112)

- Senate passed H.R. 7960, $706.5 million FY 1972 NSF authorization bill. Bill included $2.5 million for Global Atmospheric Research Program (GARP); $4 million for Arecibo Observatory in Puerto Rico; $7.7 million for Kitt Peak National Observatory; $2.5 million for Cerro Tololo Inter-American Observatory in Chile; $7 million for National Radio Astronomy Observatory; and $19 million for National Center for Atmospheric Research. Report accompanying bill described program of Research Applied to National Needs (RANN) established by NSF during FY 1971 to coordinate research in selected environmental, social, and technological problems. Major research efforts consolidated to form RANN were: weather modification, earthquake engineering, interdisciplinary research relevant to problems of our society (IRPOS), and other efforts formerly funded under scientific project support. (*CR*, 6/28/71, S9974-81; S Rpt 92-232)
- MSC announced award of $125 000, one-year, fixed-price R&D contract to NR to study safety techniques for spacecraft in earth orbit. (MSC Release 71-44)
- Bruce N. Torell was elected President of Pratt & Whitney Div. of United Aircraft Corp. at regular meeting of board of directors. Torell would succeed Bernard A. Schmickrath, who was elected a vice president. Changes would be effective July 1. (P&W *Eagle*, 7/71, 2)
- Gen. Thomas D. White Space Trophy for 1970 was presented to B/G Robert A. Duffy (USAF), Vice Commander of SAMSO, in Washington, D.C., ceremony. Citation was for efforts bringing "dramatic improvements in the effectiveness and maneuverability of advanced re-entry systems, and in both the effectiveness and variety of penetration aids designed to compound the difficulties of an enemy in intercepting re-entry systems." (*W Star*, 6/29/71, A4)
- NASA announced it had issued *Management Study of NASA Acquisition Processes*, report of team established in September 1970 to review NASA project and program planning for R&D and activities and procedures for acquiring goods and services from definition of requirements through contract award. Report covered processes used for acquisitions leading to use of formal source evaluation boards for agency-level and field center-level source selections. Study group had concluded that basic framework of NASA acquisition process was sound, but it recommended some improvements in project planning, source solicitation, and source selection. (NASA Release 71-118; Text)

June 28-30: AAS 17th annual meeting in Seattle, Wash., was devoted to exploration of outer solar system.

JPL astronomer Dr. Donald G. Rea said Jupiter was biggest enigma and challenge to most space scientists. It "emits about three times as much energy as it receives from the Sun. It has a magnetic field 20 times that of Earth, intense radiation belts, and apparently rotates at three different speeds. An important characteristic for flight projects is its gravity which will allow us to send a spacecraft to study Jupiter and then continue on" to investigate outermost planets of solar system.

JPL scientist James E. Long recommended spacecraft be launched in 1981 for insertion into Jupiter orbit early in 1984 after 750-day trip from earth. "Depending on Jupiter's radiation hazard, the spacecraft will have an active life in orbit of six months to a year." It was "imperative that we get direct measurements well into Jupiter's atmosphere over long periods . . . to determine the dynamic processes involved."

Technology for Thermoelectric Outer Planet Spacecraft (TOPS), sustained by self-repairing computers and capable of 10-yr scientific missions to outer planets, was described by JPL engineers Carl C. Wertz, Paul O. Chelson, and Richard A. Easton, JPL computers being developed and tested were STAR (self-testing and repairing) and CATS (computer-aided telemetry system). Both had built-in redundancy, with backup parts for each unit that might fail. Easton said STAR computer was spacecraft's brain and CATS computer its nervous system. Self-repairing computers were needed for outer planetary mission because "potentially hazardous environments, such as the asteroid belt between Mars and Jupiter, or radiation from Jupiter itself, require that Grand Tour spacecraft subsystems be self-adaptive to either transient or permanent failures. Another consideration is the eight-hour roundtrip communications delay from Neptune or Pluto."

Leonard Jaffe, NASA Deputy Associate Administrator for Applications, received Lloyd V. Berkner Award, presented annually by AAS to persons making significant contributions to commercial space utilization.

AAS Flight Achievement Award was presented to Astronaut John L. Swigert, Jr., for *Apollo 13* crew.

LaRC Director Edgar M. Cortright received AAS 1970 Space Flight Award "in recognition of his contribution to space flight through early planning for the organization of NASA, his direction of the Lunar and Planetary Program, his influence while Deputy Associate Administrator of the Office of Space Science and Applications, and later the Office of Manned Space Flight and his current position as Director of NASA Langley Research Center."

In speech on technology crisis Cortright said: "It is ironic that after 10 years of 'delivering the goods' in an unparalleled manner, culminating in manned exploration of the Moon, the space program is fighting for its life with a host of other priority programs. Already the national team of Government, industry, and university scientists, engineers, and technicians has been reduced to about $\frac{1}{3}$ of its peak of over 400 000. If we are not to see the progress of a decade of dedicated national effort squandered away, we had better make our case for space loud and clear. I think we have a compelling case—even in these days of relevancy." (JPL Releases 577, 578, 579; *NASA Activities*, 7/15/71, 117; 8/15/71, 129–34; *Langley Researcher*, 7/19/71, 1; Boeing Release S–0826)

June 29: Mariner 9, en route to Mars, was 7 900 000 km (4 909 000 mi) from earth at 9:00 am EDT after traveling 82 500 000 km (51 263 000 mi) and receiving over 500 commands since launch May 30. Spacecraft was expected to accomplish its basic 90-day scientific mission objectives, but was using more nitrogen gas than predicted for maintaining attitude stabilization. Current predictions indicated nitrogen gas supply would be depleted in early August 1972, nine months after spacecraft went into orbit around Mars. Increased rate of consumption was due to circuit design error in attitude-control-system electronics. (NASA Proj Off; NASA Release 71–120)

June 29

- Tenth anniversary of launch of SNAP-3A, first radioisotope thermoelectric generator—launched from KSC aboard USN *Transit 4-A* satellite by Thor-Able-Star booster June 29, 1961, in joint NASA–DOD–AEC program to develop use of nuclear power in space. SNAP-3A was still transmitting signals earthward but was no longer in use. It had orbited earth more than 50 000 times and traveled more than 2.4 billion km (1.5 billion mi). (AEC Release O-103; *A&A*, 1961)
- Senate, by vote of 82 to 5, passed H.R. 7109, $3.281-billion FY 1972 NASA authorization bill, after rejecting by vote of 64 to 22 Mondale amendment No. 233 that would have deleted from bill $138 million for space shuttle program.
    Senate also passed by voice vote H.J.R. 742, making continuing appropriations for FY 1972, after rejecting by vote of 63 to 24 modified Proxmire amendment limiting ceiling for defense expenditures to $68 billion. (*CR*, 6/29/71, S10126-62, S10177, S10272-98)
- Plans for 1974 World Weather Program experiment GATE (GARP Atlantic Tropical Experiment) were announced by NOAA. Part of Global Atmospheric Research Program (GARP), project would study behavior of cloud clusters and their role in larger circulation of atmosphere in Atlantic Ocean area from west coast of Latin America eastward to Ethiopia, with more intensive study of cloud clusters in smaller ocean area. GATE project was one of succession of experiments to increase understanding of atmosphere in order to develop numerical models and program computers for long-range weather prediction. (NOAA Release 71-82)
- MSFC announced award of $97 996 contract to Brown Engineering Co. for development of systems simulator to assess life support and habitability requirements for future space applications. Simulator was scheduled for Oct. 1 delivery. (MSFC Release 71-112)
- President Nixon transmitted to Congress third annual report of National Science Board, *Environmental Science: Challenge for the Seventies*. Report found environmental science "unable to match the needs of society for definitive information, predictive capability, and the analysis of environmental systems as systems. Because existing data and current theoretical models are inadequate, environmental science remains unable in virtually all areas of application to offer more than qualitative interpretations or suggestions of environmental change that may occur in response to specific actions." Situation constituted "crisis for the Nation." Report recommended establishing national program for advancing science of environmental systems. (*PD*, 7/5/71, 1019; Text)

June 30: Soyuz 11 Cosmonauts Georgy T. Dobrovolsky, Vladislav N. Volkov, and Viktor I. Patsayev were posthumously awarded titles of Hero of the Soviet Union for "heroism and courage shown during the test of the new space complex, the orbital station Salyut and the transport ship Soyuz-11." Cosmonauts had died during reentry from June 6–30 mission. (FBIS–Sov–71–127, 7/1/71, L1)
- Soviet government ordered formation of special commission to investigate deaths of *Soyuz 11* cosmonauts during reentry from 24-day mission. (Mills, B *Sun*, 7/1/71, A1)
- President Nixon sent message to Nikolay V. Podgorny, Chairman of Presidium of Supreme Soviet, on death of *Soyuz 11* cosmonauts: "The American people join me in expressing to you and the Soviet people

our deepest sympathy on the tragic death of the three Soviet cosmonauts. The whole world followed the exploits of these courageous explorers of the unknown and shares the anguish of their tragedy. But the achievement of cosmonauts Dobrovolsky, Volkov and Patsayev remains. It will, I am sure, prove to have contributed greatly to the future of space and thus to the widening of man's horizons." (*PD*, 7/5/71, 1010)

- Donald K. Slayton, MSC Director of Flight Crew Operations, sent cable on behalf of U.S. astronauts to Soviet Academy of Sciences: "We share with the world community deep sorrow for the loss of Cosmonauts Georgy Dobrovolsky, Vladislav Volkov and Viktor Patsayev, and we mourn with all the Russian people the deaths of three brave countrymen." (Reuters, B *Sun*, 7/1/71, A2; MSC PIO)

- U.S. manned space flight program would continue on schedule despite *Soyuz 11* tragedy, Dr. Robert R. Gilruth, MSC Director, told press in Houston.

    Dr. Charles A. Berry, MSC Director of Medical Research and Operations, said cosmonauts could have been killed by a toxic gas that accidently escaped, either by a chemical reaction or by a substance being heated inadvertently. He said odds were 1000 to 1 against one cosmonaut dying from effects of 24 days of weightlessness and it was almost impossible for all three to die simultaneously from weightlessness.

    Dr. George M. Low, NASA Deputy Administrator, told press cosmonauts' deaths might have been caused by failure of spacecraft's oxygen-supply system or by physical rupture of spacecraft. Deaths, he said, were "a terrible tragedy. They were pioneers in their achievements in space—in establishing the first manned space station. Our hearts go out to their families and to their colleagues. The cause of their death is not yet known to us. But I would speculate that the fault was with the spacecraft and not with the men. In all of our experience man has readily adapted to new conditions, while machines have sometimes failed." Dr. Low said he did not expect changes in NASA's Skylab program because of cosmonauts' deaths: "All of our experience in 24 manned flights suggests we can and should move ahead." (UPI, W *Star*, 7/1/71, A3; *NASA Activities*, 7/15/71, 107)

- Aerobee 150 sounding rocket was launched by NASA from WSMR carrying GSFC solar astronomy experiment. Rocket and instruments functioned satisfactorily. (SR list)

- Discovery of two discrete sources of very-high-energy gamma rays in southern Milky Way by team of scientists from Case Western Reserve Univ. and Univ. of Melbourne, Australia, was announced by NSF. Rays had been detected by instruments aboard three balloon flights made at altitude of 40 km (25 mi) in stratosphere on Feb. 6 and 27 and Nov. 27, 1969. All three balloons had been launched from Australia. Newly discovered sources also emitted x-rays. Scientists hoped that fact sources emitted both gamma and x-rays might provide additional clue as to how their energy was generated. They might be related to other unusual objects discovered during last decade such as quasars and pulsars. (NSF Release 71–171)

- Breaking-in process of world's most powerful atom-smasher—500-bev accelerator at Batavia, Ill.—began with first firing of protons around 6.4-km (4-mi) ring. Accelerator would build up to original design energy of 200 bev and was expected ultimately to produce seven times the energy of Soviet accelerator at Serpukhov, near Moscow. (Sullivan, *NYT*, 7/2/71, 24)

- International law had been "enriched" by draft of international convention on responsibility for damaging space objects, Tass reported from Geneva. Document had been completed by juridical subcommittee of U.N. Committee on the Peaceful Uses of Outer Space. Tass said convention, which would be submitted to September meeting of U.N. space committee, provided "reliable and acceptable . . . juridical foundation for settling problems of material responsibility which . . . may arise due to the rapidly expanding research programmes and the use of space." (FBIS–Sov–71–126, 6/30/71, L8)

- Soviet progress in science and technology was discussed in *Sovetskaya Rossiya* article by M. D. Millionschikov, Vice President of Soviet Academy of Sciences: "In recent years there has been an immense improvement in the development of astronomy and astrophysics. Our science has made discoveries of fundamental significance and has shifted considerably the limits of phenomena in the universe known to man. Research into the nature of powerful sources of radiowaves discovered some years ago has led to the conclusion that the objects noted are located almost on the edge of the observable area of the universe at a distance of 5 billion light years. The power of radiation exceeds all currently known scales. In order to explain the colossal energy they radiate from the standpoint of the most powerful known mechanism for emitting energy—thermonuclear synthesis—it is necessary to suppose that its source must be simultaneous thermonuclear explosion of a mass equivalent to a hundred million stars the size of the sun." (FBIS–Sov–71–134, 6/30/71, L1)

- Miss Baker—13-yr-old South American squirrel monkey launched and returned from 483-km (300-mi) altitude on USA Jupiter missile May 28, 1959—arrived at Alabama Space and Rocket Center, where she would reside permanently. Monkey was transferred from U.S. Naval Aerospace Medical Institute. (*Huntsville Times*, 6/29/71)

- Award by DOT of $3.5-million contract to Grumman Aerospace Corp. for construction of tracked air cushion research vehicle was announced by Secretary of Transportation John A. Volpe. Vehicle, 15.5 m (51 ft) long and weighing nearly 27 000 kg (60 000 lbs), would accommodate four persons including operator and test engineer. DOT would test vehicle at its High Speed Test Center in Pueblo, Colo., on its completion, scheduled for March 1972. (DOT Release FRA 971)

- FAA announced award of $390 000 contract to Saphier, Lerner, Schindler—Environetics (Div. of Litton Industries) for proposed two-phased study of offshore jetport [see May 21] to serve metropolitan New York City. (FAA Release 71–99)

- NASA leaders were "realistic enough to sense that their case for the [space] shuttle must be built on more than chauvinistic appeal to national prestige, which sold the Apollo in the early 1970s," Jonathan Spivak said in *Wall Street Journal*. They were "working hard to trim the shuttle's prospective costs, demonstrate its economic payoff and convince the scientific community that it offers attractive new opportunities for research." But NASA's "primary problem" was "convincing the public and the politicians that it still makes sense to go into space at all. The moon is no longer a glamorous goal; Mars is too far distant and expensive, and orbiting space stations to date have made few influential converts." (*WSJ*, 6/30/71, 10)

During June: Skylab was described in article by Dr. George M. Low, NASA Deputy Administrator, in *Astronautics & Aeronautics*: "By far the largest manned spacecraft we have ever built, Skylab will be the forerunner of permanent stations in space. In orbit 435 km [270 mi] above Earth, the Skylab cluster will be 36 meters [118 ft] long and will weigh 82 238 kilograms [181 304 lbs]. One of its major components will be a workshop with 362 cubic meters [474 cu yds] of work area. Other parts will include a Multiple Docking Adapter, Airlock Module, Apollo Command and Service Module, and an Apollo Telescope Mount, the last of these a full-fledged solar observatory capable of observing, monitoring, and recording phenomena on the Sun's surface that cannot be seen from Earth." Skylab would carry from earth orbit "about 50 scientific, medical, applications, and solar-astronomy experiments. Twenty of these will be in the life sciences, to determine how man performs and adjusts under conditions of prolonged space flight. The study of materials and manufacturing will be another area of interest. By means of an electron-beam generating device and an electric furnace, metals will be melted, molded, and welded, and other metallurgy and crystal tests performed in a work chamber exposed to zero-gravity and the vacuum of space." (*A&A*, 6/71, 20–1)

- NASA released two publications for sale by Superintendent of Documents, GPO. *Apollo 14 Preliminary Science Report* (NASA SP–272), prepared by MSC, described *Apollo 14*: mission and summarized scientific results.

  On the Moon With Apollo 15, guidebook to Hadley Rille and Apennine Mountains, had been prepared by Dr. Gene M. Simmons, MSC Chief Scientist. Publication would aid in following progress of Apollo 15 mission by giving timeline for mission, describing and illustrating lunar surface scientific experiments, and defining terms. (Texts)

- NASA released *Goals and Means in the Conquest of Space* by R. G. Perel'man (NASA Technical Translation F–595 dated May 1970). Translation of *Tseli i Puti Pokoreniva Kosmosa*, "Nauka" Press, Moscow, 1967, concluded: "The time will come when space liners will be sent across interstellar routes, but the achievements of Soviet science and technology will never be forgotten, since they first made it possible for man to set foot in the universe, and since the labor of every Soviet citizen is involved." (Text)

- Stanford Univ. and MIT experiments to determine internal structure of protons and neutrons by use of 21-bev beam at Stanford's Linear Accelerator Center were described in *Scientific American* by Dr. Henry W. Kendall of MIT and Dr. Wolfgang K. H. Panofsky of Stanford Univ. Scientific team had smashed what they thought were basic atom particles, to find that "the way ultrahigh-energy electrons are scattered by protons and neutrons suggests that these 'elementary' nuclear particles have a complex internal structure consisting of pointlike entities." New entities had been named "partons" because they seemed to be parts within parts. Scientists had concluded, "It is still too early to say whether the parton model will lead to an understanding of the nucleon's structure or whether entirely new ideas may be required." (*Scientific American*, 6/71, 61–77)

July 1971

July 1: Bodies of dead *Soyuz 11* cosmonauts, killed during return from June 6–30 mission, lay in state in flower-decked biers in Central House of Soviet Army in Moscow. Thousands of persons waited in miles-long line to pay last respects. Top Communist Party and U.S.S.R. government leaders appeared shortly after hall was opened. Bodies would be cremated in preparation for July 2 funeral. (Gwertzman, *NYT*, 7/2/71, 14; B *Sun*, 7/2/71)

• Newspaper editorials commented on death of *Soyuz 11* cosmonauts:

Chicago Tribune: "Any one of a million things could have gone wrong, because there are a million things that have had to go right for each Soviet and American success in space. It is no discredit to Russian scientists that something went wrong at the last moment. It is testimony to the good fortune and skill and care of the United States space program that something of this kind hasn't happened yet to us." (*C Trib*, 7/1/71)

Christian Science Monitor: ". . . those men who fly such missions are taking the risk for the aspirations of mankind to break the limitations of earth's gravitational and atmospheric fields. They depend on the accumulating technological knowledge of mankind to protect them. As a result, the Soyuz 11 event argues the more determinedly that nations—for the moment chiefly the United States and the Soviet Union—should share their space research efforts." (*CSM*, 7/1/71)

Washington Post: "The special sorrow that must be felt at this event does not proceed from their numbers—three deaths next to, say, thousands among the world's more recently savaged refugees. Nor does it proceed alone from their conspicuous bravery, their willingness to take the ultimate risk in pursuit of a national mission: daily, men in combat make the same commitment and do so minus the glory and unambiguous sense of purpose that cosmonauts and astronauts share. So it is more than the death of three men on a high risk venture that must move the earthbound to a special sorrow: it is the death of these men in pursuit of the ultimate adventure, exploration of the outer limits of human knowledge and endurance—and beyond." (*W Post*, 7/1/71)

New York Times: ". . . it should be apparent that the staggering dangers of space can be met most effectively by mobilizing all the knowledge and talent available irrespective of nationality. It is time to abandon the wasteful irrationality of prestige-motivated competition. The real lesson of this tragedy is . . . that space exploration ought to be a cooperative endeavor of all nations." (*NYT*, 7/1/71)

Baltimore *Sun*: "The three cosmonauts lived in space for almost 23 days, and by all indications the station's operation came up to its designers' hopes. Only in the final brief stage, a stage often successful before, did disaster intervene. It is tragedy, but it is not defeat." (B *Sun*, 7/1/71)

- MSC announced award of four, four-month, fixed-price extensions to contracts for space shuttle preliminary design. McDonnell Douglas Corp., North American Rockwell Corp., and General Dynamics Corp. would each receive $2 800 000, and Lockheed Aircraft Corp. would receive $1 400 000. (MSC Release 71-46)
- NASA issued invitation to scientists to participate in definition phase for Planetary Explorer missions to Venus. Emphasis would be on low-cost approach using spin-stabilized Explorer spacecraft launched by Delta booster. From respondents to invitation NASA would select scientists to play active role in defining typical payload for initial missions in 1976 and 1977 and to make recommendations for subsequent orbiter and probe missions in 1978 and 1980. (Text; NASA Release 71-124)
- *Apollo 11* Astronaut Edwin E. Aldrin, Jr., retired from NASA to become Commandant of Aerospace Research Pilot School at Edwards AFB, Calif. He had logged 289 hrs 53 min in space during astronaut career. (AP, *W Post*, 7/2/71, A3)
- LaRC Director of Aeronautics Laurence K. Loftin assumed for one year post of Special Assistant for Aeronautics to Assistant Secretary of the Air Force for R&D Grant L. Hansen. (*Av Wk*, 5/31/71, 23)
- Robert F. Froehlke was sworn in as Secretary of the Army. (USA PIO)

July 2: Ashes of *Soyuz 11* Cosmonauts Georgy T. Dobrovolsky, Vladislav N. Volkov, and Viktor I. Patsayev, killed during reentry from June 6–30 mission, were buried in Kremlin Wall in Moscow's Red Square. Military funeral, climaxing national day of mourning, was watched by millions of TV viewers and attended by leading Soviet and foreign officials and thousands of other onlookers. Condolences were sent by world leaders. Astronaut Thomas P. Stafford represented President Nixon at funeral.

Eulogy at funeral was presented by Andrey P. Kirilenko, head of funeral commission: "On a small island of the motherland traveling at a fantastic speed across the expanses of the universe they were filled with courage and the conviction that the tasks of the party and people would be fulfilled; and they fulfilled these tasks, successfully completing a complex and varied program of scientific research and technical experiments. The results of their observations are invaluable for science, for the future of space technology, for mankind. For 24 days we saw the valorous cosmonauts on the television screens and listened to their voices. With unbated attention we followed their work. Millions of people, as it were, became intimately linked with the heroes of the cosmos. We all impatiently awaited the cosmonauts' safe return to home territory. Their lives tragically came to an end when the long cosmic journey was close to completion."

Letter from 19 surviving cosmonauts printed in *Pravda* said: "Today we are paying the last tribute to our talented and courageous comrades. There is not only grief in our hearts, but also pride for what they have done for the homeland in outer space. We know that our road is a difficult and thorny one, but we never doubted the correctness of our choice and were always ready for any difficult flight. We express firm confidence that what happened cannot stop the further development and perfection of space engineering and man's striving for space, striving for knowledge of the mysteries of the universe."

(FBIS-Sov-129-3, 7/6/71, L1; Kingston, Canada, *Whig-Standard*, 7/2/71, 1; AP, W *Star*, 7/1/71, A2)

- Unofficial sources in Moscow said deaths of *Soyuz 11* cosmonauts had been caused by embolism which occurred because of improperly sealed hatch, according to press reports. London *Evening News* Moscow correspondent Victor Louis reported cosmonauts had failed to close hatch securely during undocking from *Salyut 1*; as spacecraft reentered atmosphere it had lost pressure and crew had been deprived of oxygen. Official commission formed to investigate deaths had not yet released findings. (Mills, B *Sun*, 7/3/71, A1; Auerbach, *W Post*, 7/3/71, A1)

- Soviet Embassy in Washington, D.C., was opened to public for signing of condolence book for *Soyuz 11* cosmonauts. (*W Post*, 7/3/71, A12)

- NASA launched series of Nike-Cajun sounding rockets carrying GSFC grenade experiments to obtain temperature, pressure, density, and wind data between 35 and 95 km (22 and 59 mi). First rocket, launched from Churchill Research Range, carried 19 grenades to 122.4-km (76.1-mi) altitude, with all 19 grenades exploding as planned.

  Second rocket was launched from Point Barrow, Alaska, with 31 grenades, but electrical power to DOVAP ground transmitter was lost and no trajectory data were recorded beyond 4.5 sec GET. All grenades exploded and were recorded, but data were not useful without trajectory data. Third rocket, launched from Point Barrow as backup, successfully carried 19 grenades to 94.5-km (58.7-mi) altitude, with all grenades detonating and recorded as planned.

  Fourth rocket, launched from Wallops Station, carried 19 grenades to 122-km (75.8-mi) altitude. All 19 grenades detonated as planned and 14 sound arrivals were recorded. (NASA Rpts SRL)

- NASA announced award of $2.3-million contract to Sperry Rand Corp. Flight Systems Div. for design and construction of STOLAND. Advanced STOL avionics system based on digital computer would provide navigation and control information to pilot via advanced electronic displays and drive aircraft's control surfaces in response to inputs from pilot or computer. STOLAND would be installed in C-8 Buffalo aircraft being modified as jet STOL aircraft in NASA-DOT-FAA program to develop data base for systems concepts, design criteria, operational procedures, and certification criteria for STOL aircraft, STOL runways, and STOL air traffic control system. (NASA Release 71-125)

- Decline of science in America was discussed by Arnold Thackray, Univ. of Pennsylvania professor of history and sociology, in *Science*: "That some new social accommodation and organization of the scientific enterprise is in process of formation is, I think, evident from the present restlessness in the larger scientific and political community . . . reflected in a host of articles, meetings, discussions, and resolutions. From a historian's perspective the central question would seem to be whether specialist societies will reorganize in ways that help accommodate broader social concerns, or whether such societies will undergo a relative decline in importance. It could well be that generalist societies, like the AAAS and the National Academy of Sciences, are better adapted to pioneer these new roles, which also cut across traditional disciplinary boundaries and concerns. If so, we may be on the edge of a new era in the life of scientific societies as social organisms, an era in which both general and scientific politics feature unashamedly in the *raisons d'etre* of reinvigorated and reorganized generalist scientific societies." (*Science*, 7/2/71, 27–31)

- NSF published *Unemployment Rates for Scientists, Spring 1971*. Rate was 2.6% in 1971, jumping from 1.5% in 1970. National unemployment rate for all workers for first quarter 1971 averaged 6.5%. Doctorates experienced 1.4% unemployment rate in 1971 and 0.9% in 1970, while nondoctorate scientists were reported at 3.5% rate in 1971 and 2.9% in 1970. Defense and space constituted largest areas of activity, with 11% and 4% of unemployed. Approximately 45% of 1971 unemployed scientists reported last science-related job had been supported to some degree by Government funds. (NSF *Highlights*, 7/2/71, 1)
- John R. Schaibley, Executive Assistant to Apollo Program Director, OMSF, died in Virginia after extended illness. He had been with NASA since 1959 and had received NASA Exceptional Service Medal. (NASA Ann)

July 3: U.S.S.R.'s *Lunokhod 1* lunar rover was activated following eighth lunar night since vehicle was landed on moon by *Luna 17* Nov. 17, 1970. Temperature in instrumentation compartment was 295 K (72°F), pressure was 100 375 newtons per sq m (753 mm mercury), and all systems were functioning satisfactorily. (Tass, FBIS–Sov–129–3, 7/6/71, L1)

July 4: Boris Petrov, chairman of Soviet Intercosmos Scientific Council on Space, said in *Pravda* that deaths of *Soyuz 11* cosmonauts would not halt U.S.S.R.'s manned orbital space station: "New flights into space, the building of new manned orbital stations of the Salyut type, lie ahead. Undoubtedly, larger and more complex multipurpose and specialized manned space stations will be built. One can say with confidence that the 1970s will become an epoch of the development and broad use of long-term manned orbital stations with changing crews which will make it possible to switch from occasional experiments in space to a regular vigil by scientists and experts in space laboratories." (UPI, *NY News*, 7/5/71, 7)

July 5: L/G Samuel C. Phillips, SAMSO Commander, had been appointed to NASA Space Systems Advisory Committee, *Aviation Week & Space Technology* reported. (*Av Wk*, 7/5/71, 9)

- *Aviation Week & Space Technology* editorial commented on deaths of Soviet cosmonauts: "Space technology has already proved some of its greatest gains for the common good on earth are in medical areas. If the deaths of the Soyuz 11 cosmonauts focus more attention not only on the need for this research but also for the need to share it with all mankind, they will have made a contribution far beyond the specific achievements" of their voyage. (Hotz, *Av Wk*, 7/5/71, 7)

July 6: Apollo 15 Astronauts David R. Scott, James B. Irwin, and Alfred M. Worden began three-week preflight medical isolation to minimize exposure to disease or illness that could delay planned July 26 launch to moon. Crew would be confined to KSC crew quarters, training building, and launch pad until liftoff. (*W Post*, 7/7/71, A13)

- Award of $1.4-million cost-plus-fixed-fee contract to Boeing Co. Vertol Div. for fabrication and test of hingeless, low-disk loading tilt rotor was announced by NASA. Contract was part of joint USA–NASA program to provide technology for design of research aircraft. Although actual flight had not been scheduled, rotor would be fabricated consistent with flight application. (NASA Release 71–126)

- MSC announced selection of McDonnell Douglas Corp. to receive $380 000, 10-mo, firm-price extension to contract for design study of space shuttle auxiliary propulsion system (APS). McDonnell Douglas would define oxygen-hydrogen system compatible for both booster and orbiter vehicles. (MSC Release 71–47)

- Astrologer Jeane Dixon in *Washington Daily News* predicted "tremendous adventures in space for the United States . . . provided we keep an eye on platform placements and docking attempts above earth." She also said that three U.S. satellites had been intercepted "and deadened" by U.S.S.R. (*W News*, 7/6/71, 7)

July 7: NASA announced it had issued RFPs for automated HEAO to study high-energy radiation from space. Proposals were due Aug. 27 at MSFC and contract would be awarded early in 1972 if HEAO became approved project in FY 1972. Single contract for two spacecraft would continue through launch and mission operations support for seven years. Contract called for system engineering of HEAO payload; design and development of spacecraft; procurement and integration of orbit adjust stage and shroud; integration of experiments; design, development, and delivery of one set of HEAO ground support equipment; and launch operations support. Winning contractor also would provide mission operations support for observatory for up to two years for each mission. (NASA Release 71–127)

- Institute of Rehabilitation Medicine at New York Univ. Medical Center had accepted for evaluation and testing motorized wheelchair controlled by sight-switch, NASA announced. Switch, worn on head, enabled user to open and close circuit controlling chair's wheel movements with his eyes. It had been developed under NASA direction to help astronauts in training to operate equipment under extreme flight conditions. Institute's director, Dr. Howard A. Rusk, had estimated that more than 100 000 quadriplegics—persons with no use of arms or legs—could achieve mobility if chair could be made available. (NASA Release 71–122; *A&A 1970*)

- *Pravda* article said prospects were good for U.S.–U.S.S.R. agreement in SALT, which resumed in Helsinki July 8, but stressed that agreement must provide "equal security" for both nations. "Action is met by counteraction. If one side strengthens its defense, then the other feverishly looks for new, more powerful means to pierce the defensive shield of the opponent. This process, if it is not stopped with the help of reasonable agreements in curbing the arms race, is endless." (*NYT*, 7/8/71, 3)

July 8: NASA launched NRL's *Solrad 10* (*Explorer 44*) (Solrad-C) solar radiation satellite from Wallops Station at 6:58 pm EDT by Scout booster. Satellite entered orbit with 630.3-km (391.7-mi) apogee, 436.3-km (271.1-mi) perigee, 95.3-min period, and 51° inclination. NASA mission objective was to place satellite in orbit where it could monitor sun's x-ray and UV emissions.

Solrad 10, 115-kg (253.5-lb) 12-sided cylinder, carried 14 experiments to monitor solar electromagnetic and UV radiation continuously and to measure, on command, stellar radiation from other celestial sources. Information gained by satellite was expected to contribute to better understanding of physical processes in solar flares and other solar activities and potential effects of this activity on shortwave communications and on manned space travel. Satellite was third in series of solar radiation satellites launched from Wallops Station in joint NASA–NRL program. Two previous satellites were *Explorer 30* (*Solrad 8*, launched Nov. 19, 1965) and *Explorer 37* (*Solrad 9*, launched March 5, 1968). *Solrad* program was sponsored jointly by OSSA and Naval Air Systems Command, with NRL providing project management for mission. (NASA Proj Off; NASA Release 71–115; WS Release 71–13)

- Technological intelligence concerning SALT was reaching U.S. through Soviet scientist who had defected to West under pseudonym of Anatoly Fedoseyev [see May 26] but had been identified subsequently as Ignatiy A. Nikitrine, deputy director of Soviet space program in charge of electronics and cybernetics, Don Cook of *Los Angeles Times* reported in *Washington Post*. Meanwhile, AP said, British officials insisted scientist's real name was Fedoseyev and that he was expert in field of "powerful magnetrons" whose research had applications to Soviet space effort. (*W Post*, 7/8/71, A1; AP, *W Post*, 7/9/71)

- U.S. and Soviet astronomers had concluded joint observations of quasars to determine their size and structure, Tass reported. Observations had been conducted from three locations—in Crimea, in eastern, and in western parts of U.S.—about 10 000 km (6200 mi) apart. Soviet scientists had used 22-m (72-ft) radiotelescope. (FBIS–Sov–71–132, 7/9/71, L1)

- MSC announced selection of Lockheed Electronics Co., Inc., to receive $20-million, one-year, cost-plus-award-fee contract for general electronic, scientific, and computing center support services, at MSC. (MSC Release 71–48)

- MSFC announced award of $325 000 contract to North American Rockwell Corp. for 12-month study of inspace propellant logistics and safety. (MSFC Release 71–116)

- Lockheed Aircraft Corp. had released market analysis that showed it could sell at least 400 new L–1011 TriStar jet airbuses—about 60% more than needed for firm to break even, Dow Jones reported in Washington *Evening Star*. (*W Star*, 7/8/71, B9)

July 9: NASA began distribution to 24 U.S. scientists of lunar soil samples returned by Soviet *Luna 16* spacecraft and exchanged by U.S.S.R. for *Apollo 11* and *12* samples June 10. About 50% of total three grams (one tenth ounce) of material would be distributed to principal U.S. experimenters in stages. First lot was about 600 mg (9 grains), or about 20% of total sample. Dr. Paul Gast, chief of MSC Planetary and Earth Sciences Div., had said Soviet scientists had implied samples from Sea of Fertility differed from U.S. lunar samples because regolith was thinner and closer to underlying strata than soil of other lunar mare areas. Dr. Gast said Soviet samples contained wider grains than did U.S. samples. (NASA Release 71–128)

- Aerobee 170B sounding rocket was launched by NASA from WSMR to 191.5-km (119-mi) altitude on flight performance mission. Rocket and

- instruments functioned satisfactorily and payload was recovered. (SR list; Proj Scientist)
- President Nixon, in Proclamation 4066 proclaiming Oct. 24 United Nations Day, said: "Through the UN, we all share stewardship over the planet Earth: together we face the challenges of coordinating measures to heal and protect the world's fragile ecosystems; of ensuring that the new resources of the sea are developed for the benefit of all mankind; of promoting international cooperation in the use of outer space." (*PD*, 7/12/71, 1044)
- Appointment of Robert E. Bower, Director of Advanced Development for Grumman Aerospace Corp., as Director for Aeronautics at LaRC had been announced by LaRC Director Edgar M. Cortright, *Langley Researcher* reported. Bower succeeded Laurence K. Loftin, Jr., who had become Special Assistant for Aeronautics to Assistant Secretary of the Air Force for R&D [see July 1]. (*Langley Researcher*, 7/9/71, 1)
- Award to Global Associates of $4-million, one-year, cost-plus-award-fee contract to provide institutional support services at MTF was announced by MSFC. Selection had been made following competitive negotiations during which eight firms had submitted proposals. Contract included maintenance of rocket stage testing facilities in standby status and support services for other Federal agencies using MTF. (MSFC Release 71-117)
- International Galabert Prize, awarded annually by French for outstanding successes in exploration of outer space, had been presented to Soviet Academy of Sciences for *Luna 16*, Tass reported. (FBIS–Sov–71–133, 7/9/71, L2)
- Sen. Hugh Scott (R-Pa.) said at Washington, D.C. press conference, "I hope to see the day when our astronauts go up in Russian spacecraft and the Russian cosmonauts in ours." He would explore idea with Soviet Ambassador Anatoly F. Dobrynin before he left for visit to U.S.S.R. in August, "if our own government will permit me to do so." (UPI, *W Post*, 7/10/71, A3)

July 10: Soviet defector Anatoly Fedoseyev told London *Daily Telegraph* he had left U.S.S.R. because life had become unbearable and "sooner or later I would have finished up in a prison or concentration camp." He had laughed at reports he was space expert comparable in importance to NASA's Dr. Wernher von Braun and that his name was Nikitrine [see July 5]. Fedoseyev told *Telegraph* he had "nothing to do with space projects" and was not ABM specialist. (AP, *W Star*, 7/10/71)

July 11: Dr. James C. Fletcher, NASA Administrator, commented on *Salyut 1–Soyuz 11* docking in AP interview published by *Chicago Tribune*. Docking indicated U.S.S.R. was "more or less on schedule in their declared program for developing a major experimental station in space, at least two years ahead of us." In answer to question Dr. Fletcher said prospect "frightens me only because I don't know what they're doing there. They may have some military purposes that we don't know about." On U.S.S.R.'s two vehicles en route to Mars he said that "we don't know what these vehicles are going to do, but it's conceivable that there will be attempts to land." Dr. Fletcher said NASA had been "shocked and saddened" by loss of *Soyuz 11* cosmo-

nauts but "we find no reason to make any major changes in our plans for manned missions." (AP, *C Trib*, 7/11/71)
- Jeffrey T. Hamilton, Special Assistant to NASA Assistant Administrator for Industry Affairs and Technology Utilization, became Director of Technology Utilization, succeeding Ronald J. Philips, whose resignation had been announced June 26. (NASA Ann, 7/13/71)
- Dr. William H. Pickering, JPL Director, announced appointment of Dr. Aaron Finerman of State Univ. of New York at Stony Brook to manage JPL's newly established Office of Computing and Information Systems. (JPL Release 583)
- Boeing Co. was reactivating 767 project that was "simmering on a back burner" while SST was alive, William Hines said in Washington *Sunday Star*. Project "borrows liberally" from NASA aircraft development work. Boeing 767 subsonic jet would be about same size as Boeing 707, with same passenger capacity, but would be more than 160 km per hr (100 mph) faster. Cruising speed would be 1167 km per hr (725 mph) or mach 0.98, "about the greatest achievable without generation of a sonic boom."

  LaRC Director Edgar M. Cortright had said aircraft's speed was attributable to its supercritical wing that could operate at speeds near mach 1 in shockproof condition. Boeing spokesman had said aircraft could be in service by end of decade. Hines said Boeing also was considering production of stretched version of 727. (W *Star*, 7/11/71, D4)

July 12: U.S.S.R. announced findings of special commission which investigated deaths of *Soyuz 11* cosmonauts during reentry from June 6–30 mission. Official report distributed by Tass said spacecraft's flight had proceeded normally up to descent trajectory, with cosmonauts Georgy T. Dobrovolsky, Vladislav N. Volkov, and Viktor I. Patsayev performing according to flight program. About 30 min before landing "there occurred a rapid drop of pressure within the descent vehicle, which led to the sudden death of the cosmonauts. This is confirmed by the medical and pathologico-anatomic examinations. The drop in pressure resulted from a loss of the ship's sealing. An investigation of the descent vehicle, which made a soft landing, showed that there are no failures in its structure." (Tass, FBIS–Sov–133–3, 7/12/71, L1)
- NASA launched series of three Nike-Cajun sounding rockets carrying GSFC grenade experiments to obtain temperature, pressure, density, and wind data between 35 and 95 km (22 and 59 mi) by detonating grenades and recording their sound arrivals on ground. Rocket launched from Churchill Research Range carried 19 grenades to 124.9-km (77.6-mi) altitude with 18 grenades detonating as planned and 16 sound arrivals recorded. Rocket launched from Point Barrow, Alaska, carried 31 grenades to 108.8-km (67.6-mi) altitude with 31 explosions and 30 sound arrivals recorded. Rocket launched from Wallops Station carried 19 grenades to 135.8-km (84.3-mi) altitude, with 19 explosions and sound arrivals recorded. (NASA Rpts SRL)
- NASA Associate Administrator for Manned Space Flight Dale D. Myers had selected MSFC to integrate two proposed earth orbital projects, space station and research and applications modules (RAM), by supervising definition, design, and verification of design concepts, MSFC announced. Development of space station would follow development of

space shuttle. RAM was family of space payload carrier modules to be delivered to earth orbit by shuttle. (MSFC Release 71-118)
- NASA announced availability of two new publications. First, *Implantable Biotelemetry Systems* (SP-5094), listed biomedical telemetering instruments developed by NASA for monitoring physiological functions in human beings and animals and described devices and techniques for radiotelemetry physiological monitoring. *Analysis of Apollo 10 Photography and Visual Observations* (SP-232) described *Apollo 10* mission and contained miniature halftones of photos taken by *Apollo 10* astronauts, which could be obtained for study. (NASA Release 71-30; NASA Hq *WB*, 7/12/28)
- FAA announced it would lower base of area-positive-control airspace (APC) from 7300 m (24 000 ft) to 5500 m (18 000 ft) over southwestern U.S., effective Aug. 19, to provide operating aircraft with additional protection against midair collisions. (FAA Release 71-105)
- President Nixon forwarded to Congress nomination of L/G Benjamin O. Davis, Jr. (USAF, Ret.), Deputy Assistant Secretary of Transportation for Safety and Consumer Affairs, to be Assistant Secretary of Transportation (Safety and Consumer Affairs). (*PD*, 7/19/71, 1060)
- AFSC announced reassignment of officers: M/G Lee V. Gossick, Deputy Chief of Staff, Systems, would become Chief of Staff for AFSC on departure of M/G Clifford J. Kronauer; M/G Kenneth W. Schultz, Deputy for Minuteman, SAMSO, Norton AFB, Calif., would succeed M/G Gossick as Deputy Chief of Staff, Systems; B/G Abner B. Martin would succeed M/G Schultz as Deputy for Minuteman; B/G Herbert A. Lyon, Assistant Deputy Chief of Staff, Systems, AFSC, would succeed B/G Martin as Deputy for Reentry Systems; B/G William W. Gilbert, Vice Commander of USAF Flight Test Center, Edwards AFB, Calif., would command European Communications Area, with additional duties as Deputy Chief of Staff, Communications, USAF, Europe. (AFSC Release 216.71; AFSC PIO)

July 13: Apollo 15 completed simulated liftoff at 9:34 am EDT in preparation for launch from KSC July 26. Spacecraft was unmanned and fully loaded with propellants. Astronauts would practice countdown in spacecraft without fuel July 14. (*SBD*, 7/14/71, 56)
- Selection of NR Rocketdyne Div. for negotiations leading to award of $500-million, cost-plus-award-fee contract for development and delivery of 35 space shuttle engines by 1978 was announced by NASA. Selection followed 12-mo competition during which NASA had accepted proposals from three firms on April 21. Program would be managed by MSFC and would support space shuttle orbital flights beginning in 1978. (NASA Release 71-131)
- NASA launched two Nike-Tomahawk sounding rockets from Wallops Station carrying Univ. of Texas experiments to study ionosphere. Rockets and instruments functioned satisfactorily. (SR list)
- Sleep analyzer developed for NASA by Dr. James D. Frost of Baylor Univ. and built by SCI Electronics, Inc., Div. of SCI Systems, Inc., as Skylab experiment had been made available as research device to U.S. medical institutions, NASA announced. Device would record quality of sleep of Skylab astronaut during 28-day stay in space by combining and evaluating inputs of electro-encephalogram and electro-occulogram. Medical application of device could be to treat insomniacs and other patients

experiencing "sleep neuroses." FAA doctors were considering study of sleeping patterns of air controllers and pilots after stressful duty. (NASA Special Release)

- Senate Committee on Banking approved by vote of 10 to 5 revised bill authorizing Government guarantee of loans to companies whose failure would "adversely and seriously affect the economy of the nation." Bill would permit Government to guarantee up to $2 billion in such loans, with limit of $250 million for one company. Bill represented first legislative test of Administration's efforts to save Lockheed Aircraft Corp. from bankruptcy, *New York Times* said. (Witkin, *NYT*, 7/14/71, 1)
- President Nixon announced appointments to Citizen's Advisory Committee on Environmental Quality, including *Apollo 8* Astronaut Frank Borman. (*PD*, 7/19/71, 1056)
- FAA announced award of $530 125 contract to Cornell Aeronautical Laboratory, Inc., for analytical and experimental program to develop new microwave landing system (MLS) for civil-military common use, to replace VHF/UHF instrument landing system (ILS) developed in early 1940s. Microwave system would provide more precise electronic guidance to aircraft on approaches and landings than current system; selectable flight paths to permit greater number of aircraft operations at airports; closer spacing of parallel runways; and procedures to ease noise over surrounding communities. (FAA Release 71-109)
- U.S. faced no dire threat from current "numbers or accuracy" of Soviet SS-9 missiles, but must extend its own antimissile system as "bargaining chip" with U.S.S.R. at SALT, Deputy Secretary of Defense David Packard said in testimony before House Committee on Foreign Relations' Subcommittee on Arms Control. (*NYT*, 7/14/71, 7)
- Secretary of Transportation John A. Volpe announced FAA was considering major policy change to permit establishment of FAA facilities and services at privately owned airports open to public. Such facilities and services were restricted to publicly owned airports. FAA Administrator John H. Shaffer said policy change would be "in accord with FAA's assigned mission and responsibility of assuring safety and efficiency of all civil aircraft operations and of promoting air commerce and civil aeronautics." (FAA Release 71-110)
- Dr. Glenn T. Seaborg, AEC Chairman, was formally presented with membership in Soviet Academy of Sciences by Soviet Ambassador Anatoly F. Dobrynin in ceremony at Soviet Embassy in Washington, D.C. (AEC Release O-120)
- Cornell Univ. and EDP Technology, Inc., announced that contract for sale of Cornell Aeronautical Laboratory by Cornell to EDP had been allowed to expire by mutual consent. (Van Gelder, *NYT*, 7/14/71, 58)

July 14: Apollo 15 completed final practice countdown at KSC in preparation for launch to moon July 26. (Wilford, *NYT*, 7/15/71, 16)

- Discovery of hydroxyl molecule in interstellar dust between stars of M-82 and NGG-253 galaxies 97 billion km (60 billion mi) from earth by use of big dish radio telescopes, was reported by Dr. Leonid N. Weliachew, visiting fellow at Cal Tech from Meudon Observatory in France. Discovery implied that chemical evolution was proceeding throughout universe. Hydroxyl was considered a chemical essential to evolution of life. (LATNS, *W Post*, 7/15/71, A3; *B Sun*, 7/19/71)

- MSFC announced award of $250 000, 10-mo, "phase A" contract to NR Space Div. to study feasibility of developing low-cost, reusable, chemical-propulsion stage that could be launched from earth on space shuttle booster and refueled in space for up to 10 missions. Vehicle could be used for high-lift capability to low earth orbit, for placing large payloads in geosynchronous orbit, and for lunar and unmanned planetary missions. (MSFC Release 71–120)
- MSC announced award of three seven-month, firm, fixed-price contracts for development of new surface materials for space shuttle orbiter stage. McDonnell Douglas Corp. would receive $325 000; GE Aerospace Group, $319 200; and Lockheed Missiles and Space Co., $322 500. (MSC Release 71–50)
- Sen. Hubert H. Humphrey (D-Minn.) and Sen. Edmund S. Muskie (D-Me.) proposed U.S. and U.S.S.R. freeze deployment of nuclear weapons while they sought agreement to control strategic arms at SALT. Joint proposal was made during hearings of Senate Committee on Foreign Relations' Subcommittee on Arms Control. (Finney, *NYT*, 7/15/71, 1)

July 15: Senate Committee on Appropriations favorably reported H.R. 9382, FY 1972 HUD-space-science appropriations bill that contained $3.325-billion NASA appropriation. Committee added $24 million for NERVA to NASA request of $2.518 billion for R&D, making total of $2.542 billion, and rejected House recommendations for cuts in construction of facilities appropriation, including $20 million for space shuttle funding. (*CR*, 7/15/71, D703; *SBD*, 7/16/71, 70)
- LaRC awarded $3 934 000 contract to Carey General Contractors, Inc., and Metropolitan Construction Co. of Missouri for joint construction of 4200-sq-m (45 500-sq-ft) Aircraft Noise Reduction Laboratory. Research facility, to be completed by November 1972, would include acoustics physics laboratory, applications area with anechoic room and reverberation room, and external effects simulation laboratory. (NASA Release 71–133)
- Dr. James C. Fletcher, NASA Administrator, presented LM similar to those used for *Apollo 11, 12,* and *14* lunar landings to Michael Collins, Director of Smithsonian Institution's Air and Space Museum and *Apollo 11* astronaut. LM would remain on permanent display at museum. (Smithsonian Release 111–71; NASA PAO)

July 16: USAF launched unidentified satellite from Vandenberg AFB by Thorad-Agena booster. Orbital parameters: apogee, 506.9 km (315 mi); perigee, 487.6 km (303 mi); period, 94.5 min; and inclination, 75°. (Pres Rpt 72; *SBD*, 7/26/71, 109)
- U.S.S.R. launched *Meteor 9* from Plesetsk into orbit with 642-km (398.9-mi) apogee, 615-km (382.1-mi) perigee, 97.2-min period, and 81.1° inclination. Tass said satellite carried meteorological equipment to obtain pictures of clouds and snowcaps on day and night sides of earth and data on thermal energy reflected and radiated by earth and atmosphere. (GSFC *SSR*, 7/31/71; *SBD*, 7/19/71, 78; FBIS–Sov–71–137, 7/16/71, L1)
- Findings on atmosphere of Mars from UV spectrometer observations made during NASA's 1969 Mariner flyby missions were reported in *Science* by Univ. of Colorado scientists Dr. Charles A. Barth and Dr. Charles W. Hord: "Mars . . . reflects sunlight in the ultraviolet, but it is the

atmosphere, not the surface, that is responsible for the reflected light. Even though there are atmospheric scatterers in addition to the molecular scatterers, it is possible to relate the intensity of the scattered radiation with the atmospheric pressure. The variation of the pressure over the planet reveals the topography to vary over 7 kilometers [4½ miles] in height and to be correlated with visible features. The carbon dioxide polar cap, in addition to being a cold trap for volatile gases in the atmosphere, may also be a very efficient absorption trap for nonvolatiles. This last property may make the cap a repository for gases produced by geological or biological activity." (*Science*, 7/16/71, 197–201)

- Lowell K. Zoller, Chief of Program Control Group in MSFC Office of Program Development, had been appointed manager of task team for research and applications modules (RAM) by MSFC Director, Dr. Eberhard F. M. Rees, MSFC announced. RAM was proposed family of space payload carrier modules to be delivered to earth orbit by space shuttle. (MSFC Release 71–122)

- White House announced President Nixon would accept resignation of Dr. William D. McElroy as Director of National Science Foundation. Dr. McElroy would remain in position until February 1972. (*PD*, 7/26/71, 1078)

July 16–18: Lewis Research Center celebrated 30th anniversary. Groundbreaking had been on Jan. 23, 1941, but anniversary celebration had been postponed for summer weather. Ceremonies honored former LeRC employees who had transferred to other NASA centers, retired employees, and visiting NASA officials, including Dr. George M. Low, Deputy Administrator. Each employee received 30th anniversary coin and special anniversary edition of *Lewis News*. Anniversary exhibition included F–106 aircraft, *Apollo 8*, *Gemini 7*, and Mercury spacecraft. Center Director Bruce T. Lundin described progress of LeRC—which had been dedicated May 28, 1943, as aircraft engineering research laboratory by Dr. George W. Lewis, NACA Director of Aeronautical Research, for whom it was later named. Center now occupied 3400 hectares (8500 acres) with 4000-member staff and plant valued at $311 million. Open House was held for general public July 17 and 18. (LeRC Release 71–30; *Lewis News*, 7/16/71, 1; NASA Hq *WB*, 7/5/71, 1)

July 17: U.S.S.R.'s *Lunokhod 1* lunar rover completed experiments planned for ninth lunar day on moon's Sea of Rains and was preparing for lunar night, Tass announced. Experiments included stereoscopic photography of crater bed and slopes, chemical analysis of rock, studies of crater 190 m (208 yds) in diameter, and measurements of cosmic radiation. Vehicle had traveled 10.2 km (6.3 mi) over lunar surface since landing on moon Nov. 17, 1970. (FBIS–Sov–138–3, 7/20/71, L1)

- Nine Soviet scientists visited AEC's Nevada test site near Las Vegas during meeting of U.S. and Soviet scientists on peaceful uses of atomic energy. (*NYT*, 7/14/71, 41; AEC PIO)

July 19: Apollo 15 astronauts would wear pressure suits—instead of flight coveralls as originally planned—during LM jettison, NASA announced. Requirements for crew to wear suits had been reevaluated after deaths of three cosmonauts during reentry from June 6–30 *Soyuz 11* mission. (NASA Release 71-134)

- U.S.S.R. sent telegram to American Science and Engineering, Inc., designer of Apollo 15 x-ray detector, accepting invitation to cooperate with U.S. in studying x-ray sources during Apollo 15 mission. Soviet scientists would train 2591-mm (102-in) telescope on same region of sky that Apollo 15 crew would explore during return to earth and would compare notes on how strongest x-ray source in sky appeared through telescope and through x-ray detector. (McElheny, *W Post*, 7/21/71, A10)
- Clare F. Farley, Executive Officer in Office of NASA Administrator since 1968, became Assistant Administrator for Technology Utilization. Farley was succeeded as Executive Officer by Henry E. Clements, formerly of USAF Hq. Directorate of Space Research and Development. (NASA Ann, 7/12/71)
- Sen. B. Everett Jordan (D–N.C.) introduced, for himself and cosponsors, S. 2302 "to establish an Office of Technology Assessment for the Congress as an aid in the identification and consideration of existing and probable impacts of technological application; to amend the National Science Foundation Act of 1950; and for other purposes." (*CR*, 7/19/71, S11346)
- President Nixon submitted to Senate nomination of Stanford Univ. chemist John D. Baldeschwieler to be Deputy Director of OST. He would succeed Dr. Hubert Heffner, who had resigned to return to duties at Stanford. (*PD*, 7/26/71, 1066, 1080)
- Apollo program's contribution to lunar science was praised in *Aviation Week & Space Technology* editorial: "Apollo has taken lunar exploration out of the hands of tree-full-of-owls theorists and put it into the care of the lunar explorers, who will eventually provide sufficient data to unravel some of the major mysteries of the moon and our universe." (Hotz, *Av Wk*, 8/19/71, 7)
- Los Angeles Harbor Commission granted one-year extension to lease with Hughes Tool Co. for storage area for *Spruce Goose,* mammoth wooden flying boat designed by Howard Hughes and flown once, on brief taxi test in 1947 [see Feb. 10]. (UPI, *W Post*, 7/21/71, C8)

July 20: Final countdown for July 26 launch of Apollo 15 manned lunar landing mission began at KSC at 6:30 am EDT. (AP, B *Sun*, 7/21/71, A1)

- More than 20 hrs of live TV coverage from Apollo 15 spacecraft and lunar surface would be carried via orbiting Intelsat comsats, ComSatCorp announced. Coverage would include liftoff, transposition and docking, intravehicular transfer, landing site approach, lunar surface exploration, first color broadcast of lunar liftoff, rendezvous and docking, film transfer in space, lunar eclipse, press conference from space, and splashdown. (ComSatCorp Release 71-39)
- U.S.S.R. launched *Cosmos 429* from Baykonur into orbit with 256-km (159.1-mi) apogee, 177-km (110-mi) perigee, 88.8-min period, and 51.7° inclination. Satellite reentered Aug. 2. (GSFC *SSR*, 7/31/71; 8/31/71; *SBD*, 7/26/71, 109)
- President Nixon issued Proclamation 4067 designating day as National Moon Walk Day, "to commemorate the anniversary of the first moon walk on July 20, 1969, and to accord recognition to the many achieve-

ments of the national space program." He said, "Two years after the first landing . . . other brave men are following in the footsteps of [Astronaut Neil A.] Armstrong and [Astronaut Edwin E.] Aldrin to explore the unknown and advance scientific knowledge for the benefit of all mankind." (PD, 7/26/71, 1068)

- U.S.S.R., in first status report on *Salyut 1* (launched April 19) since deaths of *Soyuz 11* cosmonauts June 30, said *Salyut 1* was functioning normally after 1490 orbits of earth. Tass said station was in orbit with 262.3-km (163-mi) apogee and 223.7-km (139-mi) perigee and all systems were functioning normally. (UPI, C *Trib*, 7/21/71)
- Senate by unanimous vote of 87 passed H.R. 9382, FY 1972 HUD-space-science appropriations bill that included $3.321 billion NASA appropriation and $647-million NSF appropriation (after adopting amendment increasing NSF research funds by $25 million). (*CR*, 7/20/71, S11478–502)
- Senate passed H.R. 9388, FY 1972 AEC authorization bill that contained additional $37 million in operating costs for space nuclear propulsion program, bringing total authorization for space nuclear propulsion to $52 million. (*CR*, 8/20/71, S11502–65)
- Members of 28-nation U.N. Scientific Subcommittee on Space, meeting in New York, had agreed to establish study panel to examine uses of satellites for earth resources surveys, *New York Times* reported. Agreement represented compromise. U.S. and U.S.S.R. had opposed earlier Swedish proposal that panel consider establishment of operational systems within U.N. framework. U.S. position, expressed by NASA Assistant Administrator for International Affairs Arnold W. Frutkin, had been that practical application of using satellites for remote sensing of earth would be demonstrated adequately only after testing of NASA's ERTS in spring of 1972. (Teltsch, *NYT*, 7/20/71, 12)
- Wapakoneta, Ohio, birthplace of *Apollo 11* Astronaut Neil A. Armstrong, celebrated second anniversary of first moon walk by Armstrong with display of flags and model of LM on concrete blocks on street corner. For third anniversary, in 1972, Armstrong expected to return to home town for dedication of $1-million Armstrong Museum being constructed at edge of town. Daniel R. Porter, Director of Ohio Historical Society, had said museum would feature Ohio's contributions to aerospace age from early balloon ascents through achievements of Wright brothers, who had lived in Ohio, to space exploits of Ohioan Astronaut John H. Glenn. Exhibits would include experimental aircraft flown by Armstrong as test pilot, gallery of flight artifacts, "NASA Room" containing space travel exhibits, and "infinity cube," mirrored room with lighting to give effect of projection into infinity. (Wilford, *NYT*, 7/21/71, 22)
- FAA announced award of two cost-sharing research contracts totaling almost $6 million to Boeing Co. to determine feasibility in design hardware of acoustical treatment techniques in reducing noise levels of United Aircraft Corp. Pratt & Whitney Div.'s JT8D and JT3D engines. (FAA Release 71–118)

July 21: Senate-House conferees on H.R. 7109, FY 1972 NASA Authorization bill, filed H Rpt 92–368 proposing compromise bill authorizing $3.355 billion. Report authorized $702.7 million for space flight operations, $30 million above NASA request of $672.7 million, and

recommended $15 million be spent on Skylab rescue capability and $15 million on space shuttle. (H Rpt 92-368)
- Black Brant VC sounding rocket was launched by NASA from Wallops Station on test and support mission to evaluate flight performance. Vehicle developed aerodynamic instability and payload broke off. (SR list; Proj Scientist)
- President Nixon accepted resignation of Dr. Glenn T. Seaborg as AEC Chairman, effective on date to be determined, but asked him to remain as head of U.S. delegations to atomic energy conferences in U.S.S.R., Geneva, and Vienna during 1971. President also asked Dr. Seaborg to serve as consultant after meetings. He said: "As a world famous chemist, scholar, and administrator, you have contributed in a unique and meaningful way to far greater understanding and application of the miracles of the atom."

  President submitted to Senate nomination of James R. Schlesinger, Assistant Director of Office of Management and Budget, to be AEC member and successor to Dr. Seaborg as AEC Chairman. (PD, 7/26/71, 1069, 1080)
- "Apollo Day" was celebrated in State of Indiana. Gov. Edgar H. Whitcomb attended ceremonies at Spring Mill State Park honoring Indiana-born astronaut, the late Virgil I. Grissom who was killed in Jan. 27, 1967, Apollo fire. (MSFC Hist Off)
- Dr. Dudley G. McConnell, Assistant Executive Secretary in NASA Office of Administrator, became Director of Office of Scientific and Technical Information in NASA Office of Industry Affairs and Technology Utilization. (NASA Release 71-142)

July 22: NASA and AFCRL scientists would measure plume from Apollo 15 during launch July 26 in attempt to solve mystery of *Apollo 12* lightning strike during Nov. 14, 1969, launch, AFSC announced. Measurements were part of Lightning Strikes to In-Flight Missiles program to test conductivity and electrical field breakdown strength on Minuteman and Atlas missiles. Scientists theorized that exhaust plume acted as electrical conductor and might have influenced electrical fields in thunder clouds and provoked lightning stroke.

AFCRL project scientist John L. Heckscher said USAF tests showed that "the rocket plume acts as a good conductor of electricity—like a wire . . . stretched from the rocket to the ground. And, should the rocket enter a thunder cloud, it would act like a lightning conductor. In the Apollo 15 launch we're primarily interested in the effective length of the conducting plume. The rocket is 363 feet [110 m] long, and the visible portion of the plume, which contains incandescent carbon particles and other burning matter, is perhaps another 500 ft [152 m] of good conductor. However the trailing invisible portion may also be electrically conductive and that's what we want to find out." (AFSC Release 180.71)
- *Mariner 9*, launched by NASA May 30, had covered 142 600 000 km (88 600 000 mi) in 167-day journey to Mars. Spacecraft was 15 300 000 km (9 500 000 mi) from earth and had received more than 600 commands. (NASA Release 71-136)
- ARC-developed oximeter—device to measure oxygen content of blood by light absorption of blood circulating through ear—was being used by National Cancer Institute's Leukemia Service for early detection of

shock, NASA announced. Device had been introduced to Cancer Institute by Biomedical Application Team (BATEAM) of Research Triangle Institute. Team was employed by NASA to exchange space-developed technology between physical and medical sciences. Oximeter had been developed in 1960s to study effects of rapid acceleration in centrifuge on subjects, to determine reactions to simulated manned space flight conditions. (NASA Release 71-132)

- Tass said Communist Party Central Committee and Soviet Council of Ministers had decided to unveil busts of *Salyut 1–Soyuz 11* Cosmonauts Georgy T. Dobrovolsky, Vladislav N. Volkov, and Viktor I. Patsayev in their home towns and to erect obelisk in their honor at landing site of *Soyuz 11* spacecraft. Memorial plaques would be placed on buildings at cosmonaut training center and at cosmonauts' schools. Cosmonauts' names would be given to their schools and to streets in Odessa, Moscow, and Aktyubinsk, cosmonauts' home towns. (FBIS–Sov–71–142, 7/22/71)

- Japan would launch scientific observation satellite from Uchinoura Space Center Sept. 20, *Space Business Daily* reported. Satellite design and planned orbit would be similar to that of Japan's *Tansei*, launched Feb. 16. (*SBD*, 7/22/71, 96)

- SES–100B Surface Effect Ship, built for USN by Textron Bell Aerospace Div., was launched at Michoud Assembly Facility. Ship was 23.7 m (78 ft) long, weighed 90 700 kg (100 tons), and traveled at 41 m per sec (80 knots) on air cushion maintained by eight lift fans and contained by catamaran-style side hulls and flexible bow and stern seals. Test and evaluation program would be conducted in Lake Pontchartrain and Gulf of Mexico beginning July 23. (Bell Aerospace Div Release)

July 23: U.S.S.R. launched *Cosmos 430* from Plesetsk into orbit with 265-km (164.7-mi) apogee, 187-km (116.2-mi) perigee, 89-min period, and 65.4° inclination. Satellite reentered Aug. 5. (GSFC *SSR*, 7/31/71; 8/31/71; *SBD*, 7/26/71, 109)

- Battery and companion unit in Apollo 15 LM failed during tests at KSC and were replaced. Investigation revealed battery had been contaminated with magnesium chips. Problem did not delay countdown, which continued on schedule toward 9:34 am EDT liftoff July 26. (Wilford, *NYT*, 7/24/71, 29)

- NASA's M2–F3 lifting-body vehicle, piloted by NASA test pilot William H. Dana, successfully completed seventh flight from FRC, reaching 18 300-m (60 000-ft) altitude and mach 0.83. Objectives were to expand flight envelope to mach 0.85, obtain stability and control data at mach 0.8, and evaluate reaction control system. (NASA Proj Off)

- Appointment of John P. Donnelly, Vice President for Corporate Communications of Whittaker Corp., as NASA Assistant Administrator for Public Affairs effective Aug. 15, was announced by NASA. He would be responsible for development and direction of public affairs activities, including public information, public services, and educational programs. Acting Public Affairs chief since resignation of Julian W. Scheer March 22 had been Alfred P. Alibrando, Director, Public Affairs Div. (NASA Release 71–137; NASA Ann, 2/23/71)

- NASA announced it had signed agreement with New York Times Special Features, syndicate division of *New York Times*, for one byline article each by Apollo 15 Astronauts David R. Scott, James B. Irwin, and

Alfred M. Worden after mission and three articles by Scientist-Astronaut Harrison H. Schmitt while crew was on lunar surface. Agreement, extending to 30 days after mission's end, provided for payment by Special Features of 50% of gross proceeds from syndication of articles. It also permitted still photos in Apollo 15 astronauts' homes and interviews of family members. Agreement did not infringe on NASA policy of free flow of information to all media nor limit *NYT* coverage of all aspects of space program. (NASA Release 71-138)

- *The Decision To Go to the Moon* by John M. Logsdon of George Washington Univ. was reviewed in *Science* by L. Vaughn Blankenship, State Univ. of New York at Buffalo political scientist: Book pulled together "most of the publicly available data and commentary on the events surrounding President Kennedy's decision in May 1961 to commit the United States to effecting a lunar landing before 1970." Logsdon seemed to conclude that "because the decision process somehow corresponded to his theory of how decisions are, or should be made . . . the decision must have been a good one." It was "legitimate to choose to present a history of this decision within the limited framework adopted in this book" but "less legitimate to tout this decision as evidence of a political system operating at its best. Or if it is its best, we may well worry for the future." Dr. Logsdon replied to review Sept. 17. (*Science*, 7/23/71, 317-8; 9/17/71, 1079-80)

- GAO published *Comparison of Military Research and Development Expenditures of the United States and the Soviet Union*. Report quoted Organization for Economic Co-operation and Development (OECD) as saying much Soviet R&D was likely to have been directed toward military and space programs. Spin-off to civilian industry in U.S.S.R. was considered to have been small because all invention and innovation of military application, while in principle available for civilian use, was surrounded by impenetrable security blanket. Consequently, strain of R&D effort had been much greater for Soviet economy than for that of U.S. (*CR*, 7/31/71, E8607-81)

July 24: President Nixon issued statement supporting legislation before Congress to provide emergency loan guarantees for major business enterprises confronted with temporary financial stringencies: "The administration originally sought legislation only to help the Lockheed Aircraft Corporation. That support is still needed very badly. But I have instructed Secretary of the Treasury Connally . . . to accept the broader legislation. It would be most useful in providing a systematic procedure for helping any major business enterprise with temporary financial problems whose failure would adversely affect the economy of the Nation or a region thereof." (*PD*, 7/26/71, 1078)

July 25: President Nixon telephoned Apollo 15 astronauts to extend good wishes on eve of launch. (*PD*, 8/2/71, 1099)

- KSC launch complex where Apollo 15 spacecraft and Saturn V booster stood ready for liftoff had been struck by lightning 11 times during past six weeks, *Washington Post* reported. First strike, June 14, had been equivalent to 98 000 amps—almost three times as powerful as average lightning strike and two thirds as strong as biggest lightning strike ever recorded in Florida. Two more strikes had followed within 15 min of each other and a 31 000-amp strike had hit launch pad one day later. On July 20 five separate lightning bolts with total amperage

July 26–August 7: Apollo 15 carried three astronauts on the fourth successful lunar landing mission, returning them with 77 kilograms (170 pounds) of samples after 66 hours 55 minutes on the moon's surface. During prelaunch activities in Firing Room 1 of the Kennedy Space Center's Launch Control Center, Dr. James C. Fletcher, NASA Administrator (right), spoke with NASC Executive Secretary William A. Anders; L/G Samuel C. Phillips (USAF), former Apollo Program Director; and Dr. Wernher von Braun, Deputy Associate Administrator for Planning (left to right).

of almost 100 000 amps had been recorded. Ninth strike of 22 000 amps had occurred June 25, and last—weakest to date at 6600 amps—had occurred July 19. Neither spacecraft nor launch vehicle had been damaged. Apollo 15 was first spacecraft to have even a near miss from lightning strikes while being prepared for launch. (*W Post*, 7/25/71, A16)

July 26–August 7: NASA's *Apollo 15* (AS-510) carried three-man crew on fourth successful lunar landing mission. LM-10 *Falcon* landed on moon's Hadley-Apennine region and two astronauts conducted experiments, rode first manned lunar roving vehicle (LRV) on moon, and explored lunar surface. After 66 hrs 55 min on surface, *Falcon* rejoined orbiting CSM-112 *Endeavor* and astronauts transfered for safe return to earth with lunar samples.

July 26–28: Spacecraft—carrying Astronauts David R. Scott (commander), Alfred M. Worden (CM pilot), and James B. Irwin (LM pilot)—was launched from KSC Launch Complex 39, Pad A, on time at 9:34 am EDT July 26 by Saturn V booster. Launch was watched by 2000 invited guests—including 185 members of British Astronomical Society and Royal Astronomical Society and 41 Congressmen—and estimated 1 million other viewers.

Spacecraft and S-IVB combination entered parking orbit with 171.4-km (106.5-mi) apogee and 169.4-km (105.3-mi) perigee. Translunar injection (TLI) was achieved at 2:56 GET and CSM separated from LM/S-IVB/IU at 3:22 GET. Docking of CSM with LM at 3:34 GET was shown clearly by onboard color TV.

S-IVB APS burns were conducted at 5:48 GET and 10:00 GET to send stage toward moon. Separated S-IVB/IU impacted lunar surface at 79:25 GET (4:59 pm EDT July 29) at 1° south latitude and 11.9° west longitude, 188 km (117 mi) northeast of *Apollo 14* landing site and 355 km (221 mi) northeast of *Apollo 12* landing site. Impact was detected by *Apollo 14* seismometer 37 sec after impact and by *Apollo 12* seismometer 55 sec after impact.

First midcourse correction (MCC-1) was canceled because spacecraft trajectory was near nominal. Shortly after CSM–LM docking, telemetry data indicated electrical short in service propulsion system (SPS). Troubleshooting isolated problem and MCC-2 at 28:41 GET used SPS bank A to analyze apparent intermittent short. Data indicated bank A could be safely operated manually. Modified procedures were developed for using bank B alone for remaining midcourse corrections except lunar orbit and transearth insertions, which would be dual bank burns.

Scott and Irwin entered LM at 33:56 GET, 50 min earlier than planned, to check out LM communications and other systems. TV pictures of CSM and LM interiors were shown between 34:55 and 35:46 GET. During checkout crew discovered range/range-rate exterior glass cover had broken, removing helium barrier. Crew began LM housekeeping 1½ hrs earlier than scheduled and vacuumed broken glass. At 61:13 GET, during preparations for water chlorination, water leak developed in CM chlorination septum gland but was stopped by crew, following repair instructions from ground, and water was absorbed with towels. Door of scientific instrument module (SIM) was jettisoned at 74:06 GET.

July 29–August 4: Apollo 15 entered lunar orbit with 314.8-km (195.6-mi) apolune and 107.5-km (66.8-mi) perilune after LOI burn at 78:32 GET (4:06 pm EDT July 29). "This is really profound ... it's fantastic," Scott said of view from moon. Apennine Mountains looked unreal. "They stand out in tremendous relief. They appear to be smooth or rounded. But they are cratered and in many places rough in texture. We don't see any jagged peaks. They don't look like ... any other mountains we've seen on earth."

CSM–LM undocking and separation maneuver was initiated during 12th lunar revolution, on far side of moon at 100:14 GET, but spacecraft did not undock. Worden entered tunnel to inspect CSM–LM umbilical and found plug was loose. After he reconnected plug and adjusted spacecraft attitude, undocking and separation were achieved at 100:39 GET—about 25 min later than planned.

LM *Falcon* touched down in moon's Hadley-Apennine region near Salyut Crater at 104:42 GET (6:16 pm EDT July 30) about 600 m (656 yds) north-northwest of planned target. "OK Houston," Scott said after touchdown, "the Falcon is on the plain at Hadley." He said LM's engine had stirred up so much dust that landing site had been completely obscured from 15-m (50-ft) altitude to touchdown: "We flew IFR from then on down."

Stand-up extravehicular activity (SEVA) to observe and photograph landing site and surrounding area began after cabin depressurization at 106:43 GET. Scott put his head out of upper hatch and described and photographed area for 33 min. Scott said he could describe site

July 29–August 4: Apollo 15 *entered lunar orbit July 29 and the* LM *Falcon touched down in the moon's Hadley-Apennine region July 30 for three* EVA *periods of exploration. The lunar roving vehicle Rover, deployed during the first* EVA *July 31, was photographed by Astronaut James B. Irwin near Hadley Rille (right center above), with the Hadley Delta in the background and St. George Crater partially visible at the upper right edge. Astronaut David R. Scott was working at the Rover. In the photo at right Irwin scooped a trench, with Mount Hadley in the background. A gnomon indicated the local vertical and the sun angle. Meanwhile* CM *pilot Alfred M. Worden, orbiting the moon in* Endeavor, *completed lunar and astronomic photography.*

for hours. Area was covered with craters, but strewn with very few large boulders; it was hilly, but not rocky or rugged.

First EVA began July 31 at 9:04 am EDT. Scott climbed down ladder, deployed MESA containing camera which recorded his descent, and stepped on lunar surface. Irwin followed and, while Scott removed TV camera from MESA and deployed it on tripod, Irwin collected contingency sample. Crew had difficulty deploying LRV Rover; during checkout they found that front steering mechanism was inoperative and that there were no readouts on LRV battery No. 2 amp-volt meter. Battery was found to be operating satisfactorily and decision was made to perform EVA–1 without LRV front-wheel steering. Crew mounted LRV and proceeded on EVA–1 traverse at 121:45 GET. Crew described EVA as "exploration at its greatest," and repeatedly used words "fantastic," "breathtaking," and "spectacular" to describe features as they rode LRV at about eight km per hr (five mph). Scott said Rover handled quite well: "It negotiates small craters quite well, although there is a

lot of roll. The steering is quite responsive . . . and I can maneuver pretty well. . . . There is no accumulation of dirt in the wire wheels." Stops at stations where crew collected lunar samples and took photos were broadcast on TV with excellent transmission. At end of traverse ALSEP was deployed. Scott and Irwin reentered LM after 6 hrs 33 min exploring surface, 28 min sooner than planned because Scott used up more oxygen than anticipated.

After resting inside LM, recharging and repairing portable life support system (PLSS), and reviewing plans for second EVA, astronauts left LM at 142:15 GET (7:49 am EDT Aug. 1). LRV was powered up, circuit breakers were cycled, and LRV front steering was found to be completely operational. EVA-2 traverse included stops at Spur Crater, Dune Crater, Hadley Plains, and area between Spur and Window craters. Numerous samples and photos were obtained and TV transmission was good. Crew completed heat flow experiment initiated during EVA-1, collected core sample, and deployed U.S. flag. After stowing sample container and film in LM, astronauts reentered LM at 149:27 GET, after 7 hrs 12 min on surface during EVA-2.

Third EVA began at 163:18 GET (4:52 am EDT Aug. 2), 1 hr 45 min later than planned because of cumulative changes in surface activities timeline. Late start and requirement to protect nominal liftoff time required shortening EVA. Alternate EVA plan was devised with traverse west from LM to Hadley Rille. Astronauts traveled to Scarp Crater, "The Terrace" near Rim Crater, and Rim Crater, collecting samples and photographing lunar surface features. Scott tripped over a rock

July 29–August 4: *The command and service modules were photographed in lunar orbit from the lunar module, giving a view of the scientific instrument module (SIM) bay, housing cameras and sensors to record data from the moon's surface and atmosphere.*

and fell, but experienced no difficulty in getting up. Astronauts retrieved drill-core-stem sections and samples near ALSEP, returned to LM, off-loaded LRV and stationed it for TV coverage of LM liftoff, and reentered LM at 168:08 GET, after 4 hrs 50 min—bringing total EVA time to 18 hrs 35 min.

While LM was on moon Worden, orbiting moon in *Endeavor*, completed lunar and astronomic photography and prepared for rendezvous. His observations included discovery of fields of cinder cones made by volcanic eruptions, delineation of landslide or rock glacier on northwest rim of crater on lunar farside, interpretation of ray-excluded zone around crater Proculus as result of fault system, and discovery of layers on interior walls of several craters, suggesting volcanic collapse craters of calderas in maria.

Scott and Irwin depressurized LM, discarded excess equipment, repressurized LM, and lifted off lunar surface with 77 kg (170 lbs) of lunar samples at 171:34 GET (1:11 pm EDT Aug. 2)—66 hrs 55 min after landing on moon. Liftoff, accompanied by USAF anthem "Off We Go Into The Wild Blue Yonder" taped and broadcast by astronauts from LM, was photographed in color by camera on LRV left on moon and was seen by millions of TV viewers. Spacecraft docked successfully at 173:36 GET, as TV viewers watched. Scott and Irwin transferred from LM to CSM with samples, equipment, and film. CSM–LM separation and LM jettison were delayed one revolution to verify that

CSM and LM hatches were completely sealed. LM ascent stage was jettisoned and intentionally crashed onto lunar surface at 26° 22′ north latitude and 15′ east longitude at 181:30 GET, 93 km (59 mi) west of *Apollo 15* ALSEP site. Impact was recorded by *Apollo 12, Apollo 14,* and *Apollo 15* seismometers.

Orbit-shaping maneuver was performed during 73rd lunar revolution in preparation for subsatellite launch. Scientific subsatellite was launched at 222:39 GET (4:13 pm EDT Aug. 4) into lunar orbit with 141.3-km (87.8-mi) apolune and 102.0-km (63.4-mi) perilune. Hexagonal, 36-kg (80-lb) satellite 79 cm (31 in) long and 36 cm (14 in) in diameter carried three experiments and was housed in SIM bay with seven other new experiments. Satellite extended three 1½-m (5-ft) booms, one with magnetometer to measure interplanetary and earth magnetic fields near moon and two for stabilization. Satellite also carried charged-particle sensors and equipment to detect variations in lunar gravity caused by mascons.

Transearth injection maneuver at 223:49 GET put CSM on trajectory for home after 85 hrs 18 min (74 revolutions) in lunar orbit.

August 5–7: At 241:58 GET (11:32 am EDT Aug. 5) Worden left CSM for inflight EVA to retrieve panoramic and mapping camera film cassettes from SIM on SM. He made three trips to SIM bay, two to retrieve cassettes and one to observe condition of instruments, reentering CSM after 38-min 12-sec EVA.

CM *Endeavour* separated from SM at 294:44 GET. Drogue and main parachutes deployed but one of three main parachutes partially closed during descent, causing harder landing than expected. CM splashed down in mid-Pacific about 10.1 km (5.5 nm) from recovery ship U.S.S. *Okinawa* at 295:12 GET (4:47 pm EDT Aug. 7)—12 days 7 hrs 12 min after liftoff. Astronauts, wearing fresh flight suits, were carried by helicopter to biomed area on recovery ship for post-flight examinations. After being declared in very good shape, astronauts were flown on following day to Hickam AFB, Hawaii, and to Ellington AFB, Tex. CM was retrieved and placed on board recovery ship. Lunar samples, data, and equipment were flown to Ellington AFB, and CM was off-loaded at San Diego.

Mission achieved primary *Apollo 15* objectives: to make selenological survey and sampling of materials in preselected area of Hadley-Apennine region; emplace and activate surface experiments; evaluate capability of Apollo equipment to provide extended lunar surface stay time, increased EVA operations, and surface mobility; and conduct inflight experiments and photographic tasks from lunar orbit. Launch vehicle and spacecraft systems performance were near nominal except for intermittent short circuit in Delta V thrust switch A, CSM–LM failure to undock properly, and increase in CM tunnel pressure preceding LM jettison. Flight crew performance was excellent. Accomplishments included first use of LRV, lunar surface navigation device, direct lunar communications without LM relay, and ground-controlled remotely operated TV camera on moon; largest payload in earth orbit (140 310 kg; 309 330 lbs) and largest payload in lunar orbit (33 803 kg; 74 522 lbs); longest lunar surface stay time, lunar surface EVA, distance traversed on lunar surface (28 km; 17.4 mi), and lunar orbit time (74 orbits); first subsatellite launched in lunar orbit; and largest amount of lunar samples brought to earth.

Apollo 15 was 12th Apollo mission to date, 9th manned Apollo mission, and 4th successful lunar landing mission. *Apollo 14* mission had been conducted Jan. 31–Feb. 3. Apollo program was directed by OMSF; MSC was responsible for Apollo spacecraft development, MSFC for Saturn V launch vehicle, and KSC for launch operations. Tracking and data acquisition was managed by GSFC under overall direction of OTDA. (NASA Proj Off; NASA Release 71–245; *NYT*, 7/26–8/7/71; *W Post*, 7/27–8/8/71; GSFC *SSR*, 7/31/71; NASA Special Release; NASA Release 71–119K; Lannan, W *Star*, 7/30/71, A1; AP, W *Star*, 8/1/71, A10)

July 26: President Nixon issued statement on successful *Apollo 15* launch: "The flight of Apollo 15 is the most ambitious exploration yet undertaken in space. Even as it reflects man's restless quest for his own future, so it also reenacts another of the 'deeper rituals of his bones'—not only the compulsion of the human spirit to know where we are going, but the primal need in man's blood to know from what we have come. We hope, by this journey, to know better the origins of Earth, the moon, and their other planets. We hope to understand something more of the mysteries of God's great work. And . . . we hope to understand more of man himself." (*PD*, 8/2/71, 1088)

August 7: Safe splashdown—with an extra bump. The Apollo 15 *command module, completing its 12-day lunar mission, descended to the waves on two of its three parachutes. The third parachute collapsed, increasing impact velocity to 35 km per hr (22 mph), but did not endanger the astronauts or the spacecraft.*

- House and Senate conferees submitted H. Rpt. 92–377 on H.R. 9382, FY 1972 HUD-space-science appropriations bill. Report recommended total NASA appropriation of $3.298 billion. Recommended NASA appropriation included $2.523 billion for R&D instead of $2.518 billion proposed by House and $2.542 billion proposed by Senate; $52.7 million for construction of facilities instead of $33.8 million proposed by House and $56.3 million by Senate; and $722.6 million for research and program management as proposed by Senate, instead of $720 million as proposed by House.

  Report also recommended $622-million NSF appropriation, $1.150 million below NSF budget request. (H Rpt 92–377)

- Rep. Joel T. Broyhill (R-Va.) introduced H.R. 10065 "to redesignate the Washington National Airport as 'Neil Armstrong Airport.'" Bill was referred to House Committee on Interstate and Foreign Commerce. (CR, 7/26/71, H7164)

- Union of Concerned Scientists—Boston area coalition of scientists, engineers, and other professionals—issued report saying AEC tests had confirmed unreliability of emergency cooling system designed to supply water to nuclear reactors but that AEC continued to license operation and construction of reactors containing system. Report said if accident occurred "resulting catastrophe and loss of life might well exceed anything this nation has seen in time of peace." (AP, NYT, 7/27/71, 53)

July 27: U.S.S.R.'s *Mars 2* (launched May 19) and *Mars 3* (launched May 28) were continuing to operate satisfactorily en route to Mars, Tass announced. By 6:00 am Moscow time (11:00 pm EDT July 26) *Mars 2* had completed 43 communications sessions and was 17 580 000 km (10 924 000 mi) from earth; *Mars 3* had completed 38 sessions and was 16 400 000 km (10 190 000 mi) from earth. Data from *Mars 2* and *Mars 3* measurements of corpuscular beams of solar and galactic cosmic radiation had confirmed data from same measurements conducted by *Lunokhod 1* lunar rover on moon. (Tass, FBIS–Sov–144–3, 7/27/71, L1)

- House agreed to conference report on H.R. 7109, FY 1972 NASA authorization bill, and sent measure to the Senate for further action. Report had recommended total authorization of $3.355 billion [see July 21]. (CR, 7/27/71, H7185–8, D768)

- Aerobee 170 sounding rocket was launched by NASA from WSMR carrying Univ. of Michigan experiment to study airglow. Rocket and instruments functioned satisfactorily. (SR list)

- UPI compared $400-million cost of *Apollo 14* mission (Jan. 31–Feb. 9) with $445-million cost of *Apollo 15* mission. Breakdown of *Apollo 15* cost was reportedly: CSM, $65 million; LM, $50 million; Saturn V booster, $185 million; ALSEP, $25 million; LRV, $8 million; lunar orbit science laboratory, $17 million; and operations, $95 million. Total estimated cost of entire Apollo lunar exploration program through Apollo 17 was $26.5 billion. (UPI, NYT, 7/27/71, 14)

- NASA's *Apollo 15* was 5351st man-made object launched since U.S.S.R.'s *Sputnik 1* Oct. 4, 1957, according to Air Defense Command Hq. figures reported by AP. Center was tracking 2454 objects still in orbit. (NYT, 7/27/71, 14)

July 28: U.S.S.R. launched *Molniya 1–18* comsat from Plesetsk to ensure "operation of a system of long-distance telephone and telegraphic radio communications, and also the transmission of programs of USSR central television to points of the 'Orbita' network . . . in the far north, Siberia, the Far East, and Central Asia." Orbital parameters: apogee, 39 340 km (24 444.7 mi); perigee, 995 km (618.3 mi); period, 707.2 min; and inclination, 65.3°. (GSFC *SSR*, 7/31/71; Tass, FBIS–Sov–71–145, 7/28/71, L1; *SF*, 1/72, 30)

- Senate agreed to conference report on H.R. 7109, FY 1972 NASA authorization bill that recommended $3.355-billion authorization. Bill was cleared for President's approval. (*CR*, S12328–9)
- NASA announced it would start Life Scientist Program, recommended by NAS, to increase cooperation between university and NASA life scientists in advancing disciplines related to NASA mission. Disciplines included medical, biological, behavioral, bioengineering, and life support engineering sciences as related to support of living systems in aeronautic and space operations, and exobiology—inquiry into existence of life outside universe, scientific origin of life, and planetary ecology. Program would start with selection of five scientists from different universities by evaluation of proposed investigations and their relevance to NASA's needs and interests. Scientists selected would spend one third of time with their graduate students at MSFC, ARC, or LaRC. Each would be awarded three-year grant on step-funded basis. (NASA Release 71–140)
- USN had been forced to drop plans for purchase of F–14B jet fighter aircraft—more powerful version of F–14 Tomcat fighters—because of continuing financial problems, which had already pushed estimated price to $16 million per aircraft, *Washington Post* reported. (Getler, *W Post*, 7/28/71, A1)
- Soviet biochemist Zhores Medvedev, in book smuggled to West, had said that, despite space achievements, science in U.S.S.R. was being throttled by bureaucracy and political ideology, AP reported. In *Plight of Soviet Science Today* Medvedev had said system willing to let only certain scientists travel abroad harmed U.S.S.R. (AP, *NYT*, 7/29/71, 37)

July 29: USAF X–24A lifting-body vehicle had completed flight-test program and would be reshaped and renamed X–24B, FRC announced. Decision to end flight operations and reshape vehicle had been made by joint NASA–USAF flight research team.

X–24A was one of three lifting-body vehicles in joint NASA–USAF program managed since 1967 by FRC. Vehicle had made 28 flights—10 glide and 18 powered—since beginning test program April 17, 1969, reaching 1687 km per hr (1048 mph) and 22 000-m (71 000-ft) altitude. New X–24B would be built around existing basic X–24A structure, but outward appearance would be changed dramatically, with sharply pointed flat-iron shape replacing bulbous one. New shape would also have improved hypersonic lift-to-drag ratio that would permit investigation of how well vehicle designed for good hypersonic performance would perform at low supersonic, transonic, and subsonic speeds, with emphasis on landing. Other changes would include 4.6-m (15-ft) extension of nose, removal of rear center fin, 1.5-m (5-ft) widening in rear, and 680-kg (1500-lb) increase in weight. Modifications

would be completed by Martin Marietta Corp. in about one year. (FRC Release 18–71)
- NASA announced selection of Martin Marietta Corp. to receive $12-million, award-fee contract to support JPL in design and development of major subsystems of Viking Orbiter. Martin Marietta would assign engineers and technicians to JPL and provide major subassemblies for propulsion system. (NASA Release 71–143)
- House, by vote of 362 to 30, agreed to conference report submitted July 26 on H.R. 9382, FY 1972 HUD-space-science appropriations bill that recommended NASA appropriation of $3.298 billion. (*CR*, 7/29/71, H7363–72)
- U.K. Aerospace Minister Frederick Corfield told Parliament Black Arrow launch vehicle would be abandoned after final launch in 1971. Project, which cost $27.6 million and took six years to develop, was U.K.'s only satellite launching program. (AP, W *Star*, 7/30/71)
- Moscow *Trud* published account of test flight by Soviet pilot Oleg Gudkov of new supersonic fighter aircraft to ascertain whether aircraft could be brought out of spin. Test had been successful. (FBIS–Sov–71–149, 8/3/71, M1)
- DOD spokesman Jerry W. Friedheim said next round of 48 F–14 Tomcat fighter aircraft for USN probably would not include improved "B" version but denied July 28 *Washington Post* report that USN had dropped plans to purchase F–14Bs. Friedheim said in press interview that new engine to power F–14B had fallen behind schedule and that lag in production had affected decision on next 48 aircraft. When engine became available, "assessment will be made on proceeding with the F–14B." (*W Post*, 7/29/71, A7)
- First two DC–10 jet transport aircraft were delivered by manufacturer McDonnell Douglas Corp. to American Airlines and United Airlines at Long Beach, Calif. (Bedingfield, *NYT*, 7/30/71, 54)
- House, by vote of 306 to 98, agreed to Senate amendment to H.R. 9667, $2.9-billion FY 1972 DOT appropriations bill that included $58.5 million to repay airlines for money they invested in SST. (*CR*, 7/29/71, H7373–87)
- *Washington Post* editorial criticized NASA agreement with New York Times Special Features, syndicate division of *New York Times*, for byline articles by *Apollo 15* astronauts [see July 23] and reprinted excerpt from Sept. 19, 1963, *New York Times* editorial that had scored earlier NASA agreement with publications. *Post* said: ". . . we would like to record . . . some reservations about the practice of commercializing the space program by permitting any of its official participants to sell personal accounts of their experiences. Strictly in legal terms, there is not much doubt that such material is subject to copyright protection. But, leaving aside regulations, there is a question of propriety, first raised back in 1963, when the Field Enterprises Educational Corp. and Life Magazine nailed down contracts totaling $1 million for the personal stories of 16 astronauts over a four-year span." *New York Times* editorial in 1963 had said sale of astronauts' stories represented "stain of commercialism on the record of the space program," and that astronauts "should not be permitted to reap enormous private profits from outside sources on the basis of their participation in a great national effort." (*W Post*, 7/29/71, A14)

- Dept. of Commerce issued *Metric America* (SP-345). Report said U.S. was only major nation that had failed to convert to metric system and urged Congress to effect conversion within 10 yrs. (AP, B *Sun*, 7/30/71, A1; DOC PIO)

July 30: President Nixon telephoned congratulations and good wishes to ground and flight crews of *Apollo 15* following moon landing. (*PD*, 8/2/71, 1099)

- U.S.S.R. launched *Cosmos 431* from Baykonur into orbit with 284-km (176.5-mi) apogee, 165-km (102.5-mi) perigee, 89-min period, and 51.7° inclination. Satellite reentered Aug. 11. (GSFC *SSR*, 7/31/71; 8/31/71, *SBD*, 8/2/71, 148)

- Apollo missions plus Surveyor, Orbiter, and Soviet missions had permitted first detailed study of formative stages of planetary body, UCLA scientist Dr. George W. Wetherill said in *Science* article. "Future work will require sampling distinctly different regions of the moon in order to provide data concerning other important lunar events, such as the time of formation of the highland regions and of the mare basins, and of the extent to which lunar volcanism has persisted subsequent to the first third of lunar history. This work will require a sufficient number of Apollo landings, and any further cancellation of Apollo missions will jeopardize this unique opportunity to study the development of a planetary body from its beginning. Such a study is fundamental to our understanding of the earth and other planets." (*Science*, 7/30/71, 383-92)

- House, by vote of 192 to 189, passed H.R. 8432, to authorize emergency loan guarantees to major business enterprises. Bill would provide for $250-million loan to Lockheed Aircraft Corp. to stave off bankruptcy, *New York Times* said. (*CR*, 7/30/71, H7453-520; Shanahan, *NYT*, 7/31/71, 1)

- B/G Benjamin N. Bellis (USAF), project manager for F-15 jet fighter aircraft and for F-14 and F-15 joint engine development, said in interview with Washington *Evening Star* that USN had instructed him to drop plans for construction of F-14B engine in FY 1972 and 1973 and to assume USN would not want new engines until at least FY 1974. DOD spokesman Jerry W. Friedheim later told *Star* that top DOD officials insisted no final decision against proceeding with advanced version of F-14 had been made. (Kelly, W *Star*, 7/29/71, 4)

- White House announced Administration had requested $112-million supplemental appropriation for DOD to pay for 12 additional F-111 fighter bombers. Washington *Evening Star* later said this would be sufficient to keep production line at General Dynamics Corp. Fort Worth, Tex., plant open. It had been scheduled to close in spring of 1972. (W *Star*, 7/31/71, A13)

- President Nixon formally changed name of Presidential Boeing 707 aircraft from *Air Force One* to *The Spirit of '76*. Change was made at request of David J. Mahoney, Chairman of American Revolution Bicentennial Commission. *Air Force Two* would retain its name. (AP, W *Post*, 7/31/71, A2)

- FAA Administrator John H. Shaffer announced FAA was installing new computer-assisted weapon detection system at one gate of Dulles International Airport in Virginia to keep potential hijackers from boarding commercial flights. System automatically screened passengers at rate of 20 per min as they passed through electro-magnetic fields that detected metal. Information was evaluated by computer that could discriminate between different metal objects on basis of weight. Alarm sounded when excess metal or gun was detected. (FAA Release 71-126)

- *New York Times* editorial commented on resignation of Dr. Glenn T. Seaborg as AEC Chairman and appointment of Dr. James R. Schlesinger to succeed him: "In an earlier era the nation looked to the A.E.C. primarily to produce the nuclear weapons on which national defense policy rested. Today the emphasis has shifted to the task of generating vast amounts of nuclear energy for present and future needs, to do so economically and in a manner that safeguards the environment, and it is in this area particularly that Dr. Schlesinger will be expected to throw light and show leadership." He could "take another hard look at the prospects for controlled thermonuclear fusion as an energy source. Substantially increased research funds invested in this area might enable this country in the decades ahead to increase power production and consumption greatly at minimum cost to the environment." (*NYT*, 7/30/71, 30)

July 31: President Nixon, at ceremonies marking dedication of Rathbun Dam in Centerville, Iowa, referred to *Apollo 15* astronauts: "At this very hour, if you were watching television you could see that fantastic ride in a golf cart on the moon." *Apollo 15* astronauts "look down on earth. What does it look like? Well, we've had descriptions from other astronauts and often they have remarked about the fact that from far out there the earth looks like a very peaceful planet." (*PD*, 8/9/71, 112-5)

- *New York Times* editorial commented on *Apollo 15* lunar landing: "Today, tomorrow and Monday will come the payoff for the years of work by the astronauts and thousands of others who made possible last night's success. The trek across the moon's surface these next three days will be the longest both in time and in distance to date, partly because of the availability of the lunar rover to permit moon rides as well as moon walks. The vast variety of geological conditions in the neighborhood of the Falcon's landing site suggests that these excursions may be extremely fruitful." (*NYT*, 7/31/71)
- NASA launched two Nike-Apache sounding rockets from ESRANGE, Kiruna, Sweden, to sample noctilucent clouds and particulate matter in upper atmosphere. First payload was launched to altitude of 114 km (70.8 mi) and recovered within 2 hrs 15 min after launch in good condition. Mission was considered 100% successful.

  Second payload attained 106 km (65.9 mi) and was recovered within 2 hrs of launch. Two of eight collector doors opened prematurely and caused slightly reduced apogee, but experiment results were not affected and mission was judged 100% successful. (NASA Rpts SRL)

During July: Sen. Howard W. Cannon (D-Nev.) praised U.S. space shuttle program in *Aerospace* article: "Logic dictates that we should move ahead with the shuttle program. Experience tells us that the aerospace industry can respond to even seemingly impossible challenges—and that nations that reject the major challenges of their times do not remain major powers. And a hard fact of international economics is that the United States can remain prosperous in this world only by maintaining and utilizing its advanced technology." (*Aerospace*, 7/71, 3–7)

- First book taken from earth to moon was not microfilmed Bible carried by Astronaut Edgar D. Mitchell on *Apollo 14* (launched Jan. 31, 1971) as had been reported by UPI, Robert H. Goddard Library publication *Goddard Biblio Log* said. Astronaut Edwin E. Aldrin, Jr., had carried miniature copy of *Robert Hutchings Goddard—Father of the Space Age* on *Apollo 11* (launched July 16, 1969, 45 years after Goddard's first liquid-fueled rocket flight) but had not been allowed by NASA to leave book on moon. *Apollo 11* astronauts had given three-by two-inch book, published by Achille J. St. Onge, to Mrs. Esther Goddard, who had donated it to Goddard Library at Clark Univ. (*Goddard Biblio Log*, 7/71, 11–2)

- After more than quarter century of dominating international aviation scene, U.S. was "well on its way to becoming, if not an also-ran, just another contender for the trophies of the world's aviation markets," John F. Loosbrock said in *Air Force Magazine* article. "For indicators, one only had to stroll the display line, watch the flying demonstrations, and eavesdrop at the chalets" at Paris Air Show. Bitterest pill was "lack of American presence in the supersonic-transport competition, while the Soviet TU-144 and the Anglo-French Concorde were impressing customers, the press, the public. . . ." Equally important was "dazzling array of commercial and military aviation products" displayed by European suppliers. For U.S. to forge ahead again in aviation and advanced technology generally it was going to require "new ways of looking at the problem, new ways of doing things. We are pricing ourselves out of the market, not only for what we hope to sell abroad, but even for what we would like to sell to ourselves." (*AF Mag*, 7/7, 6)

August 1971

August 1: Pope Paul VI told crowd assembled at Castel Gondolfo that he had sent confidential message to *Apollo 15* astronauts on their July 26–Aug. 7 lunar mission. Astronauts had replied: "Our journey is for all men and we hope that the talents which God has given us will enable us to satisfy those who so kindly support our undertaking." (AP, W *Star*, 8/2/71, A2)

- Number of accredited newsmen covering *Apollo 15* mission had dwindled to 402 from high of 1200 for *Apollo 11*, UPI noted. *Apollo 12* had been covered by 700 newsmen and *Apollo 14*, by 600. (W *Star*, 8/1/71, A10)
- Cuban internal radio newscasts gave detailed account of *Apollo 15* mission. Radio RELOJ, all-news station in Havana, read reports of astronauts' activities on moon. (*NYT*, 8/3/71)
- Twenty-fifth anniversary of AEC. In statement commemorating event Dr. Glenn T. Seaborg, outgoing AEC Chairman, said, "The primary mission of the Atomic Energy Commission to develop and use atomic energy to improve the public welfare, to promote world peace and to assure the common defense and security has been a success." (AEC Release 0–131)

August 2: Twin eight-cent stamps commemorating decade of U.S. achievement in space were issued during ceremonies at KSC, MSC, and MSFC in conjunction with *Apollo 15* mission. Stamps, depicting *Apollo 15* lunar landing and LRV carrying *Apollo 15* astronauts, later went on sale throughout U.S. (MSFC Release 71–132; *NYT*, 7/18/71, 24; W *Star*, 8/1/71, B7; MSFC PAO)

- Interview of Dr. Rocco A. Petrone, Apollo Program Director, was published by *U.S. News & World Report*. Dr. Petrone said: "Many people think Apollo had only a singular goal—that is, land a man on the moon and return him safely to earth. That was not the only goal of Apollo. Very clearly, in President Kennedy's address to the Congress in May of 1961, a landing on the moon was chosen as a focal point. The intention was that the U.S. would become pre-eminent in space—learning to operate, I believe as President Kennedy said, on a new ocean, a new frontier." Space research was "definitely relevant. Many of the problems we're trying to solve have to do with the environment. We've already made a head start in that field with our work on spaceships, which requires that we carry the environment with us. We have got to purify the gases we have, we've got to remove odors, we have to restore the oxygen we use. . . . We are learning how to rehabilitate an environment. These tools, these techniques, these understandings will have relevance here on earth." (*US News*, 8/2/71)
- Newspapers commented on *Apollo 15* mission, while astronauts were exploring lunar surface.

Christian Science Monitor editorial praised *Apollo 15*: "Once again, a superbly synchronized lunar mission has proven man's capacity to focus the knowledge of hundreds of specialists to achieve a singular goal. Apollo 15 is one more grand historic achievement. It testifies to the benefit that science and technology, properly harnessed, can reap for mankind." (*CSM*, 8/2/71)

French novelist and TV critic Jean Dutourd commented on *Apollo 15* TV coverage in *France Soir*. It had made him walk away from TV set. "It must be confessed," he wrote, "that these experiments are an unspeakable bore." (Reuters, *W Post*, 8/3/71, A11)

New York Times editorial defended manned space flight: "After Apollos 16 and 17 next year, there are no known plans of any nation to send any more men to the moon. A vast and complex technology developed at the cost of many billions of dollars over the last decade is being abandoned even as its vast potentialities are being demonstrated. If American resources do not permit manned lunar exploration beyond the end of next year, the United Nations should take over the task, appealing to all governments for funds and soliciting contributions as well from private individuals and private business enterprises. The dusty planet that is the moon could become the planet of human cooperation, a new frontier of hope for all mankind." (*NYT*, 8/2/71, 22)

- Senate approved conference report submitted July 26 on H.R. 9382, FY 1972 HUD-space-science appropriations bill that included $3.298-billion NASA and $619-million NSF appropriations. Bill was cleared for President's approval. (*CR*, 8/2/71, S12832–4; *PD*, 8/16/71, 1157)

- Louis H. Brennwald, former Vice President of Northrop Aircraft Corp. in charge of Puerto Rico operations, became Director of Administration for Ames Research Center. (ARC *Astrogram*, 9/2/71, 1; ARC PIO)

- Senate by vote of 49 to 48 passed and cleared for President's approval H.R. 8432, authorizing $250 million for emergency loan guarantees to major business enterprises. Legislation ended immediate threat of bankruptcy to Lockheed Aircraft Corp. and assured Lockheed adequate funds to complete construction of L–1011 TriStar jet transport aircraft, *Washington Post* later said.

President Nixon issued statement following Senate action: "The Senate's decision to approve a government loan to the Lockheed Corporation is in the best interests of all the American people. I greet this vote—as I greeted the vote in the House of Representatives last week—with gratitude and deep satisfaction. This action will save tens of thousands of jobs that would otherwise have been eliminated. It will have a major impact on the economy of California, and will contribute greatly to the economic strength of the country as a whole. It will help ensure that the Nation's largest defense contractor, and its largest airframe manufacturer, will continue serving the Nation's needs. It will also help ensure that this country will continue to play a leading role in the development of aerospace technology." (*CR*, 8/2/71, S12823–63; Samuelson, *W Post*, 8/3/71, A1; *PD*, 8/9/71, 116–7)

August 2–6: U.S. and Soviet scientists met in Moscow to recommend joint projects and exchanges in near-earth space research, investigations of moon and planets, development of space meteorology, and space applications related to natural environment. Recommendations of three working groups representing NASA and Soviet Academy of Sciences would be forwarded to Academy President Mstislav V. Keldysh and to Dr. George M. Low, NASA Deputy Administrator, for consideration. U.S. participants were led by Dr. John E. Naugle, NASA Associate Administrator for Space Science and Applications; Deputy Associate Administrator (Science) Leonard Jaffe; and NOAA's Director of National Environmental Satellite Services David S. Johnson. U.S.S.R. participants were headed by Dr. Boris N. Petrov, Chairman of Council for International Cooperation in Space Research of Soviet Academy of Sciences; Vice President Aleksander P. Vinogradov of Soviet Academy; and Deputy Chief L. A. Aleksandrov, Deputy Chief for Technology, Main Administration of Hydrometeorology Services, U.S.S.R. Council of Ministers. (NASA Release 71-147; NASA Int Aff)

August 3: Astronaut-geologist Harrison H. Schmitt praised *Apollo 15* astronauts in MSC briefing for their completion of EVA geological activities: "I don't think there is any question that we had a fantastic exploration mission. There's just no question in my mind we sent two very competent observational scientists to the moon." Samples collected, Schmitt said, "sounded very much like . . . crystalline rocks . . . that may well, and I emphasize may . . . represent the early lunar crust. I am not sure until we see the samples that we're much closer to understanding the origin of the rille than when we started, except that there probably are some origin theories that can be eliminated now.

"I think that the net result of the mission was to indicate that our preliminary thinking about the moon . . . still looks like a valid picture, it's a valid foundation [on] which to attempt to further select sites and to further approach the analysis of rocks and soils that are brought back." (Transcript; AP, B *Sun,* 8/4/71, A6)

- *Apollo 15* LM *Falcon's* liftoff from moon was described by Tom Wicker in *New York Times*: "The first televised liftoff of a manned spacecraft from the moon had none of the beauty and power of those moments when a huge Saturn rocket rises majestically into space from Cape Kennedy. The quick pop that sent off the Falcon in a shower of sparks looked more like something held over from the Fourth of July, and the part of the landing craft left on the moon was all too suggestive of an auto graveyard." Two years after first moonwalk, "Americans do not . . . seem nearly so interested as they were in these prodigious feats of modern man. Even in living color and with that great American love object—an auto, of sorts—to liven up the scene, the latest moon show seemed to need something. Unfortunately the LEM liftoff didn't turn out to be it." (*NYT,* 8/3/71, 27)

- U.S.S.R.'s *Lunokhod 1* lunar rover (landed on moon by *Luna 17* Nov. 17, 1970) was powered up to begin experiments for 10th lunar day. (UPI, *W Post,* 8/4/71, A6)

- *New York Times* editorial commented on *Apollo 15*: "All the evidence suggests that this is the most productive scientific expedition ever carried out in space. Even before the astronauts' own return, the information they have sent back has already forced reconsideration of some key ideas about the natural history of the moon. In ultimate perspective, however, it may turn out that the dividends from the Falcon's stay on the moon will be as substantial in the realm of changed human ideas and aspirations as in scientific knowledge. For millions who watched the superb television pictures of the two astronauts working comfortably on the lunar surface, the image that came across was that of a friendly moon, a place where men could walk and ride and work with ease. This is a far cry from the traditional picture of the moon as a hostile, alien environment totally inhospitable to man." (*NYT*, 8/3/71)

- GAO investigation of NASA selection of NR Rocketdyne Div. for prime contract for final design and manufacture of reusable space shuttle [see July 13] was requested by Bruce N. Torell, President of Pratt & Whitney Div. of United Aircraft Corp., in telegram to Elmer B. Staats, U.S. Comptroller General. Bases for protest were that source selection had disregarded objective of the RFP, NASA had failed to conduct "written or oral discussions" on proposals, Pratt & Whitney felt its proposal was "clearly entitled to a superior technical evaluation to the technical proposal of Rocketdyne," and NASA had failed to give proper consideration to "Pratt & Whitney's test-proven flight weight design and greater experience."

  In letter to Dr. James C. Fletcher, NASA Administrator, nine senators requested that no award of space shuttle engine contract be made until GAO investigation had been completed. (*CR*, 8/6/71, S13789)

- Arcas sounding rocket was launched by NASA from Wallops Station carrying Univ. of Pennsylvania Ionospherics experiment. Rocket and instruments functioned satisfactorily. (SR list)

- Twenty-fifth anniversary of invention of computer by Dr. J. P. Eckert and Dr. John W. Mauchly of Univ. of Pennsylvania. *New York Times* editorial Aug. 9 commented on anniversary: "Man's invasion of space, up to and including Apollo 15's latest feats, would have been impossible without computers. So would much of the rest of modern science and technology." (Smith, *NYT*, 8/4/71, 43; *NYT*, 8/9/71, 28)

- Dr. Georgy N. Babakin, prominent Soviet space scientist who had designed equipment for unmanned Soviet spacecraft, died at age 56. *Izvestia* obituary said Dr. Babakin had been recipient of Hero of Socialist Labor, highest Soviet civilian title, and corresponding member of Soviet Academy of Sciences and had made "a great contribution to the development of Soviet space technology and the study of the moon and Venus." (*NYT*, 8/5/71, 34)

- Of 80 new underground missile silos being constructed by U.S.S.R., at least 6 had been started after U.S. and U.S.S.R. announced May 20 that they would seek to limit construction of nuclear weapons, *Washington Post* quoted sources as saying. (Getler, *W Post*, 8/3/71, A1)

August 4: NASA issued RFPs to major airframe manufacturers for design and fabrication of two experimental transport jet STOL aircraft to start Quiet-STOL program to relieve airport noise and congestion. Aircraft, incorporating propulsive lift principles, would provide technological data and experience for developing environmentally acceptable, economical, and safe fan-jet STOL transport system in NASA–DOT–FAA–USAF program. Program would implement major recommendations of joint NASA–DOT Civil Aviation Research and Development (CARD) Study by providing industry with basis for STOL development, Government regulatory agencies with rules for STOL certification and operation, and USAF with modernized tactical airlift capability. (NASA Release 71–146)

- Committee for the Future—group of wealthy private citizens—was seeking to finance return trip to Hadley Rille, *Apollo 15* landing site, with surplus Apollo spacecraft, Reuters reported in Baltimore *Sun*. Group hoped to recover contemplated $400-million investment by selling moon rocks, TV film, and photographic and literary rights, and by charging for scientific experiments conducted. Group had announced proposal to operate mission "Harvest Moon" after final Apollo flight in 1972 because it seemed incredible that man could invest $25 billion and 10 yrs to reach moon and then "stop without . . . looking at the moon's value to man and his future." Members of committee had spoken informally with NASA officials, Reuters said. Officials had said privately that plan stood little chance of approval. (B *Sun*, 8/4/71, A6)
- NASA, National Weather Service and Environmental Data Service spokesmen, questioned by *Washington Post*, said their agencies had received complaints from public that *Apollo 15* mission had caused heavy rainfall over Washington, D.C., area. They denied that mission was responsible and that rains accompanying *Apollo 13* and *14* missions had been anything more than coincidence. (Cohen, *W Post*, 8/5/71, B1)
- Senate Committee on Armed Services unanimously approved H.R. 8687, $21-billion FY 1972 military procurement authorization, decrease of $100 million over House-passed figure of $21.1 billion. Committee, at Administration's request, restored $801 million for procurement of 48 F–147 carrier-based fighter aircraft built for USN by Grumman Aerospace Corp. (*CR*, 8/4/71, D824; Finney, *NYT*, 8/5/71, 11)

August 4–13: ARC's specially equipped Convair 990 jet aircraft *Galileo* carried 29 scientists and 4500 kg (10 000 lbs) of instruments on series of flights from Hickham AFB, Hawaii. Purpose of flights was to investigate Mars' invisible infrared light radiations while flying above 99% of earth's occluding atmospheric water vapor during Mars' closest approach to earth since 1671. Flight path was 869 km (540 mi) south of Hawaii along 13° north latitude meridian. Mars was about 56 million km (35 million mi) from earth. *Galileo* also carried vector magnetometer to investigate continental drift and sea floor spreading in ARC experiment. Expedition was sponsored by ARC Airborne Science Office. (NASA Release 71–145; UPI, *NYT*, 8/7/71, 8; ARC *Astrogram*, 8/19/71, 1)

August 4–20: Dr. Glenn T. Seaborg, outgoing AEC Chairman, and group of U.S. nuclear scientists toured peaceful nuclear energy facilities in U.S.S.R. at invitation of Chairman Andronik Petrosyants of Soviet State Committee for Nuclear Energy. (AEC Release 0–126; AEC PIO)

August 5: U.S.S.R. launched *Cosmos 432* from Baykonur into orbit with 252-km (156.6-mi) apogee, 200-km (124.3-mi) perigee, 89.1-min period and 51.7° inclination. Satellite reentered Aug. 18. (GSFC *SSR*, 8/31/71, *SBD*, 8/9/71, 179)

- ESRO and FAA announced in Madrid agreement on program for use of satellites to control civil air traffic. At least four satellites would be launched over Atlantic and Pacific beginning in 1975. System would be used for air-to-air communications and as navaid to avoid midair collisions. Committee to administer program would be appointed from ESRO, FAA, and representatives of Australia, Canada, Ireland, Philippines, Japan, and Portugal. (Reuters, B *Sun*, 8/6/71, A2)

- British aviatrix Sheila Scott arrived at London Airport in Piper Aztec aircraft to complete 55 000-km (34 000-mi) solo flight during which she participated in NASA *Nimbus 4* experiment [see June 1]. Miss Scott claimed seven records. (*W Post*, 8/6/71, A6)
- First commercial flight of McDonnell Douglas DC-10 jet transport aircraft was flown by American Airlines between Los Angeles and Chicago. More than 80% of seats were occupied during 3 hr 18 min flight. (*CR*, 8/6/71, E9096–7; Yarborough, W *Star*, 8/18/71, D10; AAL PRO)
- Soviet engineer O. Sashin described in *Sotsialisticheskaya Industriya* activities of U.S.S.R.'s *Lunokhod 1* lunar rover, landed on moon by *Luna 17*, Nov. 17, 1970: At time of launch "it was supposed that the whole . . . program could be fulfilled in 3 lunar work days. . . . Over these three days Lunokhod-1 . . . visited a multitude of craters, repeatedly surmounting rock fields. . . . In all, during the 3 working days envisaged by the initial program, 42 radio communications sessions were held with Lunokhod-1, during which 8,924 radio commands were issued and fulfilled. . . . the route covered by the apparatus amounted to 3,551 meters [3883 yards], the television cameras transmitted to earth more than 80 high-quality panoramas of the surrounding locality and the scientists obtained vast information about the moon. The fourth working day passed, then a fifth, and a sixth. . . . Each was succeeded by a 2-week lunar night with all its 'charms' including an outside temperature which drops to below minus 150 degrees. But the next morning came and Lunokhod-1, roused by a radio command from earth, obediently set about its work.

 "The work of the self-propelled lunar scout . . . continues. . . . 132 radio communications sessions have been held . . . in which 22,792 radio commands have been given. A route 10,226 meters [11 183 yards] . . . over the lunar surface has been covered. The 10th working day on the moon has arrived. And again the indefatigable scout [is] continuing its search." (FBIS–Sov–158–3, 8/16/71, L1–3)
- Results of Harris survey of 1614 U.S. households on priority areas for cuts in Federal funding were published in *Washington Post*. Top three candidates for cuts were Vietnam war, foreign aid, and space program; 50% of those polled favored cut in space program; 13% would cut program least of Federal spending areas listed. Support for spending in education, urban aid, and environmental areas, Harris said, "appears to be on the rise." (Harris, *W Post*, 8/5/71)

August 6: USAF launched nine satellites from Vandenberg AFB by one Atlas booster at 5:11 pm PDT (8:11 pm EDT). *Ov 1–20* Orbiting Vehicle—carrying energetic proton analyzer, particle energy and flux thermal detector, and UV solar radiation experiment—entered orbit with 1932.8-km (1201-mi) apogee, 136.8-km (85-mi) perigee, 105.9-min period, and 92° inclination and reentered Aug. 28. *Ov 1–21*—carrying velocity mass spectrometer, atmospheric composition sensor, and ELF/VLF antenna effects transceiver—entered orbit with 917.3-km (570-mi) apogee, 788.6-km (490-mi) perigee, 101.9-min period, and 87.6° inclination. *Cannonball 2* (OAR 901) carried accelerometers to model atmospheric densities into orbit with 1794.4-km (1115-mi) apogee, 130.4-km (81-mi) perigee, 104.2-min period, and 92° inclination. It reentered Jan. 31, 1972. *Musketball* (OAR 907) carried instruments to measure density variations into orbit with 653.4-km (406-mi) apogee, 130.4-km (81-mi) perigee, 92.3-min period, and 87.6° inclination. It reentered Sept. 19.

Five remaining satellites were ejected from *Ov 1-21*. RTD 701, which would serve as calibration target, entered orbit with 915.7-km (569-mi) apogee, 774.1-km (481-mi) perigee, 101.7-min period, and 87.6° inclination. Four AVL 802 satellites would gather aerodynamic data for use in design and development of orbiting vehicles. Orbital parameters: apogee, 917.3 km (570 mi); perigee, 762.8–791.8 km (474–482 mi); period 101.6–101.9 min; and inclination, 87.5°–87.8°. (Pres Rpt 72; *SBD*, 8/10/71, 187; GSFC *SSR*, 1/31/72)

- President Nixon approved H.R. 7109, $3.384-billion NASA FY 1972 authorization bill that became P.L. 92–68. Total authorization was $84 million above budget request of $3.300 billion and $86 million above FY 1971 new obligational authority (NOA) of $3.298 billion. Bill authorized for R&D $2.603 billion, increase of $85.5 million over budget request of $2.518 billion and decrease of $90 million from FY 1971 authorization of $2.693 billion; for construction of facilities, $58.4 million, up $2.1 million from budget request of $56.3 million and $24 million above FY 1971 authorization of $34.5 million; and for research and program management, $723 million, down $4 million from budget request of $727 million but $40 million above $683.3 million authorized in FY 1971.

    Bill included $29.3-million budget amendment to cover Federal pay increases. It added $30 million to budget request for Skylab and Space Shuttle programs, $43 million for nuclear power and propulsion, $2.5 million for earth resources survey, and $12.5 million for aeronautical research and technology. It reduced requested authorization for lunar and planetary exploration by $10 million. (*PD*, 8/9/71, 1136; NASA Budget Off; *A&A 1970*)

- Final communications session of 10th lunar day was held by Soviet ground controllers with *Lunokhod 1* lunar rover on moon, Tass announced. Vehicle had traveled 215 m (235 yds) during 10th lunar day, bringing total distance covered since landing on moon Nov. 17, 1970, on board *Luna 17* to 10 452 m (11 430 yds). (FBIS-Sov-159-3, 8/17/71, L1)

- NSF announced $51 000 grant to Emory Univ. to develop and evaluate experimental program to reorient former defense-aerospace scientists and technologists into research on societal problems. (NSF Release 71-177)

- Daniel J. Haughton, Chairman of Lockheed Aircraft Corp., told press in London that he hoped L-1011 TriStar airbus and Rolls-Royce RB-211 engine program would be revived in "matter of weeks." Lockheed needed more orders to reach estimated breakeven figure on TriStar sales of from 255 to 265 aircraft. (AP, *NYT*, 8/7/71, 28)

- Senate confirmed nomination of Dr. James R. Schlesinger and William O. Doub as members of AEC. (*CR*, 8/6/71, D845)

August 7: President Nixon issued statement following successful splashdown of *Apollo 15* spacecraft: "Apollo 15 has returned safely to Earth. It brings with it new proof of man's invincible will to master the moment of his own destiny. It brings new tribute to man's unquenchable thirst for understanding. It brings the seeds of scientific understanding which may open a new chapter in the history of our planet. We salute Astronauts Scott, Irwin and Worden, who have stood poised on behalf of mankind on the brink of man's new world." (PD, 8/16/71, 1154)

- Delegation of congressmen watched Apollo splashdown from onboard recovery ship for first time. Congressmen—prevented from attending previous manned space splashdowns by limited space on ships and expense of flying them from Washington, D.C., to mid-Pacific—were flown to U.S.S. *Okinawa* from Hawaii by helicopter. Congress was not in session. Watching *Apollo 15* splashdown were Rep. Olin E. Teague (D-Tex.), Chairman of House Committee on Science and Astronautics' Subcommittee on Manned Space Flight; Rep. Edward P. Bolane (D-Mass.); Rep. J. Edward Roush (D-Ind.); Rep. Richard T. Hanna (D-Calif.); Rep. Joseph M. McDade (R-Pa.); and Rep. Charles A. Mosher (R-Ohio). (*W Post*, 8/8/71, A10)

- U.N. Secretary General U Thant said in message to President Nixon that *Apollo 15* was "another brilliant chapter in man's peaceful exploration of outer space." (Reuters, *NYT*, 8/8/71)

- *Philadelphia Inquirer* editorial praised *Apollo 15* crew: "Their workmanlike approach to arduous tasks, combined with fascinating descriptive commentary, gave Scott and Irwin good credits for showmanship as well as technical expertise. And the superb orchestration from mission control in Houston with Joseph Allen directing the score with the precision of a maestro on the podium, added an esthetic touch to a performance that should win raves from drama critics as well as space buffs. All concerned with Apollo 15 rate the nation's applause for an almost flawless performance thus far." (*P Inq*, 8/7/71)

- Arcas sounding rocket launched by NASA from Ft. Greely, Alaska, carried GSFC experiment to obtain ozone measurements in conjunction with *Nimbus 4* satellite overpass. Because of problems with ground tracking equipment no telemetry signal was received from payload until rocket reached 27-km (16.8-mi) altitude on descent side of apogee. Radar did not acquire rocket or ejected payload and payload was not recovered. Flight, first in three-flight series, was rated partially successful. Second and third rockets in series would be launched Aug. 9 and 14. (NASA Rpt SRL; SR list)

August 8: U.S.S.R. launched *Cosmos 433* into orbit with 299-km (185.8-mi) apogee, 112-km (69.6-mi) perigee, 88.6-min period and 49.4° inclination. Satellite reentered Aug. 9. (GSFC *SSR*, 8/31/71)

- *Apollo 15* Astronauts David R. Scott, James B. Irwin, and Alfred M. Worden were greeted by crowd of 3500 persons during brief stopover in Hawaii en route to Ellington AFB, Tex. On arrival in Texas astronauts were cheered by 2500 persons who waved banners and flags while band played "Fly Me to the Moon." Astronauts were first to return from moon without being required to undergo quarantine. (Stevens, *NYT*, 8/9/71, 1; AP, *C Trib*, 8/10/71)

- Soviet President Nikolay V. Podgorny had sent congratulations to President Nixon on successful completion of *Apollo 15* mission, Tass reported. Podgorny had asked President Nixon to convey congratulations and good wishes to "courageous" *Apollo 15* astronauts. (FBIS–Sov–71–153, 8/9/71, G1)

- Newspapers published preliminary summary of *Apollo 15*'s scientific and technical achievements. Thomas O'Toole said in *Washington Post*: "The voyage of Apollo 15 did far more than extend man's reach into the heavens. It showed dramatically that man belongs in space. At the end of their three days on the moon and six in lunar orbit, the astronauts . . . had gathered more scientific information about the moon than all the manned and unmanned spacecraft that had flown before them."

    Cameras and instruments in SIM had made first extensive observations from lunar orbit. One of first findings was confirmation that moon was not perfectly round and was dimpled. Evidence from seismometer was backed up by bistatic radar in which spacecraft signals were used to penetrate lunar surface to core. Seismometer's reception of LM's impact on moon indicated that even small explosion or impact could be recorded at extreme range and verified earlier hypothesis that meteorite impacts were being recorded daily from all over lunar surface. Addition of *Apollo 15* station to existing seismic net completed triangular array of instruments for determining focal points of moonquakes.

    Data from cold cathode gauges at Apollo landing sites and mass spectrometer in orbiting *Endeavour* confirmed presence of thin lunar atmosphere. Spectrometer picked up pockets of neon and argon and scooped up carbon dioxide that had apparently burst from fissure near where sun was lighting lunar surface at daybreak. Other data obtained by *Endeavour*'s instruments indicated lunar highlands were rich in aluminum and deficient in magnesium. Waterless basins were rich in magnesium and deficient in aluminum. Most highland regions contained three times as much aluminum as lowlands, and one mountainous region had five times as much aluminum as typical lunar basin. Magnetometer on subsatellite launched into lunar orbit discovered moon had electromagnetic boundary layer. Camera system recorded data for comprehensive map of areas never before flown over, with details and resolution previously impossible.

    Report by Astronauts David R. Scott and James B. Irwin of terracing along sides of Hadley Rille identical to terracing noted on sides of two tallest mountains in Apennine group—Hadley Delta and Mt. Hadley—suggested mountains and Hadley Rille had been thrust up from moon by internal pressures that built up inside moon when object the size of state of Rhode Island struck moon and created Sea of Rains 4 billion yrs ago. Scientists had previously thought Hadley Rille to be younger than mountains and independent of their formation. Samples might hold clues to sun's history and its effects on earth climate, past and future.

    Walter Sullivan said in *New York Times*: "Sun-determined changes in the earth's climate have controlled man's history, from the ice ages that plagued the cave men to the desiccation that buried Middle Eastern civilizations, not under invading armies but invading sands. By understanding the role of the sun in these past events, man may be able to assess the effect our parent star will have in determining climates of the future." (*W Star*, 8/8/71, A10; *W Post*, 8/8/71, B1; *NYT*, 8/8/71)

- Newspaper editorials praised U.S. exploration.

    New York Times: "The Apollo program was born a decade ago as an element in Soviet-American rivalry for prestige and political advantage. But the program's brilliant success has made obsolete the nationalistic considerations that seemed so important in the early 1960's. Now it is evident that lunar exploration and eventual settlement offer rewards and challenges for all mankind. Do Washington and Moscow have the courage and imagination to begin making that prospect a reality?" (*NYT*, 8/8/71)

Washington Post: "There seems to be very little doubt that the mission completed yesterday by astronauts Scott, Irwin and Worden has been the most productive in the history of the Apollo program. The yield of scientific data has already been large and much more is sure to come as laboratories around the world analyze the rocks, pictures and other material. . . . The trip has been a bonanza . . . and it may well go down in history as one of the most useful expeditions ever conducted in helping mankind understand the universe in which it lives." (*W Post*, 8/8/71)

- Responsibility for NASA space tug activities was assigned to MSFC by Dale D. Myers, Deputy Administrator for Manned Space Flight, in letter to Dr. Eberhard F. M. Rees, MSFC Director. MSFC would manage "in-house and contracted activities" and provide liaison with DOD and European groups interested in participating in development and use of space tug. Dr. Rees later said Space Tug Project Group would be established within Advanced Systems Analysis Office of Program Development. (MSFC Release 71–178)
- Joint NASA-U.S. Geological Survey project CARETS (Central Atlantic Regional Ecological Test Site) was described in *Washington Post* article. NASA high-altitude aircraft were flying from Wallops Station over Maryland and Virginia to obtain data for ecological analysis. Information gleaned would be included in land use maps, water use and pollution maps, and research reports. (Pfeiffer, *W Post*, 8/8/71, C10)
- U.S.S.R. had in past year "become the potential target of a vastly increased number of nuclear bombs and missile warheads," Washington *Sunday Star* noted. While U.S. was menaced only by Soviet nuclear weapons, U.S.S.R. had become "first to be put at a disadvantage in the game of nuclear proliferation." France and China had begun deploying missiles that could reach targets within U.S.S.R. but not targets in U.S. France's first squadron of nine intermediate-range nuclear missiles had become operational Aug. 4. Another nine were to be added. Missiles carried 100-kiloton warhead—five times as powerful as atomic bomb dropped on Hiroshima—and sufficient range to reach some major industrial cities of U.S.S.R. Communist Chinese had deployed small number of medium-range nuclear missiles and U.S. experts thought they were not concentrating on development of intermediate range and intercontinental missiles.

U.S. had 100 Minuteman missiles, 50% of which would carry up to three warheads; 54 Titan missiles; 41 submarines carrying 16 missiles each, with 31 submarines being converted to carry 10-warhead Poseidon; and 470 strategic bombers. U.K. had 3 Polaris submarines; 40 Canberra V–8 bombers; and 2 aircraft carriers with 80 Buccaneer strike aircraft capable of delivering nuclear weapons. French also had 36 Mirage IV–A aircraft capable of low-level delivery of 60-kiloton nuclear bombs. NATO forces in Europe had 7000 tactical nuclear weapons available, with about 2250 missiles and aircraft to deliver them. (Kelly, W *Star*, 8/8/71, A15)

August 9: Dr. Charles A. Berry, MSC Director of Medical Research and Operations, told press at MSC that preliminary examination of *Apollo 15* crew had revealed astronauts were in good health, but were not returning to normal as rapidly as expected. Examination had also revealed that Astronaut David R. Scott had ruptured blood vessels under four fingernails—apparently during first EVA on moon. Ruptures had resulted from increased pressure on fingers from spacesuit as Scott moved his arms to conduct experiments. Although Scott had mentioned some discomfort to his hands after EVA-1, officials had assumed discomfort was typical hand cramp experienced by previous astronauts. Dr. Berry said Scott's injury was not serious enough to warrant spacesuit modifications: "It's just one of those things that you've got to put up . . . with, and we keep trying to make suits better all the time, and I think this current suit is probably as far along the line to being comfortable and providing mobility as any suit we've ever had. . . ." (Transcript)

- NASA's M2–F3 lifting body, piloted by NASA test pilot William H. Dana, reached mach 0.93 during eighth flight from FRC in joint NASA–USAF program. Purposes of flight were to evaluate effects of moving center of gravity slightly forward and of decreasing launch mach number by launching at 12 200-m (40 000-ft) altitude instead of 13 700 m (45 000 ft). (NASA Proj Off)

- Arcas sounding rocket launched by NASA from Ft. Greely, Alaska, carried GSFC payload to obtain ozone measurements in conjunction with *Nimbus 4* satellite overpass. Because of problems with ground tracking equipment, no usable telemetry signal was received from payload until rocket reached 45-km (28-mi) altitude on descent side of apogee. Payload recovery was not successful; telemetry data were late and quality of signal was poor. Experiment was rated partially successful. Launch was second in three-flight series. First had been conducted Aug. 7. (NASA Rpt SRL)

- Newspaper editorials praised *Apollo 15*.

  Washington *Daily News*: "The record harvest of rocks and other data brought back by Apollo 15 may turn out to be one of the biggest steps ever toward understanding the solar system, the dominant influence on earthly existence. This has been the most remarkable decade of research and development in history, and the space program has been the core of it." (*W News*, 8/9/71)

  Philadelphia *Evening Bulletin*: "The pictures taken by Endeavour and the data being supplied by the tiny satellite the astronauts left orbiting the moon are providing the most exact and comprehensive intelligence yet about minerals on the moon." It was "becoming obvious that what we spend in space we are investing in the future of mankind." (P *Bull*, 8/9/71)

- Prague newspaper *Rude Pravo* commented on *Apollo 15*: "The scientists eagerly anticipate the results of the work of the astronauts and instruments which were sent on such a long trip. People of goodwill throughout the world who follow these flights will always wait long into the night for the two words from Houston, Baykonur, and eventually from other places on earth: 'Mission complete.' " (FBIS–Czechoslovakia–71–155, 8/11/71, D1)

- President Nixon had sent letter to daughter of *Soyuz 11* Cosmonaut Georgy T. Dobrovolsky after seeing photo of girl at cosmonauts' funeral, *Time* reported. Letter had said: "I know your father would want you to face his loss with a strength and courage to match his own. As your sister Natasha grows older I know you will help her understand how very proud of her father you both should be." (*Time*, 8/9/71, 31)

August 10: President Nixon approved H.R. 9382, FY 1972 HUD-space-science appropriations bill that contained $3.298-billion NASA FY 1972 appropriation and $619 million NSF appropriation. Bill became P.L. 92–78. NASA total included $2.523 billion for R&D, $52.7 million for construction of facilities, and $722.6 million for research and program management. Bill appropriated $39 million for NERVA program, $13 million for space shuttle main engine test facilities, and $5.5 million for space shuttle thermal-protection facilities.

President also signed H.R. 9667, $2.9-billion DOT FY 1972 appropriations bill that included $58.5 million to repay airlines for money they invested in SST but eliminated funding for SST program. Bill became P.L. 92–96. (*PD*, 8/16/71, 1157; NASA Budget Off; H Rpt 92–377)

- Scientists at LRL in Houston began opening bags of rocks returned from moon by *Apollo 15* crew. Before and after day-long debriefing Astronauts David R. Scott and James B. Irwin assisted geologists in identifying samples. Material included football-size piece of black, glass-like rock; brick-size chunk of volcanic rock pitted with small cavities (vesicular basalt); and 9½-kg (21-lb) rock, largest single sample ever returned from moon, from rim of Hadley Rille. Scientists were anxiously waiting to examine crystalline rock in sample bag No. 196, scheduled to be opened Aug. 12, that Scott had described as possible chip of moon's original crust. (*W Post*, 8/11/71; AP B *Sun*, 8/10/71, A1)

- Lagos, Nigeria, *Morning Post* congratulated *Apollo 15* astronauts for successful mission, which they attributed to ingenuity and resourcefulness of NASA. Lunar explorations, *Post* said, would strengthen faith in and appreciation of God's work. (FBIS–Nigeria–71–153, 8/11/71, W6)

- NASA launched three sounding rockets from WSMR. Two Aerobee 170 sounding rockets, one carrying NRL experiment to study stellar spectra and one carrying GSFC experiment to conduct x-ray studies, were unsuccessful. Boosted Arcas II carried GSFC experiment to 70-km (43.5-mi) altitude to measure mobility of charged particles in lower D region. Rocket functioned satisfactorily, but some anomalous data were received. Data would be correlated with data from two Nike-Cajun launches scheduled for Aug. 31. (SR list; NASA Rpt SRL)

- Australian pilots Trevor Brougham and Bob Dickeson landed twin-engine Beechcraft Baron aircraft at Darwin, Australia, after 125-hr 27-min flight round the world which was believed to have set world record for light aircraft. (Reuters, *W Post*, 8/11/71, A6)

August 11: Dr. Paul W. Gast, Director of Planetary and Earth Sciences at MSC, explained at MSC briefing significance of photos taken by *Apollo 15* astronauts: Scientists had always wondered whether earth and moon had a gradual or cataclysmic kind of history. "What we've sort of seen in the Apollo 15 site is that indeed the lunar history is in part cataclysmic." Dr. Gast said photos of Apennine Mountains confirmed that mountains were formed out of same collision that created Sea of Rains north of mountain range, involving another body as big as state of Rhode Island. "We theorized that this would be the case," Dr. Gast said, "but until now it had just been a hypothesis." Dr. Gast and his colleagues had examined about 24 of the more than 1400 photos taken on lunar surface, but they were able to make conclusions because pictures were clear and detailed. (Transcript; O'Toole, *W Post*, 8/12/71, A13)

August 12: USAF launched unidentified satellite from Vandenberg AFB by Titan IIIB-Agena booster. Satellite entered orbit with 402.3-km (250-mi) apogee, 133.6-km (83-mi) perigee, 89.8-min period, and 110.9° inclination and reentered Sept. 3. (Pres Rpt 72; *SF*, 2/72)

- U.S.S.R. launched *Cosmos 434* from Baykonur into orbit with 11 798-km (7331.0-mi) apogee, 187-km (116.2-mi) perigee, 228.2-min period, and 51.5° inclination. (GSFC *SSR*, 8/31/71; *SBD*, 8/13/71, 205)

- *Apollo 15* Astronauts David R. Scott, James B. Irwin, and Alfred M. Worden held press conference at MSC, during which they showed films and described mission. Irwin and Scott praised performance of LRV Rover and said it had exceeded their expectations. Showing film taken from Rover during traverse, Irwin said he hoped simulated ride didn't make viewers seasick. Ride on LRV, he said, "is kind of a combination of a small rowboat in a rough sea and a bucking bronco."

  Scott said only change he would recommend for Rover was "some manner of strapping yourself in the vehicle other than just pure seat belts." He also said moon should be explored to far greater extent than was currently planned, with "a whole base of scientists" and "a Rover that can carry 6 to 10 men. I think we can . . . establish bases on the moon similar to those in the Antarctic and explore it and discover a vast amount of data that we don't even know we have up there."

  Core sample drilling, Scott said, was "one of the best things we ever did on the moon. It was very difficult, very time consuming and at times it was rather perplexing . . . because we did not expect the regolith or the surface material to be quite as well integrated or packed as it was. There was nothing wrong with the equipment, it was just that we encountered the unknown. And we had to compensate for that and . . . that's why man goes to the moon. His subjective mind can evaluate the situation and come up with a fix far better than a machine. The machine would have stopped. It would have never gotten the drill."

  Scott said in closing, "We went to the moon as trained observers . . . to gather data, not only with our instruments on board, but with our minds, and I'd like to quote a statement from Plutarch . . . : 'The mind is not a vessel to be filled, but a fire to be lighted.'"

  Astronauts later revealed they had left plaque and tiny silver figure of a fallen man on moon as memorial to 14 astronauts and cosmonauts who had died in pursuit of space exploration. Plaque, bearing names of eight astronauts and six cosmonauts, had been left in small crater about 6 m (20 ft) north of rille. (Transcript; AP, B *Sun*, 8/13/71, A1)

- Aerobee 170 sounding rocket was launched by NASA from WSMR carrying NRL experiment to conduct solar studies. Mission was partial success. (SR list)
- LRL geologists Dr. William C. Phinney and Dr. Paul W. Gast described anorthosite rock from *Apollo 15* sample bag No. 196 to press at MSC lunar sample briefing. Dr. Phinney, praising crew's descriptions of samples, said "as in all cases so far where we've looked at what the rocks are, it matches exactly what the astronauts said it was. It's about fist size, 269 grams—½-lb—anorthosite . . . essentially pure plagioclase." Grain was quite coarse, 1-cm (0.4-in) grains, "about the coarsest grain size of any lunar rocks that we've seen so far." Its milky-white color suggested "that it has probably undergone a fair amount of stress, maybe a shock type of deformation."

    Rock, popularly called "Genesis rock," was significant because it supported hypothesis that moon had been extensively molten with anorthosite crystallizing from liquid, floating to top, and producing anorthositic crust. Although discovery did not prove hypothesis, Dr. Gast explained, it was "a long step in the direction" of determining moon's origin. "If results on the dating of this anorthosite and some of the chemistry on it . . . turn out that all point to the fact that it is essentially 4.6 or 4.7 billion years old, which is the time that we think the moon formed, . . . it gets to be rather difficult to argue with the hypothesis." (Transcript)

August 13: Dr. Charles A. Berry, MSC Director of Medical Research and Operations, described *Apollo 15* crew's medical status to press at MSC. Astronauts James B. Irwin and David R. Scott had suffered brief periods of irregular heart rhythm on moon, probably as a result of extreme fatigue. Irwin also had dizzy spells in space. Arythmic heartbeats and dizziness had not been disclosed to press earlier, Dr. Berry said, because tests were still being conducted and complete information was not known. Until Aug. 13 Irwin was still suffering from effects of weightlessness; he still had mild dizzy spells and, when lying prone, felt as if his bed were tilted downward about 30°. Astronaut Alfred M. Worden's medical data during transearth coast EVA had been close to predicted, and Worden was adjusting better than Scott and Irwin.

   None of the three, however, had returned to preflight norms, but there was "no medical danger" as far as crew was concerned. "Things from past space flights were all going in a pattern and now suddenly these guys don't fit the pattern." In assessing astronauts' return to normal it was important to note how *Apollo 15* differed from previous manned space flight missions. Important factors were longer time spent on lunar surface, longer lunar orbit stay time, and longer time spent in 100% oxygen atmosphere.

   Heartbeat irregularities, which had lasted about 10 heartbeats, were usually associated with fatigue, low potassium level, and increased adrenalin flow—all of which were present in astronauts. Because of fatigue levels of which astronauts were often unaware, flight surgeons were considering recommending that first EVAs on Apollo 16 and Apollo 17 be shortened. (Transcript; AP, *NYT*, 8/14/71, 1)

- NASA announced selection of Apollo 17 prime and backup crews. Prime crewmen were Eugene A. Cernan (commander), Ronald E. Evans

(CM pilot), and scientist-astronaut Harrison H. Schmitt (LM pilot). Backup crewmen were David R. Scott, Alfred M. Worden, and James B. Irwin. Apollo 17 would be launched in December 1972 on 12-day lunar-landing mission with 68-hr lunar surface stay time. (NASA Release 71-149)

- *New York Times* editorial noted that U.S. planned to cut back on Apollo program to fund space station and shuttle while U.S.S.R. *Salyut 1* was already in orbit: "In effect, the unique American capability to land men on the moon is being abandoned ahead of the original schedule so that this country can concentrate on an attempt to compete with what already exists under Russian auspices. To make it even more bizarre, when and if both countries have rival manned space stations in orbit, each is likely to suspect the other of using these installations for military as well as civilian purposes. This is a prospect that literally borders on lunacy." (*NYT*, 8/13/71, 26)

August 13-14: New York Times published series of three articles by Apollo 15 astronauts describing the July 26–Aug. 7 lunar-landing mission.

Astronaut David R. Scott said excitement of being on moon came in waves: "They are gentle swells, the kind that stop you, buoy you, but never sweep you off course. The wave enveloped me the moment our lunar module, Falcon, cleared the peaks of the Apennines and pitched over for our final descent to the moon." Greatest thrill had been discovery of anorthosite rock: ". . . I radioed to mission control, 'I think we found what we came for.' I must confess that the sight of that rock, sitting there millions of years waiting for our arrival, set off another wave of excitement for me. It triggered an emotion much like finding The Easter Egg. You know, there are lots of eggs spread across the lawn, but there's only one golden one, and there it was—The Golden Egg."

Astronaut James B. Irwin said, although others might describe moon as barren and desolate dead world, "it can be a beautiful one to anyone who loves the mountains of earth. . . . When I first climbed down the ladder . . . I was really taken back by those mountains [that] almost encircled Hadley Base. They seemed so close and so tall. But the real surprise was that the mountains, at first sight, were not grey or brown as I had expected. They were golden. Looking to the south and east, I saw the reflection of the early morning sun glancing off the mountains, especially Hadley Delta, giving them a glow of gold. It seemed like a friendly place, which surprised me. . . . you would have thought we would have felt a little lost, a little out of place. But no, I felt that I was where I should be. I knew exactly what to do, after more than a year of training, and I felt right at home."

Astronaut Alfred M. Worden compared orbiting moon in CSM with "riding in a free air balloon, floating over the countryside. You are suspended. You have no sensation of motion unless you look down and see the lunar landscape sweeping below you. At times, I thought of myself as a bird, soaring without sound, seemingly detached from all that is below. . . . one of the most important observations from Apollo 15, as far as I'm concerned, is the apparent evidence of a great deal more volcanic activity in the moon's past. In our training we studied and discussed volcanoes . . . but I guess I wasn't really prepared for the magnitude of the volcanic evidence I saw on the moon." (*NYT*, 8/13/71, 1; 8/14/71, 1)

August 14: Arcas sounding rocket launched by NASA from Ft. Greely, Alaska, carried GSFC payload to obtain ozone measurements in conjunction with *Nimbus 4* satellite overpass. Ozone payload and ground telemetry tracking equipment functioned satisfactorily. Payload was recovered and good data were collected. Launch was last in three-flight series that began Aug. 7. (NASA Rpt SRL)

- *Chicago Tribune* editorial commented on *Apollo 15* crew: "Out of all they saw, collected and recorded by camera they hope to make a significant contribution to science and the understanding of the universe, and they hope from this to apply what they learned to the betterment of life on earth. If there is one regret, it is, as Scott said, that after the vista from this peak of experience, what comparable mountain remains to be searched?" (*C Trib*, 8/14/71)

August 15: John P. Donnelly, former Vice President for Corporate Communications for Whittaker Corp., became NASA Assistant Administrator for Public Affairs. Appointment had been announced July 23. (*NASA Activities*, 8/15/71, 135)

- Baltimore *Sun* editorial praised NASA decision to include scientist-astronaut Dr. Harrison H. Schmitt in Apollo 17 crew: "Exploration of the moon can be scientifically justified as a great and extremely costly project only by the faith that it will produce evidence on which the world's whole scientific community can agree regarding the origins and age of the moon, the earth, perhaps the entire solar system." Even though Apollo program was to end with Dr. Schmitt's flight, "it may very well take a new turning by reason of his presence." (B *Sun*, 8/15/71)

August 16: France's *Eole* (CAS-1) Cooperative Applications Satellite was launched by NASA from Wallops Station at 2:39 pm EDT. Four-stage Scout boosted spacecraft into orbit with 906.1-km (563.6-mi) apogee, 677.8-km (421.2-mi) perigee, 100.7-min period, and 50.2° inclination.

Primary NASA objectives were to place spacecraft into appropriate earth orbit to obtain meteorological data from balloons and to analyze meteorological data acquired from constant-density surface balloons for study of characteristics and movements of air masses. Secondary NASA objective was to acquire technology of satellite range and range-rate system for locating free-moving balloons.

The 85-kg (187-lb) satellite—developed by France's Centre Nationale d'Études Spatiales (CNES) and named for mythical God of Winds—would collect information on wind, temperatures, and pressures from up to 500 instrumented balloons flying at 11 900-m (39 000-ft) altitude in Southern Hemisphere. Balloons, to be launched by CNES and Argentine personnel from three sites in Argentina, could be interrogated by *Eole* day and night, individually, in sequence, or in programmed group of up to 64 at a time.

Eole's meteorological objectives were to study circulation of atmosphere from tracks of balloons, provide basis for standard reference system of pressure and temperature to be used in World Weather Program, and learn more about local winds. Mission would prove feasibility of modified Doppler system plus range measurements to determine accurately location of each balloon and prove out superpressure balloon design and lightweight, frangible balloon-borne electronics.

Eole was launched under May 27, 1966, Memorandum of Understanding between CNES and NASA with no exchange of funds. CNES designed, built, and tested spacecraft and balloons and would manage balloon launchings, spacecraft tracking, data acquisition, and data processing. NASA provided launch vehicle, prelaunch support, launch operations, and initial tracking and data acquisition. Data would be analyzed simultaneously by CNES and by Eole Data Interpretation Group comprising representatives of NASA, CNES, NCAR, NOAA, and UCLA. (NASA Proj Off; NASA Release 71-144)

- Gerald J. Mossinghoff, former Director of Congressional Liaison Div. in NASA Office of Legislative Affairs, became Deputy Assistant Administrator (Policy). Robert H. Hood, former Manager of Spacecraft and Launch Systems in Washington, D.C., office of McDonnell Douglas Corp., became Deputy Assistant Administrator (Programs). (NASA Ann, 8/13/71)

August 16-20: Working Group on Remote Measurement of Pollution met at LaRC to investigate remote sensing in identifying and monitoring contaminants. Group reported trace gases were amenable to remote sensing, water pollutants could be measured in limited number, and remote measurement of specific particulate pollutants would follow only after improved understanding of their physical, chemical, and radiative properties. Remote sensing could provide essential information in all three categories that could not be obtained from other sources. Satellites and aircraft could obtain overall picture of global or regional pollution from gases but were inappropriate for local measurements because of inability of instruments to measure gases in upper atmosphere.

Group recommended global measurement from satellites of gases contributing to ozone photochemistry and radiative equilibrium; study to establish feasibility of measuring these gases with current techniques in stratosphere and troposphere; acceleration of research on application of remote sensing to water pollution; coordinated physiooptical research program to advance level of understanding of optical properties of real atmosphere; immediate measurement of earth's albedo, reflected electromagnetic radiation; and establishment of global network of atmospheric turbidity-monitoring sites to determine any changes in worldwide turbidity due to man-made particulate matter. (*Remote Measurement of Pollution*, NASA SP-285)

August 17: U.S.S.R.'s *Lunokhod 1* lunar rover completed experiments and was powered down for 11th lunar night on moon. Vehicle had landed on moon on board *Luna 17* Nov. 17, 1970. (FBIS-Sov-71-137-3, 8/13/71, L1)

- High-power radar transmitter was being added to 3000-m-dia (1000-ft-dia) radio telescope near Arecibo, Puerto Rico, in joint NASA-NSF program, NSF announced. Modifications would make telescope most powerful antenna in world for radar studies of planets and satellites. (NSF Release 71-178)

- Completion of arrangements for exchange of four computer programs between NASA and Japan's Hitachi Central Research Laboratory was announced by NASA. Exchange, negotiated by NASA OART and Computer Software Management and Information Center (COSMIC) operated at Univ. of Georgia under NASA contract, would make two Japanese programs available to U.S. users in exchange for two NASA programs that had been available to U.S. public for several years. (NASA Release 71-148)

- OST released *Final Report of the Ad Hoc Supersonic Transport Review Committee of the Office of Science and Technology, March 30, 1969.* Release of previously secret report was prompted, OST said, "by continued public interest and certain impressions that the government may be concealing factual data on the SST program." Report recommended "termination of the development contracts and the withdrawal of Government support from the SST prototype program." Reasons given were: SST operating costs would exceed those of then-available subsonic aircraft; "attendant surcharge makes the airline market uncertain"; program was unattractive to private financing and "substantial government involvement is likely to be sought in the supply or guarantee of some $3–5 billion of capital for the certification and production of a U.S. SST"; there was "substantial uncertainty regarding the range and payload and the environmental effects" of SST and "costs and duration of the program are both likely to increase" in attempt to develop production aircraft; there was "substantial doubt that the present configurations of the Concorde and the TU–144 will become commercially viable aircraft"; balance of payments argument was not sufficiently strong to warrant Government investment in SST; U.S. leadership in aviation did not depend upon affirmative SST decision in near future; and, "when the right combination of technology and market demand appears, the U.S. aircraft industry may well decide on its own to proceed with the development and production of an SST." (Text)

- Precision stepping-drive mechanism to position scale models of spacecraft accurately about horizontal axis in wind tunnels and other testing situations had been used by clock manufacturer to replace gears and other continuous-drive mechanisms in electric clocks, NASA announced. Device, developed from NASA technology, when coupled with high-accuracy oscillator provided accurate time by advancing digital clock in increments of one second or any interval required. (NASA Release 71-153)

- U.S. Postal Service public relations officers had asked Astronauts Neil A. Armstrong and Edwin E. Aldrin, Jr., to accompany cancellation of lunar landing commemorative stamp on moon during *Apollo 11* mission with dialogue from two "casual, conversational" scripts, Julian W. Scheer, former NASA Assistant Administrator for Public Affairs, said in *Washington Post* article. "But NASA, figuring the first manned lunar landing was hairy enough without ceremonials, declined politely and, instead, Armstrong and Aldrin quietly canceled the first moon stamp and came home without postal fanfare."

    Washington Post editorial commented on article: "While we have not cast the roles and read the script aloud, timer in hand, it is our definite impression that this spontaneous little exchange might have been good for the better part of one orbit of the moon and good also for a veritable barrage of phone calls from a flabbergasted public." (*W Post,* 8/17/71, A14)

- George W. Cherry, Director of Aeronautical Operating Systems Div. in NASA OART, received AIAA Mechanics and Control of Flight Award in ceremonies at Hempstead, N.Y. Award was for "pioneering the development of the digital control system for space vehicles and the application of these techniques to the Apollo Command and Service Modules and the Lunar Modules." (*NASA Activities,* 9/15/71, 155)
- Dr. James R. Schlesinger was sworn in as AEC Chairman, succeeding Dr. Glenn T. Seaborg, whose resignation became effective Aug. 16. (AEC PIO)

August 18: NASA supercritical wing, flown on TF-8A jet aircraft piloted by NASA test pilot Thomas C. McMurtry, successfully completed ninth flight from FRC. Objectives of flight—to obtain pressure distribution data at high supersonic speeds and evaluate augmented damping characteristics at low supersonic speeds—were achieved. Flight, made at 10 700-m (35 000-ft) altitude, was first in series to obtain data for performance evaluation; previous flights had been for pilot familiarization. Pressure distribution data were obtained from mach 0.80 to 0.97 and damping characteristics were evaluated at mach 1.1 and 1.6. (NASA Proj Off)

- NASA announced formation of Transport Experimental Programs Office in OART to direct experimental STOL transport research airplane program. RFPs for design and fabrication of STOL aircraft had been issued Aug. 4, requesting proposals by Oct. 15. Program would provide data and experience for development of environmentally acceptable, economical, and safe fan-jet STOL transport systems. New office, directed by Gerald G. Kayten, would integrate activities of STOL Technology Office and Supercritical Technology Office. (NASA Release 71-155)
- NASA announced it had begun joint program with USA Air Mobility Research and Development Laboratory to develop flight research vehicle to test various rotors in actual flight under controlled conditions with provision for extensive engineering measurements during maneuvers. LaRC had issued RFPs for two parallel design studies requiring 10 mos to determine feasibility of attaining Government's research goals and provide development plan for further phases of program. Joint NASA-Army team would base detailed specifications for vehicle preliminary-design competition on results of predesign studies. (NASA Release 71-152)
- Pratt & Whitney Div. of United Aircraft Corp. filed formal protest against NASA award of $500-million space shuttle main engine contract to NR Rocketdyne Div. and asked Federal Government to reverse award. In letter to GAO, Pratt & Whitney charged selection of Rocketdyne was "illegal, arbitrary and capricious, and based on unsound decisions" and reiterated complaints cited Aug. 3 to GAO. (*W Post,* 8/19/71, C12; *SBD,* 8/19/71, 230)
- Special medallion to commemorate 1971 quadricentennial of Johannes Kepler was issued by American Museum–Hayden Planetarium. Kepler —born Dec. 16, 1571—had founded science of optics, was first to ascribe tides to moon, shaped much of comet theory, and came close to theory of gravity. His *Somnium,* highly technical "dream" of voyage

August 18

to moon, had been first work of modern science fiction. (American Museum–Hayden Planetarium Release, 8/13/71)

August 18–24: Thirteenth International Congress of the History of Science was held in Moscow.

Section on History of Aircraft, Rocket, and Space Science and Technology was held Aug. 19. Paper "The Historiography of Rocket Technology and Space Exploration" submitted by Dr. Eugene M. Emme, NASA Historian, was summarized at meeting by Dr. Lloyd S. Swensen, Jr., of Univ. of Houston. Paper covered features of evolution of rocket technology pertaining to beginning of space exploration and exploitation and discussed work of international pioneers like Konstantin Tsiolkovsky, Robert H. Goddard, and Hermann Oberth, who became interested in space exploration before deducing that liquid-propellant reactive thrust was the best technology.

Dr. Swenson, coauthor of book *This New Ocean: A History of Project Mercury*, presented his own paper "On the Mixture of Science and Technology in the Apollo 8 Circumlunar and the Apollo 11 Lunar Landing Missions." Paper examined effect of internal pressures on overall posture and achievements of Apollo project as illustrated by *Apollo 10* and *11*.

Anatoly A. Blagonravov and V. N. Sokolsky of Soviet Academy of Sciences submitted paper on "Main Lines of Research in the Field of the History of Aeronautics and Astronautics." (Program; NASA Hist Off)

August 19: Apollo 17 Astronauts Eugene A. Cernan, Harrison H. Schmitt, and Ronald E. Evans held press conference at MSC. Ph.D. geologist Schmitt defended his selection for mission and declared his readiness to "compete with anybody" in flying Apollo spacecraft. Schmitt, selected over Astronaut Joseph E. Engle, said Engle was "one of the most outstandingly qualified test pilots in the business" and speculated that Engle, transferred to space shuttle program, would "probably . . . contribute for the next 10 or 15 years a . . . lot more than maybe even we can contribute by a lunar mission." Sending a "professional observer" like himself to moon, Schmitt explained, would significantly increase knowledge gained, particularly concerning early history of earth and sun. (Transcript; Reuters, *W Post*, 8/20/71, A2)

* NASA launched series of three sounding rockets from Wallops Station to measure diurnal oxygen, density, pressure, and temperature and to obtain comparisons of pitot and active-sphere density measurements. Rockets were launched during daylight hours and data would be compared with data from similar series to be launched during darkness Aug. 20.

Viper Dart carried WSMR experiment to 135-km (83.9-mi) altitude to provide density, temperature, pressure, and wind data from 40 to 130 km (25 to 80 mi). Rocket and instruments functioned satisfactorily, with Robinette sphere ejecting and inflating as programmed. Nike-Cajun carried Univ. of Michigan and NOAA experiment to 126-km (78.3-mi) altitude to measure diurnal variation of atomic oxygen in 80- to 120-km (50- to 75-mi) region. Rocket and instruments functioned satisfactorily. Nike-Apache carried Univ. of Michigan and GSFC experiment to 171-km (106.3-mi) altitude to obtain diurnal measurements of density, pressure, and temperature and obtain com-

parisons of pitot and active-sphere density measurements. Rocket and instruments functioned satisfactorily. (NASA Rpts SRL)
- French Centre National d'Études Spatiales (CNES) announced U.S.S.R. would cooperate in launching two space probes from France's Kourou Space Center. French Veronique rocket would launch Soviet radio-frequency spectrometer in mid-December in first use of Kourou facilities by U.S.S.R. France would also track Soviet MR-12 rocket carrying French mass spectrometer after launch in December from Soviet ship near Kourou.

  CNES also announced French Centaure rocket would launch Indian equipment Oct. 8 to test recovery system for space probe instruments. (Reuters, *NYT*, 8/22/71, 35)
- Successful completion of over 2000 hrs of testing—equivalent to 80-day space mission—of prototype water vapor electrolysis system was announced by NASA. Developed at ARC, system converted moisture in air directly into hydrogen and oxygen and released reclaimed oxygen back into air. It could be used instead of heavy, bulky tanks of breathing oxygen on future long-duration space missions. It would also reduce load on humidity control equipment by removing excess water vapor from cabin atmosphere. (NASA Release 71-150)
- NASA and Stanford Univ. Biomedical Technology Transfer Team (BATEAM), fourth BATEAM to be established and first formed at leading medical school, would apply NASA-generated aerospace technology to problems in cardiology, NASA announced. Team of five medical consultants and five aerospace engineering consultants under Stanford Univ. cardiologist Dr. Donald C. Harrison would concentrate on cardiovascular medicine.

  NASA had eight special teams—four working on environmental problems and four on biomedical activities. (NASA Release 71-154)
- Data obtained during Soviet Zond missions from September 1968 to October 1970 were described by Soviet scientists, including Oleg Gazenko of Academy of Sciences, in *Moskovskaya Pravda* article, Tass reported. Zond spacecraft had carried turtles, drosophilas, onions, wheat and barley seeds, chlorella strains, intestinal bacilli, and other objects. Total dosage of radiation had been same for all flights. After return to earth turtles were active and had good appetite. During experiments they lost 10% of weight. "Some blood tests . . . and electrocardiographs did not reveal essential divergencies from control animals." Morphological and histochemical analyses of organs and tissues of turtles aboard *Zond 5* (launched Sept. 15, 1968) had revealed changes in glycogen and iron content of liver and structural changes in spleen. Subsequent experiments had not confirmed findings. Flight conditions had given rise to "comparatively large increase in the number of chromosome changes in the seeds of pine and barley, and a rise in the mutants of chlorella strains." (FBIS-Sov-71-161, 8/19/71, L1)
- NASA announced issuance of RFPs for eight-month study of applications of dual-mode nuclear rocket engine to propel spacecraft and generate electrical power for use on board. (NASA Release 71-157)
- Long-term commitment and definition of policy for storage and curatorial handling of lunar materials had been urged by Cal Tech scientist Dr. Gerald J. Wasserburg, AP reported. A principal investigator

of lunar material, Dr. Wasserburg had said lack of planning and handling could deny future generations use of moon rocks now being studied. (P *Bull*, 8/19/71)

- Washington *Evening Star* editorial commented on selection of scientist-astronaut Dr. Harrison H. Schmitt for Apollo 17 crew: "The argument for the scientist-astronaut has come. So far as we are concerned, it always made sense. Therefore, any congratulations to [NASA] for finally putting one of the flying geologists to work might be accompanied by a raise of eyebrows over the obvious tokenism of the assignment and a reproachful question as to why they waited so long to put a scientist into the richest geological treasure house that man has yet discovered." (W *Star*, 8/19/71)

August 20: No viable organism and no evidence of fossil material had been found in *Apollo 14* lunar samples, Lunar Sample Preliminary Examination Team reported in *Science*. Major findings had been: only 2 of 33 rocks had basaltic textures; samples differed from earlier lunar rocks and from their closest meteorite and terrestrial analogs in chemical content; chemical composition of soil resembled that of rocks; rocks displayed characteristic surface features and shock effects similar to those of *Apollo 11* and *12* samples; concentration of solar-wind-implanted material in soil was large, as in *Apollo 11* and *12* material, but solar wind contents of rocks ranged from that of soil "to essentially zero"; carbon content of *Apollo 14* rocks was within range of that for *Apollo 11* and *12* rocks; four fragmental rocks showed surface exposure time about an order of magnitude less than typical exposure times of *Apollo 11* and *12* rocks; broader range of soil mechanics properties had been found at *Apollo 14* site than at *Apollo 11* or *12* sites and Surveyor landing sites; and no evidence of biological material had been found in *Apollo 14* samples to date. (*Science*, 8/20/71, 681–93)

- *Apollo 12, 14,* and *15* seismometers had located source of 80% of moonquakes in area 10 km (6 mi) wide and 650 km (400 mi) below lunar surface in Ocean of Storms, Dr. Gary N. Latham, Apollo program chief seismologist, reported in interview published in *Washington Post*. (Brett, *W Post*, 8/20/71, A2)

- NASA launched series of three sounding rockets from Wallops Station to measure diurnal oxygen, density, pressure, and temperature and obtain comparisons of pitot and active-sphere density measurements. Rockets were launched during darkness and data would be compared with data from daylight series launched Aug. 19.

  Nike-Cajun carried Univ. of Michigan and NOAA experiment to 132-km (82-mi) altitude to measure diurnal variation of atomic oxygen in 80- to 120-km (50- to 75-mi) region. Nike-Apache carried Univ. of Michigan and GSFC experiment to 180-km (111.8-mi) altitude to obtain diurnal measurements of density, pressure, and temperature in 80- to 120-km (50- to 75-mi) region. Viper Dart carried WSMR experiment to 144-km (89.5-mi) altitude to provide density, temperature, pressure, and wind data from 40 to 130 km (25 to 80 mi). All rockets and instrumentation functioned satisfactorily. (NASA Rpts SRL)

- Three Nike-Apache sounding rockets were launched by NASA from Wallops Station carrying Univ. of Illinois and GCA Corp. experiments to measure electron concentration, collision frequency, and temperature in lower ionosphere before sunrise. Rockets reached 193-km (119.9-mi), 198-km (123-mi), and 200-km (124.3-mi) altitudes with all instruments functioning satisfactorily. (NASA Rpts SRL)

- U.S.S.R. launched Vertikal-2 cooperative geophysical research rocket to 463-km (287.7-mi) altitude to study solar UV and x-ray emissions, ionospheric parameters, and meteor particles. Nose section of rocket consisted of instrument compartment and recoverable container carrying scientific equipment for heliophysical research and Hungarian-Czechoslovakian-Soviet instruments to study micrometeors. Instrument compartment carried equipment for heliophysical and ionospheric research designed by Bulgarian, East German, Czechoslovakian, and Soviet scientists. Rocket and instruments functioned satisfactorily and payload was recovered. (Tass, FBIS–Sov–163–3, 8/23/71, L1)

- Permanent charter for INTELSAT was signed by 54 nations—51 of whom were members of 80-member interim INTELSAT consortium—during State Dept. ceremonies. Signatures of three more original members were required for two-thirds majority necessary to inaugurate new agreement providing for gradual relaxation of U.S. control.

Under permanent charter INTELSAT Assembly of Parties, prime organ of multitiered organization, would meet biennially and take decisions with one vote for each nation. Annual Meeting of Signatories would decide on financial, operational, and technical aspects of programs with one vote for each member. INTELSAT Board of Directors would meet at least four times a year to vote on design, development, construction, establishment, operation, and maintenance of space segment. No single Board member could cast more than 40% of total vote; decisions would be taken by two-thirds majority; and nonveto provisions had been written into charter. U.S. had agreed to relinquish unilateral veto power it had under interim arrangements. Permanent charter created new executive organ to assist Board of Governors, headed first by Secretary General and later by Director General. ComSatCorp would continue to provide technical and operational management of INTELSAT space segment under six-year contract. Net investment ceiling of U.S. $5 million was established with discretion in Board of Governors to increase ceiling by 10%. (Lydon, *NYT*, 8/21/71, 40; "Permanent Arrangements for the Global Commercial Communications Satellite System of Intelsat," IAF 22nd Int'l Astro Congress, 9/25/71)

August 21: WTTG–TV in Washington, D.C., carried one-hour special telecast "Giant Leap for Mankind." Program reviewed Apollo program and described NASA future plans. (NASA Special Ann, 8/18/71; NASA PAO)

- Appointment of Richard P. Skully, Manager of FAA's Miami Area Office, as Director of FAA Office of Environmental Quality was announced by FAA Administrator John H. Shaffer. (FAA Release 71–134)

August 22: Use of machines, rather than men, in space after final Apollo mission was advocated by Cornell Univ. astronomer Dr. Thomas Gold in *New York Times Magazine* article. For lunar or Martian exploration he suggested "roving vehicle with television eyes and one or two telefactor hands, a modern version of a centaur—half man, half automobile." Vehicle could travel longer distances on lunar surface than man and retrieve samples of different terrain. "The remote hands could pick up rocks and examine them under the television eyes, and, if desired, put them in a hopper for eventual transmission back to earth. The vehicle might then rendezvous with another that has earth-return capability and hand over the contents of the hopper to it. While this type of vehicle would allow us to explore almost all regions of the moon, manned missions have been severely limited both in the type of terrain for the landing and the distance that astronauts can travel from their spaceship." Scientists had favored use of telefactor with sample-return capability but "the very success of the Apollo program had diverted attention from it." Now that no "really purposeful" manned program was planned, Dr. Gold asked, "would not this be the right time for really learning how to do good remote control?" (*NYT Magazine*, 8/22/71)

August 23: *Apollo 15* Astronauts James B. Irwin, Alfred M. Worden, and David R. Scott addressed National Press Club luncheon in Washington, D.C., before flying to New York for official welcome by city and appearance at U.N. Scott called for greater cooperation in space and said he looked forward to flight with Soviet cosmonaut. "I know six cosmonauts personally, and I would be glad to fly with any one of them any day. And I hope we get that chance." Scott said core sample 3 m (8½ ft) long he had brought from moon contained 55 separate layers of soil that illustrated last 2.4 billion yrs of lunar history. Layers, discovered through x-ray examination, would show changes occurring over billions of years in sun's radiation as it hit lunar surface.

Later, in telephone interview with *Washington Post*, LRL scientist Dr. John F. Lindsay said layers ranged in thickness from 12.7 mm (½ in) to 127 mm (5 in) with each stratum probably representing different meteorite impact on lunar surface. While no analysis of core could be made until it was opened by geologists in January, Dr. Lindsay said x-rays had indicated that Scott had not drilled through solid rock. (UPI, *NYT*, 8/24/71, 25; Auerbach, *W Post*, 8/24/71, A3)

- NASA had laid off approximately 740 employees effective Oct. 1 to reduce to authorized FY 1972 level of 28 352, *Aviation Week & Space Technology* reported. MSC had dismissed 132; MSFC, 197; GSFC, 149; LeRC, 123; LaRC, 54; ARC, 78; FRC, 12; and NASA Pasadena Office, 2. Force reductions at Hq. and KSC had been made entirely through attrition. (*Av Wk*, 8/23/71, 21)

- U.S.S.R. had improved reliability and performance of Molniya I comsat; expanded network of Orbita system ground receiving stations to 35, including Cuban station and station at Ulan Bator in Mongolia; and would build six to eight additional ground stations annually for indefinite period, *Aviation Week & Space Technology* reported. Mongolian station had been activated in February 1970. (*Av Wk*, 8/23/71, 51-2)

- President Nixon sent memorandum and statement of Government patent policy to heads of executive departments and agencies. Improvements in 1963 policy statement would provide agency heads with authority to permit contractors greater rights to inventions where necessary to achieve use or where equitable circumstances justified, additional guidance in promoting use of Government-sponsored inventions, clarification of states' and municipal governments' rights in Federal Government-licensed inventions, and more definitive data base for evaluating patent policy's administration and effectiveness. (PD, 8/30/71, 1209)

August 23-24: Scientific results from *Oao 2* were reported during Univ. of Massachusetts symposium held in conjunction with AAS conference. GSFC, Lehigh Univ., and Univ. of Wisconsin scientists reported discovery of "black hole" in space, remnant of star that had shrunk to such density that no light could leave it. Black hole would be result of collapsing process in which material remaining from burned-out star was drawn in upon itself by its own gravity. Densities of black holes were said to be so great that globe with 3.2-km (2-mi) radius could collapse to size of pinhead and still retain its original mass. (NYT, 9/2/71, 14)

August 24: Apollo 15 Astronauts Alfred M. Worden, David R. Scott, and James B. Irwin rode in motorcade to City Hall in New York, where they received gold medals from Mayor John V. Lindsay. Astronauts presented Mayor Lindsay with flag carried on *Apollo 15* mission. Later astronauts were welcomed to U.N. by Secretary General U Thant, to whom they presented replica of plaque they had placed on moon to commemorate astronauts and cosmonauts who had died in performance of duties. (NASA Activities, 9/15/71, 147-8; Kaufman, NYT, 8/25/71, 1)

- Locked-on antenna control system on NASA's *Ats 3* applications technology satellite (launched Nov. 5, 1967) had caused satellite to stop transmitting data temporarily, NASA announced. Officials believed antenna, on top and north side of spacecraft, had overheated and stopped spinning when sun was north of equator during summer. Antenna had operated sporadically during July and August, but officials said malfunctioning was consistent with similar past problems and antenna would recover as sun moved south. (NASA Release 71-158)

August 25: NASA's M2-F3 lifting-body vehicle, piloted by NASA test pilot William H. Dana, reached 20 400-m (67 000-ft altitude) and mach 1.05 during ninth flight from FRC. Objectives were to determine aileron effectiveness at speed greater than mach 0.9, obtain stability and control data at mach 0.95 and 0.90, expand flight envelope, and evaluate speed brake. (NASA Proj Off)

- Resignation of *Apollo 11* Astronaut Neil A. Armstrong, first man to set foot on moon, was announced by NASA. Armstrong, Deputy Associate Administrator for Aeronautics, OART, would join Univ. of Cincinnati as its first professor of aerospace engineering Oct. 1. He would continue to serve NASA as consultant. Armstrong had joined NASA in 1955 and had served as aeronautics research pilot for NACA and NASA. He had piloted X-15, X-1, F-100, F-101, F-102, F-104, F-5D, and B-47 aircraft and, as B-29 "drop" aircraft pilot, had participated in more than 100 rocket aircraft flights. Armstrong had been selected by NASA as astronaut in September 1962. He had served as backup command pilot for *Gemini 5*; command pilot for *Gemini 8* (launched March 16, 1966), during which he had performed first successful docking of two vehicles in space; backup command pilot for *Gemini 11*; backup command pilot for *Apollo 8*; and commander of *Apollo 11*. (NASA Release 71-159)

- PAET Planetary Atmosphere Experiment Test, launched June 20, was adjudged successful by NASA. Test had successfully demonstrated capability of selected experiments to determine structure and composition of an unknown planetary atmosphere from probe entering atmosphere at high speeds. (NASA Proj Off)
- Appointment of Mrs. Ruth Bates Harris as Deputy Director of NASA Equal Employment Opportunity Office and Director of NASA Equal Employment Opportunity Program was announced by NASA. Appointment was effective Oct. 4. Mrs. Harris had been Director of Dept. of Human Relations of Montgomery County (Md.) Public Schools. (NASA Special Ann)

August 26: Explorer 43 (IMP-I) interplanetary monitoring platform was adjudged successful by NASA. Spacecraft, launched March 13, had exceeded objectives and was functioning nominally with 11 scientific instruments providing detailed information on galactic and solar cosmic rays, interplanetary medium, and distant magnetosphere. All systems and scientific instruments were functioning normally except GSFC plasma probe, which malfunctioned 30 days after launch, and Univ. of Chicago cosmic ray telescope, which partially malfunctioned 64 days after launch. (NASA Proj Off)
- NRL's *Solrad 10 (Explorer 44)* solar radiation satellite, launched by NASA July 8, was adjudged successful by NASA. Spacecraft was functioning nominally and all 15 scientific instruments had been turned on and were operating satisfactorily. (NASA Proj Off)
- Astronaut Alan B. Shepard, Jr., received rear admiral stars from Secretary of the Navy John H. Chaffee during Pentagon ceremonies. Shepard was first astronaut to achieve rear admiral rank. (*W Post*, 8/27/71, A1)
- Immediate report from NASA on role of Dale D. Myers, Deputy Administrator for Manned Space Flight, in award of $500-million space shuttle engine contract to NR Rocketdyne Div. was requested by Rep. William R. Cotter (D-Conn.). In statement to press, Rep. Cotter said Myers "as co-chairman of the joint NASA-Pentagon space transportation system committee, may have been instrumental in cutting off Air Force funding of Pratt & Whitney's space shuttle research under the XLR-129 program." Cotter said Pratt & Whitney Div. of United Aircraft Corp. had based its contention of superior "technical know-how" on shuttle engine development [see Aug. 3] on its 10-yr XLR-129 research. Myers had been a Rocketdyne Vice President before coming to NASA in 1970. (Text; Warden, *C Trib*, 8/27/71, 22)
- Lee B. James, former Director of Program Management at MSFC, had joined faculty of Univ. of Tennessee Space Institute, UPI reported. James' retirement from NASA had been announced April 30. (UPI, *W Post*, 8/27/71)

August 27: U.S.S.R. launched *Cosmos 435* from Plesetsk into orbit with 478-km (297.0-mi) apogee, 272-km (169.0-mi) perigee, 92.0-min period, and 70.9° inclination. Spacecraft reentered Jan. 28, 1972. (GSFC *SSR*, 8/31/71; 1/31/72; *SBD*, 9/14/71, 27)
- NASA's supercritical wing, flown on TF-8A aircraft piloted by NASA test pilot Thomas C. McMurtry, successfully completed 10th flight from FRC. Objectives of 40-min flight were to obtain additional pressure distribution and performance data at high subsonic speeds and evalu-

August 27: *Noise tests of NASA's experimental, quieter jet aircraft engine were under way in Peebles, Ohio, the Lewis Research Center announced. The project sought to develop an engine of 97 850-newton (22 000-pound) thrust that would operate at 15 to 20 decibels quieter than engines used in present subsonic air transports. In the photo the engine was being checked out before the General Electric Co. noise tests. A probe inserted in the exhaust nozzle (left) measured the smoke produced.*

ate flutter characteristics at low supersonic speeds. Aircraft reached 12 000-m (40 000-ft) altitude, with pressure distribution and performance data obtained from mach 0.80 to 0.97 and flutter characteristics evaluated at mach 1.05 and 1.10. (NASA Proj Off)

- Experimental, quieter jet aircraft engine was undergoing noise tests at Peebles, Ohio, LeRC announced. Engine was being developed under contract by GE's Aircraft Engine Group. (LeRC Release 71-38)

- Effects of levels of ionizing radiation on passengers and crew members in supersonic aircraft flying at 18 000 to 20 000 m (60 000 to 65 000 ft) were described in *Science* by USN biophysicist Hermann J. Schaefer. Level of galactic radiation per mile for SST was smaller than that for conventional jet aircraft. "Even in the extreme case of someone's spending, for a full year, 10 percent of his time at SST altitude, the integral dose equivalent would still not exceed 1 rem." Increase in risk of shortening residual life-span was "insignificant." (*Science*, 8/27/71, 780-3)

- Easier relations between Soviet and visiting foreign scientists attending 15th International Union of Geodesy and Geophysics in Moscow, Aug. 1-14, were reported in *Science* by Philip H. Abelson. "To an extent not previously possible, Americans were entertained professionally and socially and were invited to visit Soviet research facilities. Only a few Russians would be so bold as to act counter to official policy in their contacts with foreigners; therefore, the extent of the hospitality reflected national policy." (*Science*, 8/27/71, 797–800)

August 28: Oxford Univ. atmospheric physicist Dr. John Houghton told British press two-year research program at Oxford had shown "even with 500 SST aircraft flying, there would be no danger of ultraviolet radiation at ground level being increased by any significant amount." Dr. Houghton said program had been similar to research that had led Univ. of California at Berkeley scientist Dr. Harold Johnston to predict that exhaust from supersonic aircraft could have dangerous effect on stratosphere and might imperil life on earth. (London *Times* dispatch, *NYT*, 8/28/71, 41)

- *Apollo 14* Astronaut Alan B. Shepard, Jr., was awarded honorary Doctor of Science degree by Univ. of Miami in Oxford, Ohio. (*NASA Activities*, 9/15/71, 155)

August 29: *Washington Post* editorial commented on DOT–NASA report that jet engine noise was high-priority problem [see May 5]. Report "does not offer all the solutions, but it does recognize that sociological, economic and engineering considerations must be taken into account in finding them. The prescription makes good sense both for the economically strained industry and the exasperated potential passenger." (*W Post*, 8/29/71, B6)

August 30: Election of *Apollo 11* Astronauts Neil A. Armstrong, Edwin E. Aldrin, Jr., and Michael Collins to International Aerospace Hall of Fame was announced by Hall of Fame President Martin R. Engler, Jr. Awards for 1971 would be presented to astronauts in San Diego, Calif., Oct. 23. Astronauts had been nominated "in view of their epic flight and its representation as one of the greatest single scientific achievements in the history of man." (Letter to NASA Historian)

- Dr. Eberhard F. M. Rees, MSFC Director, announced establishment of temporary HEAO Office in MSFC's Program Management Directorate to plan High Energy Astronomy Observatory. Office would replace HEAO Task Team formed in spring 1971. (MSFC Release 71-148)

- FAA certification of 201 series Guppy cargo aircraft and sale of one aircraft to Airbus Industries of France, European cargo airline, were announced by aircraft's manufacturer, Aero Spacelines, Inc. New aircraft was sixth in Guppy series, created to airlift space booster components for NASA. With fuselage 7.6 m (25 ft) in diameter, new Guppy could lift 24 500 kg (54 000 lbs) of cargo and fly it for 3200 km (2000 mi) nonstop at 470 km per hr (290 mph). First 201 series Guppy would move wing and fuselage assemblies of Concorde supersonic transport and 300-passenger European airbus from points of manufacture to Toulouse, France, for assembly. (*C Trib*, 9/2/71)

- Response by Soviet Ambassador Anatoly F. Dobrynin to National Space Club letter of condolence to Soviet government on death of Soyuz cosmonauts was published in NSC *News Letter*: "We highly appreciate your expression of respect for the heroic endeavour of the space explorers who died for the progress of all mankind." (NSC *News Letter*, 8/30/71, 2)

- Yugoslavia was urging other East European countries to participate actively in INTELSAT, *Aviation Week & Space Technology* reported.

Discussions had been held with Romania and Hungary on possibility that they would use ground terminal that Yugoslavia expected to be operational in 1974. (*Av Wk*, 8/30/71, 11)

- *Time* magazine reviewed *Secret Sentries in Space* by Philip J. Klass. "No one else has written in comparable detail about spy satellites," *Time* said. Klass had described latest U.S. SAMOS (satellite and missile observations system). "Big Bird" was 10 900-kg (12-ton) spacecraft capable of operating in orbit for several months. It could transmit high-quality pictures by radio and eject capsules of exposed film, which then dropped by parachute. Big Bird also included infrared heat-sensing equipment that allowed it to "see" through ice and snow to locate Soviet underground weapons. Klass had reported heaviest concentration of long-range Soviet missiles in Siberia and behind Ural mountains in Central Asia. (*Time*, 8/30/71, 26)
- Mrs. Irene Bolam, named in book *Amelia Earhart Lives* as being missing aviatrix Amelia Earhart, filed $1.5-million libel suit against authors Joseph Klass and Joseph Gervais. Mrs. Bolam claimed book damaged her reputation by depicting her as bigamist, spy, and traitor. Book claimed Amelia Earhart had disappeared in 1937 while flying U.S. spy mission, was captured by Japanese, and might have been World War II Japanese propagandist Tokyo Rose. (UPI, *W Post*, 9/1/71, D5)
- *Washington Daily News* editorial commented on Aug. 25 resignation from NASA of *Apollo 11* Astronaut Neil A. Armstrong: "Offhand, it is regrettable to see any of these accomplished men leave the program they helped make such a smashing success. But Mr. Armstrong will be teaching engineering and it will be no surprise if some of his students wind up in future space exploration. So count his 'defection' to education as another spinoff benefit of the space program." (*W News*, 8/30/71, 22)

August 31: NASA and Soviet Academy of Sciences had confirmed results of June 21–25 Joint Working Group meetings at MSC, NASA announced. Groups had considered technical requirements for compatible systems for space rendezvous and docking, radio and optical reference, communications, life support, crew transfer, and docking assemblies. They had agreed to study experiments to test technical solutions. Experiments might include docking spacecraft of Apollo type with orbital scientific station of Salyut type and docking spacecraft like Soyuz with station like Skylab. Mission models would test suitability of technical requirements and solutions. Next working group meeting was scheduled for late November in Moscow, to complete agreement on technical requirements for compatible systems. Working Groups had been established under Oct. 28, 1970, agreement. (NASA Release 71-164)

- Denver mint held preview showing of nickel alloy Eisenhower dollar which would be released in October. Back of coin depicted *Apollo 11* lunar landing. (UPI, *W Post*, 9/2/71)
- President Nixon, from San Clemente, Calif., issued Proclamation 4078 designating Oct. 11 Columbus Day, 1971. He said, "In this present age of epic journeys in space, we can appreciate more than ever the great achievements of Christopher Columbus." (*PD*, 9/4/71, 1224–5)

During August: Gabriel Board appointed June 18 to examine and reevaluate philosophy, practices, and procedures in Centaur quality assurance and workmanship program reported findings and recommendations to House Committee on Science and Astronautics' Subcommittee on NASA Oversight: "Aggressive and creative" new policy with "thorough and conservative approach to the use of new hardware in flight" was needed to improve Atlas-Centaur quality and workmanship. Board recommended enhanced emphasis on employee motivation, manufacturing, engineering, and planning, quality assurance planning "with preventative aspects dominant, workmanship training, inspection training, quality audit, and vendor control improvements." (*Review of Recent Launch Failures*, Subcom Print, 10/71)

- USN participation in space shuttle program was urged by L/Cdr Preston E. Beck (USN, Ret.), in *U.S. Naval Institute Proceedings*. USN experience had shown close parallel between life support criteria and fuel management for nuclear submarines and that required for space shuttle. USN Office of Naval Research (ONR) and USN Systems Command should determine feasibility of using air entrainment, hydroskis, skids, and hydrofoils to permit sea-level landing point for shuttle spacecraft with one or more engine inoperable after reentry. Sea transport and barge storage of fuels and oxidizers for shuttle spacecraft would ensure safety of flight personnel and minimize interference with commercial air traffic. Space shuttle orbiter and booster structures could be moved only by flyout or by sea because of size. "The obvious answer is final assembly at the launch site using a seaborne logistic system." Naval Air and Material Center had facilities for development of space shuttle materials, testing of components and subassemblies, and supporting engineering. It seemed pertinent "that the Navy take the initiative in starting to support the space shuttle program. The key to many of the problems now being subjected to resolution is the sea." (*U.S. Naval Institute Proceedings*, 8/71, 103–5)

- Electrical power development for space was summarized by AIAA Electric Power Systems Committee in *Astronautics & Aeronautics* article: "Batteries have continued to be employed beyond Explorer I for either primary power or as an element of power subsystem. Silverzinc batteries provided 13.5 kilowatt-hours (Kwh) of electrical energy for the Mercury spacecraft, up to 63 Kwh for the early Geminis, and 65 Kwh for the Apollo Lunar Module. Specific energies have advanced to 90 watt-hours per pound (wh/lb) [198 wh per kg] in unmanned Agena vehicles. The continuing flight dependence on batteries was dramatically illustrated by the rescue phase of the Apollo-13 mission, which used the LM batteries for power after the failure of the cryogenic supply of the Service Module fuel cell system. Undoubtedly, batteries will remain a vital independent power source in manned missions and perhaps unmanned missions for many years." Committee recommended that "competitiveness of the isotope and reactor systems for powers in the tens of kilowatts be encouraged rather than inhibited and that the technology programs for both be kept viable," felt "that achieving acceptance of a reactor power system for manned Earth-orbiting missions presents a major challenge to the power community in the 1970s," and criticized lack of resources "to pursue effectively all of the advanced reactor power-system options that have been carried along most of the 1960s." (*A&A*, 8/71, 22–30)

September 1971

September 1: Preliminary results of *Apollo 15* scientific experiments were reported by principal experimenters at MSC press briefing.

Columbia Univ. scientist Dr. Gary V. Latham reported that three-station seismometer network completed on moon by *Apollo 15* had detected 39 events, including "most significant event we've recorded on the moon." Event was moonquake 800 km (500 mi) beneath surface—greater depth than for any quake recorded on earth—about 644 km (400 mi) west of crater Tycho. *Apollo 15* seismic data also had revealed that swarms of quakes occurred on moon as they did on earth. Most intense swarm recorded "contained 30 events over a period of just 2 and a half days . . . at very regular intervals, at about 2 hour spacing, and culminated in the largest moonquake yet recorded." Dr. Latham, Apollo program's chief seismic investigator, said data from missions to date had proved "that we can place long life geophysical observatories on the moon and they can exist years, which means that we can expect to maintain a network of stations with larger spacing between missions."

Two heat probes drilled almost two meters (almost six feet) into surface had shown that moon was hot, with heat concentrated deep in interior or in pockets of radioactive mineral 160 km (100 mi) below surface. Temperature of moon increased 0.56 K (1°F) each foot deeper into moon, indicating heat flow from interior was one fifth earth's heat flow. Columbia Univ. scientist Dr. Marcus E. Langseth said this evidence showed that abundance of radioactive minerals inside moon was at least comparable to abundance inside earth. It also suggested moon and earth were formed at same time from same elements.

Cal Tech scientist Dr. Leon T. Silver and U.S. Geological Survey scientist Dr. Gordon A. Swann said *Apollo 15* had discovered green rocks on moon composed of magnificently colored spheres of green glass. Dr. Paul W. Gast, Chief of MSC Planetary and Earth Sciences Div., said 75 kg (165 lbs) of *Apollo 15* lunar rocks had been logged to date and would be distributed to investigators. (Transcript; O'Toole, *W Post*, 9/2/71, A1)

- NASA announced appointment of Dr. Charles A. Berry, Director of Medical Research and Operations at MSC, as NASA Director for Life Sciences. Appointment would become effective when successor for Dr. Berry was found. Dr. Berry would succeed Dr. James W. Humphreys, Jr., who became Secretary-Treasurer of American Board of Surgery. (NASA Ann, 9/1/71; NASA Release 71-166; Am Bd of Surgery PIO)
- Second flight model of LRV Rover was delivered by Boeing Co. to KSC for Apollo 16. (MSFC Release 71-144; KSC PIO)
- MSFC had signed level-of-effort contract with NR Rocketdyne Div. for design of Space Shuttle main engine, NASA announced. The interim contract—for four months at $1 million per month—had been signed

September 1

pending GAO review of protest by United Aircraft Corp. Pratt & Whitney Div. of the contract award [see Aug. 3]. (NASA Release 71-165)
- Former F-14 project manager Capt. Lionel E. Ames, Jr. (USN), said in interview that F-14 fighters would cost $16.6 million each—almost $4 million more than $12.7 million announced by DOD Aug. 10. He said confusion over cost illustrated some of problems he had had as project manager before being replaced. (W *Star*, 9/2/71, A11)
- Finalization of order for 175 F-4 jet fighter aircraft by Federal Republic of Germany was announced by Secretary of Defense Melvin R. Laird at DOD press conference. Order amounted to more than $750 million. (Transcript)

September 2-11: U.S.S.R. launched *Luna 18* unmanned lunar probe on Proton booster from Baykonur at 6:41 pm local time (4:41 pm Moscow time, 9:41 am EDT), inserting vehicle into translunar trajectory from earth orbit. Tass said purpose of mission was to carry out further scientific research of moon and near-moon space. During flight to moon 29 communications sessions were held with spacecraft and data on parameters and systems were transmitted. Midcourse maneuvers were conducted Sept. 4 and 6 and on Sept. 7 *Luna 18* entered lunar orbit with 99.8-km (62-mi) altitude, 1-hr 59-min period, and 35° inclination. Spacecraft made 54 revolutions of moon, testing methods of automatic near-moon navigation, before braking for moon landing.

Luna 18 reached moon's surface Sept. 11 near Sea of Fertility, 3° 34' north latitude and 56° 30' east longitude. Tass announced landing was "unlucky" and communications stopped at 10:48 am Moscow time (3:48 am EDT).

Luna 18 was assumed to have crashlanded. Previous mission in series, *Luna 17*, landed successfully on moon Nov. 17, 1970, and released *Lunokhod 1*, self-propelled lunar rover designed to carry out scientific investigations. (FBIS-Sov-172-3, 9/3/71, L1; 174-3, 9/8/71, L1; 173-3, 9/12/71, L1; *SF*, 11/71, 401)
- *Oao 2*—launched Dec. 7, 1968, and exceeding all design specifications—marked its 1000th working day. (*NYT*, 9/2/71, 14)

September 2: Mariner 9 Mars probe had traveled 244 316 000 km (151 805 000 mi) in solar orbit since launch May 30 and was 39 424 000 km (24 497 000 mi) from earth, traveling at 96 540 km per hr (59 987 mph). Spacecraft was expected to reach Mars' vicinity Nov. 13, collect atmospheric and surface data, and map 70% of Martian surface during 90 days in orbit. (NASA Special Release, 9/2/71)
- Nike-Apache sounding rocket launched by NASA from Churchill Research Range, Canada, carried GSFC experiment to investigate polar cap absorption event. Rocket and instruments functioned satisfactorily; payload recovery was delayed until Sept. 3 because of adverse weather. (NASA Rpt SRL)
- U.S.S.R.'s *Lunokhod 1* lunar rover, landed on moon by *Luna 17* Nov. 17, 1970, had safely endured 10th lunar night and was functioning satisfactorily, Tass announced. (FBIS-Sov-172-3)
- Dr. James C. Fletcher, NASA Administrator, held news conference during first visit to LeRC. He said LeRC would play major role in development

of commercial engines geared toward reducing noise pollution by future jet aircraft. (*Lewis News*, 9/10/71, 1; LeRC PIO)

- Rep. Leonor K. Sullivan (D-Mo.), Chairman of Consumer Affairs Subcommittee of House Banking and Currency Committee, sent to FTC material on Franklin Mint, private corporation that produced silver medallions carried on *Apollo 14* mission Jan. 31–Feb. 9. She suggested FTC investigate promotional use by Franklin of medallions, which had been melted down into 'minimoons'—coins advertised as containing silver that had been to the moon. *Washington Daily News* said Rep. Sullivan's action had been prompted by order issued by *Apollo 14* commander Alan B. Shepard, Jr., Chief of Astronaut Office, that no more Franklin Mint medals be carried on NASA missions. Shepard had noted, *News* said, that *Apollo 14* crew had carried 200 Franklin medallions in CM (which never left lunar orbit), had distributed 150 to friends and relatives on mission's completion, and had returned 50 to Franklin with understanding they would be made into coins for members of Franklin Mint Collectors Society, but not used to promote sales. (Thomasson, Scripps-Howard, *W News*, 9/2/71, 24)

September 3: Successful free-flight development test of parachute system for Viking Mars landers to be launched in 1975 was conducted by NASA at DOT Joint Parachute Test Facility in El Centro, Calif. Test was to verify functioning of main decelerator parachute and its deployment mechanism and confirm structural adequacy of system design. Deployment mortar and parachute—16 m (53 ft) in diameter, packaged in instrumented drop vehicle 3 m (10 ft) long and 51 cm (20 in) in diameter—were dropped from NASA B–57 aircraft at altitude of 15 200 m (50 000 ft). (NASA Release 71–162; FRC PIO)

- President Nixon released statement on dedication of new USAF Museum at Wright-Patterson AFB, Ohio: "The United States Air Force Museum demonstrates in a most compelling fashion the tremendous progress which American aviation has made in a remarkably brief time. It will afford a valuable opportunity for future generations of Americans to learn about the dedication and courage and skill which have made that progress possible."

  At dedication ceremonies President said, ". . . the United States intends, and we resolve today, that having been first in aviation from the beginning, we will attempt to be first in aviation for the time to come. This should be our goal, and one that we can achieve." (*PD*, 9/13/71, 1240–2)

- MSC announced resignation of Dr. Gene M. Simmons, MSC Chief Scientist since 1969. Dr. Simmons would return to MIT as professor of geophysics. (MSC Release 71–63)

- *New York Times* editorial said scientists were pressing NASA to "take another look at the cancelled Apollo 18 and 19 flights—flights for which much equipment is available. The Soviet contribution to the new picture of the moon is still miniscule, despite the feats of the Luna space vehicles. But it would make good sense for the United States and the Soviet Union to get together and plan joint manned moon flights, perhaps by making Apollo 18 and 19 the first truly international ventures to earth's companion in the skies." (*NYT*, 9/3/71, 26)

- International Institute for Strategic Studies published *The Military*

Balance 1971–1972. Annual report of London-based independent research group said Communist China had entered nuclear missile race, U.S.S.R. had moved further ahead of U.S. in ICBM deployment, and U.S.S.R. was challenging supremacy of U.S. underwater nuclear strike force. (Reuters, B *Sun*, 9/3/71, 1)

- *Astronaut Information: American and Soviet* (Revised) (71–204 SP) was published by Library of Congress Congressional Research Service. Paper provided general reference on astronauts and cosmonauts, table of U.S. and Soviet space flights, and description of highlights of each flight. (Text)

September 4: Concorde 001, French prototype of Anglo-French supersonic transport, left Toulouse for Cayenne, French Guiana, on first leg of first transatlantic test flight, to Rio de Janeiro. Flight would promote sale of aircraft in South America and continue performance testing and long-range flying in context of commercial airports. (BAC–Aérospatiale France Release 17C/71)

September 4–5: NASA launched series of two Black Brant IIIB sounding rockets from Resolute Bay, Canada, carrying Univ. of California particles and fields experiments. Flights were partially successful. (SR list; GSFC Proj Off)

September 5: NATO intelligence had reported U.S.S.R. was flight-testing swing-wing supersonic strategic bomber designed to fly at twice speed of sound and equipped for low-level penetration of enemy defenses, *New York Times* said. Aircraft, christened "Backfire" by NATO, could become operational late in 1973. Western military intelligence specialists believed Backfire marked "emergence of entirely new Soviet air weapons system." (Szulc, *NYT*, 9/5/71, 1)

- Exhibits at new $6-million Air Force Museum near Dayton, Ohio, were described by *New York Times*: DC–6 *Independence* used by President Truman; Constellation *Columbine III* and 47J Sioux helicopter used by President Eisenhower; reproduction of Wright 1909 Military Flyer; B–70, only remaining ultrasonic bomber; Douglas World Cruiser *New Orleans*, which flew around world in 1924; Soviet MIG–15; and relics of Wright brothers' *Kitty Hawk*. Museum, "first permanent repository of Air Force memorabilia," was 244 m (800 ft) long, 91 m (300 ft) wide, and 24 m (80 ft) tall and housed 150 aircraft. It was built with funds provided by private gifts. *NYT*, 9/5/71, 28; NASA Hist Off)

- MIT—"probably the world's greatest technological university"—had committed itself to seeking change in current U.S. attitude of hostility toward technology, *Washington Post* reported. In series of statements that "can be taken jointly as a new scientists' manifesto," MIT leaders had urged change in national attitude toward "harnessing technology while not bowing to it," urged end to scientists disclaiming responsibility for the effects of their work, and pledged to make profound changes in their own teaching and research. MIT led U.S. universities in obtaining Federal research funding, with $92 million in FY 1970, more than double that of its nearest rival, Stanford Univ., which had received $39 million. (Cohn, *W Post*, 9/5/71, D2)

September 5–11: First U.S.–U.S.S.R. conference on Communication with Extraterrestrial Intelligence (CETI) was held at Byurakan Astrophysical Observatory in Soviet Armenia. Conference was sponsored jointly by U.S. and Soviet Academies of Science and organized by Soviet astronomer Dr. Yosif S. Shklovsky and Cornell Univ. astronomer Dr. Carl Sagan. Later Dr. Sagan and Cornell Univ. astronomer Dr. Frank D. Drake reported conference conclusions to NAS. Group of scientists had concluded that civilization in another solar system "a few hundred light years away" might be trying to communicate with other bodies, including earth. Man had acknowledged existence of other universes and scientific evidence had shown that amino acids —building blocks of life—could be made from materials and energy sources in space. Scientists had recommended coordinated worldwide effort to intercept messages by searching sky with powerful radiotelescopes. (Sullivan, *NYT*, 9/19/71, 4:8, Auerbach, *W Post*, 9/23/71, A4; NAS PIO)

September 6: Concorde 001, French prototype of Anglo-French supersonic transport, landed at Rio de Janeiro after successful first transatlantic test flight, piloted by French test pilot André Turcat [see Sept. 4]. Turcat told newsmen aircraft had flown at mach 2 for nine minutes during 3240-km (2015-mi) flight. (AP, *NYT*, 9/7/71, 62)

- U.S. and British space experts believed purpose of four Cosmos launches in past 10 mos was to test components for manned lunar landing, *New York Times* reported. *Cosmos 434*, launched Aug. 12, was said to have been test of variable-thrust, restartable rocket engine similar to onboard propulsion system in Apollo CSMs. Other launches identified by experts were *Cosmos 379*, Nov. 24; *Cosmos 382*, Dec. 2; and *Cosmos 398*, Feb. 26. Crucial hardware believed to be missing from launches was Saturn V-size booster. Some of these spacecraft had traveled farther in space than previous Soviet flights testing manned spaceflight equipment; at least three, including *Cosmos 434*, had demonstrated greater ability to maneuver in space than any previous Soviet manned flight. British space expert Geoffrey E. Perry, who had first announced discovery of Soviet launch site at Plesetsk in 1966, had said in telephone interview with *Times* that "the Russians have never made orbital changes to the degree that they have demonstrated on these Cosmos flights." He was convinced that "the Soviet Union is testing a large rocket engine similar to the one aboard Apollo. Even some of the velocity changes made by these Cosmos flights almost exactly match the velocity changes of Apollo for such maneuvers as lunar orbital insertion and trans-earth injection." All four Cosmos were believed to have been launched by Proton, D-type, or SL-4, nonmilitary boosters with about half the thrust of Saturn V. Some were said to have transmitted recordings of human voices as test of radio communications. (Lyons, *NYT*, 9/6/71, 1)

- Soviet supersonic transport Tu-144 flew to Bulgaria on second flight outside U.S.S.R. Aircraft had appeared at Paris Air Show in May and June. (UPI, *NYT*, 9/7/71, 62)

- Dr. Wernher von Braun, NASA Deputy Administrator for Planning, told press in Bremen, Germany, that he expected to visit 50-man U.S. research station on moon within 10 yrs, at end of 1970s or beginning of 1980s. He said, "I'll have to stay there at least eight or 10 days." (AP, *NYT*, 9/7/71, 18)

- Baltimore *Sun* editorial praised progress of NASA-Soviet Academy of Sciences plans for compatible spacecraft rendezvous and docking system: "For all the impediments, actual and potential, to genuine collaboration in an almost illimitable project which no single nation can ever claim for its own alone, America and Russia have managed, it seems, to come to terms on some major technical points." Consultations that produced agreement "indicate the growth of sensible, humane realism in both Washington and Moscow in looking to the future." (B *Sun*, 9/4/71)

September 6–16: Fourth U.N. International Conference on Peaceful Uses of Atomic Energy was held in Geneva. U.S. exhibit included NASA demonstration of applying nuclear techniques to ultra-rapid analyses of environmental materials and drugs, using satellites and computers. Signals from nuclear-powered instruments on moon were relayed to Geneva from GSFC via satellite. NASA also displayed lunar samples and SNAP–27 and SNAP–19 generators.

Mitchell W. Sharp, Canadian External Affairs Minister, said at news conference following opening session that it was time for U.S. and U.S.S.R. to negotiate underground nuclear test ban.

Dr. Glenn T. Seaborg, Conference Chairman and former AEC chairman, said in paper that plutonium 244, heaviest element, could assist in determination of earth's age. In newspaper interview, Dr. Seaborg said leading nuclear scientists agreed that nuclear-generated power was "clean, safe and necessary," but that people of world needed to be convinced. He felt that responsibility for ensuring safety in use of nuclear energy should eventually be transferred from AEC to some other agency but move would be "premature" at present.

Milton Klein, manager of AEC–NASA Space Nuclear Propulsion Office and Director of AEC Space Nuclear Systems Div., reported on NASA plans to send nuclear-powered space probes to outer planets in next five years, followed by manned missions using atomic engines in the 1980s. First mission using nuclear reactor to provide onboard electricity [Pioneer F Jupiter flyby] would be launched in 1972.

U.S.S.R. withdrew from conference program paper it had intended to read entitled: "Development of nuclear power systems for space in the U.S.S.R." and substituted paper on operation of small Topaz reactor. *Washington Post* quoted observers as attributing substitution to U.S.S.R. lag in space nuclear power field. One observer had said only other explanation might be that they were embarrassed by failure of *Luna 18* "to perform on the moon." (AEC Releases O–151, 153; Reuters, *W Post*, 9/10/71, A13; *NYT*, 9/12/71, 20; Sullivan, *NYT*, 9/16/71, 14; O'Toole, *W Post*, 9/16/71, A21)

September 7: U.S.S.R. launched *Cosmos 436* from Plesetsk into orbit with 542-km (336.8-mi) apogeee, 510-km (316.9-mi) perigee, 95.1-min period, and 74° inclination. (GSFC *SSR*, 9/30/71; *SBD*, 9/15/71, 32)

* France's *Eole* satellite, launched by NASA Aug. 16, was functioning almost perfectly after 22 days in orbit. Total of 111 balloons had been launched from three Argentina sites for *Eole* to track and acquire data from on winds, temperatures, and pressures. Five had completed one revolution around earth at 11 850-m (38 880-ft) altitude and had been tracked by *Eole*. Only five balloons had failed to respond to satellite commands and were assumed lost. Balloons would be launched at rate of 8 to 10 per day until about 400 to 450 balloons were launched;

another 50 to 100 balloons would be launched to fill in gaps in array caused by failures. (NASA Proj Off)

- Nike-Apache sounding rocket launched by NASA from Wallops Station carried GCA Corp payload to 176.4-km (109.6-mi) altitude to determine winds from vapor trail using ARC CV-990 instrumented aircraft as observing platforms, test narrow-band camera filters on chemical releases with bright sky background, and provide real-time checkout of observational equipment and procedures for BIC experiment with low-altitude barium release. Payload contained standard barium and sodium-lithium canisters. Cloud was photographed by both ground and CV-990 aircraft cameras. Flight was first of two; second rocket would be launched Sept. 16. (NASA Rpt SRL)

- Plans to store lunar rocks in vault-like facility at MSC to retain their original condition for future researchers were described by NASA officials in AP interview published by *New York Times*. LRL curatorial and research projects would be consolidated in specially air-conditioned facility covering 465 sq m (5000 sq ft) of existing laboratory space scheduled for completion in time for March 1972 Apollo 16 mission. Dr. John H. Pomeroy, NASA's Assistant Director for Lunar Sample program, said project had been "a very important aim since the very beginning, because we realize these are unique and irreplaceable samples and it may be years before we go back to these places." (*NYT*, 9/7/71)

- Patent No. 3 603 433 was granted to MSC technicians William H. Keathley and Clarence J. Wesselski for spacecraft emergency shock absorbers that might also be used in automobile bumpers. Disaster-avoidance device backed up cyclic deformation system in struts of Apollo CM couches. System, installed in spacecraft since *Apollo 11* to absorb energy from over-propulsion at launch or unusually hard landing, consisted of tightly fitted washers on rod. Pressure against end washer pushed one after another along line as force decelerated. (Jones, *NYT*, 9/11/71, 33; Patent Off PIO)

- James J. Kramer, head of Quiet Engine Program at LERC, became Chief of newly formed Noise and Pollution Reduction Branch of NASA's Aeronautical Propulsion Div. He would be responsible for planning and overseeing research at NASA Centers on ways to reduce noise and pollutants from jet aircraft engines. (NASA Release 71-170)

- NSF published statistics that showed Federal R&D funding was on upward trend. Figures were taken from soon to be published report *Federal Funds for Research, Development, and Other Scientific Activities*, Vol. XX. Estimated 9% increase in overall Federal funding for R&D from 1970 to 1972 indicated upward trend after decline that began in 1968. Total of $15.3 billion in actual obligations for 1970 established low for recent years. Although 1971 Federal R&D total was expected to be same—$15.4 billion—total $16.7 billion requested in President Nixon's 1972 budget marked new high.

  NASA had risen from 5% share of Federal R&D in 1960 to high of 34% in 1965 but was expected to fall back to 19% in 1972. Final phase-out of Apollo and reduced NERVA engine development were expected to be nearly offset by increase in other NASA programs so that 1972 would be almost a leveling off. (NSF *Highlights*, 9/7/71)

September 8: NASA's supercritical wing, flown on TF–8A jet aircraft piloted by NASA test pilot Thomas C. McMurtry, completed 11th flight from FRC. Purposes of one-hour flight were to complete evaluation of flutter characteristics at high subsonic and low supersonic speeds and obtain additional pressure, distribution, and performance data. Aircraft reached 13 700-m (45 000-ft) altitude and all objectives were achieved. Flutter characteristics were evaluated at mach 0.97, 1.05, and 1.10; pressure distribution and performance data were obtained from mach 0.80 to 0.95 and at mach 1.05 and 1.10. (NASA Proj Off)

- *Ats 3* applications technology satellite, which had temporarily stopped transmitting weather pictures because of locked-on antenna [see Aug. 24], was operating again, NASA announced. Although spacecraft was still not working properly, it was transmitting cloud-cover pictures of Western Hemisphere during the five per cent of the time when camera was aimed at earth. (NASA Release 71–171)

- Apollo Telescope Mount (ATM) prototype was flown to MSC from MSFC for extensive thermal-vacuum-chamber testing. Tests would continue through early December. (MSFC Release 71–146; MSFC PIO; MSC Hist Off)

- U.S. and U.S.S.R. had agreed to replace Washington-Moscow teletype "hot line" with comsat capable of providing instantaneous link between capitals in any crisis, *Washington Post* reported. Diplomatic sources had said agreement had been worked out by U.S. and Soviet communications experts meeting at SALT in Helsinki. System would use existing comsats orbited by both countries. (Getler, *W Post*, 9/8/71, 3)

- Initial results from U.S.S.R.'s Vertikal-2 rocket launched Aug. 20 were reported by Tass. Experiments had been conducted at altitudes between 100 and 463 km (62 and 288 mi). Photos had registered image of sun in several spectral ranges and had indicated that solar flares took place during rocket's flight. Data had been obtained on temperature, electron density, and other parameters of sun's corona and on flux of particles causing ionospheric turbulence. (FBIS–Sov–174–3, 9/8/71, L2)

- Soviet Foreign Minister Andrey A. Gromyko submitted memorandum to U.N. Secretary General U Thant expressing hope that U.N. General Assembly would call world disarmament conference on nuclear and conventional weapons. Memo also reiterated Soviet proposal for disarmament conference attended by nuclear powers—China, U.K., France, U.S., and U.S.S.R. (Tanner, *NYT*, 9/9/71, 8)

- Senate by unanimous vote of 53 agreed to ratification of Convention for Suppression of Unlawful Seizure of Aircraft [see April 15]. (*CR*, 9/8/71, 13892–95)

- Sen. William Proxmire (D-Wis.) released GAO report that showed official cost estimates for F–14 USN jet fighter aircraft had risen from $11.5 million to $16.7 million per aircraft since March 1971. Sen. Proxmire said he was "drafting amendments" to military authorization bill before Congress "which would terminate" $5-billion F–15 project. (*W Post*, 9/9/71, 2)

- NRC released *Employment of New Ph.D.'s and Postdoctorals in 1971*: Science Ph.D.s who received degrees in 1970 had found fewer jobs than had 1969 graduates and were more likely to remain unemployed or to take jobs unrelated to their training. Of 13 000 1970 Ph.D.s surveyed, 70.2% had found suitable employment, 14.3% had entered postdoctoral training, and 2.8% were jobless or had accepted inappropriate employment. In 1969, 75.3% of graduates had found appropriate jobs, 9.7% had entered postdoctoral training, and 1.6% were unemployed or underemployed. (NRC Release)

- President Nixon submitted to Senate nomination of Dr. Glenn T. Seaborg, former AEC Chairman, as U.S. Representative to General Conference of International Atomic Energy Agency in Vienna, Sept. 21–27. Among nominees for alternate representatives was Dr. T. Keith Glennan, first NASA Administrator (1958–1961). (PD, 9/13/71, 1250, 1261)
- White House announced appointment of William M. Magruder, former Director of SST Development for DOT, as special consultant to President Nixon. (PD, 9/13/71, 1260)

September 8–10: NASA and United Cerebral Palsy Research Foundation sponsored conference on neurologically handicapped at ARC. Meeting—attended by scientists, physicians, and engineers—discussed application of space age technology to neurological disorders. (NASA Release 71-161; ARC PAO)

September 9–10: NASA's Stratoscope II telescope was successfully launched by giant balloon from Redstone Arsenal Army Airfield at 7:33 pm EDT on eighth flight in series to obtain sharper astronomical pictures than possible with earth-based photography. Three prime scientific targets were Galaxies M31 (Andromeda), M32, and Planetary Nebula NGC 7662.

Stratoscope II, reused with modifications for each flight, weighed 3800 kg (8400 lbs) and had optical system with 91-cm (36-in) aperture and photographic resolution of 0.1 arc sec—equal to ability to distinguish two objects 76 cm (30 in) apart at distance of 1600 km (1000 mi). Photographs would be made in visible region of spectrum (4000–6000 Å).

Balloon-borne telescope, 198 m (650 ft) long, reached operating altitude at 25 200 m (82 800 ft) and performed flawlessly during night-long flight. Stratoscope drifted westward at 32 to 40 km per hr (20 to 25 mph) and returned to earth 96.6 km (60 mi) northeast of Little Rock, Ark., Sept. 10.

Mission objectives were accomplished and scientific results were being evaluated. Instrument package was damaged during landing, apparently striking top of oak tree, but film was reported intact. (MSFC Releases 71-53, 71-141; *Huntsville Times*, 9/10/71, 9; 9/11/71)

September 9: President Nixon honored *Apollo 15* astronauts in address on economic program before Joint Session of Congress: "Later today, in this great chamber, the Congress will pay tribute to three splendid Americans back from the moon. Theirs was a magnificent achievement, a stunning testament to their personal skill and courage, and also to what American technology can achieve.

"Let us find the means to ensure that in this decade of challenge, the remarkable technology that took these Americans to the moon can also be applied to reaching our goals here on earth." (PD, 9/13/71, 1250–6)

- Senate and House met in Joint Session to receive and pay tribute to *Apollo 15* astronauts, who reported on mission. Mission commander David R. Scott said of *Apollo 15* lunar samples, "As I approached a

fragment freshly deposited on the slopes of the Apennine, I realized it had been undisturbed since before life came out of the sea on earth." Astronaut Alfred M. Worden said first thing noticed after launch, "particularly when we got into earth orbit, was that we had a further view—we were further away from earth, and our view was expanding. We did not see any area around Cape Kennedy. What we saw were continents and oceans. . . . After we left earth orbit and for the remainder of the flight our view was one of the earth." Astronaut James B. Irwin described Hadley-Apennine campsite as "friendly, beautiful spot, like a valley in the high mountains of the earth; and it was with some reluctance that we left our valley on the moon." He told Congress: "The future of man's role in space lies in your hands. We realize very well the grave responsibilities in these days. The decisions you make will determine man's travels and explorations in space. We think that an investment in space technology is an investment in America. America needs space to grow." (CR, 9/9/71, H8224–5)
- Emergency loan of $250 million was granted to Lockheed Aircraft Corp. by Emergency Loan Guarantee Board. Board was established by President under emergency loan bill approved by Congress Aug. 2. (AP, B *Sun*, 9/10/71, A7)
- President Nixon submitted to Senate nominations of representatives and alternates to 26th session of U.N. General Assembly. They included *Apollo 14* Astronaut Alan B. Shepard, Jr., Chief of MSC Astronaut Office, as alternate. (*PD*, 9/13/71, 1257, 1261)

September 9–10: Sixth Annual Aerospace Mechanism Symposium was sponsored at ARC by ARC, Univ. of Santa Clara, and Lockheed Missiles & Space Co. Films from *Apollo 15* LRV Rover on moon were shown. Soviet scientist V. S. Syromyatnikov and MSC engineer James C. Jones presented papers on space vehicle docking mechanisms. (NASA Release 71–169; ARC *Astrogram*, 9/16/71, 1)

September 10: USAF launched two unidentified satellites from Vandenberg AFB by single Thorad-Agena booster. One entered orbit with 220.5-km (137-mi) apogee, 169-km (105-mi) perigee, 88.3-min period, and 74.9° inclination. It reentered Oct. 5. The other entered orbit with 506.9-km (315-mi) apogee, 487.6-km (303-mi) perigee, 94.5-min period, and 75° inclination. (Pres Rpt 72; *SBD*, 9/23/71, 64)
- U.S.S.R. launched *Cosmos 437* from Plesetsk into orbit with 545-km (338.7-mi) apogee, 520-km (323.1-mi) perigee, 95.2-min period, and 74° inclination. (GSFC *SSR*, 9/30/71; *SBD*, 9/15/71, 32)

September 11: French CNES satellite control station at Bretigny, France, inadvertently transmitted destruct command to balloons being tracked by *Eole* satellite (launched Aug. 16), causing 70 of 135 flying balloons to be cut down. Although error was not expected to degrade planned scientific accomplishments of mission significantly, it would delay beginning of scientific experiment by 7 to 10 days. CNES instituted new training procedures to *Eole* control station personnel on transmitting commands to satellite and designed system which would alarm when certain critical commands—such as balloon destruct, gravity gradient retraction, and activation of magnetic hysteris rods—were to be placed into *Eole*. (NASA Proj Off)

September 13: S–IC–511 1st stage for Saturn V launch vehicle that would launch Apollo 16 toward moon in 1972 was shipped by barge from

Michoud Assembly Facility, to arrive at KSC Sept. 17. (MSFC Release 71-152)

- MSC announced contract awards. Grumman Aerospace Corp. received $4 364 849 supplemental agreement for changes in Apollo LM contract. Agreement formally incorporated procedural changes previously authorized by NASA and brought total value of contract since January 1963 to $1 788 034 399.

  Boeing Co. was awarded $4.3-million, one-year, cost-plus-fixed-fee contract for reliability, quality assurance, and flight safety engineering at MSC. (MSC Releases 71-67, 71-68)

- Senate by vote of 55 to 21 rejected amendment proposed by Sen. Gaylord Nelson (D-Wis.) to H.R. 8687, FY 1972 military procurement authorization. Amendment would have frozen funds for USN F-14 and USAF B-1 and F-15 aircraft programs until final environment impact statement had been made on programs. (CR, 9/13/71, S14181-14210)

- Pressure was increasing on Nixon Administration to relax aeronautical satellite system policy to allow first preoperational system to be jointly owned by ESRO and FAA, *Aviation Week & Space Technology* reported. Original decision that system should be owned by commerical company with FAA leasing its service had become unpopular because of legal complexities. (*Av Wk*, 9/13/71, 21)

September 14: U.S.S.R. launched *Cosmos 438* from Plesetsk into orbit with 273-km (169.6-mi) apogee, 181-km (112.5-mi) perigee, 89-min period, and 65.3° inclination. Satellite reentered Sept. 27. (GSFC *SSR*, 9/30/71; *SBD*, 9/23/71, 64)

- NASA announced appointment of Lee R. Scherer, Apollo Lunar Exploration Director, as Director of Flight Research Center, NASA's facility for flight-testing high-performance aeronautical vehicles at Edwards, Calif. Appointment would be effective Oct. 11. Until then De E. Beeler, Deputy Director, would continue as Acting FRC Director. Scherer had been Assistant Director for Lunar Programs and Manager of Lunar Orbiter Program from its inception in 1963 through its successful completion in 1967. He had received NASA's Exceptional Service Medal in 1967 and NASA's Exceptional Scientific Achievement Medal in 1969. (NASA Release 71-176)

- *Apollo 15* anomalies were discussed by Apollo Mission Director Chester M. Lee during NSC luncheon and *Apollo 15* film showing in Washington, D.C.: "I have been queried and have read comments that . . . we had more anomalies than we have seen in previous missions. In fact this is not so." List of 40 anomalies had been compiled by NASA. "Twenty of these had to do with the experiments and the LRV, most of which were carried for the first time on this mission. This compares with 30 total anomalies on Apollo 14, 44 total anomalies on Apollo 12 and 43 on Apollo 10. Therefore I would conclude that we are not retrogressing in so far as the Apollo hardware performance is concerned." Lee said lunar stay of 73 hrs was planned for Apollo 16. "The additional 6 hrs will not give us more EVA time than we planned on Apollo 15, but will allow more time for EVA preps, sleep, eating, etc." (Text)

- Nonflying test model of Lockheed C-5A jet cargo aircraft developed wing crack during static testing at Lockheed's Marietta, Ga., facilities.

Crack, near fuselage, occurred as wing was subjected to stress of approximately 130%. (Levine, *WSJ*, 9/16/71, 9)

September 15: NASA successfully completed 12th and 13th flights of TF-8A jet aircraft equipped with supercritical wing, flown from FRC by NASA test pilot Thomas C. McMurtry. TF-8A flight test program was being accelerated in anticipation of adverse landing conditions when lake bed at Edwards, Calif., became wet later in year. Purposes of flights were to expand flight envelope to higher dynamic pressures, obtain pressure distribution and performance data at higher dynamic pressure, evaluate unaugmented stability and control characteristics, and investigate buffet boundary for trim flight at design condition of mach 0.99 at 14 000 m (46 000 ft). (NASA Proj Off)

- Dr. James C. Fletcher, NASA Administrator, described planned Skylab mission medical research in address before Utah State Medical Assn.: ". . . we will be looking far beyond the requirements of the immediate mission; we will be seeking answers to questions with a strong impact on the future of manned space flight during the remainder of this century." Major questions as yet unanswered after Gemini and Apollo flights were the causes of moderate loss of weight by astronauts early in flight, moderate cardiovascular deconditioning, moderate loss of exercise capacity, and minimal loss of bone density. Medical results from Gemini and Apollo missions had shown that "1. There were no major surprises. 2. As of now, we see no reason why man cannot live and work effectively in space for a long period of time. 3. Man seems to adapt to space flight more easily than he does to Earth's environment after returning from space." (Text)

- *Apollo 15* astronauts were honored by Chicago with parade, special City Council meeting, and civic luncheon. Mayor Richard J. Daley presented them with honorary citizenship medals. (Wolfe, *C Trib*, 9/16/71)

- U.S.S.R.'s *Lunokhod I* lunar rover had successfully completed 11th lunar day of work on moon since it was landed by *Luna 17* Nov. 17, 1970, Tass announced. During 11th lunar day rover had made "a detailed survey of the structure of the moon surface in the area" and "panoramic route survey of the surrounding locality," traveling 100 m (328 ft) before being powered down for 11th lunar night. (FBIS-Sov-180-3, 9/16/71, L1)

September 16: Discovery of six amino acids of extraterrestrial origin in Orgueil meteorite that fell in France in 1964 was announced by ARC team of scientists at annual meeting of American Chemical Society in Washington, D.C. Findings further supported "pattern of chemical evolution" theory adopted by Dr. Cyril A. Ponnamperuma and ARC team after their discovery, announced June 18, that identical amino acids and pyrimidines existed in Murray meteorite that fell in Kentucky in 1950 and Murchison meteorite that fell in Australia in 1969. Theory held that specific set of amino acids and other life chemicals found in three meteorites might be part of basic pattern leading to origin of life. All three meteorites were believed to have come from asteroid belt and to date from formation of solar system, about 4.5 billion yrs ago. (NASA Release 71-177)

- President Nixon was host to *Apollo 15* astronauts and families at White House dinner. (*PD*, 9/20/71, 1288)

- Najeeb E. Halaby, Pan American World Airways President and Chairman, discussed space shuttle at Symposium of Society of Experimental Test Pilots in Los Angeles: "NASA has already shown that it appreciates the contributions airlines can make to the program, by having airlines participate with space contractors in the development of the space shuttle" to "make the eventual space shuttle work more like a standard airliner than a quarterly experiment." Since it took less energy to orbit objects than to fly them across U.S., "there's no reason why the cost of Space Shuttle operations should not become as low or lower than that of jets." Pan Am's "most glamorous vision" was "Space Shuttle as an air transport, carrying passengers from New York to Tokyo—or Peking—in 45 minutes, or from Los Angeles to Rome in 40."

    "We can visualize a rocket-powered vehicle with 100 passengers . . . launched vertically in a suborbital trajectory at the precise azimuth for its intended destination, re-entering the atmosphere without excess g-forces, gliding unpowered to an altitude where its conventional jet engines will be started, then landing on a runway like a conventional airliner." (Text)

- Nike-Apache sounding rocket launched by NASA from Wallops Station carried GCA Corp. experiments. Payloads ejected trail of sodium and lithium during ascent from 80-km (50-mi) to 170-km (105-mi) altitude and then ejected barium at 170-km altitude. Both trails produced glowing, colored clouds over mideastern U.S. coastline. Data on wind conditions were obtained by photographing motion of vapor trails and clouds from cameras aboard ARC Convair 990 aircraft. (WS Release 71-18)

- Senate confirmed nominations of *Apollo 14* commander Alan B. Shepard, Jr., as Alternate U.S. Representative to 26th session of U.N. General Assembly and of Dr. Glenn T. Seaborg, former AEC Chairman, and Dr. T. Keith Glennan, former NASA Administrator, as U.S. Representative and Alternate to 15th session of General Conference of the International Atomic Energy Agency at Geneva. (CR, 9/16/71, S14433)

September 17: Age of *Apollo 15* "Genesis rock" was approximately 4.15 billion years, Dr. Liaquat Husain and Dr. John F. Sutter, State Univ. of New York at Stony Brook scientists, announced at press conference. Anorthosite sample brought back by Astronauts David R. Scott and James B. Irwin during July 26–Aug. 7 mission was 150 million yrs older than oldest *Apollo 12* rock, Husain said. "This is the oldest lunar rock found on any of the missions so far." (Fine grains of soil from earlier missions had been found to be nearly 4.5 billion yrs old and one *Apollo 12* rock had contained some fragments that had crystallized 4.5 billion yrs ago but rock itself was thought to have solidified 4 billion yrs ago.)

Although rock's age was not the hoped 4.6 billion yrs—estimated age of solar system—it supported theory that moon was once molten mass. Scientists said precise age of rock could range from 3.95 to 4.35 billion yrs. Further tests had been scheduled to narrow margin. Age had been determined by measuring relationship of radioactive potassium and argon in rock samples using rare-gas mass spectrometer and nuclear reactor. Cal Tech geologist Dr. Leon T. Silver said,

September 17: *The white anorthosite "Genesis rock" collected on the moon by Apollo 15 astronauts was about 4.15 billion years old, scientists announced. In the photo the sample, oldest lunar rock collected so far, was in the nonsterile processing line of the Lunar Receiving Laboratory at the Manned Spacecraft Center.*

"This is unquestionably one of the most important rocks ever returned from any lunar mission. It has a recording in its minerals of many series of events in lunar history—events of high temperature and high pressure." (Auerbach, *W Post*, 9/18/71, A3; AP, *B Sun*, 9/18/71, A10; *NYT*, 9/18/71, 25)

- MSC announced start of series of cold weather environment tests on performance of Skylab CM postlanding and recovery systems in Climatic Laboratory at Eglin AFB, Fla. Objectives were to determine performance characteristics and limitations of Skylab CM in cold weather, determine CM test crew response to cold and to interior environmental conditions of CM, and determine extent of ice buildup on CM exterior and effect of ice on systems performance. Tests, conducted in chamber with 274 K (33°F) water temperature and 269 K (25°F) air temperature, were necessary because spacecraft would pass over areas colder than previously experienced and, in event of emergency landing, CM and crew limitations had to be known. (MSC Release 71–70)

- Letter by Dr. John M. Logsdon of George Washington Univ., responding to criticism of his book *The Decision To Go to the Moon*, was published in *Science*. Book had been reviewed by L. Vaughn Blankenship in *Science* July 23. Logsdon wrote: "Blankenship reports that I find the Apollo decision 'a good one' and 'evidence of a political system operating at its best.' I do no such thing. I do suggest . . . that 'the decision to go to the moon was a representative American action,' and was the product of a process 'typical of the way many major decisions are reached.' " (*Science*, 9/17/71, 1079–80)

- New Orleans *Times-Picayune* editorial commented on failure of Soviet *Luna 18* mission: "The Soviet's automated designs and performances are not to be belittled, for they represent research and development toward deep-probe capability. But it is interesting that Western space observers think they have grounds for suspecting that the Soviets are in fact working up a manned lunar program. If so, more than prestige and technological display are involved; it must mean agreement that man operates on the moon with the best cost-benefit ratio." (New Orleans *Times Picayune*, 9/17/71)
- USAF's "fly-before-you-buy" approach to procurement of new equipment via prototypes was described by Gen. George S. Brown, AFSC Commander, in speech before 15th Symposium of Society of Experimental Test Pilots: "We want the confidence of demonstrated performance in our acquisition programs before deciding on production" to provide "a backlog of promising technological approaches from which future requirements can be met as they develop. While this may cost more initially, in the long run we will have more proven technology . . . and consequently better performing weapon systems at more reasonable costs." (Text)
- *September 19:* NASA announced it had granted Univ. of Denver limited exclusive patent rights to frangible-tube energy absorber conceived and developed at LaRC in 1961 and patented by NASA in 1964. Seven-year rights permitted manufacture, use, and sale of device for trailer support wheel structures, passenger automobile bumpers, and elevators. DOT was considering its use on highway guard rails. Device, made of hard metal alloy, absorbed shock by shattering when pressed against suitably shaped die by downward force. (NASA Release 71–173)

September 19–22: NASA launched series of 13 Nike-Cajun sounding rockets from Kourou, French Guiana, carrying GSFC grenade payloads. Data would be correlated with data from five French Centaure vapor flights. Objective was to determine solar-driven first and second harmonic oscillations in temperature, pressure, density, and winds in upper atmosphere in equatorial zone between altitudes of 35 and 100 km (22 and 62 mi). Each Nike-Cajun carried 31 grenades to altitudes from 114 to 125 km (70–78 mi). All but four grenades were ejected and exploded as planned, and all 13 flights were rated successful. (NASA Rpts SRL)

September 20: Joint West German and NASA barium-ion-cloud (BIC) probe was launched from Wallops Station by Scout booster at 7:31 pm EDT. Objectives of cooperative project were to study behavior of barium ion cloud released at several earth radii and to determine what broad features of earth's magnetic and electric fields could be deduced.

Barium was released into space at 11:05 pm EDT above Central America. Cloud was visible over North and South America and observations were made from all prime ground-based optical sites for one to one and one half hours. Parameters of 4th-stage burnout were 31 479-km (19 560-mi) altitude, 6.77° N latitude, and 74.26° W longitude. Barium cloud was visible to naked eye for approximately 10 min. Cloud elongated along its local magnetic field line to length of about 10 000-km (6200-mi). NASA provided vehicle support necessary to accomplish mission and to make observations at NASA sites. (NASA Proj Off; WS Release 71–19)

September 20

- *Time* magazine described Soviet swing-wing supersonic strategic bomber Backfire [see Sept. 5]. Aircraft believed to have been designed by Andrey N. Tupolev, was 39.9 m (131 ft) long and weighed 113 400 kg (250 000 lbs). Aerodynamicists believed it to be made of stainless steel with titanium to resist heat stress. Airframe had skin bonding instead of riveting. Wings were in forward position and jackknifed back about 40° for speeds of mach 2.1 at 152 000 m (50 000 ft) or mach 1 at 152 m (500 ft). Aircraft was "apparently crammed with sophisticated electronic aids to confuse radar tracking." Carrying crew of three, it had estimated payload of 50 megatons in weaponry, including parachute-dropped hydrogen bombs. "With one refueling, Backfire appears capable of striking the U.S. and returning home." Intelligence sources also had reported that U.S.S.R. was "well along in the design of the MIG-25 and MIG-27; they are potential successors to the MIG-23 'Foxbat,' one of the most advanced fighters in the world." (*Time*, 9/20/71)
- *Apollo 15* Astronauts David R. Scott, James B. Irwin, and Alfred M. Worden received Order of Leopold from King Leopold of Belgium in Brussels. (*NASA Activities*, 10/15/71, 172)
- Sen. William Proxmire (D-Wis.) issued statement that C-5A program was expected to cost $4.9 billion, increase of $300 million over 1970 figures: "Taking into account that the program has been reduced from 120 planes to 81, we find an overrun of $2.2 billion. On a unit basis, the costs have risen from $28 million originally estimated for each C-5A to $60 million, an increase of over 100%." (Text)
- FAA, American Airlines, and Aeronautical Radio, Inc., in October would begin three- to-six-month operational test and evaluation of automatic data link communications techniques, FAA announced. American Airlines 747 aircraft equipped with data link system that would print out air traffic control clearances, terminal advisories, and other information in cockpit and automatically process pilot position and weather reports to ground would test whether techniques could reduce communications workload on pilots and controllers. (FAA Release 71-152)
- Office of Management and Budget had released $98.1 million in FY 1972 funds voted for science education and institutional support, while impounding $30 million, *Aviation Week & Space Technology* noted. "The 25% holdback of funding at the basic science level is considered by some in Congress as highly destructive because it will tend to reduce the number of future scientists." It was viewed as "direct contradiction of the pro-science policy they have been led to anticipate." (*Av Wk*, 9/20/71, 15)

September 20–26: International Astronautical Federation held 22nd Congress in Brussels.

Apollo 15 Astronaut David R. Scott told opening session that analysis of 2.7-m (9-ft) *Apollo 15* lunar core sample had disclosed 58 distinct layers that represented 58 "chapters in the history of the sun." Number of "pages in each chapter" awaited further analysis. Later Scott said in interview that Dr. Liaquat Husain, State Univ. of New York at Stony Brook scientist who had analyzed *Apollo 15* "Genesis" rock, was concerned that alteration of rock by shock when it impacted with site at which it was found might have partly invalidated determination of rock's age at 4.15 billion yrs.

ARC engineer John C. Arvesen described development of method for monitoring biological productivity from air by detecting signs of chlorophyll. Monitoring device, when tested over California lakes and offshore areas, had produced results that conformed closely to readings obtained by arduous direct sampling of waters. Ultimate goal was to develop satellite-borne system that could monitor global productivity of oceans.

Dr. Charles A. Berry, NASA Director of Life Sciences, said newly completed studies of medical data from *Apollo 14* had shown radiation exposure to crew that was more than twice that of previous Apollo missions. Primary reason for higher exposure was passage of *Apollo 14* spacecraft close to Van Allen radiation belts. Dr. Berry said radiation dose was "of no hazard or biological significance." Soviet scientists said later that cosmonauts carried special antiradiation drug on all space flights but had never used it.

Dr. Berry said *Apollo 14* Astronaut Stuart A. Roosa had lost 27% of fluid within his body cells during mission. His total body water had dropped 18% and red cells in his bloodstream had decreased 9%. Losses had been far less for other *Apollo 14* astronauts and no such dramatic losses had been recorded for *Apollo 15* astronauts, indicating that individual responses to space varied widely. Heavy loss of "exchangeable" potassium had been noted in astronauts during extended missions. On *Apollo 15,* loss had been as great as 15%. Potassium was deemed exchangeable because it could move readily in and out of tissues and thus could be purged from body.

E. G. Johnson, Equipment and Facilities Branch Chief in NASA Space Nuclear Systems Office, described program for development of teleoperator systems in which remote-controlled devices could perform tasks in situations dangerous to health or safety of humans. In space, teleoperator could enter high radiation environment of space station's nuclear power plant to change fuel elements or make repairs. Johnson said optical or radioastronomy observatory on moon's far side could be run for 40 to 50 yrs using computer-controlled teleoperators whose control could be assumed from earth when necessary.

Fourth International Orbital Laboratory Symposium was held during IAF congress. Sessions discussed earth resources technology and scientific research objectives and needs of developing countries in regard to earth resources research.

IAA Fifth History of Astronautics Symposium on Sept. 23 was chaired by JPL Historian R. Cargill Hall. Soviet Cochairman V. N. Sokolsky did not attend for unexplained reason. Papers read included "From the History of the Development of Powder Rockets in the U.S.S.R.," by Soviet Academician Yu A. Pobedonostsev, which traced work on Soviet solid propellants leading to Katyusha rocket in current use; "Comparative Analysis of Projects for Jet Planes Constructed up to the 1940's," by V. N. Sokolsky; "Evolution of Spacecraft Attitude Control Concepts before 1951," by R. E. Robertson of Univ. of California at San Diego; and "Memoir: The Evolution of Aerospace Guidance Technology at M.I.T., 1935–1951," by Dr. C. Stark Draper.

Fourth International Orbital Laboratory Symposium, held during Congress, discussed earth resources technology and scientific research objectives and needs of developing countries.

Daniel and Florence Guggenheim International Astronautics Award was presented to Professor Luigi Broglio of Univ. of Rome for achievements in experiments with floating platforms to launch satellites into equatorial orbits.

Only cosmonaut attending IAF Congress was Boris B. Yegorov, only physician to travel in space (launched in *Voskhod 1* Oct. 12, 1964). (IAA Annual Rpt 1971; Sullivan, *NYT*, 9/21/71, 9/23/71, 9/27/71; Proceedings)

September 21: U.S.S.R. launched *Cosmos 439* from Plesetsk into orbit with 278-km (172.7-mi) apogee, 208-km (129.3-mi) apogee, 89.3-min period, and 65.4° inclination. Satellite reentered Oct. 2. (GSFC *SSR*, 9/30/71; 10/31/71; *SBD*, 10/1/71, 101)

- General Dynamics Corp. released F–111 program status report. Aircraft had amassed total of 115 209 hrs on 44 463 operational and test flights. Airframe static tests had been successfully completed with some standards exceeded by 20%. Tests on structural fatigue life of aircraft were continuing. Tactical Air Command versions of F–111 were being tested through 16 000 hrs, or four service lives, and would soon be assigned to service units. Strategic Air Command version of airframe had successfully passed one of four service-life tests required. To date, 356 F–111s had passed proof tests begun in 1969 and 291 of these were back in service. F–111 had accumulated better safety record at 100 000 hrs than any other Century Series fighter; 15 had been destroyed, contrasting with 40 F–104s. USAF, Boeing Co., and Convair Aerospace Div. team had completed successful series of launches of short-range attack missile (SRAM) to be used on FB–111s and B–52 bombers. During launches from WSMR, F–111 had demonstrated ability to launch five SRAMs in rapid succession at five different targets. (Text)

- Attempts by four Soviet aquanauts to break 59-day record for underwater isolation were terminated after 52 days. Men had been evacuated from oceanographic laboratory *Chernomor* at 15-m (50-ft) depth after violent storm on Black Sea broke power and communications cables of submarine's chamber. *Izvestia* said Nov. 11 that experiment had been pronounced successful because of medical and oceanographic findings, although it had failed to exceed record set by U.S. Project Tektite I April 15, 1969. (Shabad, *NYT*, 11/12/71, 10; *A&A 1969*)

September 22: ARC announced completion of acceptance tests on X–14B, VTOL aircraft that could simulate hovering flight characteristics of most existing and proposed VTOLs. New NASA research aircraft, modification of Bell X–14A, was believed to be first digital-computer-driven VTOL flight-simulator aircraft. X–14B would be used by NASA to determine optimum handling qualities for any VTOL aircraft. (ARC Release 71–47)

- LeRC received its sixth IR–100 award from *Industrial Research Magazine* during ceremony at Museum of Science and Industry in Chicago. Award, presented annually to developers of 100 most significant new products, was for development of radiation-chemistry method of preparing high-purity metals and chemical compounds. Method was developed by LeRC scientists Dr. Warren H. Philipp and Stanley J. Marsik. (*Lewis News*, 9/24/71, 1)

- Air Force Assn.'s Schilling Trophy was presented to *Apollo 15* Astronauts David R. Scott, James B. Irwin and Alfred M. Worden in recog-

nition of "their unique contribution to manned spaceflight and science." (*NASA Activities*, 10/15/71, 172; AFA PIO)
- Nomination of Sen. Howard W. Cannon (D-Nev.) for Wright Brothers Memorial Trophy for outstanding service to aviation was announced by NAA. Citation was "for his continuing energetic advocacy of and lasting contribution to the development and use of aviation both as a viable national transportation system and as an essential element in maintaining a strong military posture." Trophy would be presented in Washington, D.C., Dec. 17. (NAA Release)
- Senate rejected proposed amendment to H.R. 8687, FY 1972 military procurement bill, by Sen. George S. McGovern (D-S.D.). Amendment would have reduced funds for USAF development of B-1 bomber and related programs from $370 million to $31 million. (*CR*, 9/22/71, S14784-14847)
- Soviet aircraft designer Andrey N. Tupolev was quoted by Tass as saying Soviet designing bureau would do all it could to put Tu-144 supersonic transport in service as soon as possible. Aircraft had returned after successful visit to Bulgaria Sept. 6. During flight it had developed speed of 2300 km per hr (1400 mph). (FBIS-Sov-71-185, 9/23/71, L1)
- NASA announced availability of *Shuttle Technology Program—Fiscal Year 1972*, compilation of projects making up NASA FY 1972 Space Shuttle Technology Program. (NASA Release 71-183)

September 22-23: Command stored in *Mariner 9* computer before May 30 launch towards Mars switched spacecraft's radio transmitter from low-gain omnidirectional antenna to high-gain antenna for increased signal strength. On Sept. 23 spacecraft was 14 million km (9 million mi) from Mars and had traveled approximately 290 million km (180 million mi) in curving trajectory. Increasing distance between *Mariner 9* and earth required concentration of radio signals in narrow beam to maintain communications at required data rates during remainder of mission. Spacecraft was scheduled to reach Mars Nov. 13. (NASA Release 71-184)

September 23: Dr. James C. Fletcher, NASA Administrator, addressed annual meeting of National Security Industrial Assn. in Washington, D.C.: "I am impressed with the soundness of this new space program for the Seventies and I urge you to support it. I also believe very much that our country needs it." Among program's strengths were: it had congressional backing; it was well planned and well balanced to meet basic national needs, including national security; it stressed earth orbit as "new realm of prime importance and great opportunity, where America's capacity for world leadership will be tested not only in this decade but in the decades to follow." Program promoted economic progress based on new technology, was essential to President Nixon's peace policies based on international cooperation, and was "already part of America's destiny, as all of you can testify who watched the Apollo 15 operations on international television."

NASA had been "impressed and encouraged" by President's call for new programs to ensure maximum enlistment of technology in meeting challenge of peace and by President's promises to present those programs in next session of Congress.

"Technology, in my judgment, has made this country great and will keep it great. It is certainly true that we need adequate political con-

September 23

trols and, if possible, social controls to keep technology the servant of man, and not his master. But nevertheless new technology we absolutely need to assure jobs and a better life for all Americans in the decades ahead."

Dr. Fletcher felt recent shuttle studies had "reinforced the general belief that America's future in space in the remainder of the 20th Century depends in large measure on our skill and our determination in defining and developing the shuttle." (Edited text)

Vice President Spiro T. Agnew said in speech that U.S.S.R.'s Foxbat fighter aircraft was "now operational in Egypt and cannot be matched in performance by anything we have in operation." (Text)

- NASA conducted second successful test of Viking Mars lander free-flight, development parachute system at DOT Joint Parachute Test Facility at El Centro, Calif. First test had been conducted Sept. 3. (FRC PIO)
- Spencer M. Beresford, NASA General Counsel, testified before House Committee on Science and Astronautics' Subcommittee on Manned Space Flight during hearing on H.R. 4545, Chapel of the Astronauts bill. NASA had agreed to consider request by Chapel of the Astronauts, Inc., to purchase up to seven acres (three hectares) of land adjacent to KSC Visitor Information Center for construction of "nondenominational, nonsectarian, nonprofit public facility for worship or meditation and a memorial to the astronauts." NASA was then "actively engaged in the defense of a law suit arising from the reading of portions of the Bible by the crew of Apollo 8 while in space." Wishing to avoid a further suit, NASA had sought assurance that Chapel of the Astronauts, Inc., purpose was "truly nondenominational." Corporation had complied with NASA's request for assurance. "It is my professional opinion that . . . if present plans for the Chapel are followed and it is operated with sound discretion, the Chapel Corporation, NASA and the United States Government are as secure from successful legal attack as is possible." (Proceedings)
- *Izvestia* published interview with Soviet aircraft designer Andrey N. Tupolev and Aleksey A. Tupolev, Tu-144 test pilot Mikhail V. Kozlov, and Tu-144 navigator Georgey N. Bazhenov on Tu-144 Sept. 6 visit to Bulgaria. Kozlov said aircraft had flown regular air route, "but we flew almost three times as fast and twice as high as the other planes." Although difficulties "did not arise, we did seem to sense a dislocation, a lack of coordination between the future—which is what the TU-144 represents—and the present, even the past—those principles which are followed by the 'ground' which controls (the flight of) the plane." Tu-144 had flown air route on automatic pilot. Aleksey Tupolev said aircraft was designed for 30 000 flying hrs. "This is 10 years of work for a supersonic aircraft. Every system and every structure is designed for this." (FBIS-Sov-71-187, 9/27/71, L1)
- Miniature computer, the size of an automobile battery, was being developed by Honeywell Aerospace Div. to pilot satellite to beam educational TV to India under NASA and Indian government project to start in 1973, AP reported. Computer would use as much electricity as 30-w light bulb. (*NYT*, 9/23/71, 14)
- U.S. and U.S.S.R. ended fifth round of SALT in Helsinki. *New York Times* sources said some progress had been made toward agreement on freeze that would maintain ICBM arsenals of both nations at about

current levels. SALT would resume in Vienna in mid-November. (Hamilton, *NYT*, 9/24/71, 1)

September 23–24: Fortieth anniversary of American Institute of Physics was observed at AIP annual meeting Sept. 23 and at breakfast of AIP officers and governing board members Sept. 24. Dr. Frederick Seitz, Rockefeller Univ. President and former NAS President, spoke on "Four Decades of AIP" at annual meeting. (AIP *Newsletter*, 10/71)

September 24: U.S.S.R. launched *Cosmos 440* from Plesetsk into orbit with 789-km (490.3-mi) apogee, 271-km (168.4-mi) perigee, 95.2-min period, and 70.9° inclination. (GSFC *SSR*, 9/30/71; *SBD*, 10/1/71, 101)

- Tenth flight of M2–F3 lifting body, piloted by NASA test pilot William H. Dana, was prematurely terminated when ignition malfunction in two of four rocket engines caused shutdown of remaining engines. Small fire that started in base area of vehicle during fuel jettison was extinguished when jettisoning was stopped. Vehicle made hard landing at emergency site on Rosamond Dry Lake. Purpose of flight was to evaluate stability and control at mach 1.0 and to evaluate reaction control augmentation system. (NASA Proj Off)

- White House announced President Nixon's approval of U.S.–U.S.S.R. agreements negotiated at SALT to reduce risk of outbreak of nuclear war and to improve and modernize Washington-Moscow direct communications link. Both nations would guard against accidental or unauthorized use of nuclear weapons, arrange for rapid communication should danger of nuclear war arise, and notify each other in advance of certain planned missile launches. Two satellite circuits would be established, one by each nation, as well as multiple terminals to increase reliability of link. U.S.S.R. would provide circuit through its own satellite system and U.S. would arrange for channel through INTELSAT system. Agreements were to be signed in Washington Sept. 30. (*PD*, 9/27/71, 1318–9)

- Skylab payload shroud (nosecone) was delivered by McDonnell Douglas Astronautics Co. to MSFC. Shroud, 18.3 m (60 ft) long with 6.7-m (22-ft) diameter, weighed almost 12 000 kg (26 000 lbs), and was first major piece of Skylab hardware to be delivered to NASA. (MSFC Release 71–162; MSFC PIO)

- Senate Committee on Aeronautical and Space Sciences published *Statements by Presidents of the United States on International Cooperation in Space—A Chronology: October 1957–August 1971* (Sen. Doc. 92–40). Statements by Presidents Eisenhower, Kennedy, Johnson, and Nixon had been compiled by Dr. Eugene M. Emme, NASA Historian. In letter of transmittal to Committee Chairman Sen. Clinton P. Anderson (D-N. Mex.), Committee Staff Director James J. Gehrig said statements revealed "unanimous agreement that space research and exploration should be an international undertaking." (Text)

- Determination of crystallization ages for six *Apollo 14* lunar rocks was described in *Science* by Dr. Liaquat Husain, Dr. John F. Sutter, and Dr. Oliver A. Schaeffer, State Univ. of New York at Stony Brook scientists. Rocks, from Fra Mauro region, had been dated by measuring argon ratios in mass spectrometer. All six rocks gave age of 3.62 to 3.92 billion yrs, same as that of fragmental rocks from Fra Mauro. Scientists had concluded that Imbrium impact—one of last major events in evolution of pre-mare lunar surface—and crystallization of significant portion of pre-Imbrian basalts were "essentially contemporaneous." (*Science*, 9/24/71, 1235-6)

- Senate rejected by vote of 39 to 12 modification of amendment proposed by Sen. Hubert H. Humphrey (D-Minn.) to H.R. 8687, FY 1972 military procurement bill. Amendment called for placement of funds for MIRV in special Treasury Dept. account to be used only when President and Congress jointly determined that Soviet activity in MIRV field necessitated further U.S. testing and deployment. (*CR*, 9/24/71, S15020-68)
- Appointment of James T. Murphy as Director of FAA Office of Air Transportation Security was announced by FAA Administrator John H. Shaffer. Murphy, former FBI special agent, had been serving on acting basis since May 1971. (FAA Release 71-153)

September 24-25: Apollo 15 Astronauts David R. Scott, James B. Irwin, and Alfred M. Worden were honored by USAF Academy at Colorado Springs, Colo., with parade and special civic day. They received USAF Distinguished Service Medals from Dr. Robert C. Seamans, Jr., Secretary of the Air Force. (*NASA Activities*, 10/15/71, 172)

September 25: New York Times editorial commented on U.S.-U.S.S.R. agreements at SALT: "Improved communications, no matter how desirable, will not offset the growing danger of a nuclear holocaust unless followed speedily by substantive steps to check the headlong race in strategic nuclear arms." (*NYT*, 9/25/71, 30)

September 26: Smithsonian Institution's Langley Gold Medal for Aerodromics was presented to L/G Samuel C. Phillips (USAF), SAMSO Commander and former NASA Apollo Program Director, at ceremonies commemorating Smithsonian's 125th anniversary. Medal, presented periodically for "meritorious investigations in connection with the sciences of aeronautics and astronautics," was established in memory of Dr. Samuel P. Langley, aviation pioneer. Among recipients had been Astronaut Alan B. Shepard, Jr., in 1964 and Dr. Wernher von Braun, NASA Deputy Associate Administrator for Planning, in 1967. (Smithsonian Release SI-131-71)

September 26-30: International Aeronautical Federation Annual General Conference was held in Lucerne, Switzerland.

Gold Space Medal, Federation's highest award for space achievement, was presented to Astronaut James A. Lovell, Jr., for "his courageous achievements and leadership as the Commander of Apollo 13." In recognition of achievements of entire *Apollo 13* crew, Astronauts Lovell, Fred W. Haise, Jr., and John L. Swigert, Jr., received IAF's V. M. Komarov Diploma for 1970. Diploma had been created in memory of Cosmonaut Vladimir M. Komarov, killed during *Soyuz 1* mission April 24, 1967.

Cosmonauts Andrian G. Nikolayev and Vitaly I. Sevastyanov were awarded Yuri Gagarin Gold Medal for 1970 for achievements as crew of *Soyuz 9*, June 2-19, 1970.

Gold Air Medal, highest IAF award for aeronautics, was presented to Capt. Henry T. (Dick) Merrill for contributions to aviation over 50 yrs. Capt. Merrill, civilian pilot, had flown Atlantic in 1935 and 1937.

Concorde technical team of BAC and Aérospaciale employees received group diploma for "valuable example set by . . . effective collaboration developed since 1962 between the engineers of British Aircraft Corp., Aérospaciale, Rolls-Royce and Snecma, operating as a single industrial team."

During conference LRV was presented to Swiss Transport Museum. (Letter to NASA Administrator, 6/10/71; letter to NASA Historian, 9/29/71; *W News,* 9/30/71, 42; *Av Wk,* 10/18/71, 9; BAC PIO; NAA PIO)

September 27: Feasibility study at MSC had ruled out use of small TV camera on LRV Rover while in motion during Apollo 16 mission, *Los Angeles Times* reported. Cost would exceed usefulness and pictures might be hazy and induce nausea in viewers because of up-and-down motion of Rover over lunar surface. Necessary adjustments to provide TV from Rover would cost almost $5 million. (*LA Times,* 9/27/71)

- NASA announced assignment of Robert O. Aller, Manager of Space Station Operations, Space Station Task Force for OMSF, to Skylab Program Office as Director of Operations. Aller would be responsible for coordination and development of operationally related program and mission planning activities. (NASA Hq *WB*)

- Willis H. Shapley, NASA Associate Deputy Administrator, testified before House Committee on Foreign Affairs' Subcommittee on Africa on operation of NASA tracking station in South Africa. Subcommittee was investigating U.S. business involvement in South Africa. NASA station was manned, staffed, and operated by South African Council for Scientific and Industrial Research (CSIR). "Committed as we have been to attaining the important national objectives entrusted to us, we have thought of our stations overseas essentially in terms of their efficiency in supporting our flight missions. We have monitored station operations mainly from the standpoint of technical effectiveness and sound financial management. While we have recognized that human factors can have a direct bearing on performance, we are limited in our ability to control these factors, particularly in societies where statutes and practices differ markedly from our own." NASA wished "to do all we can to see that the NASA station in South Africa reflects to the greatest degree that local circumstances permit, the practices and ideals which govern employment at our domestic facilities." (Testimony)

- AFSC announced establishment by Aeronautical Systems Div. (ASD) of Prototype Program Office at Wright-Patterson AFB, Ohio. Office, headed by Col. Lyle W. Cameron, would manage aeronautical prototype programs for USAF. USAF had identified as candidate aeronautical projects advanced STOL transport, very low radar-cross-section (RCS) test vehicle, lightweight fighter aircraft, and quiet aircraft. (AFSC Release 240.71)

September 28–October 19: Luna 19, U.S.S.R.'s second unmanned lunar probe in less than a month, was launched from Baykonur at 3:00 pm local time (6:00 pm EDT). In earth orbit *Luna 19* was separated from last stage of booster and set on trajectory toward moon. Tass said mission would conduct scientific investigation of moon and near-lunar space from lunar orbit. During flight to moon 26 communications sessions were held with spacecraft and data on parameters and systems were transmitted. Midcourse corrections were conducted Sept. 29 and Oct. 1, and on Oct. 3 *Luna 19* entered lunar orbit with 140-km (87-mi) altitude, 122-min period, and 40° 35' inclination. Tass said Oct. 19 that all systems were operating normally and spacecraft was conducting geophysical research of moon's gravitational field and taking photographs of lunar surface.

Previous mission in series, *Luna 18*, crashed on moon while attempting softlanding Sept. 11. (FBIS–Sov–188-3, 9/28/71, L1; FBIS–Sov–192-3, 10/4/71, L1; FBIS–Sov–203-3, 10/20/71, L1; *SBD*, 9/29/71, 89)

September 28: Japan successfully launched its third satellite, *Shinsei* ("New Star"), from Uchinoura Space Center at 1:00 pm Tokyo time (4:00 am EDT Sept. 27). The 65.8-kg (145-lb) spacecraft was Japan's first scientific satellite and was launched by 22.9-m (75-ft) Mu rocket into orbit with 3892-km (2418.4-mi) apogee, 619-km (384.6-mi) perigee, 133-min period, and 31.2° inclination. Purpose of mission was to study cosmic rays and solar electric waves for one year.

Shinsei was 26-sided body, 75 cm (30 in) in diameter, equipped with 5184 silicone solar batteries and used "gravity-turn" device in lieu of conventional guidance and control techniques. Payload included solar radio receivers, cosmic-ray detectors, and ionospheric probes. Satellite functioned perfectly and relayed scientific information.

Japan had become fourth country to orbit satellite with successful launch of *Ohsumi* Feb. 11, 1970. (*Spacewarn*, 10/5/71, 2; *W Post*, 9/29/71, A14; *W Post*, 9/30/71, A6; GSFC *SSR*, 9/30/71)

- U.S.S.R. launched *Cosmos 441* from Baykonur into orbit with 266-km (165.3-mi) apogee, 207-km (128.6-mi) perigee, 89.2-min period, and 65° inclination. Satellite reentered Oct. 10. (GSFC *SSR*, 9/30/71; 10/31/71; *SBD*, 10/1/71, 101)

- Completion of tests to determine cause of *Apollo 15* parachute failure was announced by NASA. Possible causes were fuel dump, which expelled six pounds of monomethyl hydrazine (MMH) from CM reaction control system, and flaws in links connecting suspension lines to risers. Officials doubted exact cause could be determined. Dump through hot engine could cause tongues of flame that might reach parachute lines. Flaws were detected in links on one recovered chute, but the failed chute had not been recovered. Corrective action would be taken on both possible causes for Apollo 16. Dump would be eliminated and material for links changed. (NASA Release 71-188)

- Dr. Charles S. Sheldon II, Chief of Science Policy Research Division of Library of Congress, commented in interview after Sept. 28 launch of unmanned lunar probe *Luna 19* that Moscow still had plans to land men on the moon, perhaps about 1973 or 1975. Dr. Sheldon said five previous Soviet Zond circumlunar flights had been capable of carrying men and several other earth orbital flights had simulated manned lunar landing maneuvers. Dr. Sheldon also recalled that Mstislav V. Keldysh, President of Soviet Academy of Sciences, had said in 1969 that manned missions to moon were not ruled out. (Sehlstedt, B *Sun*, 10/29/71, A5)

- Selection of National Science Teachers Assn., Washington, D.C., for negotiation of $39 000, 19-mo, cost-reimbursement contract for management and operation of Skylab Student Project was announced by NASA. Purpose of project was to stimulate interest in science and technology by directly involving students in space research. NSTA would notify student and educational community of opportunity and method of participation, develop procedures for evaluation of proposals based on educational value, and develop certificates of participation and awards system for entrants. NSTA also would develop plans for entrants whose proposal ideas had been selected by NASA and conduct Skylab Student Education Conference at KSC in conjunction with Skylab launch. (NASA Release 71–183)

- U.S. patent No. 3 608 844, for self-deploying arm to serve as spacecraft boom or antenna, was granted to GSFC engineers William T. Tumulty, Jr., and Wayne P. Sours. Called Minimech, device could also retrieve objects from burning building, serve as escape chute, or erect itself as instant tent pole or instant plumbing. It could be made from various materials depending on purpose. (*NYT*, 10/2/71; Pat Off PIO)

September 29–30: NASA's *Oso 7* (OSO–H) Orbiting Solar Observatory was launched from ETR at 5:45 am EDT by two-stage Thor-Delta DSV–3N booster to study sun and its influence on earth's atmosphere. Second-stage anomaly resulted in noncircular orbit of 575.7-km (357.7-mi) apogee, 329.1-km (204.5-mi) perigee, 93.6-min period, and 33.1° inclination. Primary mission objective was to obtain high-resolution data from solar corona in particular spectral bands in EUV and in visible regions during one solar rotation.

Tetr 3 (TETR–D) Test and Training Satellite, carried pickaback on Delta 2nd stage, was ejected into orbit with 571-km (354.8-mi) apogee, 391-km (243-mi) perigee, 94.3-min period, and 33.0° inclination. Primary objective of 45-lb (20.4-kg) secondary payload was to test Manned Space Flight Network and train MSFN personnel.

Second-stage anomaly of launch vehicle separated spacecraft at pitch angle outside normal sun acquisition limits. For eight hours following orbital insertion, ground controllers manually commanded spacecraft to despin. Pitch angle was then changed to place it within automatic acquisition range. Sail section was commanded to acquisition mode and sail acquired and locked onto sun at approximately 2:00 pm EDT. Batteries were then charged. Orbit would permit observations for more than a year.

Oso 7, improved over earlier OSOs, provided for increase in experiment size and weight and power and extended capability in pointed mode, including stability and pointing during night portion of orbit. It was spin-stabilized; weighed 636 kg (1403 lbs); carried six scientific experiments; and had two main sections—wheel (lower), which carried nondirectional scanning experiments and basic support equipment, and sail (upper), which carried pointed experiments. Experiments, designed to continue and extend work of preceding OSO spacecraft, were provided by GSFC, NRL, MIT, Univ. of New Hampshire, and Univ. of California (San Diego).

Oso 7 was 8th in series of 11 OSO spacecraft designed to provide direct observations of sun during most of 11-yr solar cycle. *Oso 1* (launched March 7, 1962) and *Oso 2* (launched Feb. 3, 1965) had passed their six-month lifetimes and together provided more than 8600 hours of scientific information. OSO–C (launched Aug. 25, 1968) had failed to reach orbit when booster malfunctioned. *Oso 3* (launched Mar. 8, 1967) and *Oso 4* (launched Oct. 18, 1967) transmitted on command only. *Oso 5* (launched Jan. 22, 1969) had both tape recorders and seven of eight experiments operating satisfactorily after six months in orbit. *Oso 6* (launched Aug. 9, 1969) was

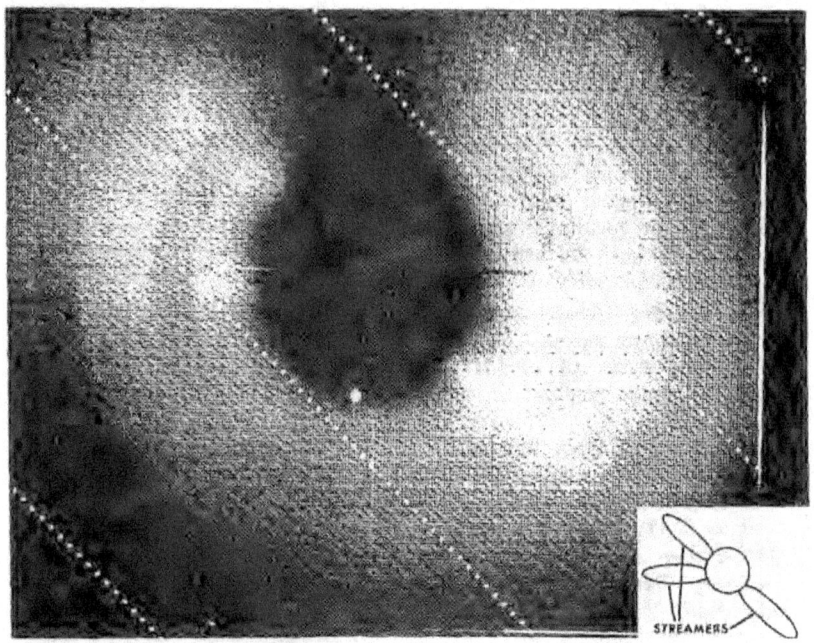

September 29–30: Oso 7 *Orbiting Solar Observatory* was launched into orbit to study the sun and its influence on the earth's atmosphere. High spatial- and spectral-resolution data were obtained by the pointed experiments on the sail and scanning experiments on the wheel. First x-ray observations from a spacecraft of the beginning of a solar flare and of solar "streamers" were made Oct. 3–6. The white light coronagraph above, from a Naval Research Laboratory experiment on Oso 7, showed three coronal streamers from the sun. The black bullseye in the center was produced by a circling occulting disc that caused an artificial eclipse. The distance from the center of the disc to the edge of the photograph was about 10 solar radii, or about 6.4 million kilometers (4 million miles).

relaying information from all seven experiments. OSO program was managed by GSFC under OSSA direction. (NASA Proj Off; NASA Release 71-163; *Spacewarn*, 10/5/71)

September 29: U.S.S.R. launched *Cosmos 442* from Plesetsk into orbit with 272-km (169.0-mi) apogee, 199-km (123.7-mi) perigee, 89.7-min period, and 72.8° inclination. Satellite reentered Oct. 12. (GSFC *SSR*, 9/30/71; 10/31/71; *SBD*, 10/4/71, 108)

- Fourteenth flight of TF–8A jet aircraft equipped with supercritical wing was completed, flown from FRC by NASA test pilot Thomas C. McMurtry in accelerated TF-8A flight test program [see Sept. 15]. (NASA Proj Off)

- *Apollo 15* Astronauts David R. Scott, James B. Irwin, and Alfred M. Worden visited KSC to thank personnel for their support during *Apollo 15* mission. KSC Deputy Director Miles Ross told crowd of about 5000 that with all talent assembled in VAB "I think this is the real Kennedy Center for the Performing Arts." (*Spaceport News*, 10/7/71, 1)
- NASA and FAA announced agreement for joint participation in flight-simulation R&D projects at ARC. NASA would make ARC simulation facilities and supporting services available to FAA. FAA would provide its own technical personnel and resident director, who would coordinate FAA R&D projects with ARC director. (NASA Release 71-185)
- Engine fell off USAF C-5A training aircraft preparing for takeoff from Altus AFB, Okla. Later, USAF officials said preliminary investigation had traced cause to structural problems in pylons that held engines to wing. Aircraft had accumulated 1300 flying hrs and more than 3000 landings. USAF said it had grounded seven C-5A's with similar flying hours until new pylons could be installed. (Getler, *W Post*, 10/7/71, 2)
- Senate by vote of 61 to 28 rejected amendment proposed by Sen. William Proxmire (D-Wis.) to terminate USN F-14 fighter aircraft program by deleting $801.6 million from H.R. 8687, $51-billion FY 1972 military procurement authorization. (*CR*, 9/29/71, S15329–15405)
- Virginians for Dulles, citizen group representing persons living near flightpaths of Washington National Airport, filed suit in U.S. District Court to prevent stretch-jet aircraft from landing or taking off at Washington National and to ban all jet traffic at airport after 11 pm because of noise pollution. (*W Post*, 9/30/71, B3)
- *Washington Post* editorial questioned need for F-14 USN fighter aircraft: "The F-14 has too many attributes of another great defense turkey [F-111B] to be approved at this time by the Senate. We believe the Senate should send the Navy back to the drawing board to come up with either an improved F-4 or a new fighter plane that is not too costly to be purchased in adequate numbers and that has a reasonable chance of performing a valid function." (*W Post*, 9/29/71, A18)
- New Orleans *Times-Picayune* editorial commented on U.S.–U.S.S.R. agreement on New York-Moscow comsat link to prevent war by nuclear accident: "The technology of satellite relay is beyond us, but we recall a treaty signing recently in Washington and Moscow at which the signers watched each other on satellite-relayed television. Yet it is reported that the link may take a long time and require building ground stations near the two cities and launching special satellites. Cannot the hot line be given a reserved channel on present facilities?" (New Orleans *Times-Picayune*, 9/29/71)

September 30: Medical instructions relayed by *Ats 1* (launched Dec. 6, 1966) from equatorial orbit 35 800 km (22 300 mi) above the earth saved lives of two women in remote native Alaska villages. Native medical aide in Chalkyitsik, northeastern Alaska, and aide in Anatuvuk Pass, north central Alaska, broke into satellite time reserved for relay of educational broadcasts to Alaskan village centers to request advice in treating hemorrhaging pregnant Indian woman and Eskimo woman suffering appendicitus attack. Physician in Anchorage, Alaska, hearing requests, broke into broadcast with instructions and alerted medical personnel nearer scenes. Univ. of Alaska officials handling transmissions telephoned NASA satellite control station at Rosman, N.C., and received permission to extend reserved satellite time during night. Sen. Mike Gravel (D-Alaska) later said in Washington, D.C.: "I am grateful for the teamwork in Alaska and at NASA that so dramatically saved two lives, and I am so proud of my fellow Alaskans who responded once again through the code of the North—but this time using a space satellite." (AP, B *Sun*, 10/8/71, A10)

- *Apollo 15* Astronauts David R. Scott, James B. Irwin, and Alfred M. Worden thanked USAF personnel at Patrick AFB, Fla., for their support during *Apollo 15* mission and presented base with U.S. flag they had carried to moon. Later, astronauts thanked MSFC employees during six-hour visit to Center. (MSFC Release 71-164; *NY News*, 10/1/71, 64; MSFC PAO)

- Formation of new Military Aircraft Programs Office within OART and appointment of Albert J. Evans, Director of Aerodynamics and Vehicles Systems Div. in OART, as its Director were announced by NASA. Office would direct ongoing support to USAF on F-15 and B-1 programs and to USN on F-14 program; USAF/NASA Transonic Aircraft Technology (TACT) Program in which NASA supercritical wing would be flight-tested on modified F-111 aircraft; USA/NASA Tilt Rotor Research Aircraft Technology Program; and other possible joint research, experimental, and prototype aircraft programs. William S. Aiken, Jr., Deputy Director of Aerodynamics and Vehicle Systems Div., would succeed Evans in former post. (NASA Release 71-186)

- *Lunokhod 1*—which landed on moon Nov. 17, 1970, on board *Luna 17*—was declared officially dead at end of lunar night because resources of its isotope heating system had been exhausted. (*SBD*, 11/3/71, 18)

- Secretary of State William P. Rogers and Soviet Foreign Minister Andrey A. Gromyko signed two nuclear control agreements in formal ceremony at State Dept. Agreements called for improved "hotline" communication between Washington and Moscow via comsats and immediate notification of any unexplained incident involving possible detonation of nuclear weapon which could cause outbreak of nuclear war. (Smith, *NYT*, 10/1/71, 1)

During September: Dr. James C. Fletcher, NASA Administrator, said in interview published by *Air Force Magazine* that NASA planned to let hardware contracts for space shuttle by spring 1972, contingent on White House approval. NASA had decided in summer to extend preliminary design contracts by four months to "look more intensively at alternative configurations," examining phased approach for building reusable upper stage first and testing and using it with interim expendable booster, instead of developing reusable booster at same time. Reevaluation of design concepts was stirred by need for additional analyses and research and by need to trim expenditures "in a given year."

NASA's viewpoint on space in this decade was "to bring the program down to earth, both figuratively and literally. We must ask ourselves the question: Is what we are doing useful in alleviating our problems on the ground?" Crucial criteria of each prospective space program were "whether it will pay for itself by doing useful work; whether it can be expected to produce valuable scientific knowledge that can only be obtained from space; whether it is important to national security; and whether it will catalyze valuable new technologies or lead to other practical results." For shuttle "answers to all these questions is 100 percent affirmative." (*AF Mag,* 9/71, 53–61)

- Communist China's technological progress was described by Bernard T. Feld in *Bulletin of the Atomic Scientists*: "The inescapable conclusion that the Chinese had mastered the difficult technologies of U–235 separation and hydrogen bomb production was somehow sublimated in the conviction that they would be much slower in mastering intercontinental ballistic missile technology. This conviction, advanced in the teeth of evidence—after all, the Chinese invented the rocket—was reinforced by reports of widespread technological disruption in the Cultural Revolution. Their two successful launchings of space satellites, the last one weighing around 480 pounds [218 kg], in March 1971, should have disabused even the most skeptical or prejudiced. In any case, there are a sufficient number of recent eyewitness accounts by Westerners of Chinese scientific achievements in various fields to dispell any notions of Chinese technical inferiority." (*Bull of Atomic Scientists,* 9/71)

October 1971

October 1: Dr. James C. Fletcher, NASA Administrator, said strong technology base was "absolutely vital" to U.S., in speech before 14th Annual Aerospace Luncheon of Los Angeles Area Chamber of Commerce. Indications from critics of lack of confidence in advancement of "greatest technological evolution in the history of mankind" was reminder that "we must do a better job of convincing the man-on-the-street of the [space] program's merits." Public needed information "to put its priorities in proper perspective." There was "something seriously wrong" when "as a nation we pay people more not to work and more in farm subsidies than we do on space research, a program which impacts our national economy and vitally affects our national security. . . . I am convinced that further erosion of our technological leadership will start us on a downhill course from which it will be difficult to recover." But, "in our zeal to forge ahead in technical and scientific fields, it is possible to lose sight of the debt we owe to the great American public that so steadfastly sustained us during the early years." Average American had his priorities. "But he needs to be informed if he is to arrive at an intelligent decision about how to rank his priorities." Burden of convincing public of necessity for "continuing well-balanced space program for the 1970's and beyond" lay with those in program.

Dr. Fletcher also said: "We want cooperation with the Soviets as well as with other nations throughout the world. We are ready to do our part to get it, but if we have learned the lessons of history, we should know that the Soviets will not be eager to cooperate with a second-rate power. . . . And that is what we will become if we do not proceed with the space shuttle in this decade. We cannot create the environment needed for space cooperation, or an era of enduring peace, by handing over a monopoly in the realm of orbital flight to the Soviet Union or anyone else." (Text)

- Subcommittee on NASA Oversight transmitted report *Review of Recent Launch Failures* to House Committee on Science and Astronautics. Subcommittee had held hearings June 15-17 on failure of Atlas-Centaur launch vehicle to place *Ats 4* in planned orbit Aug. 10, 1968, and failure to orbit OAO–B Nov. 30, 1970, and Mariner 8 May 8, 1971.

  Subcommittee concluded that NASA's decision not to attempt recovery program for unsuccessful Mariner 8 mission using proof-test model "seems justified" on grounds *Mariner 9* (launched May 30) had been conceived as backup mission for Mariner 8 from beginning; to alter Atlas-Centaur for different project would have been "challenging;" and proof-test model could better be used as *Mariner 9* backup, if necessary, or in 1973 Venus-Mercury and Viking projects. While NASA's backup mission and recovery policies after failures "seem to have been sound in the past," decision not to have OAO–B recovery program "appears . . . to have been an unwise exception." To forego OAO–B goals "which have occupied the talents and efforts of many scientists and engineers for more than a decade in order to initiate new astronomical experiments [HEAO Project] in different portions of the magnetic spectrum strikes the Committee as a decision of questionable merit." Committee recommended future use of prototype spacecraft that could be reengineered and launched as part of flight program to extract full value from expensive hardware. Manufacturing and quality control procedures in Centaur program could be improved. "It is quite clear that a great deal can and should be done short of 'man-rating' Centaur, which would significantly improve its reliability." Centaur stage problems "seem serious enough to justify . . . broad scale, independent examination by disinterested experts" so that Centaur, "which figures so importantly in the space program of the next decade, can be utilized with renewed confidence." (Text)

- McDonnell Douglas Astronautics Co. began 14-mo, $350 000 study continuation for system definition of reusable nuclear stage (RNS) for multipurpose space missions. Previous studies had determined that RNS could be launched by either Saturn or space shuttle booster. New study would focus on RNS concepts to be launched by space shuttle, but would consider two-stage, fully reusable shuttle, shuttle booster with expendable 2nd stage, and launch of RNS as complete unit or with components to be assembled in space. Study also would support NERVA program. (MSFC Release 71–174)

- Assistant Secretary of the Air Force for R&D Grant L. Hansen discussed space shuttle at Air Force Assn. meeting in Bethpage, N.Y. NASA and USAF work on shuttle engine had been helpful to shuttle program in areas of reusability and higher-pressure-engine technology. USAF had supported advanced development efforts on liquid-fuel rocket engines and had concluded hydrogen and oxygen high-performance-engine demonstration program. Work on XLR–129 reusable rocket engine had provided basis for shuttle's reusable high-pressure-rocket concept. Residual hardware of USAF XLR–129 program had been turned over to NASA for shuttle engine program. Development of shuttle main engine was to be conducted at USAF's Arnold Engineering Development Center. USAF would continue in 1972 to develop technology for high-energy orbit-to-orbit stages compatible with a shuttle to accomplish maneuvering missions for late 1970s and early 1980s. (*AF Mag*, 12/71, 34)

- More than 15 astronauts were taste-testing three meals daily for five days in MSC experiment to evaluate proposed food items for 28- and 56-day Skylab missions in 1973, MSC announced. Tests would determine astronauts' preferences and each astronaut's energy requirements. They were coordinated by Technology, Inc., and conducted in MSC Life Systems Laboratory. (MSC Release 71–74)

- Twenty-fifth anniversary of commissioning of Naval Air Missile Test Center at Point Mugu, Calif. Center, later renamed Pacific Missile Range, conducted R&D for Federal agencies, including NASA. (PMR *Missile*, 10/1/71, 1)

- Completion of DOT study of general-aviation safety was announced by Secretary of Transportation John A. Volpe. General-aviation industry consisted of 133 814 single-engine aircraft, 16 000 multi-engine aircraft, 2229 jet aircraft, 33 992 flight instructors, 2170 FAA-approved flight and ground schools, and 732 729 pilots, including 195 861 students. In 1970 general aviation had accounted for more than 25 million flight hours and nearly 6 billion km (nearly 4 billion mi) flown. Of 621 fatal general-aviation accidents in 1970, 201 had been identified as weather-caused. Study recommended biennial flight review of pilot competency, priority support of FAA General Aviation Accident Prevention Program, tightening of certification requirements for flight instructors and pilots, modernization of FAA flight service stations, improvement of FAA reporting of real-time weather data, review of FAA general-aviation structure, clarification by FAA of regulations covering general operating and flight rules, publication of FAA regulations in separate parts rather than in current volume format, and adoption of standard traffic pattern for uncontrolled airports. (DOT Release 20271)

- Impact of "leftish student movements" was described in *Science* by Harvard Univ. physicist Dr. Harvey Brooks: "Today some believe that the . . . movements, which are worldwide, form the vanguard of a new revolution, the first true revolution in advanced industrial societies. If so, it will be the first in modern history which had not attempted to ally itself with science. So far as its ideology is discernible at all, it seems to be antiscientific and antirational, more akin to the early Christians than to the modern Marxists, despite the Marxist slogans." (*Science*, 10/1/71, 21–9)

October 1–5: Fourth International Symposium on Basic Environmental Problems of Man in Space was held jointly in Yerevan, U.S.S.R., by IAA and Soviet Academy of Sciences. Papers dealt with medical problems of prolonged space missions, studies of mechanism underlying single and combined stress effects and methods of ameliorating deleterious influences, and bioastronautical contributions to general medical science and practice. Symposium reviewed results of *Soyuz 11–Salyut 1* mission, June 6–30, that ended with crew's accidental death on reentry. (IAA Annual Rpt, 8/31/71; FBIS–Sov–71–212, 11/3/71)

October 3: Poorly designed drive system controlling tilt of "world's largest movable" antenna, 100 m (330 ft) in diameter—operated by Max Planck Institute for Radio Astronomy in Bonn, West Germany—had set dish oscillating at rate of once every second, *New York Times* reported. Oscillations endangered antenna's structure and made observations "virtually impossible." Antenna, at Effelsberg, West Germany, had been cited by astronomers as demonstration of why massive and costly radio astronomy engineering enterprises should be under professional management rather than scientific direction, *Times* said. (Sullivan, *NYT*, 10/3/71, 82)

October 3–6: Oso 7, launched Sept. 29, made the first x-ray observations from spacecraft of beginning solar flare and solar "streamers" (long, tapered structures observed in solar corona). Observations would lead to new understanding of flares not possible before. Flares had many effects on earth's environment, weather, and communications. Streamers had been photographed before only during solar eclipses and occasionally by short sounding rocket flights. *Oso 7* experiment created own "eclipse" by using occulting disc.

By Oct. 6 all six scientific experiments had been turned on. NRL coronographs obtained first observations of corona in white light and EUV. Wheel section experiments obtained scanning data continuously. Only one of two tape recorders was operating but would be sufficient to achieve mission objectives. (NASA Release 71-193; *Goddard News*, 10/71, 2; NASA Proj Off)

October 4: U.S.S.R.'s *Lunokhod 1*, first self-propelled lunar vehicle, ceased operations after 321 days on moon, Tass announced Oct. 9. Eight-wheeled vehicle stopped transmitting photos and soil analyses to earth when its isotope heat reserves were exhausted during its 11th lunar night on moon. Depletion of nuclear heat source had caused temperatures on rover to drop during lunar night. Vehicle had been landed on moon in Sea of Rains by *Luna 17* Nov. 17, 1970.

Solar-powered *Lunokhod 1* had successfully conducted scientific and technical research program with all systems functioning normally. It had traveled distance of 10 540 m (34 680 ft), exploring more than 80 000 sq m (860 000 sq ft) of lunar surface and had produced more than 500 panoramas, 20 000 TV pictures, 25 chemical analyses, and hundreds of probes of physical-mechanical characteristics of lunar soil. Vehicle had been parked on level surface with French-made laser reflector pointing toward earth, to permit continued use of laser for ranging experiments. (FBIS–Sov–71–197–3, 10/12/71, L1; 212–3, 11/3/71, L5–6; *NYT*, 10/10/71, 68; *Av Wk*, 10/18/71, 17)

- NASA, FAA, and USAF began joint runway research program at Wallops Station to evaluate methods for measuring runway slipperiness at military and civil airports and improve techniques for estimating aircraft performance on runways that become slippery when wet. Tests would be followed by similar measurements at Houston Intercontinental Airport, Lubbock (Tex.) Regional Airport, Edwards AFB (Calif.), Seattle-Tacoma International Airport, and John F. Kennedy International Airport in New York. FAA Boeing 727 aircraft instrumented by LaRC would be used for flight portions of program. USAF C–141 aircraft would carry personnel and equipment. Runway surface measurements would be made by mu-meter—friction measuring device provided by USAF—and diagonally braked automobile developed by NASA at LaRC. (NASA Release 71-190)

- Award of $11 640 773 addition to cost-plus-fixed-fee contract with Chrysler Corp. for launch operations support for Skylab from Oct. 1, 1971, to June 30, 1973, was announced by KSC. Addition brought amount of contract to $25 016 708. (KSC Release 226–71)

- Two prototype x-ray devices for detecting weapons and explosives in carry-on luggage of air passengers had been successfully tested at Dulles International Airport in Virginia, FAA announced. Low radiation x-ray units—Flying Spot developed by American Science and Engineering, Inc., and Bendix Ray Inspectoscope developed by Bendix Aerospace Systems Div.—sent x-ray pulse through luggage to obtain shadowgraph, which was displayed on TV screen. Devices did not affect photographic film or recording tape. (FAA Release 71-162)

October 4–5: Representatives of ELDO and European space industries met with NASA officials at MSFC to discuss project for space tug to operate in earth orbit in conjunction with space station and space shuttle. Tug would transfer payloads in space and perform transportation services in orbit. (MSFC Release 71-168; MSFC PIO)

October 5: Apollo 15 Awards Ceremony was held at MSC. Dr. James C. Fletcher, NASA Administrator, presented NASA Exceptional Scientific Achievement Medal for "accomplishments in the Apollo 15 mission" to Dr. Larry A. Haskin, Univ. of Wisconsin; Dr. Robert O. Pepin, Univ. of Minnesota; Dr. Leon T. Silver, Cal Tech; Dr. M. Gene Simmons, MIT; and Dr. Gordon A. Swann, U.S. Geological Survey. NASA Exceptional Service Medal was presented to 22 MSC employees and NASA Public Service Award to representatives of contractors Boeing Co., Grumman Aerospace Corp., Delco Electronic Div. of General Motors Corp., and North American Rockwell Corp.

NASA Group Achievement Award was presented to MSC Extravehicular Mobility Unit Team, MSC Lunar Orbit Experiments Team, MSFC and MSC Lunar Roving Vehicle Team, MSC Lunar Traverse Planning Team, NOAA Space Flight Meteorology Group at KSC, and MSC Television and Lunar Communications Relay Unit Team.

Dr. Fletcher read telegram from President Nixon congratulating *Apollo 15* Team which, President said, deserved "highest praise for their vital contributions to man's scientific knowledge and our nation's prominence in space." (NASA Ann, 10/1/71; MSC *Roundup*, 10/8/71, 1)

- KSC announced award of $33 157 979 addition to cost-plus-fixed-fee contract with Bendix Corp. for support services for Apollo spacecraft and Skylab programs from Oct. 1, 1971, to Dec. 31, 1972. Addition brought total amount of contract to $210 081 335. (KSC Release 227-71)
- Reuters said at least 12 supersonic, swing-wing FB-111 bombers armed with six nuclear bombs of total 4.5-metric-megaton (5 megaton) yield were on constant ground alert at SAC bases in northeast U.S. News agency quoted SAC spokesman as saying portion of FB-111 force had been written into U.S. nuclear deterrent force. (*NYT*, 10/6/71, 52)
- Israel had begun to manufacture missile that could carry 450- to 680-kg (1000- to 1500-lb) warhead 480 km (300 mi) or more, according to Western intelligence reports quoted by *New York Times*. (Beecher, *NYT*, 10/5/71, 1)
- Daniel and Florence Guggenheim Space Theater at American Museum-Hayden Planetarium was dedicated at ceremonies in New York. (Invitation)

October 5-7: Huntsville, Ala., chapter of Society of Aerospace Material & Process Engineers held national technical conference, attended by experts from five NASA centers, Redstone Arsenal, other Government agencies, and private industry. Papers covered composites and high-temperature metals, high-temperature and cryogenic insulation, reuse of materials, materials compatibility with space shuttle environment, life support materials, and special materials testing. (*Marshall Star*, 10/6/71, 1)

October 6: NASA test pilots successfully completed two flights of TF-8A, modified F-8 jet aircraft with supercritical wing, flown from FRC. Flight 15, flown by Thomas C. McMurtry, and flight 16, flown by Gary Krier, were part of accelerated series to obtain data for performance evaluation of TF-8A [see Sept. 15]. (NASA Proj Off)

- Radioastronomers had obtained first "picture" of distant galaxy from radio signals received by Synthesis radio telescope near Westerbork, Netherlands, *New York Times* reported. Picture of whirlpool galaxy had led astronomers to believe radio-signal photography using antenna systems under construction or planned might rival photography from exposures at visible wavelengths. Picture from Synthesis—1.6-km (1-mi) row of 12 identical dish antennas set along east-west line— seemed to confirm hypothesis that explosive events in core of each rotating galaxy sent waves spiraling outward. Success of Westerbork antenna array was being used to encourage congressional approval for very large array (VLA) proposed for construction in southwestern U.S. (Sullivan, *NYT*, 10/6/71, 43)

- Senate by vote of 82 to 4 passed H.R. 8687, $21-billion FY 1972 military procurement authorization bill. (*CR*, 10/6/71, S15956–16004)
- Rep. Joseph E. Karth (D-Minn.) resigned from House Committee on Science and Astronautics to take seat on House Ways and Means Committee. (Off of Rep Karth)
- Rep. James G. Fulton (R-Pa.) died of heart attack in Washington, D.C. He had been ranking Republican member of House Committee on Science and Astronautics and member of House Committee on Foreign Affairs. In 1958 Rep. Fulton had volunteered at committee hearing to serve as astronaut. *Washington Post* obituary said he had fought hard for programs to keep U.S. from "dropping out of the space race." (Weil, *W Post*, 10/8/71, B8)

October 6–8: Science fiction writer and space expert Arthur C. Clarke addressed Playboy International Writers Convocation in Chicago. He said, "Much of the criticism of space expenditure is a pure excuse— a deliberate or unconscious avoidance of real problems." Politicians, administrators, and Government officials, "often too stupid, too lazy or often too corrupt to deal with the abuses of society as they should and could, are absolutely delighted to have the space program as a scapegoat." Space was now "almost a dirty word." (*NYT*, 10/8/71, 25; *Playboy* PIO)

October 7: U.S.S.R. launched *Cosmos 443* from Plesetsk into orbit with 297-km (184.6-mi) apogee, 201-km (124.9-mi) perigee, 89.4-min period, and 65.4° inclination. Satellite reentered Oct. 19. (GSFC *SSR*, 10/13/71; *SF*, 4/72, 132)

- Dr. James C. Fletcher, NASA Administrator, enumerated major factors relevant to NASA FY 1973 budget request in speech before AIAA National Capital Section: 1. "Program continuity must be maintained." NASA could not "shut down major programs for several years and then start them up again efficiently." 2. NASA programs had helped stabilize U.S. aerospace industry. "This is a very significant reason for maintaining NASA programs up to or near current levels at this time." 3. NASA would reach minimum "institutional base" requirement for efficient management of approved programs in current fiscal year. "I will strongly resist any further reductions." 4. Space shuttle was "next logical step to develop our national capabilities for using space more economically and more effectively with manned and unmanned spacecraft, for both civilian and defense purposes." 5. U.S. was lagging behind in development of new technology for commercial and military aviation. "We killed the SST while Britain and France went ahead to build the Concorde, and the Soviet Union built a competing version. The Europeans—and our Canadian neighbors—are ahead of us in Short Take Off and Landing Aircraft. The question must be asked, how long can we, as a great nation, afford to be mere spectators while others move ahead in aircraft technology?" NASA did not intend "to shift the gears of our space program into the 'park' position; we are geared up for continued steady and impressive progress in those areas of space activity which we consider most rewarding scientifically or most important to our national security." Within few years "there will be no time of day or night, year in and year out, when men are not working in orbit above the earth. And it is up to us, in decisions to be made this year in this capital, whether these men in orbit will include Americans." (Text)

- NASA announced six-month delay in development of space shuttle to give contractors more time to study in detail alternatives resulting from current studies. Under new schedule, North American Rockwell Corp., McDonnell Douglas Co., Lockheed Missiles & Space Co., and Grumman Aerospace Corp. would study designs until April 30, 1972, rather than concluding at end of October. Selection of sites for shuttle development and operational flights would be deferred until overall system characteristics were defined. (NASA Release 71–199)

- Skylab multiple docking adapter (MDA) trainer arrived at MSC from Martin Marietta Corp. facility at Denver, Colo. MDA trainer was flown to MSC aboard NASA Super Guppy aircraft. (NASA Release 71–187; MSC PIO)

- Dr. Wernher von Braun, NASA Deputy Associate Administrator for Planning, said significance of man's movement into space with both "automated, unmanned spacecraft and manned vehicles," ought to be grasped by every nation and every individual. Speaking at "Report on America" dinner sponsored in Washington, D.C., by *Time* magazine for group of European businessmen, he added: "Man has *left* Earth. There is as great an awesome potential in that simple statement as there is in the fact, man *appeared* on Earth. Human destiny now has another door opened to it that previously was firmly locked." Satellite investigations and space probes had advanced human knowledge about earth, moon, and planets. "Astronomers now are able to view across the entire electromagnetic spectrum to obtain information about the birth, evolution, and extinction of celestial objects previously beyond reach of the most powerful telescope. Space is, in fact, a vast laboratory in which to experiment with weightlessness, hard vacuum, solar plasmas, and extremely high cosmic ray particles.

    "The potential value of space technology for Earth applications alone, including long-range weather forecasting, earth resources surveys, and communications, has been estimated . . . at billions of dollars annually to the world." (*NASA Activities*, 11/15/71, 182)

- San Francisco Bay area was being surveyed and photographed by U–2 reconnaissance aircraft in U.S. Geological Survey project to gather continuous information fast and accurately on urban development, UPI reported. Project also covered Phoenix-Tucson area in Arizona and Baltimore-Washington, D.C., area as basis for more thorough, nationwide study of 27 urban regions. In 1972 NASA ERTS satellite would provide constant check on urban growth patterns and development. (*NYT*, 10/7/71, 55)

- Small "quasi-moon" 1.6 to 3.2 km (1 to 2 mi) in diameter was gravitationally linked to earth and moon in three-body team of celestial objects orbiting sun, Dr. Hannes O. Alfven of Univ. of California at San Diego said in telephone interview with *New York Times*. Although little asteroid, named Toro, was discovered in 1964, computer calculations showing it was locked with earth and moon had not been reported until September at Nobel Conference in Stockholm. Toro came within 15 million km (9.3 million mi) of earth every eight years. Discovery was first indication that another celestial object was part of earth-moon system. (Rensberger, *NYT*, 10/8/71, 24)

- DOD spokesman Jerry W. Friedheim announced that 22 additional C-5s were being grounded as safety precaution pending additional tests to find cause of Sept. 29 incident at Altus AFB, Okla., in which engine and pylon mounting fell from aircraft during engine warmup for takeoff. (Getler, *W Post*, 10/8/71, A1)
- Dr. Jerome B. Wiesner was installed as 13th President of MIT at ceremonies in Cambridge, Mass. Dr. Wiesner had been science adviser to Presidents Kennedy and Johnson. (*NYT*, 10/8/71, 25; MIT PIO)
- President Nixon issued statement on Oct. 6 death of Rep. James G. Fulton. Rep. Fulton would be "especially remembered for his unfailing support of our space program." (*PD*, 10/11/71, 1382; Weil, *W Post*, 10/8/71, B8)
- Mrs. Leatrice M. Pendray—early rocket experimenter, syndicated columnist, and public relations executive—died in Princeton, N.J., at age 66 following long illness. She had been cofounder of American Interplanetary Society in 1930. Society had been renamed American Rocket Society and was later merged with Institute of Aerospace Science to form American Institute of Aeronautics and Astronautics. (*NYT*, 10/9/71, 30; *A&A 1963*)
- Charles H. Hubbell, well-known artist-historian of air age—who in 1930s began painting winners of annual Thompson Trophy Race—died at age 72. More than 375 "Hubbells" were painted for TRW Aviation Calendar series and quarter of million lithographic reproductions had been published. Hubbell's pictures had hung in White House, museums, palaces, libraries, and homes of aviation buffs. (TRW Calendar, 1972)

October 9: India successfully launched short-range rocket for first time from its new launching facility at Sriharikota Island. Indian-made Rohini rocket was launched to test complex at Sriharikota, near Madras, from which an Indian satellite was expected to be orbited in next three or four years. (Reuters, *NYT*, 10/10/71, 15)

- Nobel Prize for Medicine was won by Vanderbilt Univ. physiologist Dr. Earl W. Sutherland, Jr., for discovery of how cyclic adenyl acid regulated work of hormones in human body. (Auerbach, *W Post*, 10/15/71, A1)

October 10–13: Offshore Airport Center Planning Conference was held by AIAA and City of New York at Harriman, N.Y., to identify problems and determine feasibility of offshore airport construction in metropolitan New York area. Conference recommended major multifaceted study to develop data for environmental impact report and comprehensive study to determine economic significance of commercial aviation to New York City and penalty of not providing adequate facilities to meet projected aviation demands. (AIAA Release, 10/15/71)

October 11: U.S.S.R.'s *Salyut 1*, first manned space laboratory (launched April 19), reentered earth's atmosphere on command and disintegrated over Pacific Ocean after completing its program, Tass announced.

Flight of orbital station, lasting about six months, had consisted of two phases. First was station's joint flight with *Soyuz 10* spacecraft (launched April 23), piloted by Cosmonauts Vladimir A. Shatalov, Aleksey S. Yeliseyev, and Nikolay N. Rukavishnikov. *Soyuz 10* linked up with *Salyut 1* for 5½ hours, undocked, and reentered April 25.

Second phase began June 6 with orbiting of *Soyuz 11*, carrying Cosmonauts Georgy T. Dobrovolsky, Vladislav N. Volkov, and Viktor I. Patsayev. Spacecraft docked with station June 7 and *Soyuz 11* threeman crew transferred to *Salyut 1* for 23-day stay. While on board, crew carried out comprehensive scientific and technical studies and experiments. Upon return June 30 after record-breaking near-24 days in space, three cosmonauts were found dead in spacecraft. Soviet authorities said deaths were caused by sudden loss of pressure in Soyuz capsule.

Salyut 1 remained in orbit until Oct. 11, when braking engine was switched on to decelerate spacecraft; station began descending trajectory and entered atmosphere. Lost with *Salyut 1* were films, tapes, biological samples, and scientific experiments left on board by *Soyuz 10* cosmonauts. During entire lifetime, orbital station's onboard equipment operated normally. Since spacecraft was not recoverable and might fall on populated area if left to deorbit naturally, it had to be destroyed. (FBIS–Sov–201–3, 10/18/71, L1; FBIS–Sov–205–3, 10/22/71, L1; Reuters, B *Sun*, 10/16/71, A3; *Av Wk*, 10/25/72, 14)

- Space program for 1970s was outlined by Dr. Wernher von Braun, NASA Deputy Associate Administrator for Planning, in speech before Society of Sigma Xi at Daytona Beach, Fla. It would be "program of transition; a program in which the exploration and technical pioneering of the Sixties will begin to pay substantive dividends; it is the logical transition to the intensive and economic use of space in the 80's and 90's. Strong evidence will emerge . . . that will refute many of the arguments advanced against science and technology as embodied in the space program." Program's primary strength was "that it is balanced to meet the basic national needs." Program "strikes a harmony among scientific exploration *in* space, the practical utilization *of* space, and the development of improved technologies *for* space." (Text)

- Molniya II earth station in U.S. for new U.S.–U.S.S.R. "hotline" comsat would be under U.S. Government control but might be owned and operated under contract by private entity, *Aviation Week & Space Technology* reported. Hotline system would consist of two duplex telephone-bandwidth circuits for secondary telegraphic multiplexing and four ground stations for transmission and reception. (*Av Wk*, 10/11/71, 17)

- U.S. scientists conceded that U.S.S.R. led U.S. in many areas of basic research and space medicine largely because U.S. had "opted for an engineering approach to manned space flight and deliberately ignored some areas of theoretical research as irrelevant to the moon landing mission," *Aviation Week & Space Technology* reported. Scientists were worried that political developments in U.S.S.R. could harm free flow of this data. (*Av Wk*, 10/11/71, 14–5)

- Satellite photos of U.S.S.R. had shown continued construction of two new classes of silos for large missile, construction of third class of silo, and doubling of production facilities for Soviet missile submarines, *New York Times* reported. (Beecher, *NYT*, 10/11/71, 9)

October 11–14: Joint U.S.–U.S.S.R. Space Biology and Medicine Working Group—cochaired by Dr. Charles A. Berry, NASA Director for Life Sciences, and Dr. N. N. Gurovsky of Soviet Academy of Sciences—met in Moscow. Statement issued by American Embassy on completion of meetings said group had begun "exchange of data and results from the Soyuz and Apollo programs and developed recommendations and procedures for expanded exchange of information in space biology and medicine." Recommendations would be forwarded to Dr. George M. Low, NASA Deputy Administrator, and to Soviet Academy of Sciences President Mstislav V. Keldysh. (NASA Release 71–204)

October 12: Training mockups of two Skylab spacecraft components—Orbital Workshop and Apollo Telescope Mount—arrived at MSC aboard NASA barge *Orion* from MSFC. Shipment also included multiple docking adapter exterior shell and portion of airlock module mockup. Trainers and hardware would be used by MSC for training prospective Skylab crewmen for missions scheduled to begin in early 1973. (NASA Release 71–187; MSC PIO)

- USAF grounded 18 more C–5 cargo aircraft as result of Sept. 29 incident at Altus AFB, Okla., in which C–5 lost engine and pylon mounting at start of takeoff. Action removed entire fleet of 47 aircraft from service. (AP, B *Sun*, 10/13/71, A19)

- *Washington Post* editorial commented on Senate's Oct. 6 passage of $21-billion military procurement bill: "The new B–1 manned strategic bomber, the F–14 Navy fighter plane, the ABM, the main battle tank or XM803—these were among the controversial items that easily survived attempts at elimination or modification by amendment." What had recently been billed as "*the* issue for the seventies, a great debate over priorities and defense spending" had "gone the way of the hula hoop. On the part of a number of senators who just a few years back had made a loud and dramatic commitment to the issue, interest was low, absenteeism high, and preparation for debate pitiful. Sen. Proxmire and those others who carried the main burden deserved much better from their colleagues." (*W Post*, 10/12/71, A18)

October 13: U.S.S.R. launched eight Cosmos satellites from Plesetsk into near circular orbits with single booster. Tass reported Oct. 14 that equipment was functioning normally. Orbital parameters were:
 Cosmos 444, 1510-km (938.3-mi) apogee, 1323-km (822.1-mi) perigee, 114.1-min period, and 74° inclination.
 Cosmos 445, 1513-km (940.1-mi) apogee, 1352-km (840.1-mi) perigee, 114.4-min period, and 74° inclination.
 Cosmos 446, 1513-km (940.1-mi) apogee, 1383-km (859.4-mi) perigee, 114.8-min period, and 74° inclination.
 Cosmos 447, 1516-km (942-mi) apogee, 1412-km (877.4-mi) perigee, 115.1-min period, and 74° inclination.
 Cosmos 448, 1518-km (943.2-mi) apogee, 1442-km (896-mi) perigee, 115.5-min period, and 74° inclination.
 Cosmos 449, 1542-km (958.2-mi) apogee, 1485-km (922.7-mi) perigee, 116.2-min period, and 74° inclination.
 Cosmos 450, 1531-km (951.3-km) apogee, 1464-km (909.7-mi) perigee, 115.9-min period, and 74° inclination.
 Cosmos 451, 1575-km (978.7-mi) apogee, 1490-km (925.8-mi)

October 12: *Training models of Skylab's Orbital Workshop and Apollo Telescope Mount arrived at the Manned Spacecraft Center from fabrication at the Marshall Space Flight Center. They would be used for training crewmen for the 1973 mission, planned to include three separate crews of three astronauts each—the first crew to stay for 28 days and the next two for 56 days each. The overhead view above shows crew quarters in the 1-g trainer model of the Workshop. At the top is the general control and working area with several experiments. At the right is the wardroom, where crewmen would prepare and eat food and spend leisure time. At the lower center is the waste management compartment, or "space bathroom." At the left is the sleep compartment.*

perigee, 116.6-min period, and 74° inclination. (GSFC *SSR*, 10/31/71; FBIS–Sov–199–3, 10/14/71, L1; *SF*, 4/72, 132)

- NASA launched series of three sounding rockets from White Sands Missile Range, New Mexico. Nike-Cajun carried ion mass spectrometer, x-ray detector, ion probe, Faraday rotation counter, magnetometer, accelerometer, solar aspect sensor, telemetry, and tone-ranging system to perform D-region ion composition studies under subsonic conditions. Payload attained 80.8 km (50.2 mi) and was recovered in good condition.

  Second Nike-Cajun carried similar instrumentation for same purpose. Subsonic portion of flight near apogee provided principal data; payload reached 80.8 km (50.2 mi) and was recovered in good condition.

  Boosted Arcas II carried electrometer, magnetometer, solar aspect sensor, telemetry, and tone-ranging system to measure ion mobility at altitudes from 65 to 10 km (40 to 6 mi), using parachute-supported instrumentation package. Payload reached 63.1-km (39.2-mi) and was recovered in good condition. (NASA Rpts SRL)

- Completion of NASA–U.S. Dept of Agriculture experiment to gather information on southern corn leaf blight from aircraft was announced by MSC. Data were being analyzed but preliminary results showed blight could be detected and its progress monitored by sensor-bearing aircraft. Although southern corn leaf blight had not been as severe as projected, its widespread presence had provided "unique opportunity to assess remote sensing as a tool for large-scale crop surveys." NASA and USDA were analyzing experiment data to determine usefulness of crop infestation surveys and application of surveys to other agricultural needs. Results of analysis would be announced within few months. (MSC Release 71-78)

- President Josip Broz Tito of Yugoslavia and Mrs. Tito visited MSC during state visit to U.S. They were shown *Apollo 9* CM, Mercury and Gemini spacecraft, and LM test article. The Titos visited Mission Control and saw demonstration of spacesuit and *Apollo 15* films. Center presented Tito with framed and autographed photo of earth as seen from moon's vicinity by *Apollo 8* astronauts. (MSC PAO)

- MSFC announced award of $1 197 000 incentive contract to American Science and Engineering, Inc., for support services in Skylab ATM project. Contract would run from Aug. 22, 1971, through June 30, 1972. (MSFC Release 71-176)

- Sen. Edward M. Kennedy (D-Mass.) introduced amendment to S. 32, Conversion Research, Education and Assistance Act of 1971, to establish New Cities Research and Experimentation Administration. Agency would employ defense and aerospace personnel. (*CR*, 10/13/71, S16259-63)

- Secretary of Defense Melvin R. Laird told Washington, D.C., press conference that Soviet buildup of land- and sea-based missiles was "far outdistancing" estimates he had offered to Congress seven months previously. He confirmed report that U.S.S.R. was expected to match U.S. strength by deploying 41 Polaris-class missile submarines by 1973. (Beecher, *NYT*, 10/14/71, 5)

- Adm. Arthur W. Radford (USN, Ret.) in letter to the editor, published by *Washington Post*, defended USN F-14 fighter aircraft against suggestion by *Post* Sept. 29 that $800 million in funds for F-14 be eliminated from military budget: "In conjunction with the full spectrum of conventional naval weapons and sensors, the design capabilities of the F-14 are credible and responsible to the perceived threat for the decade ahead. As in the case of the F-4, which has now reached the end of its growth potential against a constantly upgraded Soviet technology, the F-14 is conceived as an aircraft to be used for more than a decade into the future—it, too, will be capable of substantial performance improvement as the threat changes and as technology is advanced."

  Washington Post responded to letter with editorial comment: ". . . there is a school of thought which holds that the heavily and expensively defended [aircraft] carrier can be rendered inoperable by comparatively simple Soviet weapons and thus is inadequate to the kind of U.S.-Soviet conflict that is presupposed by the design of the [carrier-based] F-14. It is this view which we find persuasive and which therefore gives us pause concerning the usefulness of the F-14." (*W Post*, 10/13/71, A18, A19)

October 14: USAF launched unidentified satellite into orbit from Vandenberg AFB by Thor-Burner II booster. Satellite achieved 878.7-km (546-mi) apogee, 795.0-km (494-mi) perigee, 101.6-min period, and 98.9° inclination. (Pres Rpt 72; *SF*, 4/72, 132)

- U.S.S.R. launched *Cosmos 452* from Plesetsk into orbit with 280-km (174-mi) apogee, 175-km (108.7-mi) perigee, 89-min period, and 64.9° inclination. Satellite reentered Oct. 27. (GSFC *SSR*, 10/31/71; *SF*, 4/72, 132)

- NASA successfully completed 17th and 18th flights of TF-8A aircraft with supercritical wing, flown from FRC by NASA test pilots Thomas C. McMurtry and Garry Krier, and announced results in accelerated program to evaluate TF-8A performance in advance of adverse landing conditions. All flight objectives had been achieved in series to expand flight envelope to higher dynamic pressures, obtain pressure distribution and performance data at higher dynamic pressure, evaluate unaugmented stability and control characteristics, and investigate buffet boundary for trim flight at mach 0.99 design conditions at 14 000 m (46 000 ft). Pressure distribution and performance data had been obtained over mach 0.80 to 1.00 at altitudes to 14 000 m. Unaugmented stability and control characteristics had been evaluated from mach 0.70 to 0.99 at 10 700 m (35 000 ft). Buffeting had neared that predicted from wind-tunnel tests and buffet boundary had been penetrated to 22° attack angle. Buffeting had been "intense but not severe." (NASA Proj Off)

- FAA award of $229 746 planning grant-offer to New Jersey Dept. of Transportation for master plan of V/STOLport to serve New York metropolitan and northern New Jersey areas was announced by FAA Administrator John H. Shaffer. (FAA Release 71-168)

- Plans for establishment of International Institute of Applied Systems Analysis by U.S., U.S.S.R., U.K., France, Italy, Poland, and East and West Germany were described in *New York Times* by Walter Sullivan: Project, "fruit of four years of private Soviet-American negotiations" had culminated in eight-nation conference in Paris. Prominent among principals had been NAS President, Dr. Philip Handler, and Dzherman M. Gvishiani, Soviet specialist in American management techniques. (*NYT*, 10/14/71, 1)

- Assembly of solar radio telescope 10 m (33 ft) in diameter was nearing completion at Baldone, Latvia, Riga *Sovetskaya Latviya* reported. Parabolic antenna would enable Latvian Academy of Sciences Radio-astrophysical Observatory scientists to study all fluctuations of sun's radiowaves. (FBIS-Sov-70, 10/22/71, L4)

- *Wall Street Journal* editorial commented on KSC economy drive: "One of the first things that occurs to householders when a money pinch comes is to turn off all the unneeded lights to save money on electricity." This had occurred to KSC officials who "came up with a real enough saving, an estimated $75 000 a year." KSC parking lots and some buildings were "a little dimmer but everything continues to function." (*WSJ*, 10/14/71)

October 15: March 7 discovery of water vapor on moon by two suprathermal ion detectors left on lunar surface by *Apollo 12* and *14* astronauts was announced by Rice Univ. scientists Dr. John W. Freeman and Dr. H. Kent Hills. Water bursts had been found 14 min apart at two lunar sites just after seismometers at sites had picked up swarm of moonquakes. Dr. Freeman said in Houston, "I think this proves that the moon is still actively quaking and venting gas and also indicates there is a good possibility of liquid water somewhere on the moon." He said bursts were best described as "geysers," but cautioned against likening them to geysers on earth. Later, *Washington Post* reported that Apollo engineers had said water could have come from burst tank on descent stage of *Apollo 12* LM or from backpacks of *Apollo 12* crew. (O'Toole, *W Post*, 10/16/71, A1; Reuters, B *Sun*, 10/16/71, A3)

- Skylab airlock trainer was flown from McDonnell Douglas Astronautics Co. in St. Louis, Mo., to MSC aboard NASA Super Guppy aircraft. (NASA Release 71-187; MSC PIO)
- Skylab astronauts would begin extravehicular training in pressurized suits in MSFC Neutral Buoyancy Space Simulator later in year. Mockups of Skylab space laboratory modules had been submerged in water tank 12 m (40 ft) deep and 23 m (75 ft) wide, which simulated weightlessness of space environment. (NASA Release 71-205)
- Univ. of Florida was using KSC particle measurement computer system to analyze upper-air samples taken by aircraft at various altitudes, *NASA Activities* reported. System was only one of its kind in Florida. (*NASA Activities*, 10/15/72, 169)
- Grumman Aerospace Corp. demonstrated $500 000 Turbo-Mallard amphibious aircraft in Potomac River for DOT and CAB officials. Ten-seat, 20-m (66-ft) aircraft cruised at 355 km per hr (220 mph) and could land on ground or in water as shallow as 1.5 m (5 ft) and at least 3000 m (1000 ft) long. Grumman had $50 000 contract with City of New York for feasibility study of water landings and takeoffs. (Levey, *W Post*, 10/16/71, B3)
- NSF published *Enrollment Increase in Science and Mathematics in Public Secondary Schools, 1948-49 to 1969-70* (NSF 71-30). Total in 1969-70 was more than 2.5 times larger than total enrolled in 1948-49. Largest relative increases were reported for enrollments in psychology, economics, and biology. (NSF *Highlights*, 10/15/71)

October 16: Sleeping bag, handgun, photos, and papers carried by aviator Wiley Post on flight on which he lost his life Aug. 15, 1935, had been donated to State of Oklahoma for display by historical society, AP reported. Post was killed with comedian Will Rogers when aircraft engine failed after takeoff from Pt. Barrow, Alaska, on sector of contemplated flight to Siberia. (*NYT*, 10/17/71, 32)

- Lasers were finding increased applications in defense technology, *New York Times* article reported. Weapon scientists were "clearly excited over its possibilities as a deadly offensive weapon against nuclear missiles and aircraft." AEC and DOD spokesmen had conceded that laser programs were underway but had said national security barred discussion of subject. DOD had said lasers had been used in new rangfinding devices and in medical research and that, while it was primarily USAF program, "some" Army research was also going on. (Ripley, *NYT*, 10/16/71, 15)

October 17: USAF launched *Sesp 1971–2* Space Experiments Support Program satellite from Vandenberg AFB by Thorad-Agena booster. Orbital parameters: 801.5-km (498-mi) apogee, 774.1-km (481-mi) perigee, 100.5-min period, and 92.7° inclination. Objectives of mission were to test advanced development payloads—including celestial mapping IR scanner, 1500-w flexible roll-up solar cell array, secure command and control system, and energetic particles experiment. (Pres Rpt 72; *SBD,* 10/19/71, 174)

- *Eyewitness to Space* by H. Lester Cooke with James H. Dean, published by Harry N. Abrams, Inc., was reviewed by Frank Getlein in Washington *Evening Star.* Book contained more than 250 paintings and drawings by artists in NASA artist program. "What they have created is an extraordinary record . . . because it records more than the eye, even the trained eye, can see. The sketches and many of the paintings do indeed record what any eye can see, but the seeing is special. Beyond that, artist after artist at some point found himself pushed by the material itself into taking an abstract, symbolic view of the space flights themselves, painting a vision, not a view." NASA artists program, "one of the most remarkable ventures in art-government relations," had brought hundreds of artists to NASA facilities to paint or draw what they saw and they were to see whatever they wanted and depict it in whatever manner suited their perceptions." Only requirement was that all work was to be turned over to NASA, which had given it to National Gallery. Coauthor Cooke was National Gallery Curator of Painting. It was "hard to think of anything in art history that quite compares with the NASA program." Getlein recalled Italian TV coverage of first moon walk. "One big response to the event was to convene gatherings of famous Italians—non-scientists all—in studios . . . and have them talk about the moon. But at many points in the protracted coverage, against a background of Beethoven, a beautiful Tuscan voice read Dante." Qualities evoked by this "odd response, but an oddly perfect one," were "precisely those that have eluded American television coverage of the space program. They are qualities that this book reaches for constantly and often attains." (W *Star,* 10/17/71, B5)

October 18: Heart problems encountered by *Apollo 15* astronauts necessitated changes in diet and exercise regimen for Apollo 16 astronauts, Dr. Charles A. Berry, NASA Director of Life Sciences, said in speech before annual clinical meeting of American College of Surgeons in Atlantic City, N.J. All three *Apollo 15* astronauts had experienced drops in potassium levels during July 26–Aug. 7 mission. Astronaut James B. Irwin had developed cardiac arrhythmia (irregular heartbeat) for one minute after he returned to CM from lunar surface and ventricular contractions (premature heartbeats) for following hour. Astronaut David R. Scott also had developed cardiac contractions during mission. Because of findings, first Apollo 16 EVA would be shortened, astronauts would rest full eight hours between moonwalks, and more potassium would be added to Apollo 16 diet. (McGehan, B *Sun,* 10/19/71, A5)

- Scientists and engineers at MIT described planned Apollo 17 surface electrical properties experiment to probe for water on moon, in telephone interview with *New York Times.* Experiment consisted of 6.8-

October 18

kg (15-lb) radio transmitter with four 35-m (115-ft) lengths of wire which astronaut was to unreel on lunar surface in X pattern as antenna. Radio signals at varying frequencies would be beamed into lunar surface. Small receiver on LRV would receive signals bounced back from beneath lunar surface and signals would be recorded by tape recorder on LRV. Pattern, timing, and strength of reflected signals were expected to indicate presence of water, if any, and kind of material in lunar subsurface. *Times* said experiment's importance had been heightened by Oct. 15 announcement of discovery of water vapor on moon. (Wilford, *NYT*, 10/19/71, 29)

- NASA board investigating Jan. 23 helicopter crash in Indian River near KSC—in which pilot, Astronaut Eugene A. Cernan, escaped injury—had named misjudgment in estimating altitude as primary cause, NASA announced. Mitigating factors were lack of familiar objects on Indian River surface to help pilot judge altitude, possible visual focusing on false water surface because of water's smoothness, change in sun reflection on water caused by change in course, and possibility that Cernan's experience with high-speed aircraft might have contributed to altitude misjudgment in slower craft. Board said Cernan's survival training as astronaut and naval aviator was major factor in his escape from submerged wreckage and burning fuel. (NASA Release 71-206)

- NASA announced receipt of three proposals for design and fabrication of two experimental, transport, STOL research aircraft with propulsive lift—from McDonnell Douglas Corp., from Boeing Co. with Grumman Corp., and from Lockheed Aircraft Corp. with North American Rockwell Corp., Bell Aerospace Co. Div of Textron, Inc., and Cornell Aeronautical Laboratories, Inc. Aircraft would provide data and experience for developing environmentally acceptable, economical, and safe fan-jet STOL transport systems in joint NASA–DOT–FAA–USAF program to relieve noise and congestion at busy airports. (NASA Release 71-207)

- President Nixon announced appointment of Coleman duPont Donaldson as member of Air Quality Advisory Board to succeed William O. Doub, who was appointed AEC member Aug. 10. Donaldson, President of Aeronautical Research Assn. of Princeton, Inc., had been head of aerophysics section of NACA Gas Dynamics Laboratory. He had worked recently on development of techniques for computing dispersal of chemically reacting pollutants in atmosphere and generation of turbulence. Work had been supported by NASA and Environmental Protection Agency. (*PD*, 10/25/71, 1426)

- NASA wished protocol officials who scheduled President Nixon's trip to Moscow for May could be assigned to handle space cooperation talks as well, *Aviation Week & Space Technology* said. Magazine quoted NASA official as saying, "Anyone who would schedule the summer session for Houston and the winter session for Moscow just can't be serious about an agreement." (*Av Wk*, 10/18/71, 11)

- USAF Space and Missile Systems Organization planned conceptual design study by industry of spaceborne system for tracking Soviet satellites at altitudes beyond 480-km (300-mi) range of existing ground-based Spacetrack system, *Aviation Week & Space Technology* reported.

Five-month study was expected to begin in 1972, magazine said. (*Av Wk*, 10/18/71, 9)

- U.S. technological society was product of past R&D efforts, President Charles A. Anderson of Stanford Research Institute said in *Aviation Week & Space Technology* editorial. Current assault on U.S. leadership in science and technology was "both external and internal." Western Europe, Japan, and U.S.S.R. were progressing in areas traditionally controlled by U.S. "In radio astronomy, high energy physics and conventional energy research we have already been surpassed. Internally, a new national distaste for science and technology is undermining the national research and development effort." (*Av Wk*, 10/18/71, 7)

October 18–21: Lunar Science Institute held conference on lunar geophysics at MSC. Scientists said new evidence from *Apollo 15* heat flow experiment indicated moon was hotter than expected and might account for melting temperatures within moon and volcanic activity on lunar surface. Thermometers placed in two holes near Hadley Rille had shown that amount of heat from moon was one half that of heat measured from earth. Since moon was one fourth earth's size, relative abundance of radioactive isotopes which generated heat within planet might be greater in moon than in earth. Lunar scientists also were puzzled by area across Sea of Rains and Ocean of Storms several times more radioactive than rest of moon. Dr. James R. Arnold of Univ. of California at San Diego said, "It is not obvious why a whole geographic area on the moon would be different."

Dr. John W. Freeman, Jr., of Rice Univ. referred to results of *Apollo 12* and *14* suprathermal ion detector experiments that had discovered water vapor on moon [see Oct. 15]. Dr. Gary V. Latham, Apollo program chief seismic investigator, said neither he nor Dr. Freeman could locate source of tiny moonquakes that released gas from beneath lunar surface, including water vapor seen March 7. Dr. Latham described sudden and large surge from lunar surface detected by seismometer in April during series of moonquake swarms. Though he knew of no volcano on moon, he compared event—largest moonquake yet recorded—to volcanic activity on earth. (MSC Release 72–77; Maloney, *H Post*, 10/19/71; Brett, *W Post*, 10/21/71, F2)

October 18–29: Team of 10 NASA officials attended Oct. 18 ELDO Symposium on space shuttle technologies in Paris. Officials then divided into four working groups and visited industries in France, Italy, Germany, U.K., the Netherlands, and Belgium to review ELDO-sponsored space shuttle research and European technological capabilities. Team including Adelbert O. Tischler, Chief of Shuttle Technologies, OART; William A. Summerfelt, Engineering Director, Space Shuttle Program Office, OMSF; and James J. Gangler, Materials Branch Chief, Materials and Structures Div., OART. (NASA Release 71–209)

October 19: U.S.S.R. launched *Cosmos 453* from Plesetsk into orbit with 492-km (305.7-mi) apogee, 270-km (167.8-mi) perigee, 92.1-min period, and 70.9° inclination. Satellite reentered March 19, 1972. (GSFC *SSR*, 10/31/71; 3/31/72; *SBD*, 12/26/71, 195)

- Saturn IB 1st stage for first manned Skylab launch vehicle was removed from environmentally controlled enclosure at Michoud Assembly Facility after three-year hibernation. Booster, one of nine IB stages stored there in December 1968, would begin 10-mo refurbishment program in preparation for spring 1973 launch. (MSFC Release 71-183)

- Development of surgical implement to simplify cataract removal with high-frequency vibration and pump to liquify and remove cataract and lens material was announced by LeRC. Hand-held instrument—developed by four-man LeRC team in cooperative program with Dr. William J. McGannon, prominent Cleveland eye surgeon—had been successfully tested on animals. NASA was seeking patent on instrument, which would reduce cataract operation time and patient's convalescent period by minimizing stitching problem in most cases. Instrument made only small puncture. (NASA Release 71-203)
- Sen. Howard W. Cannon (D-Nev.), in letter to President Nixon, appealed for inclusion of NERVA funding in FY 1973 budget: "It seems to me . . . that this program continues to offer the nation's best chance to take the next logical step forward in space, and that the already staggering $1.4 billion investment in succesful R&D would make continuation of the program not only desirable but mandatory, since we are so close to a flyable engine." As senior member of Senate Committee on Aeronautical and Space Sciences, Sen. Cannon was "greatly disturbed that the space program in recent years has been progressively cut back." If U.S. space program was to continue "the NERVA funding issue is terribly important. . . . if we are to cast aside our earlier desire to go forward in space and subject our investment to a less-than-starvation funding level, we are only deceiving ourselves. I believe that rather than merely giving lip service to space, we ought to consider a total restructuring or delegation of NASA's role to the military." (Text)
- FCC ordered AT&T to activate transatlantic cable and satellite circuits on one-to-one ratio. Order reversed decision made in May to set ratio at five-to-one in favor of satellites. FCC said it had received "strong representations" protesting five-to-one ratio from European countries that had 50% ownership in cables but less than 30% ownership in satellites. (W Post, 10/20/71, C10)
- Aerobee 170 sounding rocket was launched by NASA from WSMR carrying GSFC payload on test and support mission. Rocket and instruments functioned satisfactorily. (SR list)
- President Nixon announced 25 appointments to new National Advisory Committee on Oceans and Atmosphere and designation of Dr. William A. Nierenberg, Director of Scripps Institution of Oceanography, and Dr. William J. Hargis, Jr., Director of Virginia Institute of Marine Science, as Chairman and Vice Chairman. (PD, 10/25/71, 1429)
- KSC announced award of $99 189, 24-mo contract to Univ. of Florida for research and preparation of narrative history of KSC Apollo launch facilities and operations. Contract included option to extend for six months. (KSC Release 245-71)
- ComSatCorp had requested permission to build two antennas 29.5 m (97 ft) in diameter as backup systems for its comsat stations in Andover, Me., and Etam, W. Va., Washington *Evening Star* reported. (W Star, 10/19/71, A14)
- Former astronaut M. Scott Carpenter, who had also been aquanaut in USN's Sealab Project, testified before Senate Committee on Com-

merce's Subcommittee on Oceans and Atmosphere. Carpenter, president of Sea Sciences Corp., said man must explore ocean floor if he is to understand it, just as he had to explore lunar surface to understand it. Subcommittee was examining ocean pollution. (*CR*, 10/18/71, D1039; Sehlstedt, B *Sun*, 10/19/71, A1)

- *New York Times* editorial commented on discovery of water on lunar surface [see Oct. 15]: "The possibility that the water vapor that has now been detected on the moon was originally brought there by Apollo astronauts cannot yet be dismissed. Nevertheless, even before Drs. Freeman and Hills announced their finding it was becoming increasingly clear that the moon is a much more complex body than had been supposed. We are still in the very infancy of the era of moon exploration, and its development for the benefit of the human race." (*NYT*, 10/19/71, 40)

October 20: Apollo 16 preliminary timeline was announced by NASA. Launch was scheduled for 1:03 pm EST March 17, 1972. Translunar insertion would occur at 3:40 pm. Lunar landing was planned for 5:25 pm March 21, with EVA to start at 9:18 pm. Second EVA was scheduled for 7:48 pm March 22, and third for 7:13 pm March 23. LM would leave moon at 6:23 pm March 24 and earth landing was scheduled for 4:14 pm March 29. (NASA Release 71–211)

- Flight path of *Mariner 9*, en route to Mars, was reported sufficiently precise for orbital insertion without scheduled second midcourse correction. Next critical event for *Mariner 9* would be Mars orbit insertion Nov. 13, when spacecraft would begin to collect atmospheric and surface data and map 70% of Martian surface during 90 days in orbit.

  At 9:00 am EDT, after 142.4 days in flight, *Mariner 9* was 88 847 850 km (55 207 500 mi) from earth. Spacecraft was traveling at 59 220 km per hr (36 800 mph) relative to earth and had traveled total distance of more than 349 452 100 km (217 139 500 mi) since May 30 launch. (NASA Proj Off)

- NASA and Soviet Academy of Sciences had agreed to rapid exchange of findings on Mars from U.S. and U.S.S.R. probes nearing that planet, NASA announced. Recommendations of Joint Working Groups of U.S. and Soviet scientists who met in Moscow Aug. 2–6 to negotiate agreement on joint projects and exchanges in space research and applications had all been approved. Included was telegraphic exchange of findings of special interest by NASA's *Mariner 9* (launched May 30) and Soviet *Mars 2* (launched May 19) and *Mars 3* (launched May 28). All three spacecraft were scheduled to reach Mars orbit in November.

  NASA and Soviet Academy also had agreed on joint seminars to facilitate analysis of data; exploration of feasibility of transmitting magnetometer data from U.S.S.R. ground-based observing stations in real-time to U.S. satellite ATS–F (scheduled for mid-1972 launch) and retransmitting data to ground with data from onboard magnetometer; and bilateral discussions on scientific results, objectives, and strategy of planetary exploration. Further agreements were to exchange information on photography from past and future missions and all lunar maps, to continue exchange of lunar samples, to conduct multipurpose aerospace and field experiments on environment, and to make joint efforts in remote sensing of ocean. Technical Note would be exchanged on temperature sounding from satellites to develop joint experiment over Western Europe in 1973 and 1974. Information on methods of microwave measurement and of data on cloud emissivity would also be exchanged. Efforts would be made to achieve similarity in U.S. and Soviet requirements for ground receiving equipment. NASA and Soviet Academy also had agreed to research topics for meteorological sounding rocket investigations by both nations, to exchange alerts of stratospheric warmings, to conduct rocket system intercomparison tests, and to present scientific results in regular seminars and symposia. (Text; NASA Release 71-210)

- GAO said USAF had not charged ComSatCorp for more than $6 million in launch service costs for comsats. Agency blamed $3.1 million of loss on USAF accounting procedures and said other $3 million was not charged because of DOD–NASA agreement. GAO could find no legal basis for changing method to determine costs chargeable to ComSatCorp but recommended ComSatCorp be assessed on full user-charge basis for future launches and Intelsat IV series that began in January. (UPI, *W Post*, 10/21/71, A15)

- Lockheed Aircraft Corp. Chairman Daniel J. Haughton told firm's annual meeting that "we are now on an upward track." He reported net income for quarter ended Sept. 30 of $2.2 million, or 19 cents per share, up from $2 million, or 18 cents per share, in 1970 period. Sales had climbed from $565 million in 1970 to $667 million. Sales for all of 1971 were estimated at close to $3 billion, although major percentage would be booked at no profit because of Lockheed's settlement with Government on C-5A military transport aircraft and Cheyenne helicopter contracts. Haughton said 1972 sales should be at 1970 level of just over $2.5 billion. (Wright, *NYT*, 10/21/71, 69)

- Shift in AEC's role from promoting atomic energy to protecting public interest in nuclear affairs was announced by Dr. James R. Schlesinger, AEC Chairman, at annual meeting of Atomic Industrial Forum and American Nuclear Society in Bal Harbour, Fla. It was AEC's responsibility to "develop new technical options and to bring those options to the point of commercial application," not to "solve industry's problems which may crop up in the course of commercial exploitation. That is industry's responsibility, to be settled among industry, Congress, and the public." AEC's role was "primarily to perform as a referee serving the public interest." (Text)

- President Nixon submitted to Senate nomination of Albert C. Hall, Vice President for Engineering and Research with Martin Marietta Corp., to be Assistant Secretary of Defense (Administration). Hall would succeed Robert F. Froehlke, who had become Secretary of the Army July 1. (*PD*, 10/25/71, 1430, 1441; USA PIO)

- NASA signed $195 400 000 supplemental agreement to Apollo LM contract with Grumman Corp. for changes to increase lunar staytime, activity support, and other features of LM. (*NASA Activities*, 11/15/71, 187)

October 21: ITOS–B Improved Tiros Operational Satellite failed to reach orbit after 4:32 am PDT launch by NASA for NOAA from WTR on two-stage long-tank, thrust-augmented Thor-Delta (DSV-N-6) booster. Satellite, weighing 306 kg (675 lbs), had been intended for sun-synchronous orbit to make regular, dependable, day and night cloud-cover observations by both direct readout and onboard storage.

Flight appeared normal through Delta first burn. At start of last period (approximately one hour coast), pitch and yaw altitude-control jets began pulsating to counteract unknown force tending to tumble vehicle. Pitch and yaw jets were able to maintain vehicle in proper altitude until gas was expended. At that time, unknown force caused vehicle to tumble. Planned vehicle functions remaining, including second Delta burn and spacecraft separation, occurred approximately on time. Planned circular orbit was not achieved and spacecraft and Delta 2nd stage impacted above Arctic Circle. Review board was being formed to investigate Delta 2nd-stage failure.

ITOS–B was second Tiros spacecraft funded by NOAA and third spacecraft in ITOS series. First, *Itos 1* (Tiros-M), had been launched Jan. 23, 1970. Second, *Noaa 1* (ITOS–A), was launched Dec. 11, 1970, and ceased operations Aug. 19, 1971. Office of Space Science and Applications had overall responsibility for mission, with Earth Observations Program Office managing project for Headquarters. Technical and administrative management was assigned to GSFC. (NASA Proj Off)

- Impact of aerospace industry slump and termination of lunar excursion module project on Grumman Corp. and Long Island, N.Y., area was described by *New York Times*. With 26 000 employees, Grumman Corp., "reaches out into every corner of the economy of Nassau and Suffolk Counties and, as one of the nation's largest aerospace combines, even far beyond those two counties that jointly claim Grumman as their own." Nearly 1000 subcontractors depended on Grumman's patronage. Firm was "largest real estate taxpayer in Nassau County, the largest customer for the Long Island Lighting Company and hundreds of smaller corporations. And in salaries for its employes alone it pumps nearly a third of a billion dollars each year into the Long Island economy." Grumman figures indicated that for every person laid off, 4½ persons somewhere in U.S. economy lost their jobs, including 2½ in Nassau or Suffolk Counties. From December 1969 to December 1970 nearly 5700 persons at Grumman had lost jobs; 14 250 elsewhere on Long Island had lost jobs as result. (Andelman, *NYT*, 10/21/71, 45)

- NASA director of Safety Jerome F. Lederer gave guidelines for noting and meeting problems of drug abuse among aerospace employees at 24th International Air Safety Seminar in Mexico City. Drug addiction was "not yet a discernible problem among NASA and its contractor employees, of which there are now about 160 000. But the work force of the future will involve many young people who have experienced hard drugs." (NASA Release 71–208)

- AFSC announced USAF development of battery-powered, expendable, remote-operating weather station (EROWS) to gather data from inaccessible terrain or battle fields and transmit data 320 km (200 mi). Dropped from cargo aircraft and helicopters, EROWS would embed itself vertically in preselected spot, turn itself on, monitor its surroundings, and send meteorological data to control recorder. (AFSC Release 237.71)

October 22: Establishment of joint NASA–DOT Office of Noise Abatement was announced by Dr. James C. Fletcher, NASA Administrator, and Secretary of Transportation John A. Volpe. Office would provide overall leadership and focal point for national program to attack noise problems of current and planned transportation systems. Charles R. Foster, head of DOT Office of Noise Abatement, would be Director of joint office. Walter F. Dankhoff, Chief of LeRC Office of Operations and Planning, would be Deputy Director. New office would be responsible to Robert H. Cannon, Jr., Assistant Secretary of Transportation for Systems Development and Technology, and to Roy P. Jackson, NASA Associate Administrator for Advanced Research and Technology. It would also integrate programs with other agencies through Interagency Aircraft Noise Abatement Program and was fulfillment of some of recommendations from Civil Aviation Research and Development (CARD) Policy Study. (NASA Release 71-213)

October 23: USAF launched unidentified satellite from Vandenberg AFB by Titan IIIB-Agena booster into orbit with 392.7-km (244-mi) apogee, 133.6-km (83-mi) perigee, 89.7-min period, and 110.9° inclination. Satellite reentered Nov. 17, 1971. (Pres Rpt 72; *SF*, 4/72, 132)

• Aerobee 170 sounding rocket was launched by NASA from WSMR carrying Cal Tech soft x-ray astronomy experiment. Rocket and instruments functioned satisfactorily. (SR list)

• Preliminary results of experiment in which two clocks were flown around the world to test Einstein theory of relativity had indicated "clock paradox" of theory was correct, Dr. J. C. Hafele, Washington Univ. in St. Louis scientist, told meeting of American Assn. of Physics Teachers in Rolla, Mo. Clocks, carried in two separate aircraft, had not kept same time as master clock at Naval Observatory in Washington, D.C. Einstein theory said that as body approached speed of light, variables in time, size, and weight changed. (W *Star*, 10/24/71, A3)

• Dangers inherent in separation of science and government were discussed in *New York Times* article by Dr. Albert Szent-Gyorgyi, Nobel Prize-winning Hungarian biochemist at Marine Biological Laboratories in Woods Hole, Mass. "Among our 450 representatives in Washington there is not a single scientist, so they do not know that problems are like equations which cannot be solved by trickery, blows or bombs, not even atomic bombs." With different assumptions different self-consistent systems could be constructed; each seemingly logical. "Ptolemy placed the earth at the center of the universe, Copernius the sun. Each system is perfectly logical and self-contained. Copernicus himself never dared to give preference to his own. The only way to decide between the two is to see where they lead. An astronaut basing his calculation on Ptolemy would end in the nowhere. Our present political system leads to poverty, inflation, unemployment, crime, drugs, war and the neglect of our priorities toward a final catastrophe. The solution is not separating science from politics . . . but making it penetrate into politics." (*NYT*, 10/23/71, 33)

October 24: Latest edition of *Jane's All the World's Aircraft,* published in London, listed series of Cosmos satellites that could intercept other spacecraft in orbit, determine their purpose, and "blow them to pieces." In report of Soviet orbital interception tests during year, publication said *Cosmos 397* (launched Feb. 25) had passed near *Cosmos 394* (launched Feb. 9) "and was subsequently destroyed in an explosion." *Cosmos 400* (launched March 19) had been "intercepted" by *Cosmos 404* on *404*'s launch date, April 3. In preface, editor John W. R. Taylor also said U.S. maintained satellites in stationary orbit over Pacific, including one with "fantastic 11-ton reconnaissance camera," to monitor launches of Soviet and Red Chinese long-range missiles. (UPI, *NYT*, 10/24/71, 4)

October 25: MSC engineers were studying use of liquid oxygen and propane as fuel combination for recoverable ballistic booster that was possible vehicle for space shuttle transportation system, *Aviation Week & Space Technology* reported. (*Av Wk*, 10/25/71, 12)

- Decisions on NASA's space shuttle program might be politically timed, *Aviation Week & Space Technology* said. Four contractors would complete study contracts in spring 1972. "At this point, President Nixon—in the midst of his re-election campaign—might publicly state his support for the project." Selection of winners "of the biggest plum in space" would not be announced until after the November election, magazine said. (*Av Wk*, 10/25/71, 11)

- Execution of FAA-ESRO Memorandum of Understanding for joint aviation satellite program had been delayed from Nov. 3 to early December because Office of Management and Budget could not complete review of proposed program in time, *Aviation Week & Space Technology* reported. Final pre-bid briefings for prospective contractors also had been delayed. (*Av Wk*, 10/25/71, 23)

- French national space agency Centre National d'Etudes Spatiales (CNES) budget request for 1972 would be approximately $150 million, up from $132.7 million received for 1971, *Aviation Week & Space Technology* reported. Budget would be augmented to cover participation in ELDO Europa 3 launch vehicle program if production of advanced launch vehicle was approved. Increase in CNES funding would be largely offset by inflation, French officials had said. (*Av Wk*, 10/25/71, 20)

- *Aviation Week & Space Technology* editorial noted "major surprises" from first four manned lunar explorations: Possibility of existence of water or ice below lunar surface; detection of heat near moon's core and of seismic activity on moon; greater magnetic field than expected; and fact moon was "mineral storehouse of tremendous magnitude with some elements in combinations rarely or never found on earth." There was also "sizable accumulation of other scientific data to be refined from the instrumentation still functioning on the moon and the nearly 400 lb [181 kg] of lunar rocks and soil samples. . . . The moon has proved to be a scientific bonanza regardless of what more material benefits may be realized eventually. All of this must be a cause for considerable chagrin and embarrassment to that segment of the scientific community that led the sneering section during the early stages of the Apollo program when flight testing of operational hardware was of necessity its primary concern." (Hotz, *Av Wk*, 10/25/71, 7)

- Mikhail K. Yangel, Soviet scientist and rocket expert, died of heart attack at age 60. Tass later quoted obituary, signed by Soviet leaders, as saying Academician Yangel had "made an inestimable contribution to the cause of development and perfection of rocket-space engineering and exploration of near-earth outer space." (FBIS–Sov–71–207, 10/27/71, L3)

October 25–28: AIAA held 8th Annual Meeting and Technical Display in Washington, D.C. Meeting Chairman, Pan American World Airways President Najeeb E. Halaby, dedicated meeting "to establishing a rapport between supporters of technology and its critics." In program introduction he said, "This is a time not for routine reading of sophisticated papers about a future that is itself in question, but for addressing ourselves to the basic issue of where we are headed—if anywhere."

Technical display included space shuttle technology exhibits by NASA and major contenders for shuttle contracts and models of Grumman F-14A fighter aircraft, McDonnell Douglas F-15 fighter, and NR B-1 bomber aircraft. On Oct. 25, designated by AIAA as "Aerospace Day," *Apollo 15* Astronaut David R. Scott narrated NASA *Apollo 15* film.

Panel discussions included "Rocketry in the 1950's" chaired by Dr. Wernher von Braun, NASA Deputy Associate Administrator for Planning; "STOL and VTOL—Where Do They Fit In?" chaired by ARC Director Hans Mark and transmitted by telephone link up to LeRC employees; and "The Aerospace Professional—What Is His Future?" with Dr. James C. Fletcher, NASA Administrator, as panel member.

Von Kármán Lecture, "Trends in the Field of Automatic Control in the Last Two Decades," was given by Dr. Irmgard Flugge-Lotz of Stanford Univ. Automation had been penetrating nearly every field of engineering. Mathematical theory and large computing equipment allowed development and extension beyond handling of linear and weekly nonlinear systems. Investigation of systems with large transport time lags had been advanced.

Major AIAA awards presented included John Jeffries Award to Richard S. Johnston, MSC Deputy Director of Medical Research and Operations, for "outstanding contributions to the advancement of Manned Space Flight and aerospace medical research through your leadership in the development of life support systems." Lawrence Sperry Award was given to Ronald L. Berry, Chief of MSC Planetary Missions Analysis Branch, for "many outstanding achievements in the field of manned space flight, particularly in lunar mission analysis and design development."

Spacecraft Design Award was presented to Anthony J. Iorillo, Manager of Special Projects for Hughes Aircraft Co., for "invention of the Gyrostat stabilized spacecraft which formed the basic stabilization and configuration for the successful Tactical Communications Satellite, Intelsat IV, and a series of other satellites now in design and production." Louis W. Hill Space Transportation Award was given to Dr. Hubertus Strughold, Emeritus Professor of Medicine at Brooks AFB, Tex. Citation said, "With his prophetic foresight and exceptional talents he established the rational biomedical foundations for manned exploration of space." DeFlorez Training Award went to Capt. Walter P. Moran of American Airlines for his "pioneering efforts in the utilization of flight simulation to increase safety and economy of airline flight training and checkout operations. . . ." Aerospace Communications Award was presented posthumously to Siegfried H. Rieger for "his personal contributions over a 10 year period, in analytical work evolving basic concepts, and in the origination and implementation of the INTELSAT Program which led to the establishment of the present worldwide communications satellite network."

AIAA History Manuscript Award was presented to Dr. Richard C. Lukas of Tennessee Technological Univ. for "best historical manuscript dealing with the science, technology and impact of aeronautics and astronautics on society." Award winning manuscript was "Eagles East: The Army Air Force and the Soviet Union."

Among 23 new Fellows honored at Honors Night Banquet Oct. 28 were George H. Hage, Boeing Co. Vice President for Product Development and former NASA Apollo Program Deputy Director; Dale D. Myers, NASA Associate Administrator for Manned Space Flight; Dr. Richard T. Whitcomb, Head of 8-Foot Tunnels Branch at LaRC; Eugene S. Love, Chief of Space Systems Div. at LaRC; James W. Plummer, Vice President and Assistant General Manager of Lockheed Missiles & Space Co.; Wilbur L. Pritchard, Comsat Laboratories scientist; and John F. Yardley, McDonnell Douglas Corp. designer. (Program; AIAA Releases; *Lewis News*, 10/22/71, 1)

October 25–30: Soviet Communist Party General Secretary Leonid I. Brezhnev paid state visit to France as guest of French President Georges Pompidou. Later Tass reported that they had discussed Soviet-French cooperation in science and technology: "They noted with satisfaction the successes achieved in this field since the time of signing the agreement of June 30, 1966, and expressed their appreciation, in particular, of the results of the sixth session of the Mixed Soviet-French Standing Commission held in Moscow from July 22 to 24, 1971. In this connection the sides referred to the important achievements in cooperation in the fields of space exploration, peaceful uses of atomic energy, as well as high energy physics, in particular the installation of a French laser reflector on the moon surface, the commissioning of the French bubble chamber Mirabelle at the Soviet proton accelerator in Serpukhov, the Soviet-French contact on enriching French natural uranium in the Soviet Union." (FBIS–Sov–71–210, 11/1/71, F4)

October 26: Identification of molecule magnesium tetrabenzporphine in outer space as forerunner of chlorophyll by Dr. Fred M. Johnson, California State College spectroscopist, was reported in *Washington Post*. Dr. Johnson had analyzed spectral lines taken of Milky Way from Lick Observatory optical telescope during 18-yr effort to identify nitrogen-bearing hydrocarbons among hydrogen clouds in interstellar space. He had succeeded after analyzing spectral line surrounding hot, young star in Orion constellation. Finding, Dr. Johnson believed, might provide clue to building blocks of life in universe. "Whatever mechanism occurs here occurs throughout the universe and you can be sure it occurs in the solar system." (O'Toole, *W Post*, 10/26/71, A3)

- NAS held meeting on molecular constitutions of interstellar space.

Dr. William J. Welch, Univ. of California at Berkeley astronomer, said 23 different molecules had been detected in interstellar space. Dr. Arno A. Penzias of Bell Telephone Laboratories said new radiotelescope at Kitt Peak, Ariz., had detected amount of carbon monoxide that "covers vast regions of space." Detected carbon seemed "to play a disproportionately large role." Dr. Benjamin M. Zuckerman of Univ. of Maryland said it might be possible that carbon monoxide was formed in shells around very young and very old stars from which it was ejected into space. (Lyons, *NYT*, 10/27/71, 51)

- *Soyuz 11* accident in which crew lost lives [see June 6–30] might have been "greatest setback in Soviet space flight activity in past several years," *Space Business Daily* reported. October, traditionally "bumper

month for Soviet flight programming" had seen "relative depression" this year. In October 1970 U.S.S.R. had recorded 12 missions. This year, "up to last week, there had been only 4 missions recorded." (*SBD*, 10/26/71, 194)
- USAF F–111 fighter-bomber wing and wing-carry-through structures had withstood test loadings equivalent to 24 000 flight hrs—four times estimated number of hours aircraft would fly in 15 years—General Dynamics Convair Aerospace Div. announced. Aircraft was undergoing program to extend service life beyond original 15-yr requirements. (General Dynamics Release 71–36)
- Rolls-Royce, Ltd., and Pratt & Whitney Div. of United Aircraft Corp. signed agreement in London for joint development of Harrier VTOL aircraft's Pegasus 11 engine. (Fairhall, Manchester *Guardian*, *W Post*, 10/29/71, D11; P&W PIO)
- USAF Aeronautical Systems Div. was issuing letter contracts for STOL aircraft, DOD announced. Fairchild Industries would receive $3 771 536 for 15 light, armed STOL aircraft; Helio Aircraft Corp. would receive $3 398 750 for 15 STOL aircraft. (DOD Release 911–71)

October 27: Dr. John S. Foster, Jr., DOD Director of Research and Engineering, and team of DOD scientists testified before congressional Joint Committee on Atomic Energy during hearings to determine whether treaty could safely be negotiated to ban underground tests without inspection. Dr. Foster and Dr. Stephen J. Lukasic, Director of DOD Advanced Research Projects Agency (ARPA), said that, while advances had been made in detecting underground tests, it was still impossible to spot some detonations of up to 100 kt. (Lyons, *NYT*, 10/28/71, 1)
- Authorization by President Nixon of underground test of five-megaton nuclear warhead on Amchitka Island in Aleutians was announced by AEC Chairman, Dr. James R. Schlesinger. Announcement came shortly before U.S. Court of Appeals for District of Columbia met to hear oral argument on petition by Committee for Nuclear Responsibility and six other environmental organizations to delay test. Opponents of test feared $200-million blast, code-named "Cannikin," could release radioactive gases and stimulate seismic sea wave. (AEC Release O–193; Kenworthy, *NYT*, 10/28/71, 1)
- Aerobee 170 sounding rocket was launched by NASA from WSMR carrying GSFC stellar spectra experiment. Mission was unsuccessful. (SR list)

October 28: U.K. successfully launched 66-kg (145-lb) *Prospero* (X–3), its first technology satellite, from Woomera Test Range, Australia, on Black Arrow booster. U.K. thus became sixth nation to launch its own satellite with its own launch vehicle. *Prospero* entered orbit with 1540-km (956.9-mi) apogee, 552-km (343-mi) perigee, 106.4-min period, and 82° inclination. Satellite, developed for $28.8 million in six years, carried experiments to test satellite technology. *Prospero* was first and last U.K. satellite scheduled for launch on U.K. booster. X–4, scheduled for launch in early 1974, would be launched by NASA Scout booster. (GSFC *SSR*, 10/31/71; *SBD*, 11/2/71, 14; Owen, *KC Times*, 11/26/71)
- NASA announced establishment of Delta Launch Vehicle System Review Board to investigate and recommend corrective action for failure of Thor-Delta 85 to place *Oso 7* in planned orbit Sept. 29 and failure of Thor-Delta 86 to orbit ITOS–B Oct. 21. Dr. John E. Naugle, Associate Administrator for Space Science and Applications, had appointed

Dr. William R. Lucas, MSFC Deputy Director, as Board Chairman. (NASA Release 71-217)

- Regional conference at ARC explored application of aerospace technology to community and public problems. Procedures for locating and obtaining technological developments were outlined to potential manufacturers and users. Dr. Joel A. Snow, NSF Deputy Assistant Director for Science and Technology, discussed technology in changing society and application of research to national needs. NASA Technology Utilization Program was described by Jeffrey T. Hamilton, Director of Technology Utilization Office. Dr. Donald C. Harrison, Chief of Stanford Univ. Medical Center Cardiology Dept., discussed NASA-Stanford biomedical technology transfer project. Conference officials noted that of 1200 problems submitted to NASA biomedical and technology application teams to date, 200 had been brought to satisfactory solution through aerospace technology. (NASA Release 71-216; ARC PIO)
- Rotating-cylinder-flap system had been installed in wing flaps of modified OV-10A Bronco aircraft provided by USN for research into STOL aircraft characteristics at ARC, ARC *Astrogram* reported. Cylinder, when rotated rapidly in direction of airflow, induced propeller slipstream to flow smoothly across wing surfaces and deflected flap. Wind-tunnel tests had shown that smoother airflow greatly increased lift necessary for STOL operations. (ARC Astrogram, 10/28/71, 1)
- Smithsonian Institution's revived plans for construction of $40-million Air and Space Museum on south side of Mall in Washington, D.C., were described by Museum Director Michael Collins, *Apollo 11* Astronaut, in interview published by *Washington Post*. Plans for building, to contain exhibit areas and galleries with displays on history of aviation and science and technology of flight, would be presented to National Capital Planning Commission in November. Collins said only best items in Smithsonian's collection would be exhibited in new museum, designed by architect Gyo Obata, who had designed previous building that would now cost $70 million to construct. Collins said Smithsonian would ask Congress to appropriate funds for new building in FY 1972. (Scharfenberg, *W Post*, 10/28/71, B1)
- Lockheed Aircraft Corp. announced sale of two L-1011 TriStar jet aircraft to British charter airline Court Line Aviation. Carrier had made "second buy" commitments to purchase three additional TriStars. New orders, first since Lockheed received U.S. Government guarantee for $250-million loan, brought total TriStar sales to 105, with 49 "second buys." (*W Post*, 10/29/71, D11)
- Rep. George P. Miller (D-Calif.) introduced H.R. 11487 for himself and cosponsors. Bill would authorize sale of up to seven acres of unimproved land adjacent to KSC Visitor Information Center to Chapel of the Astronauts, Inc. Organization was nonprofit Florida corporation interested in constructing and operating nonsectarian chapel on property. No federal funds were involved in proposed transaction. (Text; H Rpt 92-600)

October 29: NASA held *Mariner 9* pre-encounter news briefing at NASA Hq. Dr. John E. Naugle, Associate Administrator for Space Science and Applications, and Robert H. Steinbacher, Mariner Project scientist at JPL, described mysterious yellow dust storm—greatest observed since 1956—that had developed on Mars Sept. 23 and 24 and was first

October 29

detected through telescopes Sept. 26. Steinbacher said, "The question as Mariner goes into orbit . . . is, will we be doing mapping or covering some dynamic atmospheric phenomenon. . . ." It would be fortuitous "if, with the cameras and other instruments of Mariner 9, we catch the most dramatic event on Mars." (NASA Release, 10/20/71; Haughland, *W Post*, 10/30/71, A3; Sehlstedt, B *Sun*, 10/30/71, A3)

- NASA's 13th Annual Awards Ceremony was held in Washington, D.C. Dr. James C. Fletcher, NASA Administrator, presented NASA Distinguished Service Medal to Charles J. Donlan, Deputy Associate Administrator (Technical), OMSF; Vincent L. Johnson, Deputy Associate Administrator, OSSA; Bruce T. Lundin, LERC Director; Bernard Moritz, Deputy Associate Administrator, Office of Organization and Management; Oran W. Nicks, LaRC Deputy Director; and John W. Townsend, Jr., NOAA Associate Administrator.

  Group Achievement Award was presented to Apollo Recovery Communications ATS Satellite Support Team, GSFC; Ionospheric Studies Project Team; *Oso 7* Recovery Team, GSFC; Planetary Atmosphere Experiments Test Projects, ARC and Hq; and San Marco Project Team.

  Other awards included Exceptional Service Medal to 42 persons and Exceptional Scientific Achievement Medal to 28. (Program; NASA Release 71–218)

- Technique for accurately forecasting occurrence and location of solar flares by inferring solar magnetic fields from hydrogen light photos taken by earth-based optical telescopes was announced by NOAA. Technique, developed by NOAA scientist Patrick S. McIntosh, was alternative to use of magnetographs for determining solar magnetic-field data. It had enabled NOAA's Space Environment Laboratory at Boulder, Colo., to forecast region of sun where solar flares would occur, correctly 85% of time for all flares and 95% of time for major flares. NASA had contracted with NOAA to refine its solar forecasting capabilities in preparation for operating solar telescopes aboard Skylab beginning in spring 1973. (NOAA Release 71–153)

- Synthesis, without using water, of six amino acids that formed building blocks of life by exposing combination of formaldehyde, ammonia, methanol vapor, and formic acid vapor to UV light for 25 days, was reported in *Nature* by Columbia Univ. scientists Goesta Wollin and David B. Ericson. Because chemicals in combination had been identified in interstellar space, findings suggested life might arise on moon or waterless planet. (*Nature*, 10/29/71, 615–6)

October 30: UPI quoted MSC Deputy Director Christopher C. Kraft, Jr., as saying in Houston that, because of success of *Apollo 15*, NASA was considering addition to Apollo program of extra mission to orbit moon and return to earth. (*W Post*, 10/31/71, A20)

- Cracks had been found in seven engine pylons on seven C–5 cargo transport aircraft, all apparently due to "improper fabrication," USAF said in status report to Congress. Pylons would be replaced and aircraft released for flight. New pylon design was under study and might be installed within three to five months. Meanwhile C–5s could be expected to operate without difficulty with normal modifications and repairs for at least 6000 flight hours. Report said 23 C–5s had been returned to service since Oct. 12 grounding of entire fleet following Sept. 29 engine loss at Altus AFB, Okla., and subsequent discoveries of engine cracks. (Witkin, *NYT*, 10/31/71, 95)

October 31: Nixon Administration's New Technology Opportunities Program "expected to be announced early next year" was described by *New York Times*. Basic concept was to define technologies that could help solve domestic problems, establish R&D priorities accordingly, and provide incentives for industry to pursue technological goals. *Times* said program could be major step toward making Government partner for first time in industries outside aerospace and defense, result in greater Federal R&D funding, and constitute counterattack against "those who increasingly criticize science and technology as disruptive and destructive forces in society." (Wilford, *NYT*, 10/31/71, 1)

- USAF commanders believed USAF had entered "critical period in which American strategic and tactical air power is declining" while U.S.S.R.'s "is expanding," *New York Times* reported. Chief elements of problem were that USAF's basic weapon systems, B-52 bomber and F-4 fighter-bomber, were nearing obsolescence and needed to be replaced at "high cost" by B-1 and F-15; satellite-gathered intelligence indicated U.S.S.R. had established lead over U.S. in land-based ICBMs, was building emplacements for larger missiles, and had deployed FOBS to bring missiles on target from any direction; and developments were taking place against "background of budgetary stringency" in U.S. (Middleton, *NYT*, 10/31/71, 95)

- Revised NASA Policy Directive for administering Agency activities for protection and enhancement of environmental quality became effective. (NASA Pol Dir 8800.6A)

- *New York Times* editorial criticized President Nixon's decision to hold nuclear warhead test Cannikin: "Multiple anxieties" had found "responsible expression in many quarters." It was "simply not good enough" for Administration to make known its decision through AEC. "There was a time when the public accepted the father-knows-best assurance about 'overriding requirements of national security.' But that time is not now. Scrutiny and skepticism by Congress and the public have forced a healthy re-examination of many national security decisions from the manufacture of biological horror weapons to the defoliation of Vietnam to the roles of the manned bomber and the aircraft carrier. Nuclear weapons technology, although an arcane subject, is no exception." (*NYT*, 10/31/71, 16)

- London *Sunday Express* reported U.S.S.R. was building launching sites for intercontinental hydrogen bomb missiles at unprecedented speed in rocky areas. Paper said more than 60 silos had been photographed by spy satellites in recent months. U.K. Defence Ministry had said article was speculative, Reuters reported. (*NYT*, 1/11/71, 48)

During October: NAS published *Plans for U.S. Participation in the GARP Atlantic Tropical Experiment* and *Plan for U.S. Data Management in the Global Atmospheric Research Program*. Reports of panels of NCR's U.S. Committee for the Global Atmospheric Research Program called for increased preparation for major tropical observation experiment in Atlantic in summer 1974. Ad Hoc Tropical Task Group said experiment would require geostationary satellite "having both day and night imaging capability," closely spaced array of ships with calibrated radar, upper-air wind-measuring systems, and from 10 to 12 aircraft.

During October

GARP Data Management Panel recommended international data management for project, with fulltime professionals appointed by U.S. and foreign atmospheric agencies. (Texts)

- U.S.S.R. completed five-day experiment during which at least 35 meteorological sounding rockets were launched from area near Volgograd and from Heuss Island in Franz Josef Archipelago in Soviet Arctic. Purpose of experiment was to determine effect of sunspots on magnetic storms. Ground-based observatories in U.S.S.R. and Antarctic had participated in study. Rockets were said to have collected vast amount of data on impact of solar flares on earth's magnetic field, including impact on radio communications. (NYT, 11/5/71, 18; SBD, 11/9/71, 44)

November 1971

November 1: Publication of revised patent licensing regulations to accelerate commercial use of space-related patented inventions or technology was announced by NASA. Revisions would permit granting of exclusive licenses as early as nine months after announcement of invention's availability, granting of more meaningful exclusive licenses to qualified applicants to ensure prompt use of NASA inventions, and provision of royalty payments if appropriate. (NASA Release 71-220)

- NASA continued to receive proposals for using surplus Apollo hardware for "Apollo 18" lunar mission, *Aviation Week & Space Technology* reported. Dr. Robert R. Gilruth, MSC Director, had suggested mission be flown in April 1974 to map moon from polar orbit. During mission, which would cost about $190 million, LM would be converted into "massive scientific experiments bay." (*Av Wk*, 11/1/71, 15)

- Lockheed Missiles & Space Co. and 15 other aerospace firms were constructing advanced comsat model to compete for contract for INTELSAT V series, *Aviation Week & Space Technology* reported. Hardware from firms in Europe, Canada, and Japan would arrive at Lockheed for assembly before year's end. Frame had already been shipped from Contraves A. G. of Switzerland. Spacecraft would be mounted on platform to permit testing at 1 g in simulated space environment in spring 1972. (Johnsen, *Av Wk*, 11/1/71, 21)

- Airline Pilots Assn. in letter to FAA expressed concern about "decreasing margin of safety in takeoff and landing at U.S. airports." Union President John J. O'Donnell said "length of time now planned by your agency for installation of landing aids is so inadequate that it can only invite more accidents, increased inconvenience to passengers, and increased cost of operation for the airlines." Letter criticized FAA's 10-yr plan to improve airport safety and called for accelerated 5-yr program. O'Donnell said ALPA studies had shown that 573 persons had been killed in past 10 yrs in approaches to airports not fully equipped with instrument landing systems. (Text)

- NSF published "Estimated Academic R&D Direct Price Trends 50 Percent Higher Over Decade 1961-71" in *Science Resources Studies Highlights*. Price inflation accounted for estimated 50% increase in direct costs of academic R&D over 10 fiscal years ending June 1971. Most of rise had occurred in last five years, when compounded annual increase rate was 5.0%. Preliminary estimates for FY 1971 indicated that increase from FY 1970 was 5.5%, equal to change over preceding year. Increases in salaries—accounting for 65% of academic R&D direct costs—were responsible for large increase in total R&D direct costs. Payments to personnel were estimated to have increased two thirds over decade; prices of equipment, expendable supplies, and miscellaneous services, only one fifth. (NSF *Highlights*, 11/1/71)

- Yugoslavian charter airline Air Yugoslavia announced in Belgrade that it was first foreign carrier to obtain permission from Communist China to organize charter flights to Peking. (AP, *W Post*, 11/2/71, A4)

November 1-2: Dr. George M. Low, NASA Deputy Administrator, chaired symposium and workshop series sponsored by NAE in Washington, D.C. Program, "Application of Technology To Improve Productivity in the Service Sector of the National Economy," explored methods of using technology to ease social, educational, and health care problems. (NAE Release)

November 2-5: USAF Titan IIIC booster launched *Dscs 2-1 and Dscs 2-2* improved Defense Satellite Communications System satellites from ETR at 10:09 pm EST into transfer orbits and then into separate, synchronous orbits over Pacific. Orbital parameters for both spacecraft: 35 816.0-km (22 255-mi) apogee, 35 807.9-km (22 250-mi) perigee, 1436.1-min period, and 2.6° inclination. Immediately after launch telemetry difficulties were experienced during standard operational status feedback. By Nov. 5 malfunctions had been corrected and satellites were functioning satisfactorily.

The two 544-kg (1200-lb), 2.7- x 4-m (9- x 13-ft), TRW-built comsats were designed to replace smaller IDSCS satellites launched since 1966. First of improved system, new satellites would increase DOD communications volume. Each satellite had 40-w output for multichannel communications, including 1300 two-way voice conversations, TV, teletype, and data relay and could transmit between large ground terminals over long distances, as well as tactical messages between small portable ground stations. (Pres Rpt 71; AFSC *Newsreview*, 1/72; *SBD*, 11/4/71, 25; *Detroit News*, 11/3/71; UPI, *M Her*, 10/31/71)

November 2: Mariner 9—launched by NASA May 30 and expected to orbit Mars Nov. 13—experienced anomolous condition at 11:00 am EST that caused spacecraft to lose navigational lock on guiding star Canopus and lock on wrong star, Sirius. Resulting unfavorable high-gain-antenna orientation weakened spacecraft's radio signals. After extensive check showed instruments were still working and spacecraft was disoriented by only few degrees, *Mariner 9* was commanded to search for Canopus; confirmation was received that lock had been acquired at 2:51 pm EST. Radio signals returned full strength and mission operations resumed planned schedule. JPL officials speculated that spacecraft had been struck glacing blow by dust particle or Canopus sensor had been distracted by light reflected from passing dust cloud. (NASA Proj Off; *NYT*, 11/4/71, 18; *W Post*, 11/3/71, A3)

- U.S.S.R. launched *Cosmos 454* into orbit from Plesetsk with 346-km (215-mi) apogee, 204-km (126.8-mi) perigee, 90.0-min period, and 65.4° inclination. Satellite reentered Nov. 16. (GSFC *SSR*, 11/30/71; *SF*, 7/72, 179)

- Soviet scientist Boris Nepoklonov said in Tass interview that data from *Lunokhod 1* (placed on moon by *Luna 17* Nov. 17, 1970) would help define age of moon more precisely. Information showed that Sea of Rains was one of oldest lunar formations. Analysis of phototelevision panoramas had shown that large craters over 100 m (328 ft) in diameter had emerged several billion years ago, while small craters up to 10 m (32.8 ft) in diameter were millions of years younger. Study of lunar terrain showed many layers. Outer layer, from 0.2 to 0.5 mm (0.008 to 0.02 in), was loosest and had poor heat conductivity, which resulted in sharp temperature fluctuations on lunar surface within half hour. Denser layer beneath was strong enough to sustain weight of

Lunokhod 1. Nepoklonov, scientific leader of *Lunokhod 1* control team, said that information received from vehicle would be published in multivolume monograph. First book had already appeared. (FBIS-Sov-71-212-3, 11/3/71, L5-6)

- Federal Government had filed $3 690 290 suit against Bechtel Corp. in Federal District Court in San Francisco, *Wall Street Journal* reported. Suit charged that Bechtel's designs for space environment simulation chamber for MSC were "inadequate, improper and unsuited for the purpose intended." Chamber had failed in 1964 NASA test, Government said. (*WSJ*, 11/2/71)
- Nobel Prize for physics was awarded to Dr. Dennis Gabor, Hungarian-born British scientist with Columbia Broadcasting System laboratories, for invention of three-dimensional lensless photography system known as holography. Nobel Prize for chemistry was awarded to German-born National Research Council of Canada scientist Gerhard Herzberg for "contributions to knowledge of electronic structure and geometry of molecules, particularly free radicals." (*NYT*, 11/3/71, 1)
- House Committee on Science and Astronautics favorably reported H.R. 11484, bill permitting sale of land at KSC for Chapel of the Astronauts. (*CR*, 11/2/71, D1109; H Rpt 92-600)
- Rohr Corp. President Frank E. McCreery said in statement that extensive investigation had produced no evidence that Rohr Corp. had delivered cracked jet engine mountings for USAF's C-5A jet transport aircraft [see Oct. 30]. (UPI, *NYT*, 11/4/71, C38)
- Lockheed Aircraft Corp. had announced "voluntary" early retirement of Lockheed-California Div. President Charles S. Wagner, Senior Vice President Dudley E. Browne, and Corporate Vice President for Manufacturing W. A. Pulver, *Wall Street Journal* reported. Browne would remain a director and member of board's executive and finance committees. (*WSJ*, 11/2/71, 16)

November 2-4: NASA sponsored conference on "Vehicle Technology for Civil Aviation: The 70s and Beyond" at LaRC and Williamsburg, Va. Dr. James C. Fletcher, NASA Administrator, said in Nov. 2 address to airlines and aircraft industry executives and DOT representatives that environmental issue of noise provided "show case for technology" and "opportunity to apply technology directly to the solution of an environmental problem of increasing public concern." It was not enough "to think in terms of more power, more lift, more speed. If we wish the taxpayer to continue to support civil aviation, we should take as our motto, 'Fly Quiet!'" NASA could be proud of progress made in reducing aircraft noise. NASA was "well prepared" to develop supersonic technology that would "permit supersonic flights over continents as well as oceans without harmful sonic boom effects and outstrip the current foreign models in speed, capacity, range, and profitability."

Dr. Fletcher described NASA STOL aircraft program as "first research program of this magnitude focused on the needs of civil rather than military aviation." He felt U.S. should establish "clear national policy . . . to maintain our leading position in world aircraft markets" and "should resolve to achieve this goal by whatever reasonable means are necessary, with emphasis on an unequivocal decision to stay out front in aeronautical research and technology."

FRC engineer Marvin R. Barber said NASA study had predicted that by 1982 small aircraft would be "most economical mode of transportation on trips of between 40 and 450 miles [65 and 725 kilometers]."

B/G Gustav E. Lundquist (USAF, Ret.) said attempts to introduce use of STOL transport to public had been unsuccessful in Northeast Corridor between Boston and Washington. (NASA Release 71-222; Newport News *Daily Press*, 11/3-5/71)

November 3: NASA announced it had begun distribution of more than 2200 samples and polished thin sections of *Apollo 15* lunar material to U.S. and foreign scientists. Total weight of largest and most varied sample collection was about three kilograms (six and a half pounds). Material would be analyzed by 700 investigators in U.S., Virgin Islands, 15 foreign countries, and ESRO. (NASA Release 71-223)

- Results to date of study to determine minimum cockpit visibility required by space shuttle pilot to make safe horizontal ground landings were announced by FRC. Tests, begun in July with instrumented T-33 jet aircraft with radar system to judge accurate altitude to within two feet (0.6 meters), had shown pilots could land aircraft with greater visibility restrictions than expected. Visibility restriction of 5° had been extended to 75° on either side, or total of 150° in 225 approaches flown. Restrictions would be increased in future to 160°. (FRC Release 26-71)

- Prototype water-vapor electrolysis system, developed by ARC scientist Dr. Theordore Wydeven, Jr., had successfully completed more than 2000 hrs of testing—equivalent of 80-day space mission—*Valley* [Calif.] *Journal* reported. System converted moisture in air into hydrogen and oxygen and released reclaimed oxygen back into air. It would eliminate need for bulky oxygen tanks on space missions and could be used as onboard oxygen system in aircraft, portable oxygen generator for hospitals and homes, and air conditioner and freshener in mines, air raid shelters, or other places in need of oxygenation. (*Valley Journal*, 11/3/71)

- Federal Reserve System had begun supplying banks with 62 million copper and nickel dollars commemorating President Eisenhower and *Apollo 11* lunar landing, AP reported. One side of coin carried reproduction of *Apollo 11* astronauts' emblem—eagle against moon's surface. (B *Sun*, 11/3/71, A6)

- Lack of Government policy for engineering was scored by Sen. Edward M. Kennedy (D-Mass.) in speech telephoned to meeting of Institute of Electrical and Electronic Engineers in Boston, Mass. Sen. Kennedy cited communications engineer who had led team that designed antenna for *Apollo 15* LRV. Engineer had received special commendation from NASA for his leadership. "He also received a layoff notice from his firm shortly before he watched his antenna in action on the surface of the moon." Engineers possessed great potential to contribute to Nation's needs. "Engineers who can produce patents for their former employers can provide ideas which will benefit society. Engineers who can transmit data from the moon can help meet the data processing needs of our technological economy. And engineers who could have taught at our leading universities must have some lessons to offer of benefit to us all." (CR, 11/10/71, S18003-4; Off of Sen Kennedy)

- Air France announced first firm order from airline for A-300B 250-

passenger airbus being built by European consortium. (Reuters, *W Post*, 11/4/71, A2)

- Dr. Dennis Gabor, winner of 1971 Nobel Prize for physics, said in Bridgeport, Conn., that space exploration was "strictly a dead end occupation." It was "extremely uninteresting—the only interesting place is earth." Dr. Gabor made comment during interview at Univ. of Bridgeport honors convocation. (UPI, *W Post*, 11/5/71, A7; *W Star*, 11/6/71, A7)
- Fairchild Industries, Inc., had agreed to acquire Swearingen Aircraft Co., *Wall Street Journal* reported. Agreement called for Fairchild to provide up to $3 million of immediate working capital for Swearingen. New subsidiary would be formed. (*WSJ*, 11/3/71, 17)

November 4: National Capital Planning Commission approved plans for Smithsonian Institution's new $40-million National Air and Space Museum after hearing testimony from Museum Director Michael Collins, *Apollo 11* astronaut. During 90-min debate, Commission member Paul Thiry described museum building, designed by architect Gyo Obata, as "industrial type" from "elevator-air conditioner school of architecture" lacking "classic proportions." Collins said museum would possess "flavor of air and space." Commission stipulated that building's roof remain clear of mechanical equipment and that use of granite and marble be considered for construction. Final approval of plans would be considered following Nov. 17 meeting of Fine Arts Commission. (Combes, *W Post*, 11/5/71, A14)

- Successful first launch of Poseidon missile from surfaced submarine was observed by Soviet trawler at distance of 805 m (880 yds). Submarine U.S.S. *Nathaniel Greene* was surfaced 16 km (10 mi) east of Cocoa Beach, Fla. DOD later said it had not determined if Soviet vessel was intelligence-gathering ship. (*W Post*, 11/5/71, A5)
- NASA's structural analysis computer program (NASTRAN), designed to analyze behavior of elastic structures in space program, was being used by Ford Motor Co. in predicting performance of 1973 light truck components, NASA announced. Ford had reported 60% improvement in predicting behavior of components under stress and time-saving of two thirds in calculations. Program also had been used in design of people-mover system to be used at TRANSPO '72 exhibit planned for spring 1972 at Dulles International Airport, Va. (NASA Release 71-221)
- Full-size plywood model of B-1 strategic bomber, designed as replacement for USAF B-52, was displayed for first time by North American Rockwell Corp. in Los Angeles. (*W Post*, 11/5/71, 3)
- Soviet computer had calculated total weight of atmosphere at 5 quadrillion, 157 trillion tons, Tass reported. News agency said calculation was "essential for research in cosmonautics, space geodesy and gravimetry." Computer had, for first time, taken into account "all the peculiarities of the earth's relief and the distribution of temperatures on its surface," Tass said. (UPI, San Francisco *Examiner*, 11/5/71)
- *Wall Street Journal* quoted Bank of America as saying California's aerospace employment should stabilize near end of 1971 "and hold fairly steady during 1972." Bank, in annual forecast of California economy, had said overall aerospace employment in state would not show marked improvement until early 1973 "when defense contracting and demand for commercial aircraft pick up." (*WSJ*, 11/4/71, 42)

- Plans of naval historian R/A Samuel E. Morison (USN, Ret.) to retrace by air and sea 64 000-km (40 000-mi) first circumnavigation of globe by Ferdinand Magellan in 1519 were described by *New York Times*. Adm. Morison would begin two-month voyage with Colombian diplomat, aviator, and explorer Mauricio Obregon in December. (*NYT*, 11/4/71, 45)

November 5: Europa 2, ELDO (European Launcher Development Organization) multistage rocket, failed to launch test-satellite from Kourou, French Guiana, in first developmental flight. Two minutes forty seconds after liftoff, during separation of 1st stage (British Blue Streak) and 2nd stage (French Coralie), rocket acceleration dropped and rocket fell into Atlantic. Failure was setback for European space organization; Europa 2 was to have launched 363-kg (800-lb) satellite into orbit. Rocket's 3rd stage was West German Astris unit. Belgium had set up ground tracking and control stations and Netherlands had supplied station equipment. Italy had contributed test satellite. (Reuters, *NYT*, 11/6/71, 14; *SBD*, 11/8/71, 36)

- First combination of modern radar data from observations of inner planets with existing optical observations of sun and planets made between 1750 and 1970 had shown that Pluto's mass could not be determined reliably from existing data, MIT scientists M. E. Ash, Irwin I. Shapiro, and W. B. Smith reported in *Science*. Total 300 000 measurements had been made during experiment which also included available observations of asteroids Eros and Icarus. Radar data were from MIT's Millstone Hill and Haystack facilities; optical data were from original observatory reports transformed over six years into machine-readable form. Results had shown "insufficient basis for the widespread conclusion that Pluto's average density is greater than, or even comparable to, the earth's." (*Science*, 11/5/71, 551–6)

- Commercial pilot Elgin L. Long flew from San Francisco International Airport to Fairbanks, Alaska, in twin-engine Piper Navajo aircraft on first sector of 32 000-km (20 000-mi), month-long journey over North and South Poles and prime and 180th meridians at Equator—geographic crossroads of world. If successful, UPI said, Long would establish eight aviation records, including first flight around the world landing on all seven continents, first solo flight across Antarctica, and first solo from Antarctica to Australia. (*W Star*, 11/6/71, A2)

- National Geographic Society was quoted in *Chicago Tribune* as saying, Phobos, Martian satellite, was darkest body yet observed in solar system, "possibly because meteoroids had scoured it clean of reflecting dust." (*C Trib*, 11/5/71)

- Issuance by USAF's SAMSO of $20 932 400 contract to Martin Marietta Corp. for launch services for Titan IIIC launch vehicle at ETR was announced by DOD. Initial increment was $10 701 400. (DOD Release 942-71)

- *Washington Post* editorial commented on AEC's test of five-megaton atomic device on Amchitka Island in Aleutians scheduled for Nov. 6: "The fact that this particular test may have some element of risk involved would not automatically foreclose the question of whether it should be conducted. If the test does involve a risk to the environment, it must then be decided whether that risk is worth taking in view of the knowledge to be gained. In any case, the administration would be in far better shape if it had let the debate on the environmental aspect run free and then made its case for taking the risk. As it is, the administration is in the position of assuring Americans and Canadians alike that all is well—that it knows best. We sincerely hope it is right. But if it is wrong, the onus on the administration will be far greater as a direct result of the evidence we now have of the suppression of high-level reservations about the ecological risks involved." (*W Post*, 11/5/71, A26)

November 6: AEC detonated five-megaton hydrogen bomb on Amchitka Island, Alaska, at 11 am local time (5 pm EST). Supreme Court had refused to order delay in controversial $200-million Cannikin project to test prototype warhead for Spartan missiles of Safeguard ABM system. No earthquake followed detonation, but shock waves reverberating from deep earth strata rocked island severely. Dr. James R. Schlesinger, AEC Chairman, broadcast from control room on Amchitka immediately following blast: "We had fairly significant ground motion . . . but sensation was just about what we had expected." Tentative indications were that blast effects were "well within the range of projections." Dr. Schlesinger said test had permitted introducing of Spartan warhead into inventory "while minimizing the likelihood of a defective warhead. I believe the final results of Cannikin will permit the AEC to certify the Spartan warhead for introduction into the stockpile within the appropriate deployment schedule." Report from Palmer Observatory had indicated Richter scale reading of 7.0. "So it is indicated that we had a full yield test." Control board indicated "all of our radiation activity monitoring stations are reporting . . . not even a trace of radioactive release to this point. We are past the 12-minute mark." (AEC Release 0–204; Turner, *NYT*, 11/7/71, 1)

• Award by Communist Party Central Committee and Council of Ministers of U.S.S.R. of 1971 state prizes for achievements in science and technology was announced by Tass. State prize for celestial mechanics had been awarded to unidentified scientists "who have developed new methods of plotting the movement of natural and artificial space objects." Research provided "a theoretical basis for the solution of engineering problems . . . in the flight of space objects and their orientation in orbit." Later, *Pravda* published article by Soviet Academy of Sciences President Mstislav V. Keldysh describing prize-winning research. "A cycle of works by Ye. P. Aksenov, G. N. Duboshin, and others is devoted to research into the various astronomical, mechanical and mathematical aspects of formulating high precision theories of the strongly perturbed motion of artificial and natural satellites of the planets and also of asteroids." (FBIS–Sov–71–217, 11/10/71, L1; 218, 11/11/71, L1)

November 7: Apollo 15 Astronauts David R. Scott, James B. Irwin, and Alfred M. Worden arrived in London to begin two-week, five-country tour of lectures and meetings with European scientists. (UPI, *C Trib*, 11/8/71)

• Two goats had survived 30 hrs in pressure chamber at 31 atmospheres —equal to pressure at 300 m (1000 ft) below sea level—breathing same oxygen over and over, *New York Times* reported. Experiment, conducted by Dr. Jacques Chouteau, head of Laboratory of High-Pressure Physiology at Univ. of Marseilles, replaced bulky apparatus normally used to remove carbon dioxide from used air with small cylinder of potassium superoxide. When air breathed by goats passed through cylinder, water and carbon dioxide in used air reacted with potassium to release oxygen, which was returned to chamber. *Times* said process, developed by Air Liquide under French military contract, might be used in space, sealed aircraft, and tanks. U.S.S.R. was believed to use a similar technique. (Hess, *NYT*, 11/7/71, 84)

- Maverick, airborne missile guided to surface targets by TV camera in nosecone, had gone into production by Hughes Aircraft Co. under $69.9-million contract and would be delivered to USAF in 1972, *New York Times* reported. If missile proved successful, USAF believed its installation on fighter aircraft in West Germany and U.K. could help redress U.S.S.R.'s "quantitive armored superiority in Europe." (Middleton, *NYT*, 11/7/71, 57)

- Canadian Defence Ministry had announced intention to mothball hydrofoil *Bras d'Or*, classified as world's fastest warship, AP reported. Aluminum craft 61 m (200 ft) long had been developed over 13 yrs at $52-million cost. Vessel had attained speed of 116 km per hr (72 mph) in 1969. Major reasons for decision to mothball were vessel's lack of capability in Arctic waters and fact no foreign sales of craft were forecast, while production costs were up. (*NYT*, 11/7/71, 5)

November 8: Dr. Harrison H. Schmitt, scheduled to become first scientist-astronaut on moon during Apollo 17 mission, addressed 41st annual international meeting of Society of Exploration Geophysicists in Houston, Tex. Dr. Schmitt said lunar exploration "is permanently terminating for this generation and may be for future generations." Chances for collecting data from moon were "decreasing rapidly." In response to questions Dr. Schmitt said mankind's future could depend on scientific understanding of knowledge being gathered during Apollo program. (*Atlanta J&C*, 11/14/71)

- President Nixon announced appointment of Betty Crites Dillon as U.S. Representative on International Civil Aviation Organization Council and submitted to Senate nomination of Mrs. Dillon for rank of Minister. She would succeed Charles F. Butler who had resigned Oct. 1. (*PD*, 11/15/71, 1498, 1517)

November 9: Apollo 15 Commander David R. Scott told press in London that colonization of moon was "very practical proposition." He predicted eventual manned exploration of planets and said man would definitely explore Mars. "The Apollo missions have proved that man can do a worthwhile job exploring space." *Apollo 15* Astronauts Scott, James B. Irwin, and Alfred M. Worden were touring Europe to brief scientists on their mission. (Reuters, B *Sun*, 11/10/71, A3)

- House Armed Services Committee's Subcommittee on Investigations opened hearings on unscheduled Oct. 26 landing of Cuban An-24 transport aircraft at New Orleans International Airport. Hearings were to determine how Soviet-built aircraft had penetrated U.S. southern defense facilities so effectively that first knowledge of its presence was received when pilot requested permission to land, 10 min from airport. *Aviation Week & Space Technology* later said incident was second Cuban penetration of U.S. airspace in 25 mos. DOD witnesses at previous hearings had blamed October 1969 penetration on absence of radar. (*CR*, 11/8/71, D1140; *Av Wk*, 11/22/71, 13)

November 10–11: Mariner 9 (launched by NASA May 30) began taking preorbit science photos of Mars, showing planet as featureless, half-moon-like object, hazy and dust-covered. First photos, taken over 24-hr period from distances of 860 000 to 570 000 km (535 000 to 355 000 mi) from Mars, were transmitted to Goldstone tracking station and then to JPL Nov. 11. They were not as clear as photos taken by *Mariner 6* and *7* in 1969. Astronomers reported dust storm of "unprecedented" duration and scope ranging over much of planet, obscuring nearly all surface details. Dr. William H. Pickering, JPL Director, said storm could be regarded as a "premium," since opportunity to see dynamic, unusual event on another planet was rare. But protracted period of bad weather over Mars could jeopardize mapping of surface. Dr. Bradford A. Smith of New Mexico State Univ. said storm appeared to have reached its peak about Oct. 21 and there had been "a positive but very, very slow clearing of the dust. . . . We're quite clearly seeing the south polar cap." Polar cap, he said, was not as bright as expected, indicating dust had not cleared away completely in that region. Scientists believed they were able to see Deimos, tiny outer moon of Mars. Photos, made in three sequences at intervals, provided complete coverage of planet as it rotated about its axis. First two sequences each covered one Martian day; third covered about one third of day. Radio signals had reached Goldstone, Calif., tracking station, were relayed to JPL, and were fed into computers that had converted signals to TV images. *Mariner 9* was scheduled to enter Mars orbit Nov. 13. (NASA Proj Off; Wilford, *NYT*, 11/11–12/71; Sehlstedt, *B Sun*, 11/11–12/71)

November 11: Two USAF comsats launched Nov. 2 from ETR by Titan IIIC as part of improved series of Defense Satellite Communications System (DSCS) satellites were in operational modes as programmed. After insertion into synchronous orbit over Pacific, first satellite had failed to turn on its telemetry system. Second satellite had turned on telemetry, but telemetry was sporadic. Subsequent efforts of ground controllers were successful and both satellites responded to ground commands. (*SBD*, 11/11/71, 55)

- Smithsonian Astrophysical Observatory astronomers said in Cambridge that radio signals from *Apollo 15* CM *Endeavor* had passed into radio "dead spot" behind moon on Aug. 3, faded away, and reappeared on receivers of observatory's 46-m (150-ft) radio antenna two minutes later. Subsequent analyses had confirmed that signals had been transmitted through moon's body. (AP, *NYT*, 11/13/71)

- Political and Security Committee of U.N. General Assembly unanimously approved Convention on International Liability for Damage Caused by Space Objects. Treaty held space powers "absolutely liable" to pay compensation for loss of life, injury, or property damage. Amount of compensation would be determined by agreement between nation responsible for damage and country or countries suffering damage, or by three-member claims commission, whose decision would be recommendation not legally binding. Canada, Japan, Sweden, and Iran abstained from voting to signify dissatisfaction because treaty did not compel payment of compensation.

Convention would be open for signature in Washington, Moscow, and London following General Assembly approval—which was assured, *New York Times* later said. Treaty had been completed after seven-year delay caused by U.S.S.R.'s contention that claims should be decided directly by governments through diplomatic channels and without arbitration. U.S.S.R. had agreed to compromise in June. (*NYT*, 11/12/71, 1; UN PIO)

- Retirement of Jerome F. Lederer as NASA Director of Safety Jan. 14, 1972, was announced by NASA. Internationally recognized authority on air and space flight safety had joined NASA as Director of Flight Safety for OMSF June 1, 1967, after having been Director of Flight Safety Foundation since 1947. Lederer, who had received NASA Exceptional Service Medal in 1969, would return to Flight Safety Foundation as part time consultant. (NASA Release 71-224)
- Magellanic Premium of American Philosophical Society was presented to JPL mathematicians Paul M. Muller and William L. Sjogren at Society's Annual General Meeting in Philadelphia. Muller and Sjogren were cited for "their discovery of lunar mascons leading to the first detailed gravimetric map of the moon." Magellanic gold medal had been presented only 22 times since inception in 1790. Most recent recipient had been Dr. William H. Pickering, JPL Director, in 1966. (JPL Release 604)
- Mrs. Betty J. Baldwin, ARC computer programmer, had been selected one of top 10 business women of the year by American Business Women's Assn., for progress as NASA civil service employee and interests in educational and character-building programs, ARC *Astrogram* announced. (ARC *Astrogram*, 11/11/71, 1)
- Delivery of first of USAF's 96 F-111D supersonic jet bombers, equipped with Mark II avionics package, to 27th Tactical Fighter Wing at Cannon AFB, N. Mex., was announced by DOD in memorandum to correspondents. (Kelly, W *Star*, 11/18/71, A4)
- Future of Soviet airlines under new five-year plan being drafted by U.S.S.R. Supreme Soviet was described by G. Ustinov in *Izvestia* article: Draft "envisages the introduction on the civil airlines of new, comfortable, rapid, and more economic aircraft. The operation of the supersonic TU-144 . . . will begin. Almost all existing aircraft will be modified. The network of airports will be expanded. The transportation of passengers will increase considerably, and . . . 100 million hectares [247 million acres] of fields and forests will be treated from the air." (FBIS-Sov-71-222, 11/17/71, K4)

November 12: *Washington Daily News* editorial said *Mariner 9* had already accomplished "enough to win plaudits from any but the most jaded." It had "photographed . . . one of Mars' moons . . . and with luck will go into orbit around the planet itself, taking pictures of surface details no bigger than a football field.

"No one can tell where the ever-expanding horizons of man's knowledge will lead, nor the extent to which Mariner 9 will enlarge our knowledge of the solar system. But jaded we are not, and we are frank to say that we are downright thrilled by Mariner's exploits." (*W News*, 11/12/71)

- Portrait of Dr. Thomas O. Paine, third NASA Administrator (1969–1970) was unveiled in NASA Hq. ceremony attended by Dr. James C.

Fletcher, NASA Administrator; Dr. Paine; and Rep. George P. Miller (D-Calif.), Chairman of House Committee on Science and Astronautics. Portrait, by artist William Draper, would hang at NASA Hq. (Transcript)

- Cost of full operational space transportation system had been put at $12.7 billion, "although new cost-cutting efforts may reduce that figure," Univ. of Michigan engineer T. A. Heppenheimer said in *Science* article. Shuttle's scientific benefits "do not in themselves justify its development. . . . Defense, commercial, and applications benefits combine as well to furnish the justification. Nevertheless, the scientific benefits should not be overlooked. For the first time, laboratory-type investigations may be carried out in space, not by scientist-astronauts, but by scientists." (*Science*, 11/12/71, 646–7)
- *Seattle Times* quoted Dow Jones News Service as saying Italian government had approved funding of $320 million for joint Boeing Co. and Aeritalia S. P. A. development of STOL aircraft to carry 100 to 150 passengers. Aeritalia hoped to begin production in 1974 or 1975. Boeing could decide to produce aircraft in U.S. if market warranted. (*Seattle Times*, 11/12/71)
- Discovery of diamonds in 1.4-kg (3-lb) meteorite that fell through roof of storehouse on Finnish island of Havro Aug. 2 was announced by Smithsonian Center for Short-Lived Phenomena. It was sixth such meteorite to be found. First discovery of meteoritic diamonds had been made by Russian scientists in 1888 while analyzing meteorite found near Novoured. (UPI, *NYT*, 11/13/71)
- Univ. of Toronto had announced discovery by astronomer Dr. Sidney Van Den Bergh of three galaxies composed of thousands of stars, *Washington Post* reported. (*W Post*, 11/12/71, A36)

November 12–13: Scientists at P. N. Lebedev Physics Institute and Soviet Academy of Sciences' Crimean Astrophysics Observatory experimented with laser to locate reflector left on moon by *Apollo 15* astronauts during July 26–Aug. 7 mission. *Pravda* later quoted Soviet scientist A. Sukhanovsky as saying experiment was "continuation of research work on the 'earth-moon' system by the laser location method." First successful experiment to locate French reflector on *Lunokhod 1* had been conducted in December 1970. "The accuracy of the measurement of the distance to the reflector was then in the order of several meters." In current experiment "for the first time we have succeeded in measuring the distance to the U.S. angular reflector." Fact U.S. scientists had not succeeded in locating French reflector on *Lunokhod 1* was "apparently connected with differences in the methods of calculating the reflector's supposed coordinates." (FBIS–Sov–71–266, 11/23/71, L1)

November 13–15: *Mariner 9*, launched by NASA May 30, was inserted into elliptical orbit around Mars at 7:32 pm EST Nov. 13 after 15-min engine burn and became first spacecraft to orbit another planet in solar system. Orbital parameters were close to planned ones: 1398-km (868.7-mi) periapsis altitude, 12-hr 34-min period, and 64.5° inclination. Spacecraft propulsion system and attitude control systems performed flawlessly after 400-million-km (248-million-mi) journey. *Mariner 9* would orbit Mars twice daily while photographing up to 70% of Martian surface in next three months.

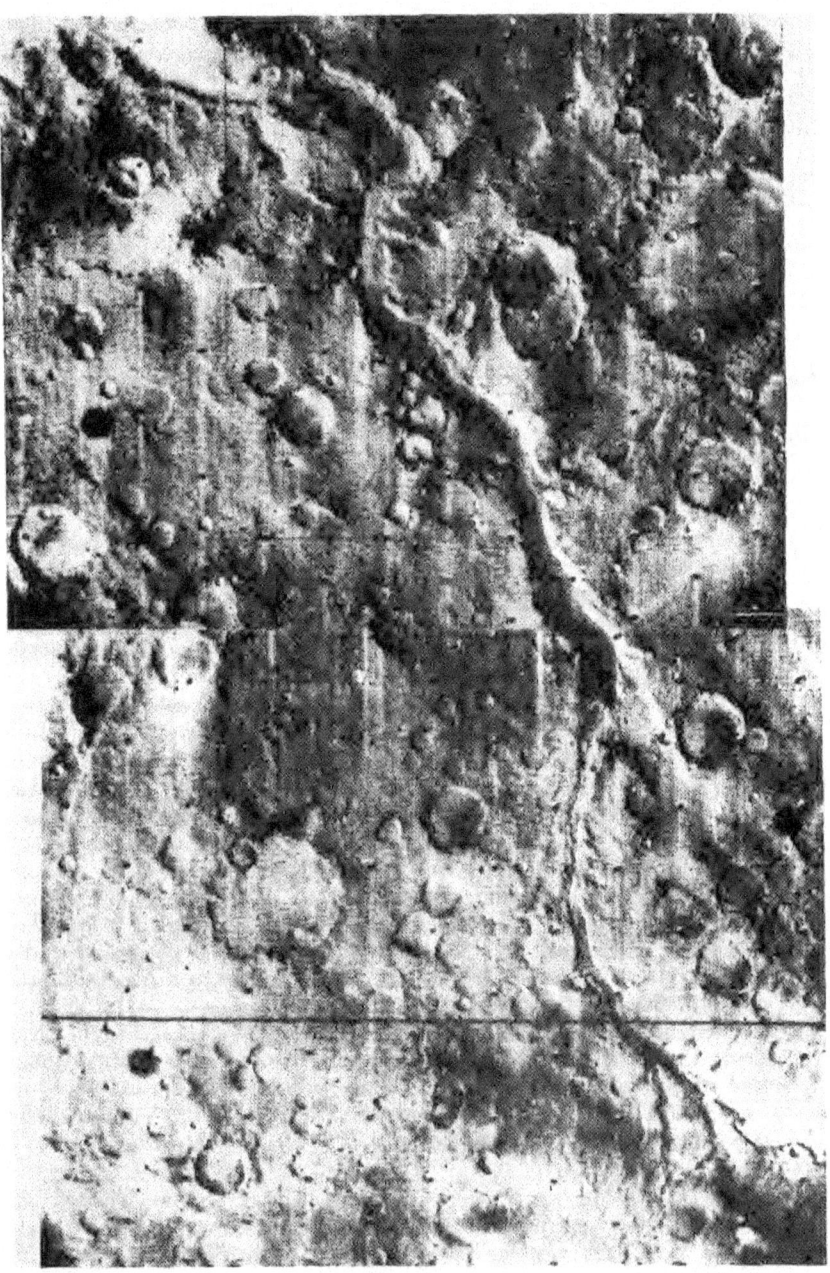

November 13–15: Mariner 9 became the first spacecraft to orbit the planet Mars. The first photos of the planet were obscured by a dust storm but later pictures revealed sinuous rilles whose origin was puzzling. The valley above, 700 kilometers (435 miles) long, was photographed Jan. 22, 1972, in the Mars Rasena region. Such rilles might have been produced by collapse of the roof over subsurface lava flows. They raised the possibility of erosion by water in ancient Martian history.

November 13–15: *Inserted into orbit around Mars Nov. 13, Mariner 9 took man's first closeup photos of the Martian moon Phobos Nov. 29 and 30. The computer-enhanced photo at the right was taken 5543 kilometers (3444 miles) from Phobos Nov. 30. The profusion of craters suggested that the moon is very old and possesses considerable structural strength. The Martian volcanic mountain Nix Olympica (below) was photographed in late January 1972 as the planet's dust storm subsided. The mountain was 500 kilometers (310 miles) across at the base, more than twice as broad as the most massive volcanic pile on the earth. Steep cliffs fell from the mountain flanks to a great plain. The complex multiple volcanic vent at the summit was 65 kilometers (40 miles) in diameter.*

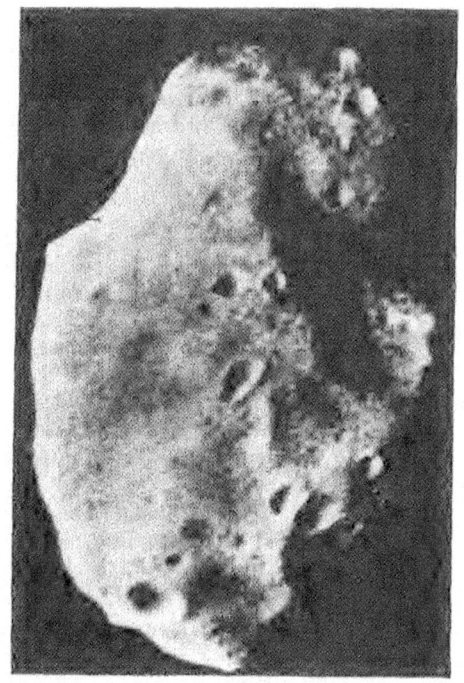

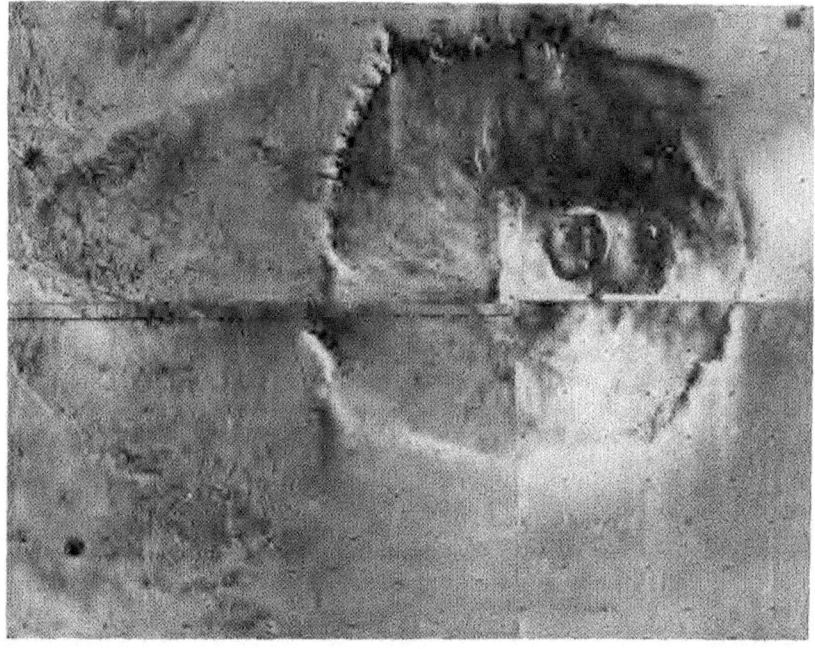

First photos were obscured by dust storm but before end of month *Mariner 9* had shown that Mars bulged at equator, was "rougher gravitationally" than expected, and probably had wide variety of rock compositions; instruments had detected small amounts of water vapor at south pole and warmer air pockets above pole [see Nov. 16–23]. Spacecraft also made man's first close photo of potato-shaped Martian moon Deimos [see Nov. 26–27] and photos of moon Phobos [see Nov. 29–30].

First Mariner photos from orbit were broadcast live by national TV network, but detail was limited by dust storms that had raged on surface since September. Dr. Robert H. Steinbacher, JPL project scientist, said "We're hopeful that as the storm dies down we can get some good pictures of the dust settling to the surface. This is an unprecedented chance to watch changes in the Martian weather." He said that if cameras worked properly pictures would be 1000 times better than any earthbound telescopes had taken.

On Nov. 14 spacecraft transmitted three sequences of TV pictures. Of more than 60 photos returned, only seven revealed surface features. Dust storm continued and atmospheric experts could see no sign of abatement. Dust obscuring planet was apparently very fine sand which high winds pushed to height of as much as 16 km (10 mi) above surface.

On fourth orbit spacecraft successfully completed orbit trim maneuver at 9:44 pm EST Nov. 15. Maneuver changed orbital period of spacecraft to make it synchronous with Mars viewing period of Deep Space Station antenna at Goldstone, Calif. New orbital parameters were 1394-km (866.2-mi) periapsis altitude, 11-hr 57-min 12-sec period, and 64.34° inclination.

Data from remote-sensing instruments confirmed earlier observations that carbon dioxide was major constituent of Martian atmosphere and also suggested possibility of "hot spots" on planet. Dr. Ellis D. Miner of JPL said sensor detected average temperature rise of 6.7 K (12°F) in area 24 km (15 mi) square below Mars equator. Temperature rise might indicate hole in cloud cover exposing mountain peak or plateau reflecting residual heat or evidence of internal source of heat on Mars. UV spectrometer data indicated dust storm was made up of fairly coarse-grained, sandlike particles. (NASA Proj Off; Wilford, *NYT*, 11/14–16/71; O'Toole, *W Post*, 11/14–15/71; *W Star*, 11/16/71, A4)

November 13: U.S.S.R. announced *Mars 2* and *Mars 3* (launched May 19 and May 28 and scheduled to encounter planet Mars within two weeks) would attempt to put first man-made objects on "red" planet. Dr. Carl Sagan, Cornell Univ. astronomer, said Soviet scientists had told him *Mars 2* and *Mars 3* were equipped to land instrument probes. He said he understood they contained flammable material that would destroy any earth bacteria when probes completed investigation. (*W Star*, 11/14/71, A6)

November 14: Gold issues of first annual United Nations Peace Medal, issued to coincide with U.N.'s 26th anniversary, had been presented to *Apollo 15* astronauts, AP reported. For collectors, medals would be in silver and bronze. (*B Sun*, 11/14/71, A3)

November 15–29: *Explorer 45* (sss–a) Small Scientific Satellite was successfully launched for NASA by Italian crew at 8:52 am local time (12:52 am EST) from San Marco launch platform off coast of Kenya. Four-stage Scout launch vehicle boosted 50-kg (110-lb) satellite into orbit with 27 034.5-km (16 798.5-mi) apogee, 220.8-km (137.2-mi) perigee, 469-min period, and 3.57° inclination. Primary objective was to measure characteristics and formation of earth's ring current and development of main-phase magnetic storms; relation between magnetic storms, aurora, and acceleration of particles within inner magnetosphere; and relative importance of various diffusion mechanisms in populating radiation zones at 4- to 5-earth-radii elliptical equatorial orbit.

By Nov. 29 all satellite household functions were nominal but spacecraft was coning at higher angle than planned, which would make data reduction slightly more difficult. Attempts were being made to reduce coning angle. Five experiments had been turned on and were operating satisfactorily. One experiment, charged particle detector, remained to be turned on.

Explorer 45, 26-sided polyhedral spacecraft 63.5 cm (25 in) in diameter, attained length of 5.8 m (19 ft) with booms deployed. It was second U.S. satellite to be launched by Italy under April 30, 1969, launch agreement. First was *Uhuru* (*Explorer 42*) Small Astronomy Satellite launched from San Marco Dec. 12, 1970. *Explorer 45* was most innovative and compact satellite in Explorer series to date; onboard data handling system could be reprogrammed by ground command to concentrate on specific sets of data.

Under NASA and Univ. of Rome agreement, NASA provided booster and satellite; Italian team conducted assembly, checkout, and launch services on cost-reimbursable basis. NASA OSSA was responsible for overall direction of program. Satellite was designed and built at GSFC, which managed project. LaRC managed Scout launch vehicle. (NASA Proj Off; NASA Release 71-212)

November 15: NASA's M2–F3 lifting body, piloted by NASA test pilot William H. Dana, successfully completed its 11th flight from FRC. Objectives were to check out fuel jettison system and newly installed engine chamber purge system. (NASA Proj Off)

- Intersputnik, international space communications organization, was formed under agreement signed in Moscow by U.S.S.R., Bulgaria, Hungary, East Germany, Cuba, Mongolia, Poland, Romania, and Czechoslovakia. Tass later quoted agreement as saying "organization is open for accession by all states of the world. It will coordinate its activities with the International Telecommunications Union as well as with other organizations whose activities are related to the use of communication satellites." (UPI, *NYT*, 11/17/71, 9)

- Full deployment of Satellite Early Warning System (SEWS), "possibly in 1972," would alert U.S. to attack from any point on globe in seconds —"time to get strategic bombers into the air and to launch a missile counterstrike before enemy warheads arrive on U.S. targets"—*U.S. News & World Report* article said. DOD was "fairly confident" that Pearl Harbor type of attack was "becoming almost impossible." Two SEWS satellites, launched May 5, were reported in near-stationary orbit

November 15

about 40 000 km (25 000 mi) above Indian Ocean. They could monitor Soviet and Communist Chinese missile launches and detect and report nuclear explosions in atmosphere. (*US News*, 11/15/71, 10)

- President Nixon submitted to Senate nomination of Dr. H. Guyford Stever as NSF Director. Dr. Stever would succeed Dr. William D. McElroy, whose resignation would be effective Feb. 1. Dr. Stever was member of National Science Board and of Advisory Panel to House of Representatives on Science and Astronautics. He had served as USAF Chief Scientist and as Chairman of President Nixon's Pre-Inauguration Task Force on Science and Technology. (*PD*, 11/22/71, 1528, 1542)
- MSC issued RFPs to develop and test polymer seal materials for space shuttle propulsion systems. Proposals, for fixed-price R&D contract, were due Dec. 3, with work to be concluded 12 mos after contract award. (MSC Release 71-82)
- House passed and sent to Senate H.R. 11487, bill to permit sale of land at KSC for Chapel of the Astronauts. (*CR*, 11/15/71, H10998, D1171)
- FAA announced award of $159 362 contract to Northwest Environmental Technology Laboratories, Inc., to measure and evaluate effectiveness of airborne commercial seeding techniques in dissipating fog at airports. (FAA Release 71-185)
- RAND Corp. President Henry S. Rowen resigned. *New York Times* later quoted sources as saying his departure had been caused partly by DOD's "dissatisfaction with Rand's role in the release of the Pentagon papers [report on background of U.S. involvement in Vietnam War prepared at request of then Secretary of Defense Robert S. McNamara and classified by DOD]." Both Rowen and RAND Board Chairman Newton N. Minow had denied relationship between Rowen retirement and Pentagon papers. (Roberts, *NYT*, 11/16/71, 22)
- Hanna Reitsch, described by Washington *Evening Star* as "Nazi Germany's famous daredevil woman test pilot and the only woman in German history ever to be given two Iron Crosses for her exploits," told of her glider research during *Star* interview. "I am searching for the right currents over the Alps. I don't have to worry about . . . borders in the air. . . . As we enter space work more, we will realize how small mankind is and how necessary it is to come together in peace." (Dean, *W Star*, 11/16/71, B7)

November 15–17: AIAA 2nd Symposium on Uranium Plasmas: Research and Applications was held in Atlanta, Ga. Symposium Chairman Robert G. Ragsdale, head of Advanced Reactor Concepts Section of LeRC Nuclear System Div., described cumulative impact of papers delivered in *Astronautics and Aeronautics*: "Significant progress in a number of research areas has been achieved along the road to determining the feasibility of gas-core reactors . . . [which] harness and control the energy released from a hydrodynamically contained fissioning uranium plasma. The major application studied to date is a gaseous-fueled rocket reactor. Other possible applications, such as nuclear-MHD power generators and breeder reactors, are also beginning to receive some attention." Developments detailed at symposium argued "the possibility of an engine with a 5000-sec specific impulse and . . . relatively high thrust-to-weight ratio." Gas-core reactors "could possibly send astronauts to Mars and return them safely to Earth in a total round-trip time of 60 days." (*A&A,* 11/71, 6)

November 15–19: Huntsville Assn. of Technical Societies (HATS) held international Congress on Space for Mankind's Benefit in Huntsville, Ala. Purpose of meeting was to communicate to nonaerospace public, in nontechnical language, practical benefits and dividends of U.S. space program.

Dr. Wernher von Braun, NASA Deputy Associate Administrator for Planning, in keynote address said tools and techniques originating from space exploration had arrived just in time to aid ecologists in saving planet earth. "Many benefits are now being realized, but we can expect far greater returns as increasingly sophisticated devices go into service." (Program; *Birmingham Post-Herald,* 11/17/71)

November 16–23: Photos from *Mariner 9* (launched May 30) showed dust storm on Mars appeared to have subsided somewhat over south pole area Nov. 16, and by Nov. 19 photos showed dust settling to surface in at least two regions of planet. Scientists believed winds might have died down in these areas.

Mariner scientists reported in news conference at JPL Nov. 19 that instruments had shown planet was flattened at poles and bulging at equator and had more irregularities in its gravitational field than moon had. Bulge was discovered in area called Tharsis, on equator, making planet's diameter 39 km (24 mi) greater at equator than through poles. Infrared spectrometer had measured reflected radiations that suggested wide range of rock compositions on Mars surface, varying from high in silicon and oxygen to low. Spectrometer had also detected small amounts of water vapor at south pole, as had been observed on previous Mariner flights.

On Nov. 23 *Mariner 9*'s infrared interferometer spectrometer (IRIS) detected pocket of warm air above south polar cap in same region as water vapor traces had been found earlier. Air pocket was about 240 K ($-30°C$). (Wilford, *NYT,* 11/20/71, 51; O'Toole, *W Post,* 11/20, 23, 24/71; Miles, *LA Times,* 11/24/71; JPL Hist Off)

November 16: NASA notified U.S.S.R. of *Mariner 9* Mars orbit in telexed message to I. A. Zhulin, Soviet space expert, in Moscow. Message, from Dan Schneiderman, *Mariner 9* Project Manager at JPL, said spacecraft had achieved desired orbit but ephemeris used in planning orbit required 60-km (37-mi) adjustment. Mars "appears brighter than expected by a factor of two. Surface features are almost entirely obscured, but the south polar cap can be seen because of its brightness. This is believed to be the result of a dust storm which covers the entire planet." Infrared instruments on the two *Mariner 9* cameras "appear to be seeing the [Martian] surface, but interpretation of their results is complicated by effects of the dust storm.

Message invited Soviet request for additional orbit information and described Nov. 10 and 11 photos taken by *Mariner 9* before Mars arrival. Photos had shown "four distinct dark features," one "possibly Nix Olympica." Nov. 12 photos had shown streaks 1000 km (620 mi) long originating at fourth spot. "Changes in appearance of these four dark areas over the coming weeks may provide unique information about the dust storm." (NASA Release 71–229)

November 16

- Pioneer-F, 250-kg (550-lb) spacecraft, was expected to be ready for launch from ETR in late February or early March 1972 and would fly by Jupiter in December 1973, NASA scientists announced at JPL press conference. Pioneer-G was to be launched in 1973 and fly past Jupiter in 1974. NASA Director of Planetary Programs Robert S. Kraemer said Pioneer flights would be "vital precursor" to future Grand Tour flights to outer planets. Pioneer spacecraft would be first to penetrate vast asteroid belt between Mars and Jupiter and first to study planet's intense radiation belts. (Wilford, *NYT*, 11/17/71, 26)
- Assistant chief designer for U.S.S.R.'s *Lunokhod 1* moon rover said in Tass interview that all goals of mission had been attained. *Lunokhod 1* had landed on moon Nov. 17, 1970, on board *Luna 17* and had exceeded expected three-month operating time and provided valuable information. Mission had resulted in acquisition of new skills and development of new techniques and control methods. Principles of remote control, reliability, and durability, as well as power supply, might find application on earth. (FBIS–Sov–71–223–3, 11/18/71, L2–3)
- NASA announced it had issued RFPs for use of space-related technology in solving domestic problems of air pollution, water pollution, solid waste management, and clinical medicine. Proposals would be reviewed by experts in each field at ARC, LaRC, and MSC. Responses were requested by Dec. 15, 1971. One or more projects in each field might be selected for negotiations and award of contract worth up to $75 000 for maximum of one year. (NASA Release 71–226)
- Dr. Cyril A. Ponnamperuma—Univ. of Maryland scientist, former Chief of ARC Exobiology Branch, and leading authority on chemical building blocks of life—commented on Oct. 29 report by Columbia Univ. scientists of "waterless" synthesis of amino acids suggesting life might arise on waterless planet. Dr. Ponnamperuma said chemical reactions in experiment were not independent of water's role, since formic acid—used in experiment—contained hydroxyl which was water molecule. Synthesis of life building blocks, amino acids, that had been identified in interstellar base had thus not been waterless. (*NYT*, 11/17/71, 13)
- House rejected amendment to H.R. 11731, FY 1972 Defense Establishment appropriations bill, that would strike out $801.6 million for F-14 USN jet aircraft procurement. (*CR*, 11/16/71, H11082–117)
- NAE issued *An Assessment of Industrial Activity in the Field of Biomedical Engineering.* Report of Committee on Interplay of Engineering with Biology and Medicine recommended creation of Government-supported national advisory body to inform and make recommendations on development and marketing of biomedical products. It also recommended establishment of separate Government agency to develop and test prototype products, issue licenses for production, and subsidize production when market for product was too limited to allow profit. (NAE memo to press)
- Elrey B. Jeppesen, Chairman of Board of Jeppesen and Co.—world's largest compiler and publisher of air navigation materials—received FAA's Distinguished Service Award "in recognition of his outstanding contributions to the progress and safety of aviation." Award was presented at Colorado Aviation Historical Society Dinner in Denver. (FAA Release 71–184)

November 17: U.S.S.R. launched *Cosmos 455* from Plesetsk into orbit with 487-km (302.6-mi) apogee, 271-km (168.4-mi) perigee, 92.1-min period, and 70.9° inclination. Satellite reentered April 9, 1972. (GSFC SSR, 11/30/71; 4/30/72; SBD, 11/23/71, 111)
- Lunar roving vehicle (LRV) which would transport astronauts John W. Young and Charles M. Duke, Jr., on moon during Apollo 16 mission had successfully completed prelaunch checks and been installed on lunar module for flight, KSC announced. Astronauts Young and Duke would make three excursions on LRV in Descartes area of moon during mission, scheduled for March 1972. Final approval of LRV flight readiness was given by Dr. Rocco A. Petrone, Apollo Program Director, after installation review. LRV was managed by MSFC and was built by Boeing Co. (KSC Release 259-71)
- Skylab flight hardware manufacturing was nearing completion, MSFC announced. Post-manufacturing checkout would soon begin at industrial and Government installations. Major Skylab spacecraft components include Workshop, Apollo Telescope Mount (ATM), airlock module (AM), and multiple docking adapter (MDA). (MSFC Release 71-212)
- Appointment of David S. Gabriel as Manager of AEC–NASA Space Nuclear Systems Office was announced by NASA. Gabriel, Deputy Manager, would succeed Milton Klein, who had accepted position as Associate Administrator for Research, Development, and Demonstration Activities in Federal Railway Administration. Gabriel had joined NACA at LeRC in 1943. (NASA Release 71-228)
- Nike-Tomahawk sounding rocket was launched by NASA from Kenya, Africa, carrying GSFC experiment to conduct stellar astronomy studies. Rocket and instruments functioned satisfactorily. (SR list)
- Fine Arts Commission—overseer of Federal architecture in Washington, D.C.—endorsed plans for $40-million National Air and Space Museum but requested architect Gyo Obata to redesign facade details. (Combes, *W Post*, 11/18/71, H3)
- FAA announced appointment of Dr. Alan J. Grobecker of Institute for Defense Analysis (IDA) to direct new DOT High Altitude Climatic Impact Assessment Program. Purpose of program was to provide basic data for projected operation of supersonic aircraft at altitudes above 14 000 m (45 000 ft). (DOT Release R-81)
- *Apollo 15* astronauts on European tour met with Pope Paul VI in Rome. Pope told them he hoped their voyage to moon would help men appreciate God and live in harmony. (AP, *W Post*, 11/18/71, H3)
- House by vote of 342 to 51 passed H.R. 11731, $71-billion FY 1972 Defense Dept. appropriations bill, after rejecting amendment to earmark $10 million to develop two prototype, light, air-superiority aircraft for USN as alternate to F-14 aircraft. (*CR*, 11/17/71, H11162-203)
- Tass quoted U.S.S.R. scientists as saying it was possible to observe long-life plasma as "light of blinding brightness, and freely soaring in space," with optical instrument. Optical plasmatron at Physics Institute of U.S.S.R. Academy of Sciences had enabled scientists to apply energy to plasma by light beam from quantum generator. Broad application of technique could be expected in physics, technology, and engineering. (FBIS–Sov–71–223–3, 11/18/71, L2)

- Cambridge Univ. astronomer Dr. David Dewhirst said in Cambridge that irregularities in orbit of Neptune, eighth planet from sun, indicated possible presence of another body. "But I must emphasize that at present this is only a notion." Dr. Dewhirst said U.S. Naval Observatory had supplied photos for study. (UPI, *W News*, 11/18/71, 15)
- Baltimore *Sun* editorial commented on quantity of new scientific data to be expected from U.S. *Mariner 9* and U.S.S.R. *Mars 2* and *3* probes of Mars—data "certain to be gathered later if not now." Whatever was learned, "the store of mystery held by the skies and even their nearest objects will remain enormous. . . . The admiration for man's persistence in pitting his intelligence, skill, courage and inventiveness against the solar system's magnificent teases is equaled by the folly of supposing we shall ever read any but a few of their overpowering riddles." (B *Sun*, 11/17/71)

November 18: Dr. James C. Fletcher, NASA Administrator, said in speech before National Space Club that Apollo program would end with Apollo 16 and 17 launches in 1972. Later, responding to question concerning possible cancellation of Apollo 17 because of FY 1973 budget restraints, Dr. Fletcher said, "The fate of Apollo 17 is beyond my control at the present time."

NASA was shifting emphasis, Dr. Fletcher said in speech: "When Apollo ends, most of our space activities planned for this decade will be in Earth orbit—and Moon exploration, for a time, will be left completely to the Soviets." Continuing effort to explore planets also was related to earth studies. "I am sure that much of what we learn will have direct and valuable application to our understanding of Earth and our current concern with environmental protection." Apollo photos of earth had helped create concern for world's environment. "I am glad to say that NASA is now well prepared to take a leading role in defining and responding to this concern." Apollo technology and equipment would be used directly in Skylab and directly or indirectly in space shuttle, Dr. Fletcher said. "Moreover, we gained invaluable experience in Earth orbital operations in the Gemini program and on the way to the Moon and back. We are well prepared for the next logical steps in near-Earth space. I wish very much we could continue lunar exploration . . . but if we have to make hard choices in this decade, then the right choice is to return to Earth orbit and defer lunar exploration."

NASA hoped to keep open option of additional Apollo missions in earth orbit. "This would enable us to carry out additional experiments in our Earth resources program, and would also facilitate our planning for cooperative flights with the Soviet Union."

Worldwide demand for applications satellites could be met only by production of space shuttle. "The key to our ability to help our country and to serve mankind is now the shuttle. The sooner we build it, the better we can serve." (Text; *Av Wk*, 11/29/71, 13)

- Bochum Observatory in West Germany said it had received what appeared to be on and off switching signals from Soviet Mars probes, *Mars 2* and *3*. (AP, *W Post*, 11/19/71, A3)
- Atomic Energy Commission announced Communist China had conducted atmospheric nuclear explosion test of low-yield (20 000 tons) at 2:00 pm local time. Explosion was 12th Chinese nuclear test and first since

Oct. 14, 1970. Test was conducted at top of Lop Nor proving ground in Sinkiang Province. Dr. Ralph Lapp, physicist, later said pattern of Chinese nuclear tests indicated Chinese had intended to explode larger hydrogen device that required small atomic blast to trigger thermonuclear reaction. Dr. Lapp believed hydrogen bomb failed to go off. Other experts said explosion could have been test of intermediate-range ballistic missile. (Auerbach, *W Post*, 11/19/71, A3; Lyons, *NYT*, 11/19/71, 1)

- AIAA Louis W. Hill Space Transportation Award was presented to Dr. Hubertus Strughold, former chief scientist of AFSC Aerospace Medical Div., during ceremonies at San Antonio, Tex. Citation was for "significant contributions indicative of American enterprise and ingenuity in the art and science of space travel." Dr. Strughold was cited for establishing rational biomedical functions for manned space exploration. (AFSC *Newsreview*, 1/72, 13)
- Moscow radio reported first Soviet sounding of outer space by radio wave. Scientists at Univ. of Kharkov had used geophysical rockets and other means to explore to 5000-km (3100-mi) altitude. Uninterrupted picture of electronic concentration "and its temporary fluctuations at any point at this height and also in the horizontal plane" had been obtained. Findings had confirmed scientists' model of the earth's ionosphere. (FBIS-Sov-71-237-12/9/71, L4)
- Senate Committee on Appropriations favorably reported H.R. 11731, $70.2-billion FY 1972 Defense Establishment appropriations bill, after decreasing House-passed figure by $805.5 million. (Sen Com Off)
- MIT had broken even financially during 1970-1971 academic year, AP reported, "but only by using funds that otherwise would have gone into endowments." In final annual report as MIT President, Dr. Howard W. Johnson had said outlook was not bright if MIT had to continue using unrestricted funds for current costs. Another way budget was balanced was by increasing tuition costs. (*W Post*, 11/19/71, B5)

November 19: U.S.S.R. launched *Cosmos 456* from Plesetsk into orbit with 287-km (178.3-mi) apogee, 175-km (108.7-mi) perigee, 89.1-min period, and 72.8° inclination. Satellite reentered Dec. 2. (CSFC *SSR*, 11/30/71; 12/31/71; *SBD*, 11/23/71, 111)

- Working Groups of NASA and Soviet Academy of Sciences on development of compatible space rendezvous and docking systems would meet jointly in Moscow Nov. 29 to Dec. 7, NASA announced. NASA representatives headed by Dr. Robert R. Gilruth, MSC Director, would meet with Soviet group led by Academician B. N. Petrov to continue joint efforts called for by Oct. 28, 1970, agreement. (NASA Release 71-231)
- Biostack, West German experiment to study biological effects of cosmic radiation during space flight, would be carried in Apollo 16 CM, NASA announced. Experiment, by Univ. of Frankfurt, would stack four layers of biological systems—spores, seeds of European watercress, bean embryos, and encysted eggs of brine shrimp—between different physical detectors of heavy particle tracks (nuclear emulsions and plastics). Measurements of effect of particles on biologic material could be compared with effect of same particles on detecting layers to provide data on strength and other characteristics of particles. Combined action of heavy cosmic ions and space environment would be evaluated by comparing results with those from balloon-borne Biostacks

and with ground-based irradiation experiments. Research on biological effects of heavy nuclei was significant because of possible relationship to biological effects of space flight on man. Experiment was sponsored by German Ministry for Education and Science. (NASA Release 71-230)

- AEC released summary report on effects of Nov. 6 Cannikin blast of five-megaton nuclear device on Amchitka Island, Alaska. Desired data had been obtained on performance of Spartan warhead. Analyses of radioactive melt, expected to be completed by spring 1972, were required to evaluate warhead performance fully. There had been "no detectable release of radioactivity to the marine or surface environment." Measurements would "continue for a number of years." No large earthquake had been triggered although "hundreds of after-shocks" had been registered in vicinity until collapse of cavity created by detonation. Ground motion had been felt "strongly on Amchitka, distinctly at Adak, nearly 200 miles [320 km] away, and less distinctly at Shemya, somewhat farther away." No large ocean wave from ocean floor movement had been recorded. Observed effects on living species on and near Amchitka Island had indicated "no permanent harm will result from the test to any population of mammals, birds, fish, other marine life, or plant life." (AEC Release O-221)

- On second anniversary of its emplacement on moon by *Apollo 12* astronauts, atomic battery designed to power *Apollo 12* experiments for one year was still generating at more than 70 w. Battery had been designed to produce 63.5-w power. *Apollo 12* instruments were still reporting data. Identical batteries, fueled by radioactive plutonium 238 and taken to moon by *Apollo 14* and *15*, were generating more than 70 w. Apollo 16 and 17 would land similar batteries on moon during 1972 missions, AEC reported. (UPI, *W Post*, 11/20/71, A3)

- Office of Management and Budget had asked NASA to cut FY 1973 budget request "quite a few million dollars" below $3.3-billion FY 1972 budget according to *New York Times*' sources, *Times* said. Dale D. Myers, NASA Associate Administrator for Manned Space Flight, had said Apollo 16 and 17 "look firm," but could "be in trouble" if there was significant cut in NASA funding. Other sources had said NASA had alternate plans that included elimination of one or both Apollo flights and curtailment, cancellation, or delay of other projects. (Wilford, *NYT*, 11/19/71)

- Daniel Guggenheim Medal was presented to Dr. Archibald E. Russell, Chief Designer for Bristol Aircraft Corp., during Washington, D.C., ceremonies marking first William Littlewood Memorial Lecture sponsored by NAC and Washington sections of AIAA and SAE. Award was for "personal devotion to aircraft engineering and design and particularly for his leadership of the Bristol team in development of the Concorde SST." Peter G. Masefield, Chairman of British Airports Authority, delivered first Littlewood lecture, "Aviation and the Environment." NAC Man of Distinction Award was given posthumously to William Littlewood, former American Airlines Vice President and aviation engineering consultant. (NAC Release)

- President Nixon had asked Congress for $9.1 million to purchase new presidential backup aircraft to replace 13-yr-old Boeing 707 used as backup aircraft for *Spirit of '76, Washington Post* reported. Presi-

dent's aircraft, formerly known as *Air Force One,* was Boeing 707–320–B delivered to USAF in 1962. (*W Post,* 11/19/71, B7)
- Senate Committee on Foreign Relations favorably reported Protocol to enlarge ICAO Council membership from 27 to 30 members. Enlarged Council would reflect expanding membership of ICAO due to establishment of increasing number of civil airlines throughout world. (*CR,* 11/19/71, D1201; ICAO PIO)
- Mrs. E. Margaret Burbidge would become first woman director of Royal Greenwich Observatory in 1972, *Washington Post* reported. British-born former professor of astronomy at Univ. of California would not inherit title of Astronomer Royal, traditionally bestowed upon Observatory Director. She would direct staff of 200, including 25 women scientists. (Nossiter, W *Post,* 11/19/71, A26)
- FAA announced availability from GPO of *FAA Statistical Handbook of Aviation* (5007–0166). Publication contained historical and statistical data for 11 yrs ending Dec. 31, 1969. (FAA Release 71–183)

November 20: U.S.S.R. launched *Cosmos 457* from Plesetsk into orbit with 1221-km (758.7-mi) apogee, 1184-km (735.7-mi) perigee, 109.4-min period, and 74° inclination. (GSFC *SSR,* 11/30/71; *SBD,* 11/23/71, 111)
- Pioneer-F, scheduled for early 1972 launch and Dec. 1973 Jupiter flyby, began its final test in TRW Inc. vacuum chamber. Spacecraft would operate for week on nuclear batteries in cold and almost total darkness to simulate Jupiter approach. Pioneer's voyage would be longest space trip to date—an estimated 660 days—and most hazardous since Pioneer would attempt for first time to fly through asteroid belt. Typical mathematical model of asteroid belt gave Pioneer 1-in-10 chance of penetrating belt without serious mishap. (O'Toole, *W Post,* 11/21/71, A3)
- NASA launched two Aerobee 170 sounding rockets carrying GSFC stellar UV experiments from WSMR. One flight was unsuccessful. (SR list)

November 21: New York Times said *Mariner 9* dust storm encounter had "made possible some unique experiments in global meteorology—experiments that may help us to understand the weather of our own planet." Storm had lasted longer than any observed so far. "Since it is hoped that Mariner will continue sending pictures for at least three months, it seems likely that it will record the gradual dissipation of the storm, which seems to have already begun, and in doing so hopefully will reveal wind patterns on the planet. This may help scientists understand why the storm is global, a phenomenon never seen on earth, and establish the science of meteorology—now a provincial one based on the weather of only one planet—on a more cosmopolitan basis." (Sullivan, *NYT,* 11/21/71, 4:8)
- CBS commentator Walter Cronkite received Bradford Washington Award of Boston Museum of Science for "outstanding contribution toward public understanding of science." Award, which included $5000 and gold medal, was presented by Astronaut James A. Lovell, Jr., at dinner commemorating museum's 35th anniversary. Cronkite was CBS anchorman for coverage of most NASA manned space flight missions. (AP, *Denver Post,* 11/22/71; Lovell Off)

November 22: Study of telemetry from unsuccessful Europa 2 European launch vehicle that failed on launch Nov. 5 had shown vehicle buckled at 1st-end stage junction 2 min 30 sec after liftoff during excessive pitch-over maneuver, *Aviation Week & Space Technology* reported. Vehicle's structural limits were exceeded; 2nd stage nozzle pierced liquid oxygen tank atop 1st stage, causing it to explode; 2nd stage separated in tumbling flight and exploded 10 sec later; and 3rd stage, 4th stage perigee motor, and payload fell into South Atlantic about 490 km (305 mi) downrange. Failure had been attributed to loss of inertial-guidance-system signals between 1 min 45 sec and 1 min 47 sec after launch. Engines on 1st stage had driven Europa 2 below programmed trajectory because corrective signals had not been sent to stage's autopilot. Exact cause of inertial-guidance-system breakdown had not been determined. If cause could not be established before Dec. 15, launch of 12th satellite in Europa 1-2 development series—scheduled for April 13, 1972—would have to be delayed. (*Av Wk*, 11/22/71, 19)

- MSC announced it had issued RFPs to 10 firms for $150 000 firm-fixed-price R&D contract to study space walking requirements in space shuttle program. Proposal called for investigation of emergency, contingency, and normal extravehicular and intravehicular activities (EVA/IVA) while outside earth's atmosphere and study of protective equipment needed to perform EVA/IVA. Proposals were due Dec. 6. (MSC Release 71-83)

- New $200-million international airport at Kansas City, Mo., was being built under NASA-developed management system, NASA announced. Midwest Research Institute, management consultant for project, was using KCI Management Information Center—fully equipped briefing room that resembled KSC's—as repository for reference data on airport project. (NASA Release 71-232)

- It was "open secret" that USAF saw space shuttle as "broad highway that will lead to manned military space operations by the United States," Universal Science News, Inc., said in *Dallas Morning News*. While "effort to keep NASA's civilian image seemed so important at one time that the military were allowed to start their own manned orbiting laboratory program, even though NASA has a similar project," recent budget squeeze had "forced NASA and the Air Force closer together in certain areas." (*Dallas Morning News*, 11/22/71)

- USAF reconnaissance satellite "Big Bird" launched in June was described in *Newsweek* article: "Streaking through its orbit, Big Bird scans broad land areas with one wide-angle camera, radios what it sees back to ground stations, and, on order, turns a giant 'narrow angle' second camera on targets of special interest for closeup pictures—a multiple function that used to require at least two less sophisticated satellites." Cameras in satellites 160 km (100 mi) above earth "can clearly photograph objects on the ground the size of small cars." (*Newsweek*, 11/22/71, 18)

November 23: Four extra manned space flight missions to use four Apollo spacecraft and one Saturn V launch vehicle left after 1973 Skylab mission were proposed by MSC officials to NASA Associate Administrator for Manned Space Flight Dale D. Myers at NASA Hq. Missions would include one lunar flight and three Saturn IB-launched earth-orbital missions. *Houston Chronicle* later quoted MSC Deputy Director Christopher C. Kraft, Jr., who had presented proposal, as saying MSC plan called for one mission of up to nine days in lunar orbit. Astronauts could map 60% of lunar surface and place Nimbus weather satellite in lunar orbit to complete photography of entire moon in one year. (Hill, *H Chron*, 11/29/71)

- Commercial pilot Elgin L. Long flew over South Pole in twin-engine Piper Navajo aircraft to become first man to make solo flight over both North and South Poles during one journey. Long, who left San Francisco Nov. 5, had flown over North Pole to Stockholm, then to London, Africa, Brazil, and Chile before crossing South Pole. (*W Post*, 11/23/71, B3)

- Dr. Wernher von Braun, NASA Deputy Associate Administrator for Planning, was first speaker in LeRC Awareness Program series. Dr. Von Braun said at LeRC: "Our agency is at the crossroads. Our success story is our greatest enemy." After four successful manned lunar landings, people had become complacent. If NASA fell behind U.S.S.R. in space, it would create another "Sputnik" environment. (*Lewis News*, 12/3/71, 1)

- NASA announced selection of TRW Systems Group for negotiations leading to $70-million contract for development of High Energy Astronomy Observatory (HEAO). Seven-year contract for two spacecraft would be awarded in 1972 and was expected to extend through launch and mission support for orbital operations. NASA planned to launch first HEAO unmanned scientific satellite from ETR on Titan III booster in 1975 and second, 15 mos. later. (NASA Release 71-233; *NYT*, 11/26/71, 30)

- Soviet defectors—writer Leonid Vladimirov, who defected in 1966, and scientist Anatoly Fedoseyev [see June 19]—said at London press conference that Soviet space technology had lagged behind that of U.S. and had produced less scientific data. Fedoseyev, not present but responding by tape recorder to questions put to him previously, said he did not think Soviet space achievements would be scientifically impressive for some time because Soviet space program was short of funds and technology level was low. Fedoseyev was said to have been responsible for bringing down U-2 reconnaissance aircraft pilot Francis Gary Powers in May 1960, Reuters later reported. (Reuters, B *Sun*, 11/24/71, A4)

- Senate by vote of 82 to 14 adopted amendment to H.R. 11731, FY 1972 Defense Establishment appropriations bill, and passed bill by vote of 80 to 5. Amendment proposed by Sen. Henry M. Jackson (D-Wash.) would provide $250 million for F-4 jet aircraft for Israel. (*CR*, 11/23/71, S19463–551)

- USN Naval Air Systems Command announced issuance of $3-million letter of offer and acceptance to U.K. Ministry of Defence for product support engineering services for Pegasus 11 engines for Harrier aircraft. Work would be performed by Hawker Siddeley Aviation, Ltd. (DOD Release 1191-71)

- *Washington Post* editorial commented on plans for new Air and Space Museum: "We are glad that skyrocketing costs have brought the proposed National Air and Space Museum down to earth. The trouble with the design architect Gyo Obata prepared several years ago was not only that it would now cost $70 million to build. It was also far too grandiose, bold and ambitious." U.S. air and space accomplishments "have a very prominent place in the hearts of the American people . . . they should also have a prominent place in the nation's capital. But that is not to say that we need or want a monumental aircraft hangar." Museum Director, *Apollo 11* Astronaut Michael Collins, had described new museum as "national center for education about the science and technology of flight and its economic, social, cultural and political meaning." New model, *Post* said, "shows a pleasing and interesting structure that is well behaved. . . . If it lacks architectural excitement, that is not what we want on the Mall."

Post also published letter from Gilmore D. Clarke, former Chairman of Commission of Fine Arts, protesting construction of Air and Space Museum on Mall, "on a site of restricted size whereon the structure cannot be enlarged in the course of years as the science of aviation and outer space technology advances. Whatever is built will, within a decade or two, prove to be too small." (*W Post*, 11/23/71, A18, A19)

November 24: U.S.S.R. launched *Molniya II-1* comsat from Plesetsk to ensure "functioning of the remote telephone-telegraph radio communications systems in the Soviet Union, for relaying central television programs to stations" in Orbita network and participating international networks. Satellite was new comsat and part of further development of Soviet satellite communications system. Orbital parameters: 39 553-km (24 577.1-mi) apogee, 516-km (320.6-mi) perigee, 712.0-min period, and 65.0° inclination. (FBIS–Sov–71–228–3, 11/26/71, L1; GSFC *SSR*, 11/30/71; UN Gen'l Assembly, A/AC.105/INF.243; *SF*, 5/72, 179)

- Tass announced that French equipment on *Mars 3* (launched May 28) for investigating solar radio emission in one-meter-waveband experiment was functioning successfully. French instrumentation was manufactured by French specialists under French-Soviet cooperative program. (FBIS–Sov–71–228–3, 11/26/71, L1)
- MSFC announced selection of Itek Corp. to perform large space telescope definition study under $400 000, 12-mo contract. High-resolution optical telescope 3 m (120 in) in diameter would be placed in 630- to 810-km (391- to 503-mi) orbit aboard research and applications module (RAM) by reusable shuttle vehicle or Titan III launch vehicle in late 1970s as national facility for operation by many astronomers. (MSFC Release 71–219)
- Merrill H. Mead, Chief of ARC Programs and Resources Office, had been named ARC Deputy Director of Administration, ARC *Astrogram* reported. (ARC *Astrogram*, 11/24/71, 1)
- *Mariner 9* and Soviet *Mars 2* and *3* race to Mars was "exciting race for knowledge and the fame of exploration, exemplifying the kind of peaceful competition the world needs," Washington *Evening Star* editorial commented. "If Mariner 9 reveals half of what it was sent out to find, it will be worth much more than the $150 million it is costing." (*W Star*, 11/24/71)
- Boris Sergievsky, World War I aviator with Imperial Russian Air Force and later leading test pilot in U.S., died in New York at age 83. He had set seaplane records with Pan American World Airways' Brazilian Clipper, flown Sikorsky amphibian aircraft to Chile in 1931, and been vice president of Helicopter Corp. of America. (*NYT*, 11/26/71, 38)

November 25: The Russian Space Bluff by Leonid Vladimirov, Soviet engineer-journalist who had defected in 1966, had been published in U.K., AP reported. Book said Soviet space program had begun as series of publicity stunts to persuade West that U.S.S.R. had reached level of advanced technology equal to that of U.S. Vladimirov had said that U.S.S.R. knew "she could not beat America to the moon because she cannot build a moon rocket. Even today the Soviet Union cannot produce any significantly big jet nozzles."

Mid-1960 flights in Voskhod series had been short because spacecraft were overloaded earlier Vostok type with some equipment removed. Last-minute modifications had been engineering disgrace. (AP, W Star, 11/25/71, H7)

November 26–27: Mariner 9 (launched toward Mars May 30) took man's first close photo of Martian moon Deimos on spacecraft's 25th revolution of Mars Nov. 26. *Mariner 9* was 10 300 km (6400 mi) from Mars and 8500 km (5300 mi) from Deimos when photo was taken. Picture told scientists that Deimos was almost 12.9 km (8 mi) in diameter and was shaped more like potato than grapefruit. Dr. Carl Sagan, director of Cornell University's Laboratory for Planetary Studies, cautioned against speculation on moon's makeup, but said "whatever its origins the Mariner mission represents a wonderful opportunity to study it." Deimos was discovered along with other Martian moon Phobos in 1877 by American astronomer Asaph Hall, who named moons after two sons of Greek mythological god Ares—Mars in Roman mythology—as given by Homer in Iliad.

On Nov. 27 *Mariner 9* failed in attempt at closeup photograph of Martian moon Phobos. JPL ground controllers said spacecraft's camera wasn't pointed correctly when attempt was made. Orbit of Phobos was not known precisely enough to ensure getting photo. Successful photo of Phobos would have shown greater detail than recent picture of Deimos, Mars' other moon, since Phobos was twice as large as Deimos and spacecraft passed 1600 km (100 mi) closer. *Mariner 9* had photographed Phobos Nov. 13 from 150 000 km (93 000 mi) out. (O'Toole, *W Post*, 11/27/71, A4; *W Star*, 11/28/71, A6)

November 26: Univ. of Michigan physicist Lawrence C. Jones was quoted in *Kansas City Times* as saying use of liquid hydrogen as nonpolluting automobile fuel was "not only technically and economically feasible" but also "desirable" and "may even be inevitable." Hydrogen could be used in conventional aircraft and automobile gasoline engines and "has the desirable feature of burning very efficiently." (*KC Times*, 11/26/71)

• *Pravda Vostoka* newspaper of Tashkent, Uzbek Republic, published account of autumn 1966 detonation by U.S.S.R. of underground nuclear charge of "unprecedented force" to extinguish natural gas fire that raged out of control for three years in Central Asia. *New York Times* said later that successful experiment was believed to be first use of nuclear explosion to quell gas-well fire, explosion was assumed to have been of low-yield, and disclosure of test had marked first Soviet pinpointing of location of peaceful nuclear blast. (Shabad, *NYT*, 12/2/71, 2)

• New initiatives in technology were discussed in *Science* editorial by Philip H. Abelson: In FY 1972, 77% of Federal R&D funding would go to atomic energy, space, and defense research. "Most citizens would agree that in view of our many domestic needs this proportion is out of line." Most "would view favorably the prospect of engaging presently unemployed resources of men and organizations in meeting social needs." Two major factors made this difficult. "First one cannot identify a single goal that has the kind of dramatic appeal that the moon venture had in the circumstances in which it was announced. Announcement of a lesser single goal would be certain to draw partisan sniping and could prove to be a political liability. Second . . . many of the possible initiatives represent a further invasion by the federal government of matters that have previously been considered the domain of private enterprise." Cynics would discount new initiatives as election year ploy. "However, win or lose, there will remain after the election a residue in the form of a further penetration of government into the management of the nation's life. There could also remain a residue of real progress, for goals once enunciated tend to carry with them a momentum of their own." (*Science*, 11/26/71, 895)

November 27: U.S.S.R.'s *Mars 2* probe (launched May 19) ejected capsule 1380 km (858 mi) above Mars on direct trajectory to planet's surface. Capsule, carrying pennant bearing U.S.S.R. hammer-and-sickle coat of arms, became first man-made object to reach surface. After capsule ejection *Mars 2* decelerated and entered Martian orbit with 25 000-km (15 500-mi) apogee, 1380-km (858-mi) perigee, 1080-min (18-hr) period, and 48.9° inclination.

Mars 2 had traveled 470 million km (292 million mi) to reach Mars. During flight it had explored interplanetary space and three midcourse corrections had been conducted, June 17, Nov. 20, and automatically on Nov. 27. Spacecraft was reported operating normally in Martian orbit. U.S.S.R. did not indicate whether capsule softlanded or hardlanded or what mission was. (*NYT*, 12/1/71, 20; O'Toole, *W Post*, 12/1/71, A23; FBIS–Sov–71–231–3, 12/1/71, L1)

- Tass reported arrival in Moscow of Dr. Robert R. Gilruth, MSC Director, and group of NASA experts "in connection with the forthcoming technical conference of Soviet and American experts in ensuring compatibility of means of approach and docking of spaceships and stations." Conference would take place Nov. 29 to Dec. 7. (FBIS–Sov–71–230, 11/30/71, L1)

- U.S. would resurrect SST program and have "production model by 1982," FAA Administrator John H. Shaffer predicted in Washington, D.C. Shaffer, who had recently returned from world tour of FAA installations, said France expected to sell three Concordes to Chinese Communists and Romania was expected to attempt purchase of several used U.S. jetliners. (UPI, *W Post*, 11/27/71, A9)

- *New York Times* editorials commented on recent astronomical events— Nov. 26 *Mariner 9* photographing of Martian satellite Deimos and Oct. 7 report by astronomer Dr. Hannes O. Alfven of discovery of asteroid Toro. Deimos' photo was "only the latest of Mariner 9's unique accomplishments. This is the first man-made artificial satellite of another major planet in the solar system and it has already journeyed more than a million miles around Mars." Barring *Mariner 9* malfunction, "lifting of the dust storm in the days ahead should permit even greater revelations of the nature of this planet that has been a mystery to man throughout the ages."

Evidence suggested asteroid Toro might crash into earth after 200 yrs, *Times* said. "Any such collision could do great damage. It may not be too early for space scientists to begin thinking about ways of nudging Toro out of a collision course with earth when, as and if the need arises. If men can send space ships across hundreds of millions of miles to Venus and Mars, it should be no great feat to visit Toro— a mere nine million miles [14 million kilometers] away at present closest approach—by way of performing whatever celestial engineering is required to prevent a mammoth cosmic traffic accident." (*NYT*, 11/27/71, 30)

November 28: Italy's *San Marco 3* (launched April 24) reentered atmosphere after more than seven months in orbit. Design lifetime had been six months. Satellite had produced excellent data on density, composition, and temperature of equatorial neutral-particle atmosphere at altitudes of 222–723 km (137.9–449.2 mi). (GSFC *SSR*, 11/30/71; NASA Proj Off)

• Senate Committee on Nutrition asked Federal Government to provide emergency food for highly skilled professionals in Seattle, Wash., who had been hit by aerospace industry unemployment. Committee, investigating Seattle situation after Dept. of Agriculture had refused request from State of Washington for free distribution of surplus food in Seattle, said many families were going hungry. (Reuters, *W Post*, 11/29/71, A6)

November 29–30: *Mariner 9*, orbiting Mars after May 30 launch on 400-million-km (248-million-mi) journey, photographed Mars moon Phobos from 14 682- and 5543-km (9123- and 3444-mi) distances. Photos, released by NASA Dec. 1, clearly showed Phobos was irregular, potato shaped chunk of rock about 21 km (13 mi) wide and 26 km (16 mi) long, with at least one crater 6 km (4 mi) in diameter. (*LA Times*, 12/2/71; Reuters, *NYT*, 12/2/71, 26)

November 29: *Oso 7* Orbiting Solar Observatory, launched by NASA Sept. 29, had completed more than 850 orbits of the earth and was officially adjudged a success. Primary objective, to obtain high-resolution data from solar corona in particular spectral bands in EUV and in visible regions during one solar rotation, had been achieved. Fulfillment of secondary objective, of investigating intensity and spectrum of solar and cosmic x-rays from nonpointed instruments beyond one solar rotation, continued. All spacecraft systems, including star-tracker— gyro night-pointing feature new to OSO—had operated satisfactorily. Tape recorder No. 1, inoperative when first interrogated on seventh orbit, had not recovered despite several attempts to operate unit. Tape recorder No. 2 had performed required data storage and readout functions satisfactorily. (NASA Proj Off)

• U.S.S.R. launched two Cosmos satellites from Plesetsk. *Cosmos 458* reached 497-km (308.8-mi) apogee, 271-km (168.4-mi) perigee, 92.2-min period, and 70.0° inclination and reentered April 20, 1972. *Cosmos 459* reached 286-km (177.7-mi) apogee, 199-km (123.7-mi) perigee, 89.4-min period, and 65.0° inclination and reentered Dec. 27 [see Dec. 27]. (GSFC, *SSR*, 11/30/71; 12/1/71; 4/30/72; *SBD*, 11/30/71, 129; *SA*, 12/13/71, 15)

November 29

- U.S.S.R.'s *Mars 3*, launched May 28, was continuing its flight toward Mars. At 9:00 pm Moscow time (1:00 pm EST) spacecraft was 138.8-million-km (86.3-million-mi) from Mars, having traveled approximately 472 million km (293 million mi). (*NYT*, 12/1/71, 20; FBIS-Sov-71-231-3, 12/1/71, L1)
- U.N. General Assembly adopted resolutions directing Committee on Peaceful Uses of Outer Space to formulate treaty by 1972 charting international course for space exploration, commending proposed liabilities convention for outer space travel [see Nov. 11], and recommending convention of working group on remote sensing of earth by satellites. (Fulton, *C Trib*, 11/30/71)
- FAA Administrator John H. Shaffer, in letter to Airline Pilots Assn. President John J. O'Donnell, rejected ALPA's Nov. 1 request for accelerated five-year plan to install landing aids at U.S. airports. Shaffer said FAA was legally unable "to accede to one limited group's requests to the prejudice of other legitimate demands." (Text)
- NASA announced publication of third annual *Research and Technology Operating Plan Summary*, RTOP-72. Publication described in capsule form hundreds of NASA research and technology efforts. (NASA Release 71-234)

November 29–December 7: U.S. and U.S.S.R. space officials met in Moscow in third of series of meetings to discuss design and testing of compatible systems for docking and linking manned spacecraft. U.S. delegation was headed by Dr. Robert R. Gilruth, MSC Director, and U.S.S.R. delegation was headed by Academician Boris N. Petrov. Joint communique Dec. 7 said progress was made in working out agreed technical requirements and planning possible joint test missions. (NASA Release 71-244)

November 30: U.S.S.R. launched *Cosmos 460* from Plesetsk into orbit with 539-km (334.9-mi) apogee, 518-km (321.9-mi) perigee, 95.2-min period, and 74.0° inclination. (GSFC *SSR*, 11/30/71; *SBD*, 12/2/71, 141)

- Assignment of MSFC as lead center within OMSF for concept verification testing (CVT)—system design and integration to verify concepts and hardware from NASA definition studies—was announced by Dr. Eberhard F. M. Rees, MSFC Director. Primary purposes of CTV program were to demonstrate representative shuttle payload concepts in research and applications modules (RAM) and space station classes, to develop shuttle-payload interface requirements from viewpoint of payload and users, to demonstrate shuttle capabilities for research and applications, to support design and development of Sortie Can (payload carrier for space shuttle), and to investigate integration and operational techniques for long-duration mission system for advanced RAMs and for eventual space station. CVT program would be part of Space Station Task Team headed by William A. Brooksbank. CVT work would be headed by Konrad K. Dannenberg. (MSFC Release 71-225)
- Meeting of NASA's Comet and Asteroid Missions Science Panel at MSFC, chaired by Dr. Ernest Stuhlinger, Associate Director for Science, discussed present and prospective efforts in planetary and interplanetary probes. Nine scientists from education field and Government attended. (MSFC Release 71-266; MSFC PAO)
- NASA and Small Business Administration announced publication of *Testing*

Methods and Techniques, summary of aerospace technology suitable for profitable adaptation by small industrial firms. Booklet, offered without charge, contained 34 abstracts of innovative processes and devices used in quality control and nondestructive testing, with emphasis on physical inspection and internal-flaw detection. (NASA Release 71–236)

- U.S. patent No. 3 624 650 was granted to JPL engineers Richard L. Horttor and Richard M. Goldstein for method of mapping planets from earth station with radar and orbiting transponder. Patent was assigned to NASA. (Jones, *NYT*, 12/4/71)
- USAF had announced that 130-member review team of specialists from USAF, Lockheed, and other aerospace companies would make year-long review of USAF C–5A program, *Wall Street Journal* reported. Object of review was to "recommend operational refinements or aircraft modifications where necessary." (*WSJ*, 11/30/71, 3)
- DOD signed $25-million contract with Lockheed Space & Missiles Corp. for definition and development phase of new-generation submarine-launched ULMS (underwater long-range missile system) with maximum 10 000-km (6200-mi) range. (Reuters, B *Sun*, 12/1/71, A12; *W Post*, 12/1/71, A14)

November 30–December 2: Apollo 15 investigator's symposium was conducted at Lunar Science Institute in Houston, Tex. Apollo Photo Science Team members Robert D. Mercer of Dudley Observatory and Lawrence Dunkelman of GSFC described photos of sun and interplanetary dust taken by *Apollo 15* CM pilot Astronaut Alfred M. Worden. Photos showed features of solar corona and zodiacal light never before observed. Preliminary examination of photos showed "light levels less than can be seen by the best telescopes on earth." Analysis of photos would permit direct, comparative measurements of solar phenomena's relative brightness to obtain information on mechanisms by which energy left sun and on distribution of particles outward from sun. (MSC Release 71–97)

November 30–December 3: Joint meeting of NASA and European Space Conference (ESC) was held in Washington, D.C., to identify candidate areas for possible European participation in post-Apollo space programs. Meeting had been proposed in Sept. 1 letter from Under Secretary of State U. Alexis Johnson to Belgian Science Minister Theo Lefevre, ESC Chairman. (NASA Release 71–235, 71–241; NASA PAO)

During November: ARC research into heat shields to protect aircraft passengers in crash fires was described in *Astronautics & Aeronautics* by ARC and Avco Corp. scientists. Concept of surrounding passenger compartment by fire-retardant shell to protect occupants long enough for fire to burn out or be extinguished had been made possible by Apollo program research into fire-retardant materials polyisocyanurate foam and intumescent paint. "Exposed to heat . . . paint expands to many times its original thickness and insulates the surface underneath it. The thermal-protection mechanisms of these materials operate on the same ablative principles . . . that protect our astronauts during re-entry of the Apollo spacecraft." (Neel, Parker, et al., *A&A*, 11/71, 18–26)

- USAF had accepted delivery of 520-kg (1150-lb) comsat from TRW Systems Group, AFSC *Newsreview* reported. Comsat was first of six being built for SAMSO use in Phase II of U.S. Defense Satellite Communications System (DSCS). (ASFC *Newsreview*, 11/71, 4)

- AFSC Aeronautical Systems Div. had established Prototype Program Office, AFSC *Newsreview* reported. Objective of prototype program was to improve performance in development and acquisition of weapon systems by placing more reliance on the performance of "hardware" and less reliance on paper analysis. (AFSC *Newsreview*, 11/71, 1)
- *Columbia Journalism Review* editorial criticized *New York Times* for Aug. 15 publication of "soap-opera style exclusive interviews with the families of Apollo 15 astronauts obtained through the newspaper's purchase of their personal stories." *Times* had "condemned this kind of exploitation, by other publishers, only eight years ago." (*Columbia Journ Rev*, 11/12/71, 6)
- Article by Paul Dickson in *The Progressive* discussed "Empire of Think Tanks." R&D in U.S. had become "vast, powerful, and well financed empire . . . replete with its own priorities, powers, and pecking order. Atop this empire is a diverse group of institutions called think tanks, which as agents of applied research and policy study have a fateful impact on the nation. The most famous of these . . . is the RAND Corporation . . . a prime mover in military research."

    R&D was "particularly critical element in the battle over national priorities because it not only reflects our current priorities but largely determines what will be technologically feasible in the future." Lunar landings had shown "how much is possible if a great deal of thought, resources, and dedication are channeled into any stated goal. The vast majority of Project Apollo money had been pumped into space R&D during the decade prior to the first landing. There is, however, a catch in invoking the Apollo/NASA example, because such a new knowledge as is developed in such a program must be applied and directed in order to get anywhere. If not conscientiously applied, well-performed research can lead to a range of ends from a simple nonuse to a more insidious ruse—a way of dodging action under the guise of studying the problem."

    Independent think tanks needed to be scrutinized along with Federally sponsored ones. "Many of them receive the bulk of their funds from the Government, and, though not officially 'sponsored' like RAND, they bring up the same questions of power, unchecked and unanswerable, evoked by the officially sponsored outfits." (*CR*, 11/3/71, S1752-5)

December 1971

December 1: NASA's M2–F3 lifting body, piloted by test pilot William H. Dana, reached mach 1.22 and 21 300-m (70 000-ft) altitude during 12th flight, launched from B–52 from FRC. Objectives of powered flight—acquisition of stability and control data above mach 1.0, expansion of flight envelope, and evaluation of reaction control system—were achieved. (NASA Proj Off)

- Quarterly review of research and applications modules (RAM) study being conducted by General Dynamics Corp.'s Convair Aerospace Div. was held at MSFC. Work on specific set of RAM payload carriers was reviewed. Preliminary design of spacecraft would be completed during remainder of study. (MSFC Release 71–217; MSFC PIO)
- Approval by NASA Project Designation Committee of designation QUESTOL for quiet, experimental, STOL transport research aircraft was announced by NASA Assistant Administrator for Public Affairs John P. Donnelly in memorandum to NASA Associate Administrator for Advanced Research and Technology Roy P. Jackson. (Text)

December 2: U.S.S.R.'s *Mars 3* probe (launched May 29) reached vicinity of Mars and ejected capsule which softlanded on planet. U.S.S.R. did not announce landing until Dec. 7 and did not release details until Dec. 19. After separation, capsule bearing Soviet coat of arms entered Mars atmosphere and parachuted to softlanding in Mars southern hemisphere between Electris and Phaetonis in area 45° south latitude and 158° west longitude. *Mars 3* continued flight into orbit with 1500-km (930-mi) periapsis and 11-day period. Tass said signals from landing capsule "were received and recorded on board . . . *Mars 3* and later . . . transmitted to the earth on December 2–5. The videosignals received from the surface of Mars were brief and suddenly discontinued." (FBIS–Sov–71–236, 12/8/71, L1)

- U.S.S.R. launched two satellites. *Intercosmos 5*, launched from Kapustin-Yar, entered orbit with 1055-km (655.6-mi) apogee, 197-km (122.4-mi) perigee, 97.2-min period, and 48.4° inclination. Satellite—joint project of U.S.S.R. and Czechoslovakia—would continue investigation of radiation and dynamics of corpuscular flux in near-earth space and investigation of nature and spectrum of low-frequency electromagnetic oscillations in natural plasma. *Intercosmos 5* reentered April 7, 1972.

  Cosmos 461, launched from Plesetsk, entered orbit with 508-km (315.7-mi) apogee, 488-km (303.2-mi) perigee, 94.5-min period, and 69.2° inclination. (FBIS–Sov–71–233, 12/3/71, L1; L2; GSFC *SSR*, 12/31/71; 4/30/72; *SBD*, 12/7/71, 164)

- Thomas O'Toole described in *Washington Post* effects of *Mariner 9* data on idea that life existed on Mars: "In the three weeks that it has spent in Martian orbit so far, Mariner 9 has found that the planet's atmosphere is even thinner than it was thought to be—about 260 times as thin as Earth's atmosphere. It also found Martian surface temperatures to be no higher than 80 degrees below zero. To date, Mariner 9's

December 2: *The U.S.S.R.'s* Mars 3 *probe softlanded a capsule on the planet Mars. The capsule transmitted signals 20 seconds before falling silent. In the Tass news service photo the descent apparatus—shown here in descent position—carried a braking cone that also served as a heat shield for the capsule. The cone fell away as the main parachute, stored in the container immediately above the capsule, opened to slow the fall.* (Photo courtesy of United Press International)

cameras have been frustrated by a storm that has covered almost all of Mars with dust for more than two months. It's almost impossible to imagine a global dust storm, and when one thinks of a global storm lasting for two months, it's even harder to imagine life surviving such cataclysmic weather conditions. . . . now the dust has begun to settle on the planet, revealing striking features that man has never seen before. Nevertheless, the idea that there might be life among those features is one that seems to be a dying one." (*W Post*, 12/2/71, A19)

- NASA selected Douglas Aircraft Co., Grumman Aerospace Corp., and Lockheed-Georgia Co. for negotiations of three separate $1.5-million, six-month contracts for design phase of QUESTOL (quiet, experimental STOL) program. After completion of design phase one contractor would be selected to fabricate two experimental transport aircraft for flight research program. (NASA Release 71-238)
- Description by Soviet professor of physics and mathematics M. Marov of *Mars 2* scientific program to Soviet journalists was reported by Tass: "Registration of various radiations from Mars, ranging from ultraviolet light to radio waves, by sputniks from a comparatively short distance will give much more complete information about Mars. Certain findings which are still quite inadequate are readily available.

It has been established that at least nine-tenths of the Martian atmosphere consists of carbon dioxide. Nitrogen is practically absent, while water vapors are very scant. The thin atmosphere of Mars is extremely heterogeneous. Its structure varies widely with seasons, the time of day and night. The structure of the surface in the areas explored is very heterogeneous too." (FBIS–Sov–71–233, 12/3/71, L3)

December 3: U.S.S.R. launched *Cosmos 462* from Plesetsk into orbit with 1782-km (1107.3-mi) apogee, 227-km (141.1-mi) perigee, 105.2-min period, and 65.7° inclination. Within hours after launch, *Cosmos 462* passed close to *Cosmos 459* (launched Nov. 29) and exploded in apparent test of satellite interceptor system at low altitude. Mission was first satellite-intercept test using target below 258-km (160-mi) altitude; previous tests had occurred at altitudes between 580 and 885 km (360 and 550 mi). (GSFC *SSR*, 12/31/71; AP, *NYT*, 1/2/71, 25; *Av Wk*, 12/13/71, 20)

- Discovery of polar caps on sun by Dr. Werner M. Neupert and team of GSFC astronomers was announced by NASA. Measurements made by *Oso 7* (launched Sept. 29) had shown black areas on sun's north and south poles with temperature of 1 000 000 K (1 800 000°F). Hot areas on sun associated with solar flares registered temperatures of 3 000 000 to 4 000 000 K (5 400 000°F to 7 200 000°F). Data from sounding rocket experiments that preceded solar polar cap discovery indicated seasonal variations in cap sizes. If data were accurate, caps were largest during time of least solar activity and might disappear near most active times. (NASA Release 71–237; NASA Special Release 12/6/71)

- NASA held *Mariner 9* press conference at JPL. Discussion centered on dust storm that had covered planet since mission began and on origin of Martian craters.

    Dr. Bradford A. Smith of New Mexico State Univ. said storm had begun 10 wks earlier, had reached peak of intensity in 5th wk, and was gradually but continuously clearing. Project scientists disagreed, however, how long dust would obscure Mars; predictions ranged from a few weeks to months. If storm continued throughout mission, primary objective might have to be changed. Dr. Smith explained prime objective "was to map 70 percent of the planet in high resolution. If the storm does not clear up immediately, it will not be possible to do that within the nominal mission. An alternative might be to focus on the phenomenon that we have on our hands; that is, the dust storm. And instead of studying it in bits and pieces and waiting for the planet to clear up . . . to rearrange our sequences to make an all out study of this atmospheric phenomenon." Failure to complete mapping as planned would not significantly affect Viking Mars landing mission because, with only slight degradation, Viking zone could be mapped in last half or last third of *Mariner 9* mission. "Also Viking has the capability to keep its landers in orbit for as long as two months while it does the mapping . . . [and] the capability to do some photographic site selection even if there were no usable pictures returned from '71."

    Dr. Harold Masursky of U.S. Geological Survey described four high points on Mars, all with large summit craters, suggesting that they might be volcanic calderas. Great alignment of relatively young topographic highs with summit craters suggested that Mars was geochemi-

December 3

cally active—that in not too distant past there had been extensive volcanic activity. Similar features might be found at lower elevations; they would be of interest because they had been, and still might be, hot. Such volcanic sources had, according to most opinions, produced terrestrial atmosphere; that is, carbon dioxide and water had come from such volcanic vents. Dr. Masursky emphasized that volcanic theory was not agreed on; project scientists' views ranged from "agreement, to skepticism, to outright disbelief."

Data on Mars moons Deimos and Phobos were described by Dr. James B. Pollack of ARC. Deimos was 12 by 13½ km (7½ by 8½ mi) and Phobos was 21 by 26 km (13 by 16 mi). Moons were "among the darkest objects in the solar system," reflecting only about 5% of sunlight that fell on them. Moons' darkness could be attributed to basalts or carbonaceous chondrites in soil. If scientists could determine reason for darkness, they would soon be able to determine whether moons were captured asteroids or remnants of material from which Mars was formed. (Transcript)

- NASA sent message to Dr. I. A. Zhulin in Moscow via telex communication system established for exchange of information on *Mariner 9* and *Mars 2* and *Mars 3* missions to Mars. Message congratulated U.S.S.R. on success of *Mars 2*; requested *Mars 2* coordinates, orbital parameters, doppler data, and data from UV spectrometer; and provided information on U.S. *Mariner 9* photos of Mars and moons, doppler data, and UV spectrometer data. (NASA Release 71-240)

- NASA announced establishment of new Office of Applications (OA). Charles W. Mathews, Deputy Associate Administrator for Manned Space Flight since 1968, was named Associate Administrator for Applications. Leonard Jaffe, Deputy Associate Administrator for Space Science and Applications (Applications) since 1966, was named Deputy Associate Administrator for Applications. Personnel, programs, and functions of Earth Observations and Communications Programs Divs. would be transferred from OSSA to new office and OSSA would be renamed Office of Space Science (OSS).

    Dr. James C. Fletcher, NASA Administrator said: "The application of space technology to solving problems here on Earth is perhaps NASA's most important new thrust. I believe it appropriate to centralize into a single office in NASA Headquarters all of the resources which we can muster to support space applications to Earth." (NASA Release 71-239)

- Joint group of NASA and European Space Conference technical experts completed four-day meeting at NASA Hq. to identify candidate areas for possible European participation in post-Apollo space programs. NASA announced Dec. 6 that areas identified included space shuttle subsystems, major responsibility for orbit-to-orbit space tug, early development of major payload elements like sortie cans and research and applications modules, and support studies in related technology. Group's report would be studied by U.S. and European authorities. Group had proposed February 1972 meeting for further technical definition and consideration of programmatic, financial, and management implications. (NASA Release 71-241)

- U.S. Embassy in Moscow released statement to press about Nov. 29–Dec. 7 NASA-Soviet Academy of Sciences meeting on compatible space docking system: "The main purpose of this third working session was to discuss the technical details of hardware compatibility and to discuss a joint American-Soviet space mission. It is intended to sign a final agreement within two months. Details will not be given out until then."

AP, in bulletin from Moscow, said: "American and Soviet scientists plan to sign an agreement on a joint manned space flight within the next two months, a U.S. Embassy official disclosed today."

Sources later described by *Washington Post* as "close to" NASA delegation in Moscow said in response to press query that it was misunderstanding to think that agreement already had been reached. NASA spokesman in Washington, D.C., told press AP report was result of misunderstanding of U.S. Embassy report. Embassy statement had stemmed from "routine programs report . . . on discussions . . . now in progress." Any understanding as to joint test mission would come after current talks had ended and been subjected to recommendations and approval.

Press later commented on misunderstanding. *New York Times* said there was "considerable confusion . . . as to how firm and far-reaching is the apparent understanding between the American and Soviet negotiators." *Washington Post* said Cosmonaut Andrian G. Nikolayev had said at October reception for U.S. governors visiting Moscow that astronauts and cosmonauts would work together in space: "We speak the common language of space." Baltimore *Sun* said "the Russians are traditionally sensitive about any publicity before official, simultaneous announcements are made on any bilateral agreement." It appeared "the Americans had inadvertently talked about the agreement prematurely." (AP News Service, 12/3/71; Wilford, *NYT*, 12/4/71, 1; Kaiser, *W Post*, 12/4/71, A1; Mills, B *Sun*, 12/4/71, A12)

- First transatlantic picture telephone call was made successfully between U.S. and Sweden via *Intelsat–IV F–2* orbiting over Atlantic. One-hour transmission was part of ceremonies in Sweden and at ComSatCorp headquarters in Washington, D.C., formally inaugurating new Nordic earth station at Tanum, Sweden, jointly owned by Denmark, Finland, Norway, and Sweden. (ComSatCorp Release 71–67)
- USAF announced issuance by Aeronautical Systems Div. of $1 781 382 fixed-price-incentive contract to General Dynamics Corp. for aerospace ground equipment for F–111 aircraft. (DOD Release 1020–71)
- Senate confirmed nomination of H. Guyford Stever as NSF Director. (*CR*, 12/3/71, S20397)

December 4: Unidentified USAF satellite and its Atlas-Agena booster were exploded by radio command when booster malfunctioned and veered off course shortly after launch from ETR. (*W Post*, 12/7/71, A7)

- U.S.S.R.'s *Mars 3*, approaching Mars, had detected possible evidence of earth's magnetic tail farther out in space than previously reported, *Pravda* article said. Both *Mars 2* and *Mars 3* had passed through area 19 million km (12 million mi) from earth, where they had detected magnetic field. When *Mars 3* passed through area it recorded streams of charged particles whose velocities differed from those observed in solar wind. Soviet scientists speculated that particles of solar wind were mixed with those of earth's magnetic tail. (*NYT*, 12/5/71, 9)

December 5: France's 90-kg (200-lb) Polaire scientific satellite, launched from Kourou Space Center, failed to enter orbit when 2nd stage of Diamant-B booster malfunctioned. (FBIS–Sov–71–234, 12/6/71, T2; *SBD*, 12/7/71, 163)

- Launch of Intelsat-IV F–3 comsat, scheduled for Dec. 6 by Atlas-Centaur booster, would be postponed indefinitely pending review of Dec. 4 failure of USAF Atlas-Agena mission, NASA announced. (*NYT*, 12/7/71, 18)
- Dr. Wernher von Braun, NASA Deputy Associate Administrator for Planning, described future benefit of space program spinoff in *Parade* article on applications of space technology to social problems. "On the 1973 flights in the Skylab experimental space station, a medical doctor will go into space with the astronauts. He will carry out a comprehensive program of experiments . . . to establish how well men live and work in weightless conditions in an enclosed space as big as a three-bedroom house." With data obtained, space benefits could be achieved in a few years and made available to people everywhere. (*Parade*, 12/5/71, 8, 12)
- Remaining "puzzles" of solar system pattern were discussed by Walter Sullivan in *New York Times* article: General assumption was that sun, planets, and moons had been formed from "swirling, flattened cloud of material." But: "Why does Mars have two tiny moons, each only a few miles in diameter, yet our moon is a quarter the diameter of the earth? Why do some of the outer moons of Jupiter go the 'wrong way'—counter clockwise—around that huge planet? Why is the earth's moon formed of a different mix of elements from that constituting the earth?" Recent information had "illuminated these questions though it had not resolved them." Earth's moon was giving off "far more heat than had been predicted on the basis of assumptions of how it was formed." *Mariner 9* photos of the two Mars moons had shown both to be irregular chunks "rather than systematically spherical bodies in the manner normally attributed to moons and planets." George Washington Univ. scientist J. Martyn Bailey had proposed that all seven of Jupiter's outer moons were asteroids captured by that planet's powerful gravity and that Jupiter's clockwise moons had been captured when planet was farthest from sun, while counter-clockwise moons had been captured when Jupiter was nearest sun. Lunar Science Institute scientists had agreed there had been sufficient sorting and layering on earth's moon to produce radioactive crust that could account for heat flow. Information on two Martian moons had indicated that Phobos and Deimos looked much like asteroidal fragments but that did not prove that they were captured asteroids. They could have been formed as spherical moons from material left when Mars was being created and later "battered into their irregular shapes." (*NYT*, 12/5/71, 4:8)
- *New York Times* editorial praised results of Apollo lunar exploration and noted there was "much more that men have to learn about this puzzling neighbor in space." It had been reported that NASA might have to cancel Apollo 16 and 17 for budgetary reasons. "Such action would confirm the widespread suspicion that the United States was less interested in the scientific results of lunar exploration than in the prestige of being the first nation to land men on the moon." Announcement that

Soviet flag had been dropped on Mars might, however, "impel the Administration to decide that this is not the time for the United States to end its scientific bridgehead on the moon. It is the promising research prospects from continued lunar study, not Soviet space competition, that give the really valid motivation for continuing the Apollo program to the extent that existing hardware, including the already-manufactured equipment for Apollos 16 and 17, permits." (*NYT*, 12/5/71, 4:10)

- Travel writer Morris D. Rosenberg described his visit to KSC in *New York Times* article: Visit had provided "whole new point of reference, more understanding of the complex drama you see on TV and read about in the newspapers, and a better appreciation of the U.S. role in space." Those opposed to space funding might be unimpressed by "avalanche of facts and figures" but it was doubtful that they could "walk away from this 2-hour and 15-minute bus tour without a strong feeling of pride." (*NYT*, 12/5/71, 1)

- Contract with NR that was expected to set pattern for aerospace industry settlements under negotiation was approved by Long Beach, Calif., Local 887 of United Automobile Workers. Proposed contract called for increases of 30% in salary over 34-mo period. (Wright, *NYT*, 12/6/71, 25)

December 5–6: NASA launched series of eight sounding rockets from Point Barrow, Alaska, to obtain temperature, pressure, density, and wind data in upper atmosphere. Eight payloads consisted of four Nike-Apache pitot density probes, two Nike-Cajun grenade payloads, one Nike-Cajun pitot atomic oxygen, and one Nike-Cajun ozone experiment. All launches were successful except one; first Nike-Apache failed when rocket motor apparently malfunctioned and payload broke off. (NASA Rpts SRL; Proj engineer)

December 6–11: U.S.S.R. launched *Cosmos 463* from Baykonur into orbit with 285-km (177.1-mi) apogee, 204-km (126.8-mi) perigee, 89.2-min period, and 64.9° inclination. On Dec. 7 satellite passed over East Pakistan at perigee on 14th revolution, British Royal Aircraft Establishment at Farnborough and British Interplanetary Society sources reported. On 16th revolution, perigee was lowered to 188 km (116.8 mi), reducing period to 89.0 min. Revolutions 30 and 46 crossed East Pakistan Dec. 8 and 9 and on revolution 48 apogee was raised, returning to original period. Satellite reentered Dec. 11. *Spaceflight* magazine reported telemetry had revealed that some consumables were used at twice normal rate for a 13-day mission.

Changes in orbit supported earlier reports of U.S.S.R.'s ability to maneuver satellites for precise target coverage. Flights of *Cosmos 463* and *Cosmos 464* [see Dec. 10–16] were made during time of Indo-Pakistan war. Paths would permit observation of events in Pakistan before early reentry; any film cassettes could have been recovered within four hours of last pass over East Pakistan on each flight. (GSFC *SSR*, 12/31/71; *SBD*, 12/7/71, 160; *SF*, 9/9/72, 351; *W Post*, 1/7/72, A16)

December 6: NASA was "in search of new, clearly defined goals, and the search is not going well," John N. Wilford said in *New York Times* article. "Self-preservation" was believed to be factor in emphasis NASA placed on current "efforts in Moscow to reach an agree-

December 6

ment with the Soviet Union that could lead to joint Soviet-American manned space missions." If NASA could establish opportunity for improved relations with U.S.S.R. through space cooperation, "this would probably create the need for a series of post-Apollo missions of historic proportions." It would be "hard" for Administration or Congress to refuse funds for such an effort. Soviet motives in current negotiations might be saving in space funds through cooperative ventures or "that the Soviet Union looks to cooperation as a way of sharing some of the advanced technologies developed by NASA." (*NYT*, 12/6/71, 30)

- JPL announced selection of Motorola, Inc., Government Electronics Div. for negotiation of $2.8-million, cost-plus-fixed-fee contract for design and manufacture of two communications subsystems—modulation demodulation subsystem (MDS) and relay telemetry subsystem (RTS)—for Viking Orbiter '75 project. (JPL Release 606)

- Dish-shaped 305-m (1000-ft) radiotelescope at Cornell Univ.'s Arecibo Observatory near San Juan, Puerto Rico, was undergoing $7-million modification program that would increase telescope's sensitivity 2000 times, Dr. Frank D. Drake, Cornell Univ. astronomer and former Arecibo Observatory Director, told San Juan meeting of American Astronomical Society. Modified telescope would search skies for unusual features and objects "we cannot even begin to predict" and organic molecules in intergalactic space that were clues to evolution of life and evidence of possible distant civilizations. "We expect to devote about 1 per cent of our time to just listening. If any civilization is advanced enough to communicate with other civilizations in the universe, we might get the message." (Goodwin, *W Post*, 12/16/71, E9)

- USAF aircraft procurement in FY 1973 would reach "historic low of fewer than 100," *Aviation Week & Space Technology* reported. DOD preliminary budget estimates had indicated cyclical dip reflecting completion of procurement of LTV Aerospace Corp. A-7D and General Dynamics Corp. F-111 aircraft. Magazine said USAF procurement might rise in FY 1974 when McDonnell Douglas F-15 fighter aircraft and NR's B-1 bomber had passed development milestones. (*Av Wk*, 12/6/71, 13)

- House defeated by vote of 203 to 172 H.R. 11624, "to amend the Military Construction Authorization Act, 1970, to authorize additional funds for the conduct of an international aeronautical exposition." (*CR*, 12/6/71, H11830-1)

- Washington *Evening Star* editorial commented on Administration's New Technologies Opportunities Program [see Oct. 31]: Much was worthwhile, "especially the attempt to mount a genuine program of conversion from defense and space-related to socially useful technological emphasis." But "purely economic objectives of the program are murkier, at least at present. Granted that this nation may be on its way, in a number of industrial fields, to losing the technological lead that for so long has been a mainstay of our economic strength. Does it follow that the answer, in high-technology fields like electronics and commercial aviation, is what one Presidential adviser has called 'an unaccustomed set' of supports and subsidies to industry?" Question contained "all the elements of explosive controversy, and the administration will be hard-pressed to make a convincing case." (*W Star*, 12/6/71, A10)

December 6–9: American Geophysical Union held annual meeting in San Francisco. MIT seismologist Dr. Mehmet Nafi Toksoz presented findings of seismic team from MIT, Columbia Univ., Univ. of Hawaii, and General Dynamics Corp. that moon had been subdivided into layers by process somewhat similar to that which produced layering within earth. Conclusion had been reached after analysis of tremors induced in moon by spacecraft impacts. Report was released simultaneously at MIT, where Dr. Frank Press, coauthor of report, told press that separation into layers, some 3.3 billion yrs ago, had brought so much radioactive material close to lunar surface that there was not enough left at depth to keep lunar interior hot. He believed deep lunar interior cooled and was quiescent compared with earth. Dr. Press said energy released in form of moonquakes was 100 million times less than in earthquakes.

Scientist Steve Korn of TRW Inc. Viking biological instrument team described instruments being developed for $11-million experiments to be conducted by two Viking landers scheduled for 1976 Mars landing. Sophisticated units for four experiments to search for microscopic life on Mars would be among most complex systems ever flown aboard unmanned spacecraft. Experiments included heat release experiment to analyze Martian soil for photosynthetic life similar to plant life on earth; experiment that would add nutrient to solid sample and measure metabolic conversion; gas exchange experiment to monitor environment surrounding sample for changes in gaseous makeup that would indicate presence of living organisms; and light-scattering experiment to provide nutrient to soil sample and detect any increases in scattering attributable to growth of microorganisms. Also, three-legged probes would snap photos; study Martian atmosphere; return meteorological data, pressures, temperatures, wind, and humidity; report on Mars' magnetic and physical properties; and check seismic events. Probes would be lowered by parachute and retrorocket after landers had obtained orbit in procedure designed to land them accurately in areas deemed most likely to be habitats for life. (Sullivan, *NYT*, 12/10/71, 29; Miles, LATNS, *W Post*, 12/19/71, A41)

December 7: Retirement of Astronaut Richard F. Gordon, Jr., from NASA and USN, effective Jan. 1, 1972, was announced by NASA. Gordon, pilot of *Gemini 11* (Sept. 12–15, 1966) and CM pilot for *Apollo 12* (Nov. 14–24, 1969), would join John W. Mecom, Jr., organization in New Orleans, La. Retirement would reduce number of astronauts to 46. (NASA Release 71–243)

• FRC announced award of $211 260, cost-plus-fixed-fee contract to LTV Aerospace Corp. for design and construction of side fairings for NASA's supercritical wing test aircraft, modified F–8 jet. Installation of side fairings would improve overall aircraft with area ruling, an aerodynamic design feature that would permit aircraft to fly more efficiently, and provide more desirable test environment for evaluating aerodynamic characteristics of supercritical wing near speed of sound. (FRC Release 29–71)

• NASA award of two contracts for feasibility studies of pressure-fed engine for water-recoverable space shuttle booster was announced by MSFC.

December 7 ASTRONAUTICS AND AERONAUTICS, 1971

TRW Inc. would receive $400 000 and Aerojet-General Corp. $367 595, for studies ending Feb. 29, 1972. (MSFC Release 71-230)

- Dr. James R. Schlesinger, AEC Chairman, announced major reorganization of AEC operating staff to "provide a coherent management structure and to improve program effectiveness." Major program areas—national security, energy and development, research, and production—would be directed by assistant general managers. New Div. of Controlled Thermonuclear Research would deal with development of fusion as new energy source. New position of Assistant General Manager for Environment and Safety would increase emphasis on environmental matters and safety research. New Div. of International Security Affairs and Div. of Applied Technology were being established. (AEC Release O-258)
- RAND Corp. specialists Robert L. Perry and Arthur J. Alexander testified before Senate Committee on Armed Services that France, Sweden, U.K., and U.S.S.R. had been turning out excellent fighter aircraft for fraction of what U.S. paid in money and manpower. Cost difference was largely due to simpler design, strict "fly before you buy" policy, and reliance on small teams of experts rather than on huge number of men and machines. Perry said government offices monitoring development of new aircraft in France, Sweden, and U.K. rarely contained more than 30 specialists. In U.S., figure was normally five times that. Within European countries developing aircraft, skilled workers committed to program rarely exceeded 700. In U.S. "from two to ten times as many comparable specialists are employed," Perry said. (Getler, *W Post*, 12/8/71, A25)
- U.K. and French governments agreed on undisclosed pricing policy for Anglo-French supersonic transport Concorde. *New York Times* later said policy favored 16 airlines that had taken options on 74 Concordes. (Giniger, *NYT*, 12/8/71, 5)
- AFSC announced development of optical aimpoint guidance (OPTAG) system, to enable bomb or missile to pinpoint specific target within large complex. System used a correlatron to compare previously made photos with actual scenery. By turning photographic and optical scenes into electronic images, system cued control fins to keep bomb or missile on course during flight. (AFSC Release 245.71)

December 7-9: Skylab multiple docking adapter Crew Compartment Fit and Function Review was held at Martin Marietta Corp. facility in Denver, Colo. Purpose of review was to familiarize astronauts with MDA equipment and storage requirements. (*Marshall Star*, 12/1/71, 2; MSFC PAO)

December 8: NASA's TF-8A jet aircraft with supercritical wing, piloted by NASA test pilot Thomas C. McMurtry, successfully completed 25th and 26th flights from FRC. Objectives of flights were to investigate buffet boundary further at low supersonic speeds and to determine effects of aileron position and angle of attack on wing pressure distributions. Buffet boundary was penetrated at mach 1.15 and 15 200-m (50 000-ft) altitude. Pressure distribution data were obtained from mach 0.9 to 1.0 at 14 000-m (46 000-ft) altitude. (NASA Proj Off)

- *Apollo 15* Astronauts David R. Scott, Alfred M. Worden, and James B. Irwin received NASA Distinguished Service Medal from Vice President Spiro T. Agnew in Washington, D.C., ceremony. Citation was for

"most complex and carefully planned scientific expedition in the history of exploration." (NASA Release 71-242)
- Portrait of Dr. Robert H. Goddard, U.S. rocket pioneer, was unveiled by artist Peter Stevens and Mrs. Goddard and presented to Rep. George P. Miller (D-Calif., Chairman of House Committee on Science and Astronautics, by National Space Club. Ceremonies were held in Committee's main hearing room, where portrait would hang. (NSC *News Letter*, 11/23/71)
- Spartan ABM intercepted one of several reentry vehicles in successful test of Spartan's ability to distinguish actual nuclear warhead from flock of dummies. DOD later said test, above atmosphere over Pacific, was first in which Spartan, its associated radar, and its computerized radar guidance system were required to make this distinction. (Farrar, *C Trib*, 12/10/71, 12)
- Fiftieth anniversary of first transatlantic transmissions by American amateur radio "hams" on 200-m frequency. Bouncing radio transmissions off ionosphere over long distances was later to prove seminal in evolution of concepts for passive reflector satellites. (*QST*, 12/71, 54-7)
- Baltimore *Sun* editorial commented on softlanding on Mars of instrumented capsule by Soviet *Mars 3*: "For the first time man has been able to reach Mars with implements that have proved capable both of surviving a landing and of working, for a time at least, after they have been landed. This of course is only one approach to the new studies of brother objects in space. The United States Mariner 9 may by photographic means add valuable pictorial knowledge to that which Mars 3 has reported by signal. And even if neither of these latest raids on the unknown returns as much information as men want, they have together shown that the means by which more and more can be learned are now in our hands, even if in only a primitive and imperfect state so far." (*B Sun*, 12/8/71, A18)
- *Kansas City Times* editorial commented on "ambiguous" announcement of progress in NASA-Soviet Academy of Sciences meetings to develop compatible space docking system [see Dec. 3]: "The first interpretation was that agreement soon would be reached on a joint mission involving astronauts of the two countries in an orbital linkup. In hasty clarification, NASA explained that the mooted accord covered only the technical matters which could make such a mission possible. . . . Either way, the conclusion is obvious: The two space powers are moving at last, and seriously, toward the sort of shared space planning that logic and economics have long indicated but which competition and secretiveness have prevented." (*KC Times*, 12/8/71)

December 9: ESRO Council meeting in Paris agreed on future program based on applications satellites. Priority would be given to use of ELDO launch vehicles provided ELDO's price did not exceed that of launch vehicles available elsewhere. ESRO budget would be equivalent of $283.6 million from 1972 to 1974 and $300 million for following three years. From 1974 to 1980 at least $70 million would be spent on construction of applications satellites. (AP, *NYT*, 12/12/71, 9)

December 10-16: U.S.S.R. launched *Cosmos 464* from Plesetsk into orbit with 311-km (193.3-mi) apogee, 180-km (111.9-mi) perigee, 89.4-min period, and 72° inclination. British Royal Aircraft Establishment and British Interplanetary Society sources reported that drift of about 5°

per day brought satellite over East Pakistan on revolution 44 Dec. 13 at height of 230 km (142.9 mi). On revolution 46, two-impulse maneuver lowered both apogee and perigee, with 89.0-min period and 182-km (113.1-mi) perigee. Westerly drift rate was reduced to less than 1° per day, permitting three further daylight passes over Pakistan area before reentry Dec. 16.

Orbital changes supported earlier reports of Soviet maneuvering capability. *Cosmos 463* [see Dec. 6–11] and *Cosmos 464* were launched during Indo-Pakistan war; altered paths permitted observation of Pakistan before early reentry. (GSFC *SSR*, 12/31/71; *SBD*, 12/14/71, 199; *SF*, 9/9/72, 351; *W Post*, 1/7/72, A16)

December 10: MSFC announced completion of study of offshore oil and gas operations by team of MSFC, MTF, and Michoud Assembly Facility experts. Study, made for U.S. Geological Survey, was confined to federally managed outer continental shelf in Gulf of Mexico but team's recommendations, in report to USGS, applied to all outer shelf operations. Team had suggested measures to produce offshore resources with safety and protection from pollution of marine and coastal environment. (MSFC Release 71–232)

• Development of lift-distribution system to alleviate stress problems in C–5A aircraft by spreading aerodynamic loads through wings rather than concentrating stress on small areas was announced by Lockheed-Georgia Co. Vice President Robert B. Ormsby. Modification could be installed during regular modification and inspection periods and would not affect aircraft's handling. (AP, *W Post*, 12/12/71, A3)

• U.K. successfully tested experimental hovertrain in 805-m (880-yd) run at 32 km per hr (20 mph) near Cambridge. Train, developed by Tracked Hovercraft, Ltd., to reach maximum 483-km-per-hr (300-mph) speed, traveled on air cushion astride concrete track and was controlled from trackside. (Scott, *NYT*, 12/11/71, 54)

December 11: U.K.'s *Ariel 4* (UK–4) satellite was successfully launched by NASA on Scout booster from WTR at 12:47 pm PST. Orbital parameters: apogee, 587 km (364.8 mi); perigee, 472 km (293.3 mi); period, 95.2 min; and inclination, 83°. Primary objective was to investigate interactions between electromagnetic waves, plasmas, and energetic particles in upper ionosphere. *Ariel 4* weighed 99.6 kg (220 lbs) and carried five experiments—four provided by U.K. groups and one by Univ. of Iowa—to study radio noise and low-frequency radio waves, measure electron temperature, and count low-energy charged particles.

Ariel 4 was fourth in series of U.S.–U.K. cooperative launches that began with *Ariel 1* April 26, 1962. British Science Research Council (SRC) was responsible for satellite design, fabrication, and tests including all but one experiment. NASA was responsible for low-energy proton and electron analyzer, technical support, Scout booster, launch operations, and tracking and data acquisition. *Ariel 2* had been launched March 27, 1964, and *Ariel 3*, May 5, 1967. (NASA Proj Off; NASA Release 71–277)

• President Nixon accepted resignation, for personal reasons, of David Packard as Deputy Secretary of Defense effective Dec. 13. President also redesignated Secor D. Browne as CAB Chairman for term expiring Dec. 31, 1972, and announced intention to nominate Whitney Gillilland for reappointment as CAB member for six-year term and, upon Senate confirmation, as Vice Chairman of CAB for term expiring Dec. 31, 1972. Gillilland appointment was submitted to Senate Dec. 13 and confirmed Feb. 29, 1972. (PD, 12/13/71, 1640–1; 12/20/71, 1672; CR, 2/29/72, S2916)

- U.S. Senate approved Protocol enlarging ICAO Council membership from 27 to 30. (CR, 12/11/71, S21400–2)

December 12: USN manned spacecraft recovery forces in Atlantic and Pacific would cachet and cancel philatelic mail for March 17 launch of Apollo 16, Washington *Sunday Star* reported. Covers would be processed through coordinators at Norfolk, Va., and Honolulu. From Hawaii covers would be forwarded to recovery ship. Since Atlantic Recovery Force would not include ship with post office, covers would be processed by Norfolk post office. (W *Star*, 12/12/71, C11)

- NOAA, after one year of functioning, was "hardly a wet NASA," Thomas O'Toole said in *Washington Post* article. While it had been predicted that Government would spend up to $2 billion a year on NOAA by 1970, spending "never went higher than $500 million and it's almost all spent on surface vessels." NOAA expected to spend no more than $700 000 for undersea research in 1972. (W *Post*, 12/12/71, A1)

- David Sarnoff, honorary board chairman and former president of RCA Corp., died at age 80 after long illness. Sarnoff, often called "electronics prophet," had played major role in development of radio and TV. (NYT, 12/13/71, 1)

December 13: Apollo 16 spacecraft was rolled out from KSC Vehicle Assembly Building to Launch Complex 39, Pad A, in preparation for manned lunar landing mission scheduled for launch March 17, 1972. Rollout was watched by about 10 000 persons at KSC. (UPI, NYT, 12/14/71, 12; KSC Release 263–71)

- First photos of solar flare spouting from far side of sun were taken by *Oso 7* (launched Sept. 29, 1971). Photos also showed for first time effects of massive eruption of sun traveling through its outer corona. Tongues of ionized gases were shown shooting up to 6½ million km (4 million mi) from solar surface. NRL Director Capt. Earle W. Sapp said later that OSOs were increasingly helpful in forecasting sun's weather to predict effect of solar storms on navigation and communications equipment. (Lyons, NYT, 1/11/72, 26)

- NASA's TF–8A jet aircraft equipped with supercritical wing, flown by NASA test pilot Thomas C. McMurtry, successfully completed 27th flight from FRC. Purposes of flight were to obtain wing pressure distributions at low angles of attack and to obtain stability and control characteristics at design conditions. Pressure distributions were obtained from mach 0.90 to 0.99 at 14 000-m (46 000-ft) altitude. Stability and control data were obtained at mach 0.99 at 14 000-m altitude. (NASA Proj Off)

- NASA announced establishment of two new positions in OART. Dr. Seymour C. Himmel, LeRC Director of Rockets and Vehicles, was appointed Deputy Associate Administrator for Technology. George W. Cherry, Director of OART's Aeronautical Operating Systems Div., was appointed Deputy Associate Administrator for Programs. (NASA Release 71–246; NASA Ann, 12/15/71)

- President Nixon referred to Anglo-French Concorde supersonic airliner at dinner with Prime Minister Marcello Caetano of Portugal and

President Georges Pompidou of France during meetings of French and U.S. heads of state at Angra do Heroisimo, The Azores: "When I arrived at the airport on the Spirit of '76, a Boeing 707, I saw parked in front of me a Concorde which had carried the President of France. Our Ambassador to France, Mr. Watson, pointed out that he had come from France at a speed three times as fast as we had come from the United States. I do not speak in envy; I only wish we had made the plane ourselves." (*PD*, 12/20/71, 1661–4)

- Deputy Secretary of Defense David Packard, whose resignation had been accepted by President Nixon on Dec. 11, said at DOD news conference that new orders for military equipment were "leveling off" after declining sharply following 1967–1968 peak period due to Vietnam war. Total defense spending, Packard said, was less than 7% of current GNP and probably would remain so for several years. *Wall Street Journal* said later that Packard would become Chairman of Hewlett Packard Co., electronics firm he had helped to found in 1941. (Transcript; *WSJ*, 12/14/71, 14)
- U.S.S.R.'s largest and most advanced space tracking ship, 40 800 000-kg (45 000-ton), 230-m (760-ft) *Cosmonaut Yuri Gagarin*, had been unveiled in Odessa Harbor, *New York Times* reported. Ship was equipped with 100 antennas, including four large parabolic antennas for deep space communications, and could operate at sea for up to one year. (Shabad, *NYT*, 12/14/71, 11)

December 14: USAF launched four unidentified satellites from Vandenberg AFB by Thorad-Agena booster. Orbital parameters: 999.4-km (621-mi) apogee, 983.3-km (611-mi) perigee, 104.9-min period, and 70° inclination. (Pres Rpt 72; *SBD*, 12/16/71, 210)

- Team of ARC astronomers headed by Dr. Charles P. Sonett had found evidence that sun's spin rate 4.6 billion yrs ago was one revolution every three hours rather than current one rotation every 27 days, ARC announced. Study of melting histories of meteorites recovered on earth and observations of young stars in constellation Taurus led astronomers to believe high rotation rate and other mechanisms typical of newly formed suns had forced flow of electric gases from sun, causing sun to lose one third of its original mass in a few million years and stripping away primordial atmosphere of Mercury, earth, and Mars. Complete melting could have occurred in asteroids. Radioactive dating had established that many meteorites were melted about 4.6 billion yrs ago. ARC astronomers believed inertia of gas molecules thrown out by sun and tied to sun by solar magnetic field, with interplanctary magnetic field itself, would have slowed sun's rotation over 4.6 billion yrs. (ARC Release 71–71)
- House Committee on Science and Astronautics published *For the Benefit of All Mankind: A Survey of the Practical Returns From Space Investment*. Report updated and expanded study of tangible benefits and practical returns from space investment based on material from Government, industry, and press—including extensive NASA research in technology utilization at Committee's request. Committee concluded that illustrations in report were "*extra* dividends which are a fallout of ingenious application of space experience by business, industry, commerce, science, government, the medical profession and the academic community. Those dividends already paid, coupled with those in sight

for the near-term future, affect practically every facet of human convenience and concern. They promise continuing and increasing return on the space investment for the benefit of mankind on earth today." (H Rpt 92-748)

- FAA Administrator John H. Shaffer, in speech before Aviation/Space Writers Assn. in Washington, D.C., said Anglo-French Concorde supersonic jet transport would be quiet enough to land at U.S. airports. Concorde would be used over North Atlantic and would not cause potentially destructive sonic booms over U.S. Shaffer predicted U.S. would build SST eventually. (Reuters, *W Post* 12/15/71, E10)

- President Nixon inspected Concorde Anglo-French supersonic airliner at Lajes Field, The Azores, before departing for U.S. aboard Presidential aircraft *Spirit of '76* after meeting with French heads of state. (*PD*, 12/20/71, 1670)

- DOD Project MAST (military assistance to safety and traffic) was example of "major role" of transfer of technical knowledge from DOD laboratories to civilian scientific community in "technological revolution," Secretary of Defense Melvin R. Laird said at meeting of Domestic Action Council in Washington, D.C. Since July 1970 inception of program, military helicopters had flown 767 missions and evacuated 983 civilian patients for civilian agencies. (Text)

December 15: U.S.S.R. launched *Cosmos 465* from Plesetsk into orbit with 1011-km (628.2-mi) apogee, 969-km (602.1-mi) perigee, 104.8-min period, and 74° inclination. (GSFC *SSR*, 12/31/71; *SBD*, 12/16/71, 210)

- U.S.S.R.'s *Mars 2* and *Mars 3* satellites were "successfully continuing the scientific exploration of Mars and the space near the planet in substantially different orbits," Tass announced. To date, 153 radio communication sessions had been held with *Mars 2* and 159, with *Mars 3* FBIS-Sov-71-241)

- USAF's X-24A lifting body was shipped by air from FRC to Martin Marietta Corp.'s Denver plant, where it would be modified and redesignated X-24B. Modified vehicle, of new shape with improved hypersonic lift-to-drag ratio, would be delivered to FRC in early fall 1972. X-24A had completed 28 flights in joint NASA-USAF research program, reaching 1676 km per hr (1048 mph) and 21 600-m (71 000-ft) altitude. (FRC Release 30-71; *SBD*, 12/17/71, 215)

- Purchase price of $31.2 million for Concorde supersonic transport was announced at Paris press conference by Henri Ziegler, head of French government-owned Aérospatiale. Announcement was made on behalf of Aérospatiale and BAC, manufacturers of Anglo-French aircraft. Ziegler said first six Concordes would be ready for service by October 1974. *New York Times* later said price was record for civilian aircraft. (Hess, *NYT*, 12/16/71, 93)

- List of Federal agencies using MTF for scientific activities was inserted in *Congressional Record* by Sen. John C. Stennis (D-Miss.). Agencies included Environmental Protection Agency, U.S. Geological Survey and other Interior Dept. activities, USA Corps of Engineers, NOAA and NASA. Mississippi State Univ. and Louisiana State Univ. had programs at MTF; state governments of Mississippi, Louisiana, and Arkansas had established full-time liaison with MTF; and Dept. of Agriculture and

December 15

other Federal agencies were investigating possible use of facility. (CR, 12/15/71)

- Conversation between Mary Ann Harbert, U.S. citizen released Dec. 14 after nearly four-year internment by Chinese Communists, and escort Col. Leonard W. Johnson, Jr., Commander of 9th Aeromedical Evacuation Group at Clark AFB in Philippines, was reported by AP. Miss Harbert had asked, "Have you put any men on the moon yet?" Col. Johnson had responded with description of U.S. space program that had put eight men on moon. Miss Harbert had said, "I really don't know, I hadn't heard about it. Many people in China don't know that. The last thing we knew about space was that Russia had something circling the moon." (B Sun, 12/15/71, A9)

- NSF published *Industrial R&D Spending,* 1970 (NSF 71-39). In 1970 total R&D performance in industry amounted to $17.9 billion—3% below 1969 level of $18.3 billion but more than double 1958 amount. In constant dollars, 1969–1970 decrease was nearly 8%. Entire drop in industrial R&D spending between 1969 and 1970 was accounted for by decrease in Federal R&D funds to industry from $8.4 billion to $7.8 billion. Decline of 8% represented lowest level of Federal R&D dollars in industry since 1965. (NSF *Highlights,* 12/10/71, 1)

December 15–16: Space Shuttle Aerothermodynamics Technology Conference at ARC discussed technology for manned spacecraft that could fly to aircraftlike landing. More than 30 papers were presented on space shuttle design, operation flight mechanics, flow fields, and heat transfer. (ARC *Astrogram,* 12/23/71, 3)

- European Launcher Development Organization (ELDO) Council, meeting in Paris, adopted $79.3-million budget to cover Europa I and II development, studies and preliminary work on Europa III, construction of F–15 and subsequent launchers, and launchers for Symphonie. Gen. R. Aubiniere of France succeeded Ambassador R. di Carroba of Italy as Secretary General. Gen. H. Hoffman was appointed Technical Director, G. Van Reeth Administrative Director, and J. P. Causse Deputy Director General, to act for ELDO in preliminary talks with U.S. on European participation in post-Apollo program. (SF, 2/72, 8; ELDO PAO)

December 16: Cosmos 466 was launched by U.S.S.R. from Baykonur into orbit with 375-km (233-mi) apogee, 175-km (108.7-mi) perigee, 90-min period, and 64° inclination. Satellite reentered Dec. 27. (GSFC SSR, 12/31/71; SBD, 12/17/71, 217)

- NASA's M2–F3 lifting body, piloted by test pilot William H. Dana, reached mach 0.8 and 13 700-m (45 000-ft) altitude during 13th flight, air-launched from B-52 from FRC. Objectives of powered flight were to obtain stability and control data at speeds above mach 1.0, expand flight envelope, and evaluate reaction control system. Evaluation of reaction control system was only objective achieved because engine shut down prematurely. (NASA Proj Off)

- Launch of Intelsat-IV F-3 by NASA for ComSatCorp, postponed from Dec. 6 following failure of USAF Atlas-Agena booster, had been rescheduled for Dec. 18, KSC announced. NASA project officials recommended that sensor probe be added to Atlas sustainer system to provide information on engine temperature between gas generator and turbine inlet and that temperature readouts be monitored during Atlas ignition sequence before liftoff. (KSC Release 274-71)

- AP quoted NASA Assistant Administrator for Public Affairs John P. Donnelly as saying NASA was not releasing some information on Soviet space program received through official channels. Clampdown was being made at U.S.S.R's request and through "bilateral agreement." AP said Donnelly had forbidden NASA engineers who had returned from NASA-Soviet Academy of Sciences meeting on compatible docking system Nov. 29–Dec. 7 in Moscow to talk to press. Text of message telexed by Soviet scientists to NASA at JPL during exchange of *Mars 2* and *3* and *Mariner 9* information would be withheld at Soviet request. Soviet scientists had sent three such messages. (*NYT*, 12/16/71, 27)

- Australia's decision to accept delivery of 24 U.S. F-111C aircraft was announced by Dr. Robert C. Seamans, Jr., Secretary of the Air Force, at USAF news briefing. Matter had "been under discussion" for some time, Seamans said. "I am very gratified that the decision has finally been made and we can put the aircraft in shape for them and get them over to Australia." (Transcript)

December 16–18: KSC added new displays to outdoor exhibit of space hardware at Visitor Information Center. Spacecraft facility verification vehicle—Apollo spacecraft mockup 25 m (82 ft) tall used to check out assembly and test facilities for early Apollo missions—was placed on temporary display with Juno 2 and Jupiter C rockets, Gemini Titan and Mercury spacecraft, Apollo LM mockup, and three kinds of rocket engines. Full-scale model of *Apollo 15* LRV would remain on display for several weeks. Scale model of *Mariner 9* spacecraft had been added to indoor exhibits. (KSC Release 273-71)

December 17: U.S.S.R. launched two satellites from Plesetsk. *Cosmos 467* entered orbit with 465-km (288.9-mi) apogee, 266-km (165.3-mi) perigee, 91.8-min period, and 71° inclination and reentered April 18, 1972.

Cosmos 468 entered orbit with 808-km (502.1-mi) apogee, 786-km (488.4-mi) perigee, 100.7-min period, and 74° inclination. (GSFC SSR, 12/31/71; 4/30/72; SBD, 12/20/71, 226; 12/21/71, 230)

- *Mariner 9* photographed Martian plateau from 6500-km (4000-mi) altitude. Photo, released by JPL Dec. 29, showed area 120 km (75 mi) square in Phoenicis Lacus region, covered with fault lines and resembling wrinkled elephant hide. Faults and scarcity of craters suggested region was volcanic and relatively young geologically. (AP, B *Sun*, 12/30/71, A5; Miles, *LA Times*, 12/30/71)

- Flight version of multiple docking adapter (MDA) for Skylab was accepted by MSFC at Martin Marietta Corp.'s Denver facility. MDA—a major Skylab module being readied for 1973 launch—had been designed and fabricated by MSFC. Martin Marietta had installed, integrated, and tested MDA subassemblies and functional equipment. (MSFC Release 71-241)

- Rep. William R. Cotter (D-Conn.) said in release to press that preliminary GAO audit of NR Rocketdyne Div. $24-million Saturn contract had shown "as much as $5 million . . . was diverted to underwrite crucial research on Rocketdyne's winning design in the $500 million space shuttle main engine contract [awarded Sept. 1]." Cotter said audit,

December 17

performed at his request, showed "that work performed under the Saturn contract influenced NASA's Source Selection Board, the group which evaluated the space shuttle designs submitted by Rocketdyne, Pratt & Whitney and Aerojet General." NASA Assistant Administrator for Industry Affairs and Technology Utilization Daniel J. Harnett later told press NASA audit of space shuttle engine contract had shown "no unfair competition advantage was given to Rocketdyne." (Text; AP, W Star, 12/19/71, E3)

- Aviation Hall of Fame installed 1971 honorees in ceremony at Dayton, Ohio: Aviatrix Jacqueline Cochran for "outstanding achievements in modern aviation"; Boeing Co. Board Chairman William M. Allen, who had backed R&D that led to Boeing production of B-47, B-52, 707, and 747 aircraft; Gen. George C. Kenney (USAF, Ret.), who had organized and commanded SAC from 1946 to 1948, commanded Air Univ. at Maxwell AFB, Ala., and been World War I combat pilot in France; and late Harry F. Guggenheim, who had helped establish schools of aeronautical engineering at U.S. universities and had supported Dr. Robert H. Goddard in rocket development and jet propulsion development. (AF Mag, 1/72, 15–16)

December 18: President Nixon signed H.R. 11731, $70.5-billion, FY 1972 DOD defense establishment appropriation bill. Bill became P.L. 92–204. (CR, 12/17/71, D1339)

December 19–20: Intelsat-IV F–3 comsat was launched by NASA for ComSatCorp on behalf of INTELSAT. Satellite, launched from ETR at 8:10 pm EST by Atlas-Centaur booster, entered elliptical transfer orbit. Primary objective was to place satellite into transfer orbit accurate enough for spacecraft onboard propulsion systems to place it in planned synchronous orbit.

Apogee kick motor was fired at 12:49 pm EST Dec. 20, and *Intelsat-IV F–3* entered orbit with 36 052.9-km (22 402.2-mi) apogee, 35 745.5-km (22 211.2-mi) perigee, 1442-min period, and 0.47° inclination. Satellite would drift from position at 284.2° east longitude to permanent station over Atlantic at 19.5° west longitude by end of January 1972.

Intelsat-IV F–3 was second comsat in Intelsat IV series. Satellite was 238 cm (93.7 in) in diameter and 528 cm (208 in) high and weighed 1412 kg (3090 lbs) at launch. It had 12 transponders, providing 12 TV channels and 3000–9000 telephone circuits, and was capable of multiple-access and simultaneous transmissions. Expected lifetime was seven years. *Intelsat-IV F–2* had been launched Jan. 25 and was operating satisfactorily over Atlantic. (NASA Proj Off; ComSatCorp Release 71–66)

December 19: Detailed Tass description of *Mars 2* and *3* missions with diagrams of spacecraft and landing procedures appeared in Soviet newspapers. Enroute to Mars, spacecraft had measured solar wind particle streams, measured parameters of interplanetary magnetic fields, determined electron concentrations in interplanetary medium, studied solar radio emissions with Soviet-French Stereo-1 experiment, and studied spacial structure, directivity, and mechanism of radiation processes. In Mars orbit spacecraft had investigated Mars solar wind flow and its interaction with Mars ionosphere, recorded spectrum of charged particles and variations in magnetic field, and measured refraction of radiowaves.

Mars 3 descent package had transmitted pictures only briefly after landing on Mars surface Dec. 2, Tass said: "At the computed time, the transmission of video pictures was begun and continued for about 20 sec. During that period a small part of a panorama was transmitted in which no discernible details were detected because of [the lack of] contrast. Now, it is still difficult to say what caused the termination of the transmission. It is possible that it was associated with the local peculiarities in the landing area, which are completely unknown, or with the dust storm raging at the time. Apparently the dust storm acted as a 'veil' covering the surface details during the panoramic exposure. The sudden cessation of signals made it impossible to obtain information from the scientific equipment."

Orbiting vehicle carried infrared radiometer, two TV cameras, photometer, and instruments to measure CO_2 and determine water vapor content, surface radio-brightness temperature, and atmospheric density. (*Sov Rpt*, 12/31/71, 1–14)

December 20: U.S.S.R. launched *Molniya 1-19* comsat from Plesetsk into orbit with 39 146-km (24 324.2-mi) apogee, 493-km (306.3-mi) perigee, 703.3-min period, and 65.4° inclination. Satellite would transmit telephone, telegraph, radio, and TV signals to link far north, Siberia, Central Asia, and Far East. (GSFC Operations Off; *Spacewarn*, 12/28/71, 1; *SBD*, 12/21/71, 230)

- Italy's *San Marco 3* (launched April 24 by NASA) was adjudged successful. Satellite had transmitted excellent data on density, composition, and neutral-particle atmosphere at altitudes of 222–723 km (127.9–449.2 mi). Only failure had been in sun sensors, which were not essential to mission success. *San Marco 3* had reentered Nov. 28 after more than seven months in orbit and had taken good measurements during reentry phase. (NASA Proj Off)

December 21: First complete details on *Mars 2* and *3* instruments were telexed by Soviet scientists to NASA scientists at JPL over direct line installed for exchange of *Mariner 9* and *Mars 2* and *3* data. Message said Soviet Mars spacecraft each had two radiometers, four photometers, one magnetometer, one ion trap, one spectrometer, and device to measure cosmic radiation. Analysis of data from Mars probes had been delayed while Soviet scientists investigated interruption of TV transmission from capsule dropped on Mars by *Mars 3* Dec. 2. Message did not state cause of interruption or whether transmission had resumed. JPL officials later said Soviet message had given location of *Mars 2* spacecraft but Soviet scientists had asked that information not be publicized. (AP, Long Beach, Calif, *Independent-Press Telegram*, 12/22/71)

- Flight version of multiple docking adapter (MDA) for Skylab was transported from Denver to St. Louis, Mo., aboard Super Guppy aircraft for mating with airlock module (AM) being manufactured by McDonnell Douglas Astronautics Co. After mating, all systems would be tested and mated modules would be placed in large altitude chamber where manned and unmanned simulated missions would be conducted. (MSFC Release 71-241)

- Gunman disarmed and held captive four Univ. of Maryland security guards outside chemistry building during alleged attempt to break into laboratories where lunar samples were kept. Later Dr. Cyril A. Ponnamperuma, former ARC scientist, said he believed unsuccesful attempt had been made to steal lunar samples. Dr. Ponnamperuma was analyzing samples at Univ. of Maryland for traces of amino acid, hydrocarbons, or other building blocks of life. (Meyer, *W Post*, 12/22/71, C1)
- Identification of galactic x-ray source GX3+1 with known star by British astronomers using lunar occulation—new technique employing coordinated observations from two sounding rockets—had been reported by U.K. Science Research Council, *New York Times* said. Observations—from Skylark rockets launched from Woomera, Australia, Sept. 27 and Oct. 24, when GX3+1's approximate position was known—had indicated two times during which x-ray source had disappeared behind moon. Data had enabled Royal Greenwich Observatory astronomer L. V. Morrison to pinpoint area in which x-ray source lay by using photographic plate of region that contained star. Attempts to confirm identification would be made in next two months but report conceded possibility that x-ray source could be associated with fainter star not visible on existing plates. (*NYT*, 12/21/71, 22)

December 22: NASA announced flight schedule for 1972. Apollo 16 manned lunar landing mission would be launched in March and Apollo 17 in December. Pioneer-F would be launched in February on two-year mission to Jupiter, and ERTS would be launched into polar orbit during the first half of 1972 to obtain earth surface imagery. Other launches would include Intelsat-IV F–4, F–5, and F–6 comsats for ComSatCorp; HEOS–A2 Highly Eccentric Orbit Satellite for ESRO; MTS Meteoroid Technology Satellite; TD–1 scientific satellite for ESRO; ITOS–C, –D, and –E Improved Tiros Operational meteorological satellites for NOAA; OAO–C Orbiting Astronomical Observatory; IMP–H Interplanetary Monitoring Platform; ESRO–4 scientific satellite for ESRO; Aeros scientific satellite in cooperation with West Germany; SAS–B Small Astronomy Satellite; Telesat-A comsat for Canada; Nimbus-E R&D satellite; and SMS–A Synchronous Meteorological Satellite.

In addition, NASA schedule would include broad range of aeronautical research and cooperative exploration with other countries. Aircraft noise abatement would be high-priority goal, with development of QUESTOL aircraft, construction of noise-reduction laboratory at LaRC, and mating of acoustically treated nacelle to quieter jet engine. Supercritical wing test aircraft would be fitted with side fairings to increase flight efficiency and wing would be fitted to modified F–111 aircraft for tests in 1973. X–24A lifting body would be modified and redesignated X–24B and would begin flight tests in fall 1972. (NASA Feature Release)

- NASA announced it had authorized MSFC to extend for one month the level-of-effort contract with NR Rocketdyne Div. for design of Space Shuttle main engine. Action had been taken pending completion of GAO review of United Aircraft Corp. Pratt & Whitney Div.'s protest of contract award to Rocketdyne [see Aug. 3]. (NASA Release 71–249)

December 23: Intelsat-IV F–2, launched by NASA for ComSatCorp on behalf of INTELSAT Jan. 25, was adjudged successful by NASA. Satellite had been placed in desired synchronous orbit Jan. 27 and was now in service at 24° west longitude over Atlantic. (NASA Proj Off)

- France's *Eole* Cooperative Applications Satellite, launched by NASA Aug. 16, was adjudged successful by NASA. Initial data tape had been received by GSFC from CNES for reduction and analysis of balloon-temperature, pressure-sensor, and wind-velocity information. Preliminary results indicated that balloon sensors, interrogation method, and range-rate system for position-locating balloons had been operating within design limits. Satellite position-location technique had been able to locate fixed balloon electronics package to within 1.2 km (0.8 mi) of its surveyed position. Location of free moving balloons had been within 3 km (1.9 mi). (NASA Proj Off)

- Joint West German and NASA barium-ion-cloud (BIC) probe launched Sept. 20 was adjudged successful by NASA. Barium had been released at planned altitude, latitude, and longitude, and cloud generated had been recorded by special cameras for about 75 min. Cloud elongated to 12 000 km (7500 mi) during observation period and good data were obtained during cloud's deceleration. Striations observed would provide information on plasma instabilities and on electric and magnetic fields in magnetosphere. (NASA Proj Off)

- NASA announced appointment of John W. King, KSC Chief of Public Information, as MSC Public Affairs Officer. Appointment would be effective in late January 1972. King had participated in more than 200 launches; voice of "Jack King" was well known to Apollo launch viewers. (NASA Release 71-250)

- GSA had said it would take bids on towers and equipment from two Saturn launch complexes at KSC, AP reported. Equipment from Complexes 34 and 37, to be sold in one lot, included cranes, boom hoists, 298 km (185 mi) of copper and steel cable, and five towers weighing total of more than 8.2 million kg (9000 tons). NASA had said complexes, which cost $147 million to build, were obsolete. Complex 34 was scene of Jan. 27, 1967, Apollo fire that took the lives of Astronauts Virgil I. Grissom, Roger B. Chaffee, and Edward H. White II. It was also launch site of *Apollo 7* (Oct. 11-22, 1968), first manned mission in Apollo program. Complex 37 had been used for pre-Apollo unmanned missions. *NYT*, 12/24/71, C7; KSC Hist Off)

- Discovery of new planetoid by Soviet astrophysicist N. Chernykh was reported by Moscow publication *Sotsialisticheskaya Industriya*. Planetoid had been described as celestial body of 16th magnitude but had not been identified by permanent number and name. Preliminary calculations of planetoid's orbit had been made by F. Khanina of Soviet Academy of Sciences. Chernykh and M. Diritis of Latvian State Univ. had called planetoid "Riga." (FBIS-Sov-71-231, 12/30/71, L4)

- DOD had decided to postpone production decision on Lockheed Aircraft Corp's Cheyenne attack helicopter while USA flight-tested aircraft against two competitors, *Wall Street Journal* reported. Competitors were United Aircraft Corp. Sikorsky Aircraft Div.'s Blackhawk and Textron, Inc., Bell Helicopter Co.'s helicopter, KingCobra. (Levine, *WSJ*, 12/23/71, 5)

- Chinese Communists were building missile tracking ship that could be used in Pacific for tracking and analyzing test-firing of ICBMs capable of application against European portion of U.S.S.R. and U.S., *Christian*

Science Monitor reported. *Monitor's* sources in Washington, D.C., had said ship could be used to spy on U.S. and Soviet missile work and to expand Chinese satellite program into wider areas. (Ashworth, *CSM*, 12/23/71, 1)

December 25: U.S.S.R. launched *Cosmos 469* from Baykonur into orbit with 262-km (162.8-mi) apogee, 252-km (156.6-mi) perigee, 89.6-min period, and 64.9° inclination. (GSFC *SSR*, 12/31/71; *Sov Aero*, 1/17/71, 15)

- U.S.S.R. and France had successfully completed joint program of sounding rocket launches from Soviet ship *Professor Zubov* off coast of Kourou, French Guiana, Tass announced. Purpose of launches was to study upper atmosphere. (FBIS–Sov–71–251, 12/30/71, L2)
- U.S.S.R. would transfer to new system of world coordinated time at midnight Dec. 31, Tass reported. (FBIS–Sov–71–251, 12/30/71, L1)

December 26: NASA was studying method of obtaining samples of comets by sending spacecraft to rendezvous with comet to fly through comet's head or tail, *Washington Post* reported. OSS Planetary Programs Director Robert S. Kraemer, had said that "best shot of all might be going right through the middle of a comet, to find out what's going on inside the comet." Not all comets were reachable by spacecraft, but some would come close enough to earth in next 20 yrs to make rendezvous feasible, *Post* said. Comet d'Arrest would fly by in 1976 and and 1982, comet Encke in 1983, comet Giacobini-Zinner in 1985. Comet "most astronomers would like to probe" was Halley's, due close to sun in February 1986. (O'Toole, *W Post*, 12/26/71, C5)

- USAF had ordered new flying procedures to reduce structural wear and tear on C–5A cargo aircraft, *New York Times* reported. "Highly unusual program for extending a plane's useful life" was additional to "more routine work of beefing up the wings in areas that show signs of being the least durable." There were indications that C–5A's structural needs could grow beyond usual level for new aircraft. "If so, huge new outlays could be added to the already massive costs that have made the aircraft a focus of debate." C–5A had already overrun initial total cost estimates by $1.2 billion. Latest figure of $4.6 billion was for diminished fleet of 81 instead of 120 aircraft, which were to have cost $3.4 billion. (Witkin, *NYT*, 12/26/71, 46)

December 26–31: AAAS held 138th Annual Meeting in Philadelphia. Activists protesting U.S. policy in Vietnam war tossed tomato at former Vice President Hubert H. Humphrey and interrupted speech of William P. Bundy, former State Dept. Assistant Secretary in charge of East Asian affairs. *Science* reported later that despite "well publicized" incidents, observers agreed level of intensity of disruptions was lower than in recent years. *New York Times* attributed relative mildness of meeting to fact proceedings were not televised, thus providing "less tempting platform for confrontations."

Symposium on "Astronomy from a Space Platform" held Dec. 27–28 included papers by astronomers, space program planners, electronics specialists, optical designers, and researchers. NASA Skylab Program Director William C. Schneider, presented plans for Skylab solar astronomy experiments. Dr. George H. Ludwig, Chief of GSFC Information Processing Div., analyzed tradeoff between earth-based and space-based data processing in astronomy missions. Dr. William F. Hoffman of NASA Institute for Space Studies discussed next-generation infrared space astronomical instruments and discoveries already made in infrared region. Panel chaired by Dr. John E. Naugle, NASA Associate Administrator for Space Science, reviewed symposium on Dec. 28. Panel also included Dr. Karl Henize, astronaut-astronomer. (Program; *Science*, 10/15/71, 11/19/71; 1/7/72; *NYT*, 1/2/72, 3:9; NASA Hist Off)

December 27: U.S.S.R. launched two satellites from Plesetsk. *Cosmos 470* entered orbit with 257-km (159.7-mi) apogee, 190-km (118.1-mi) perigee, 88.9-min period, and 65.4° inclination. It reentered Jan. 6, 1972.

Oreol, joint Soviet-French experiment in Arcade program to study upper atmosphere at high latitudes and determine nature of polar lights, entered orbit with 2477-km (1539.2-mi) apogee, 400-km (248.6-mi) perigee, 114.6-min period, and 73.9° inclination. Satellite carried instruments to study spectrum of protrons and electrons, to measure integral intensity of protons, and to determine ionic composition of atmosphere. (GSFC *SSR*, 12/31/71; 1/31/72; Tass, FBIS–Sov–71–249, 12/28/71, L1; *Sov Aero*, 1/17/72, 15)

- U.S.S.R.'s *Mars 2* and *Mars 3* had discovered presence of oxygen and atomic hydrogen in Mars' upper atmosphere, Tass announced. Hydrogen corona was located at altitude of 10 000–20 000 km (6200–12 400 mi) above surface, and oxygen at 700–1000 km (435–620 mi). Spacecraft also measured temperature changes on Martian surface. Individual areas on diurnal side did not exceed 258 K (−15°C, or 5°F). On nocturnal side, temperature in some regions exceeded that of environment by 20–25 K (20–25°C, or 68–77°F). (FBIS–Sov–71–249, 12/28/71, L2)

- NASA and Soviet Academy of Sciences had agreed on substance and mechanics of future exchanges of medical and biological data on their experience in manned space flight, NASA announced. Agreement's details were contained in recommendations of joint working group on space biology and medicine which met in Moscow Oct. 9–13. Group—cochaired by Dr. Charles A. Berry, NASA Director for Life Sciences, and Dr. N. N. Gurovsky of U.S.S.R. Ministry of Health and Academy of Sciences—recommended that meetings be held at least once annually, with next meeting held in U.S. during May 1972; working sessions be held to consider ways to predict state of human organism during and after space flights, response of certain body systems to space environment, techniques of dysbarism prevention, methods of preflight and postflight medical observations, and methods and results of biological experiments; and cochairmen explore possibility of an exchange of one or two specialists to work in laboratories doing similar biomedical research in space programs of U.S. and U.S.S.R. (NASA Release 71-251)

- NSF interdisciplinary conference at Franklin Institute in Philadelphia celebrated 400th anniversary of birth of Johannes Kepler. Kepler—astronomer, mathematician, and physicist—had devised laws of planetary motions in early 17th century. Laws were "just as valid today as at the time of their announcement," *Science* said. (*Science*, 10/15/71, 325; NSF PIO)

- Rising volume in aerospace industry production was predicted by David A. Loehwing in *Barrons* magazine: "NASA is readying some new programs, notably the space shuttle. . . . Airline revenues are climbing, and some pick-up in orders for transports is anticipated." DOD procurement policies "which nearly threw a few companies into bankruptcy" had been revised. "Most important there is a growing awareness that the Soviet arms buildup poses a real threat to U.S. security, which new weapons must be developed to counter." Persuasive but "cynical" argument for lift in aerospace business in 1972 was political expediency. "Over half-a-million people have been laid off in the industry since 1968, some 200,000 in California," state with largest number of electoral votes. *Barrons* quoted political observer as saying, "Starting pretty early next spring, you're going to see a massive injection of funds into aerospace." (*Barrons*, 12/27/71)

December 28: U.S. Patent No. 3 631 294 was awarded to Steven R. Hofstein, President of Princeton Electronics, Inc., for Lithicon, first electronic storage tube to record on silicon and silicon dioxide. Tube, with target area consisting of more than 10 million silicon storage elements, had been used by JPL to record *Mariner 9* Mars photos, by airlines and Federal agencies to inspect parcels that might contain bombs; by hospitals for brief x-ray and sonar exposures, and by libraries to retrieve microfilmed publications. (Jones, *NYT*, 1/1/72, 25)

- International Institute for Strategic Studies was quoted as saying in London that 30 to 70 new Soviet Sukhoi Su-11 delta-wing advanced fighter aircraft had been delivered to Soviet Air Force in Egypt. Aircraft were believed to be capable of mach 2.5 speed and were armed with air-to-air missiles. (Reuters, *CSM*, 12/28/71, 2)

December 29: U.S.S.R. launched *Meteor 10* from Plesetsk to "obtain meteorological information necessary for an efficient weather forecasting service." Orbital parameters: apogee, 890 km (553 mi); perigee, 876 km (544.3 mi); period, 102.6 min; and inclination, 81.2°. (GSFC SSR, 12/31/71; Tass, FBIS–Sov–71–251, 2/30/71, L1; *Sov Aero* 1/17/72, 15)

- NASA telexed thanks to Soviet scientist Dr. I. A. Zhulin for "very informative transmission" of *Mars 2* and *3* data that included description of instruments [see Dec. 21] and forwarded *Mariner 9* UV spectrometer results. Message, signed by *Mariner 9* Program Manager Dan Schneiderman, continued exchange between JPL and Moscow of information from U.S. and Soviet Mars probes. NASA requested more information on Soviet experiments and spacecraft on "detailed level . . . described in our ICARUS publication previously transmitted to you," and said *Mariner 9* photo of *Mars 3* lander site area was being dispatched by diplomatic pouch. New value for direction of Mars polar axis based on dynamic considerations was included in NASA message to Moscow. NASA noted, "Although this will probably cause a revision of the Mars coordinate system at a later date, we will continue to observe the existing convention in describing the positions of visible surface features." (NASA Release 71–254)

- Failure Review Board appointed to study malfunction of Delta 85 and Delta 86 launch vehicles had completed report, NASA announced. Failure of Delta 86, which had failed to orbit ITOS–B Oct. 21, had apparently been caused by leak in oxidizer system's vent valve, which caused 2nd stage to move from correct attitude. To compensate for disturbance, attitude thrusters using stored nitrogen gas fired continuously to keep rocket properly oriented, depleting nitrogen gas and causing rocket to tumble and crashland after reignition.

Apparent cause of Delta 85 malfunction while orbiting *Oso 7* Sept. 29 had been nitrogen gas leak in 2nd-stage hydraulic reservoir system. Although control of spacecraft had been lost before spacecraft separation, satellite had entered slightly elliptical orbit and had accomplished its mission objectives. (NASA Release 71-253)

- In 21 major tests since April 1970, Safeguard ABM system had had 17 successes, two partial successes, and two failures, *Washington Daily News* reported. USA officials had termed record highly successful for complex weapon system during testing phase. Tests had included intercepts of missiles fired against Safeguard test network and firing of ABMs at predetermined points in space. Most significant test had been firing of long-range Spartan ABM from USA's Meck Island test complex in Pacific. Missile had intercepted incoming enemy warhead after picking it out of group of decoys. Spartan had been designed to knock down incoming warheads outside atmosphere at 322 km (200 mi) from launch site. Superfast Sprint missile was being developed to intercept warheads penetrating atmosphere at 40-km (25-mi) distance. About 35 more major Safeguard tests were planned through mid-1974, including tests of missile's ability to launch and guide large numbers of ABMs simultaneously against clusters of simulated warheads, decoys, electric jamming devices, and other missile penetration aids. (*W News*, 12/29/71)

December 30: Mars orbit of *Mariner 9* was changed by 17-sec engine firing to compensate for effects of dust storm on mission objectives. Periapsis was raised from 1388 km (862 mi) to 1650 km (1025 mi) to expand area covered in mapping photos and to provide required overlapping of pictures. Mapping would be completed in three 20-day cycles with fourth cycle available to fill in open areas where photos did not overlap. Period was increased by 1 min 19 sec to place periapsis in middle of Goldstone viewing period for maximum data return. (NASA Release 71-252; Pasadena *Star-News*, 12/31/71)

- Two new clean rooms for checkout of Apollo Telescope Mount (ATM) in Skylab program were being constructed at KSC. Control of temperature, humidity, and air in rooms would help prevent contamination of delicate lenses and other critical components of ATM experiments. (KSC Release 279-71)

- NASA announced award of $1.6-million, cost-plus-fixed-fee contract to Aerospace Corp. for studies of advanced space programs. Studies, which would advance work performed under previous contract, would include space shuttle mission and payload capture analysis, payload analysis for space shuttle applications, analysis of advanced program operations and logistics analysis, and advanced missions safety. (NASA Release 71-255)

During December: U.S. manned lunar missions *Apollo 14* and *15* were voted fifth most important news story of 1971 in AP poll of member newspapers. Communist China was voted number one newsmaker for admission to U.N., changed diplomatic relations with U.S., and emerging position in world affairs. (W *Star*, 12/27/71, C9)

- Bell Aerospace *Rendezvous* articles described GSFC-devised position-location and aircraft communications experiment (PLACE) to test feasibility of transoceanic air traffic control system that would use earth-orbit satellites. Experiment was designed following earlier mission analyses by ERC. "Scheduled for 1973, the year-long experiment will determine the technical and operational feasibility of using communications satellites to give ground control centers precise and uninterrupted readings and control of air traffic over the North Atlantic 24 hours a day." Experiment would employ two satellites—Ats 5 launched Aug. 12, 1969) and ATS-F (to be launched in 1973)—and three ground stations. PLACE would be able automatically to pinpoint to within one mile [1.6 km] actual positions of up to 250 aircraft, update information once a minute, and predict locations between updates. Operational PLACE system could reduce flight-corridor widths to 48 km (30 mi) and aircraft spacings from 15- to 5-min intervals, permitting same corridors to accommodate 12 times as much traffic without affecting flight safety. In bad weather, PLACE could simultaneously reroute every aircraft within flight corridor. Although PLACE was geared only to North Atlantic air traffic, "it is the forerunner of what could become a worldwide satellite air traffic and surface control network." (*Rendezvous*, Summer/Fall, 71, 9–11; *Aerospace Technology*, 2/26/68, 33)

- Dr. Charles S. Sheldon II, Chief of Science Policy Research Div. of Library of Congress' Congressional Research Service, reviewed Soviet space program in *AIAA Student Journal*. Soviet program's physical size and upward pace "suggests it exceeds in real terms the U.S. program at its previous maximum." In contrast to U.S. program—which had been eroded as programs ended and technical manpower teams broke up—Soviet "application of resources to a broad program of space flight has continued fairly steadily upward toward an ultimate level which has not yet been publicly defined."

  Soviet program began its space flights with adaptation of its first ICBM, SS-6 Sapwood. "This vehicle, with improved upper staging, is still the mainstay of the present program while our corresponding, but very small, Redstone and Vanguard first stages have long since disappeared into history."

  Largest single element in total U.S.S.R. program was Cosmos reconnnaisance satellite, which stayed in low-circular orbit a few days and was then recovered and which represented "rather passive military support flights." Other elements with direct weapon implications were special uses of SS-9 Scarp ballistic missile carrier with space versions F-1-r carrying fractional orbital bombardment system (FOBS) and F-1-m carrying highly maneuverable payloads. Some payloads launched by F-1-m "seem almost certainly aimed at developing the techniques of rendezvous with uncooperative spacecraft to inspect them and, if need be, destroy them. In a series of flights . . . such inspectors have made close passes on other Soviet payloads and then have themselves been blown into clouds of debris. At least one, instead of being destroyed . . . was deliberately redirected to plunge into the atmosphere and ocean." (*AIAA Stu J*, 12/71, 14–28)

- Safety problems of space shuttle were discussed by Director I. Irving Pinkel of LERC Aerospace Safety Research and Data Institute, in *Astronautics & Aeronautics* article. "The rocket engine must have an airplane-like endurance of 10 years (100 missions). The airplane systems must operate where rockets do, subject to space vacuum, space radiation, and reentry heating, and still carry a fan engine with high cycle temperature." (A&A, 12/71, 28–35)

- Attributes of Q-fan engine for STOL propulsion were described in *Aeronautics & Astronautics* article by George Rosen, United Aircraft Corp. Hamilton Standard Div. engineer: "Q–Fan, with its lightly loaded rotor, operating at low tip speeds and incorporating controllable-pitch blading, is now essentially available as a new propulsor uniquely matched to the stringent performance, noise, and operational requirements of the advanced STOL transport. It offers the aircraft designer flexibility in meeting the specific requirements of a given aircraft with a limited number of available core engines. By varying the size of the Q–Fan, the bypass ratios available . . . could range from 15:1 to 30:1, or more if needed." (A&A, 12/71, 50–5)

- Slackening of U.S. R&D was scored by Eberhardt Rechtin, Principal Deputy Director of Defense Research and Engineering in DOD, in *Astronautics & Aeronautics* article: "The image of world technical leadership earned by our electronics and aerospace industries in the 1960s will predictably be eroded in the 1970s by our own lack of planned 'firsts' in contrast with the continuing achievements of the Soviet Union, France, and probably Germany and Japan." Major source of future wealth "must be the special skills of our population—yet we are presently pricing ourselves out of the world markets, including our own. We are also disseminating our science, our inventions, our management expertise, and our manufacturing technologies around the world in a way which, though it raises the standards of living elsewhere, also diminishes the competitive advantages of our own people." While U.S. had been world leader in R&D less than 25 yrs, "you hear discussions about reordering national priorities as if we could make unilateral decisions, independent of the international competition —as if international interactions were not a significant, much less a driving, factor—concerning defense R&D, the NASA space program, tax incentives to industry, or a wage/price freeze." (A&A, 12/71, 22–7)

- *Goals and Means in the Conquest of Space* by Soviet author R. G. Perel'man (NASA Technical Translation F–595) was reviewed in *Aerospace Historian* by Clarke G. Reynolds of Univ. of Maine. Reynolds quoted Perel'man conclusion: after "conquering" Solar System, Communist society "will take a further bold step, i.e., to the stars of our galaxy, and then other galaxies; to visit and study the planetary systems and civilizations of other worlds in order to bring other islands of intelligence into the system of the grand circle that is characteristic of the space age in the evolution of intelligent life." (*Aerospace Historian*, 12/71, 216–7)

During Winter: Only "small fraction" of defense, space, and atomic energy funding during 1950s and 1960s had been spent on science and technology *per se*, Gabor Strasser, former technical assistant to Presidential Science Adviser Edward E. David, Jr., said in *Science Policy Reviews*. DOD, NASA, and AEC funding had accounted for "about two-thirds of our entire Federal budget" but bulk of "these huge expenditures" had gone for "such 'unscientific' things" as plants, materials, labor, tooling, and production. With approach of mid-1970s "emphasis is shifting . . . toward alleviation of domestic social and environmental problems, or . . . improvement of our 'quality of life.' These goals are not so science-and-technology sensitive as going to the moon, as MIRV's or as phased-array radars have been." Major change affecting current scientific and technical activities was "approach required to tackle today's emerging problems." Advancement of state of the art was less important than "integration of existing science and technology with what is already known and yet to be found out in sociology, economics, politics, management and institutional arrangements." (Battelle *Sci Pol Rev*, Fourth Quarter 71, 3–12)

- NASA published *Mariner-Venus 1967, Final Project Report* (NASA SP-190). In foreword, Project Manager Glenn A. Reiff said *Mariner 5* (launched June 14, 1967, toward Oct. 19, 1967, flyby of planet Venus), apart from significant contributions to knowledge of deep space, had achieved secondary objective, "to gain engineering experience by converting a spare Mariner-Mars 1964 spacecraft into one that could be flown to Venus," with substantial savings. Cost of mission had been 10% less than anticipated. (Text)

Summary

During 1971: U.S. orbited 51 spacecraft in 29 launches—including *Apollo 14* and *15* CSMs *Kitty Hawk* and *Endeavor* with LMs *Antares* and *Falcon* and their S–IVB stages intentionally crashed onto moon. U.S.S.R. orbited 97 payloads in 83 launches; Japan, 2; U.K., 1; France, 1; Italy, 2 (1 for U.S. on U.S. booster); and Communist China, 1. U.S. total included 33 orbited by DOD in 17 launches and 19 by NASA in 13 launches. NASA's total included 1 launched for USAF on behalf of NATO, 1 for France, and 1 for U.K.

NASA's manned space flight program continued with launches of *Apollo 14* and *Apollo 15* manned lunar landing missions. Mobile equipment transporter, used for first time on *Apollo 14* to carry equipment and samples, allowed astronauts to explore 3.5 km (2.2 mi) of lunar surface and collect 43 kg (94.8 lbs) of samples. On *Apollo 15*, astronauts traveled 28 km (17.4 mi) over lunar surface on four-wheeled, battery-powered lunar roving vehicle (LRV) Rover and collected 77 kg (170 lbs) of lunar samples, including one white anorthosite rock about 4 billion yrs old.

Unmanned program was highlighted by *Mariner 9* Mars probe, which traveled 400 million km (248 million mi) to Mars and began transmitting photos of major Mars dust storm and, later, of details of Martian surface and moons Deimos and Phobos. Mariner 8 launch vehicle malfunctioned at launch. *Oso 7* Orbiting Solar Observatory—rescued from initially incorrect orbit by ground command—made first x-ray observations of beginning solar flare and revealed that sun had "polar caps." Other unmanned missions included *Intelsat-IV F–2* and *Intelsat-IV F–3* for ComSatCorp and *Explorer 43, Explorer 44,* and *Explorer 45* (the last launched for NASA by Italy). Launched as secondary payloads were *Subsatellite* on *Apollo 15* and *Tetr 3* on *Oso 7*. All elements of Skylab flight system reached final manufacturing and checkout in preparation for 1973 launch and intensive engineering studies on Space Shuttle were conducted throughout year.

In joint NASA–USAF lifting-body program, M2–F3 and X–24A lifting bodies made 17 flights, including first supersonic flight by M2–F3. X–24A reached 1687 km per hr (1048 mph) before completing program and beginning modification for new flight program as X–24B.

In aeronautics NASA TF–8A jet aircraft with supercritical wing made 27 successful flights. USN F–14 fighter aircraft successfully demonstrated first automatically programmable variable sweep wing. USA CH–54B helicopter set new world altitude record of 5120 m (16 798 ft) while carrying 10 000-kg (22 000-lb) payload. New V/STOL Wind Tunnel at LARC became operational, providing greatly improved testing capability for scaled models. GE quiet experimental engine successfully completed initial tests in NASA program to reduce jet engine noise, and NASA let competitive design contracts for development of quiet, experimental, short takeoff and landing (QUESTOL) transport

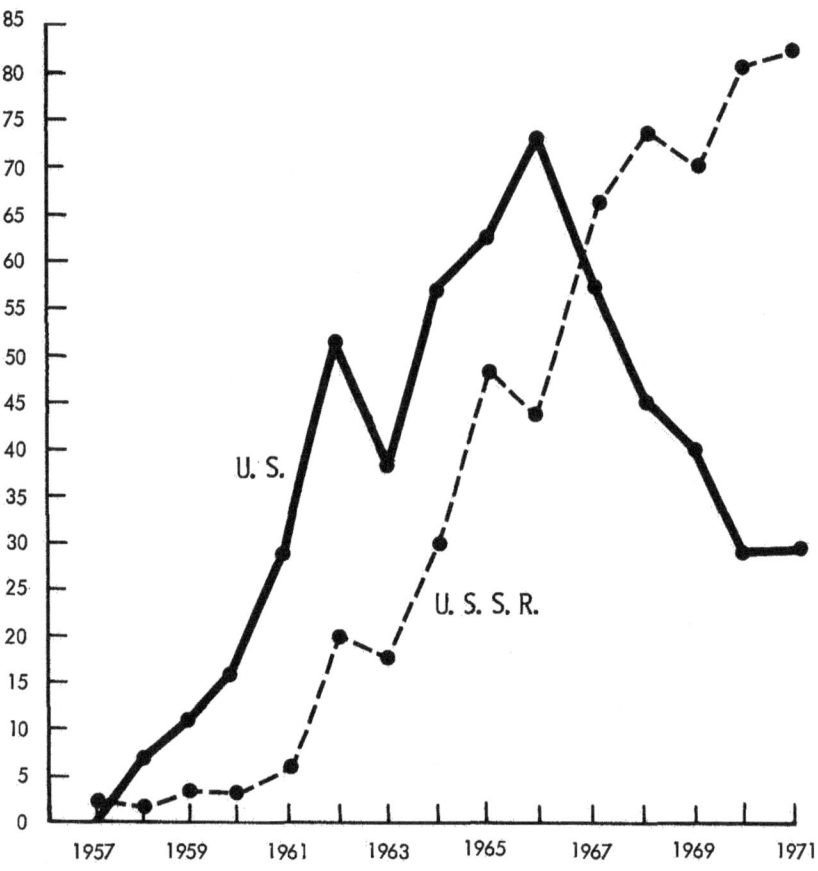

U.S.–U.S.S.R. Space Activity, 1957–1971
Successful Launches to Earth Orbit or Beyond

aircraft. SST program ended, with Congress voting to terminate program.

In sounding rocket program, 126 successful flights were made to study atmosphere, ionosphere, auroras and airglow, geomagnetic storms, meteor streams, and trapped radiation fluctuations and to make astronomical observations in x-ray, UV, and radio regions of electromagnetic spectrum.

DOD's 33 payloads included orbiting of 2 Defense Satellite Communications System (DSCS–2) military comsats, 2 OV–I scientific satellites, and *Cannonball 2* and *Musketball.*

U.S.S.R.'s 83 launches set an all-time record in number of flights. The 97 payloads included 81 Cosmos satellites, 2 Luna, 2 Soyuz, 1 Salyut, 2 Mars, 1 Intercosmos, 4 Meteor, 2 Molniya I, 1 Molniya II, and *Oreol* (launched for France). *Soyuz 11,* carrying three-man crew,

Summary of Soviet and United States Space Payloads by Mission Category

	Soviet Union					1957–1971 total	United States					1957–1971 total
	1967	1968	1969	1970	1971		1967	1968	1969	1970	1971	
Earth orbital science	5	10	3	9	7	59	12	16	14	4	11	137
Earth orbital engineering							6	1	5	1	10	50
Communications	4	4	2	5	3	23	19	11	6	6	6	83
Weather	4	2	2	5	4	22	6	4	3	5	4	49
Navigation/ferret	4	6	6	16	27	84	3	1		1		25
Geodesy							1	1	1	1		17
Military observation:												
Low orbit recoverable	22	29	32	29	28	202	19	16	12	9	7	198
Low orbit nonrecover	8	10	12	12	12	67	7	7	11	4	6	77
Intermediate orbit						4						10
Synchronous or higher							2	1	3	5	1	18
Fractional orbit bombard	9	2	1	2	1	17						
Military inspector/destruct	2	5	2	4	8	25						
Earth orbit man-related	3	4		1	2	20	1		1			11
Earth orbit manned	1	1	5	1	3	19						14
Lunar man-related	2	3	1	2		8	1	2		1	3	13
Lunar manned								2	8	2	4	16
Moon-unmanned programs		1	3	3	2	25	8	1				21
Venus	2		2	2		17	1					2
Mars					5	10			2		1	5
Interplanetary							1	1				5
Vehicle tests												13
Subtotal	66	77	71	91	102	602	87	64	66	39	53	764
Orbital launch platforms	6	8	6	10	7	59						
Total	72	85	77	101	109	661	87	64	66	39	53	764

From Senate Committee on Aeronautical and Space Sciences Staff Report, *Soviet Space Programs, 1971*, 92d Cong, 2d Sess, April 1972, p. 5.

successfully rendezvoused and docked with unmanned *Salyut 1* to form first orbital scientific station, but cosmonauts died during earth reentry because of improperly sealed hatch. *Mars 2* and *Mars 3* dropped instrumented capsules on Mars surface. *Mars 2* capsule crashlanded and *Mars 3* capsule transmitted 20 sec. *Lunokhod 1* lunar rover carried to moon on board *Luna 17* Nov. 17, 1970, continued to explore lunar surface until October 1971, when isotope heat reserves were exhausted during its 11th lunar night.

Japan launched two satellites—*Tansei* and *Shinsei*. China launched *Chicom 2*, France launched *Tournesol*, U.K. launched *Prospero*, and Italy launched *San Marco 3*. (NASA Release 71-245; NASA Lifting Body Off; GSFC Sounding Rocket Br; NASA Exec Secretariat; Sheldon, *United States and Soviet Progress in Space*, Library of Cong, 1/22/72)

- NASA's scientific and societal achievements advanced national goal of alleviating environmental problems and expanding basic research on behalf of humanity, as public pressed for immediate, usable returns from research in all fields. Apollo manned missions swelled storehouse of lunar information. Unmanned astronomical missions gathered unique data, including new clues to Martian atmosphere. NASA-developed inventions saw increased use beyond space exploration—in environmental sciences, communications, medicine, industry, and agriculture. U.S.–U.S.S.R. space meetings set stage for future joint space missions. And balanced program of space and aeronautics, despite low funding level, continued to raise U.S. prestige as aerospace pioneer while it worked to improve man's future. Some observers, however, warned of higher rate of technological advances abroad than in U.S. and of ever-growing U.S.S.R. space program and technological capability. Soviet expenditures on space were estimated at level about double U.S. peak expenditure of 1% of GNP when combined NASA–DOD space program was at highest level and four times present level.

Dr. James C. Fletcher was nominated new NASA Administrator by President Nixon. President of Univ. of Utah and former aerospace executive, Dr. Fletcher took office April 27. He succeeded NASA's third Administrator, Dr. Thomas O. Paine, who had resigned Sept. 15, 1970, to return to industry. Dr. George M. Low, Deputy Administrator, had served as Acting NASA Administrator in the interim.

Operating at lowest budget level since 1962, NASA in 1971 continued phase-out of Apollo program with successful completion of *Apollo 14* and *15*. Man's fourth landing on moon, in *Apollo 15* mission, proved to many that man had an effective scientific role in space. At year's end, Apollo 16 had been scheduled for March 1972, with Apollo 17 to complete program later in year.

Reduced funding forced additional reduction of personnel as well as program cutbacks, stretching out remaining Apollo missions, delaying first Skylab flight, and slowing studies of space shuttle. NERVA nuclear rocket engine program was reduced to holding action. Bioscience funding was eliminated. Aerospace employment in industry throughout nation also fell, reflecting cuts in DOD spending as well as in NASA's.

During year, however, Skylab hardware manufacture neared completion and space shuttle work continued. Shuttle preliminary design contracts were extended so that NASA could examine alternative configurations, contingent on Administration approval of shuttle development. In ERTS program, aircraft tested remote sensing devices for gathering earth resources data and first satellite for global monitoring of resources was scheduled for 1972. During 1971, instrumented aircraft measured air pollution, investigated corn blight infestation, and sought hidden natural resources. Pioneer-F spacecraft for planned 1972 launch and December 1973 Jupiter flyby went into final manufacturer's tests. Increased emphasis on aeronautics included research into aircraft noise reduction, increased air transport efficiency with use of supercritical wing, and STOL aircraft for transportation.

To emphasize practical goals of space science, NASA established new Office of Applications, increased technology utilization budget from $1 million to $5 million, and issued request for proposals for additional uses of space-developed technology to solve problems of society. Among NASA projects were NSF contract to investigate use of solar cells to electrify homes and buildings and HUD contract to study application of space technology to sewage recycling. Joint NASA–DOT study resulted in recommendations for civil aviation R&D policy (CARD).

Research achievements at NASA centers included ARC discovery of amino acids—building blocks of life—in second meteorite. Discovery strengthened case for chemical evolution of life elsewhere in universe.

NASA FY 1972 appropriation of $3.298 billion was $27 million more than requested $3.271 billion and $29 million above FY 1971 appropriation of $3.269 billion—halting downward trend but far below 1965 peak of $5.250 billion. (*A&A 1971*; O'Toole, *W Post*, 1/2/72, A1; Sheldon, *Soviet Space Programs, 1966–70*, Senate Doc 92–51)

- In international cooperation program, NASA successfully launched NATO's *Natosat 2* military comsat for USAF and NATO, *Eole* Cooperative Applications Satellite for France, *Isis 2* cooperative U.S.–Canada ionospheric satellite, and *Ariel 4* ionospheric satellite for U.K. Italian crew trained by NASA launched NASA's *Explorer 45* Small Scientific Satellite and Italy's *San Marco 3* from Italian platform off coast of Kenya.

Joint NASA and Soviet Academy of Sciences meetings on compatible spacecraft rendezvous and docking system progressed to discussion of possible joint test missions. Data obtained by NASA's *Mariner 9* and Soviet *Mars 2* and *3* probes were exchanged under agreement that covered broad range of U.S.–U.S.S.R. cooperation in space science and applications. Agreement on future exchanges of medical and biological manned space flight data called for meetings between U.S. and Soviet experts at least annually.

U.S. and U.S.S.R. exchanged *Apollo 11* and *12* and Soviet *Luna 16* lunar samples and Soviet scientists were among 750 experts attending Second Annual Lunar Science Conference at Lunar Science Institute in Houston. First U.S.–U.S.S.R. Conference on Communications with Extraterrestrial Intelligence was held in Soviet Armenia.

Joint meeting of NASA and European Space Conference (ESC) identified candidate areas for possible European participation in U.S. post-Apollo space programs, including design and development of space shuttle subsystems.

International Workshop on Earth Resources Survey at Univ. of Michigan was cosponsored by NASA, Dept. of State, USN, and AID to inform representatives of 51 countries and international organizations of latest techniques for interpreting earth resources data acquired by aircraft and satellite remote sensing systems.

NASA's C-130 aircraft equipped with remote-sensing devices participated in U.S. survey of Jamaica's natural resources undertaken at request of U.N. and Jamaican government. NASA signed agreement with West Germany for series of barium-ion-cloud (BIC) probe launches, and new sounding rocket agreement was signed with France. Under sounding rocket agreement, 20 French Centaure and NASA Nike-Cajun rockets would be launched from Kourou Center, French Guiana, in cooperative France-U.S. upper-atmosphere research project.

By end of year NASA had entered into cooperative agreements with 39 countries, including agreements on lunar samples, tracking and data acquisition, and geodetic satellite program. Scientists of 87 countries besides U.S. were participating in cooperative ground-based programs and personnel exchanges and 62 countries were exchanging scientific and technical information. Since the beginning of 1962, NASA had launched 21 satellites under international agreements, in addition to flying 21 international experiments on NASA satellites. Since the beginning of the sounding rocket program, 683 cooperative rockets had been launched for research. (NASA Release 71-245; NASA Int Aff; "Selected Statistics on Int'l Affairs," 2/17/72; *A&A, 1971*)

- Aerospace industry sales continued anticipated decline in 1971, to $23.3 billion from $24.9 billion—decrease of 6.3%. Commercial aerospace sales, primarily jet transports, increased 7.1% in 1971, from $4.578 in 1970 to $4.903 billion. Increase reflected increased deliveries of wide-bodied jet transports.

  Space sales continued decline, to $3.220 billion from $3.580 billion in 1970. Aerospace industry remained nation's largest manufacturing employer during 1971, despite continuing drop in employment from 1 069 000 workers in December 1970 to estimated 931 000 in December 1971. (*Aerospace*, 12/72, 3–5)

- Number of aircraft landing facilities in U.S. passed 12 000 for first time, with 12 070 airports, heliports, and seaplane bases in operation by Dec. 31 despite abandonment of 354. Figure at end of 1970 had been 261. Texas kept lead with 1128 landing facilities, Alaska was second with 762, and California was third with 746. Of national total, 4418 facilities were classified as publicly owned and 7652 as privately owned. (FAA Release 72-50)

Appendix A

SATELLITES, SPACE PROBES, AND MANNED SPACE FLIGHTS

A CHRONICLE FOR 1971

The following tabulation was compiled from open sources by Leonard C. Bruno of the Science and Technology Division of the Library of Congress. Sources included the United Nations Public Registry; the *Satellite Situation Report* issued by the Operations Control Center at Goddard Space Flight Center; public information releases of the Department of Defense, NASA, NOAA, and other agencies, as well as those of the Communications Satellite Corporation. Russian data are from the U.N. Public Registry, the *Satellite Situation Report*, translations from the Tass News Agency, statements in the Soviet press, and international news services reports. Data on satellites of other foreign nations are from the U.N. Public Registry, the *Satellite Situation Report*, governmental announcements, and international news services reports.

This tabulation lists payloads that have (a) orbited; (b) as probes, ascended to at least the 4000-mile altitude that traditionally has distinguished probes from sounding rockets, etc.; or (c) conveyed one or more human beings into space, whether orbit was attained or not. Furthermore, only flights that have succeeded—or at least can be shown by tracking data to have fulfilled our definition of satellite or probe or manned flight—are listed. Date of launch is referenced to local time at the launch site. An asterisk by the date marks dates that are one day earlier in this tabulation than in listings which are referenced to Greenwich Mean Time. A double asterisk by the date marks dates of Soviet launches which are a day later in this compilation than in listings which are referenced to Greenwich Mean Time.

World space activity increased for the second straight year. There was an increase in the total succesful launches—120 against 114 in 1970—and a marked increase in total payloads orbited—151 against 130 in 1970. The difference between launches and payloads is of course accounted for by the multiple-payload launches (DOD, the principal user of this system in the past, made 5 multiple launches in 1971, orbiting 21 payloads; NASA made only 2 multiple launches, totaling 4 payloads; the U.S.S.R. made 2 multiple launches, orbiting 16 payloads).

Of the 1971 world total, the United States launched 29 boosters carrying 47 payloads—up from 28 and 36 in 1970 and marking the first U.S. increase of both categories in one year since 1966. Of the 1971 totals, DOD was responsible for 17 launches and 33 payloads. Five of NASA's 12 launches were non-NASA missions—*Intelsat-IV F–2 and Intelsat-IV F–3* for ComSat-

Corp, *Natosat 2* for the USAF and NATO, *Eole* Cooperative Applications Satellite for France, and *Ariel 4* ionospheric research satellite for the U.K. The Soviet Union again dominated world totals, launching 97 payloads (a high for the U.S.S.R.) with a record 83 launches. It had launched 88 payloads with 81 launches in 1970.

The year was marked by success and tragedy. The U.S. accomplished two spectacular manned Apollo lunar flights, but the Soviet record flight of the world's first manned earth orbiting laboratory returned to earth with its crew dead. *Apollo 14* made the third successful U.S. manned lunar landing and returned with 43 kg (94.8 lbs) of lunar samples. *Apollo 15* returned with 77 kg (170 lbs) of lunar material and employed the lunar roving vehicle for the first time on the moon. The U.S. also successfully photographed Mars and its moons with *Mariner 9* and made the first x-ray observations from a spacecraft of a beginning solar flare and of solar streamers with *Oso 7*.

Early in the year, the Soviet Union orbited *Salyut 1*, a large unmanned station. The three-man crew of *Soyuz 10* docked with the station but encountered some difficulties and returned quickly to earth. More than a month later the three-man crew of *Soyuz 11* docked with *Salyut 1*, transferred to it, and conducted experiments from the linked spacecraft before undocking and returning in the *Soyuz 11*. Although the *Soyuz 11* mission set a new world endurance record for the longest manned space flight—nearly 24 days—the three Soviet cosmonauts died just before reentry when an imperfect hatch seal vented their air supply. *Mars 2* orbited the planet Mars and placed the first man-made object—a capsule bearing the Soviet coat-of-arms—on the surface. A capsule ejected five days later from *Mars 3* sent back video signals for 20 seconds from the planet's surface before shutting down. *Luna 18* orbited the moon but crashed on its landing attempt. *Luna 19* went into lunar orbit and remained there to conduct research and take photographs.

Also during 1971, the United Kingdom became the sixth nation to orbit a satellite (*Prospero*) with its own launch vehicle. Japan launched two of its own satellites (*Tansei* and *Shinsei*), and the People's Republic of China launched its second satellite (*Mao 2*). In addition to launching its own satellite (*Tournesol*), France participated in a cooperative launch with the Soviet Union, which orbited *Oreol*. Italy launched NASA's *Explorer 45* and its own *San Marco 3* on NASA vehicles from its facility off the coast of Kenya.

As we have cautioned in previous years, the "Remarks" column of these appendixes is never complete, because of the inescapable lag behind each flight of the analysis and interpretation of results.

Launch Date	Name, Country, International Designation, Vehicle	Payload Data	Apogee in Kilometers (and st mi)	Perigee in Kilometers (and st mi)	Period in Minutes	Inclination in Degrees	Remarks
Jan. 12	Cosmos 390 (U.S.S.R.) 1971-1A Not available	Total weight: Not available. Objective: "Continuation of Cosmos scientific satellite series." Payload: Not available.	270 (167.8)	202 (125.5)	89.2	65.0	Reentered 1/25/71.
Jan. 14	Cosmos 391 (U.S.S.R.) 1971-2A Not available	Total weight: Not available. Objective: "Continuation of Cosmos scientific satellite series." Payload: Not available.	796 (494.6)	266 (165.3)	95.2	70.9	Reentered 2/21/72.
Jan. 20	Meteor 7 (U.S.S.R.) 1971-3A Not available	Total weight: Not available. Objective: Photograph cloud cover and snow and study atmospheric thermal energy radiated by earth. Payload: Cylindrical body with 2 large solar paddles attached; 3 data-collection systems, TV cameras, infrared sensors, and actinometric scanners; 3-axis attitude-control system.	655 (407)	629 (390.8)	97.5	81.2	Meteor 7 meteorological satellite still in orbit.
Jan. 21	Cosmos 392 (U.S.S.R.) 1971-4A Not available	Total weight: Not available. Objective: "Continuation of Cosmos scientific satellite series." Payload: Not available.	276 (171.5)	203 (126.1)	89.2	64.9	Reentered 2/2/71.
Jan. 21	DOD Spacecraft (United States) 1971-5A Titan IIIB-Agena	Total weight: Not available. Objective: Develop space flight techniques and technology. Payload: Not available.	392.7 (244)	130.4 (81)	89.2	110.8	Reentered 2/9/71.
Jan. 25*	Intelsat-IV F-2 (United States) 1971-6A Atlas-Centaur	Total weight: 1397 kg (3080 lbs) at launch; 718.5 kg (1584 lbs) after apogee motor fire. Objective: Place satellite and apogee motor into proper transfer orbit; provide tracking and telemetry and backup calculations through transfer orbit so satellite can be injected into synchronous orbit for commercial communications.	36 410.4 (22 624.4)	35 740.0 (22 207.8)	1450	0.59	Launched by NASA for ComSatCorp into good transfer orbit. Apogee motor fired 1/26, stationing satellite in synchronous orbit over Atlantic Ocean at 24°W longitude to service North America, South America, and Western Europe. First satellite in improved Intelsat IV series and first launched by Atlas-Centaur. Largest commercial communications satel-

373

ASTRONAUTICS AND AERONAUTICS, 1971

Launch Date	Name, Country, International Designation, Vehicle	Payload Data	Apogee in Kilometers (and st mi)	Perigee in Kilometers (and st mi)	Period in Minutes	Inclination in Degrees	Remarks
		Payload: 5.4 m × 237.5-cm-dia (17.7 ft × 93.5-in-dia) cylindrical satellite capable of carrying 3000–9000 telephone circuits simultaneously or 12 color TV channels or combinations; spin-stabilized; 12 communications repeaters (transponders); 6 antennas (2 transmit horns, 2 receive horns, and 2 steerable 127 cm [50 in] dish spot-beam antennas); 42 240 solar cells.					lite ever launched, both in weight and communications capacity. Conducted first transatlantic picture telephone call 12/3/71 between U.S. and Sweden. Still in orbit.
Jan. 26	Cosmos 393 (U.S.S.R.) 1971–7A Not available	Total weight: Not available. Objective: "Continuation of Cosmos scientific satellite series." Payload: Not available.	417 (259.1)	281 (174.6)	91.3	70.9	Reentered 6/16/71.
Jan. 31	Apollo 14 (United States) 1971–8A Saturn V	Total weight: 137 268.8 kg (302 626 lbs) at earth orbit insertion, including S-IVB stage, instrument unit, spacecraft LM adaptor, LM, and CSM. Objective: Perform selenological inspection, survey, and sampling of materials in a preselected region of Fra Mauro formation; deploy and activate Apollo lunar surface experiments package (ALSEP); develop man's capability to work in lunar environment; obtain photographs of candidate exploration sites. Payload: 34.8-m (114-ft)-long S-IVB/IU/spacecraft LM adaptor/Block II command and service modules/lunar module.	188.9 (117.4) Lunar orbit, 314.1 (195.2)	183.1 (113.8) 108.2 (67.2)			Apollo 14, 3rd successful manned lunar landing mission, carried Astronauts Alan B. Shepard, Jr., Stuart A. Roosa, and Edgar D. Mitchell. After translunar injection and CSM separation from S-IVB, difficulties were encountered with CSM–LM docking. After 5 unsuccessful attempts, docking was achieved on 6th try and LM extracted from S-IVB 1 hr 54 min later than planned. Two midcourse corrections were performed during translunar coast. Spacecraft entered lunar orbit 2/4, LM undocked, initiated descent, and touched down at planned Fra Mauro site at 4:17 am EST 2/5. Crew performed 2 EVAs. During 1st EVA on 2/5 (4 hrs 49 min), equipment and experiments were deployed and about 19.5 kg (43 lbs) of lunar samples collected. During 2nd EVA on 2/6 (4 hrs 35 min), crew employed mobile equipment transporter (MET) and proceeded toward

374

Date	Name	Description	Weight kg (lb)	Perigee/Apogee	Period (min)	Inclination	Remarks
							Cone Crater, obtaining samples, photographs, and terrain descriptions. Crew nearly reached Cone Crater rim but steep slopes and late timeline necessitated return. Total astronaut EVA time, 9 hrs 24 min. Total weight of lunar rocks collected, 43 kg (94.8 lbs). LM liftoff from moon at 142:25 GMT 2/6. Total lunar stay time 33 hrs 31 min. After CSM–LM docking and crew transfer, LM ascent stage deorbited and impacted near landing site. Impact was recorded by *Apollo 12* ALSEP and *Apollo 14* ALSEP. Transearth injection initiated and one midcourse correction required during coast. CM separated from SM and splashed down in mid-Pacific 4:05 pm EST 2/9. Total flight time 9 days 2 min.
Feb. 2*	*Natosat 2* (U.S.–NATO) 1971–9A DSV-3M	Total weight: 242.7 kg (535 lbs). Objective: Place spacecraft into transfer orbit of sufficient accuracy to allow onboard propulsion systems to place spacecraft into acceptable synchronous orbit which does not impose unacceptable degradation to spacecraft performance or expected orbital life. Payload: 81.3 cm x 137.2-cm-dia (32 in x 54-in-dia) cylindrical satellite constructed of 2 concentric cylinders with apogee motor within inner cylinder; solar cells mounted on outside surface of outer cylinder; despun antenna system and redundant X-band communications systems with power output of 3.5 w.		37 712.2 (23 433.3) After apogee motor firing, 37 053.0 (23 023.7)	272.9 (169.6) 34 493.6 (21 433.3)	673.9	25.9	NATO military communications satellite launched into good transfer orbit by NASA long-tank thrust-augmented Thor-Delta for USAF, which acted as agent for NATO. Apogee-kick motor fired 2/4 to place spacecraft in synchronous orbit at 26°W longitude. Still in orbit.
Feb. 9	*Cosmos 394* (U.S.S.R.) 1971–10A Not available	Total weight: Not available. Objective: "Continuation of Cosmos scientific satellite series." Payload: Not available.		613 (380.9)	572 (355.4)	96.5	65.8	*Cosmos 394* was employed as target for maneuverable interceptor satellite *Cosmos 397* (launched 2/25/71), which made inspection during fast pass. Still in orbit.
Feb. 16	*Tansei* (Japan) 1971–11A Mu-4S	Total weight: 63 kg (139 lbs). Objective: Conduct engineering test for launching of scientific satellites. Payload: 26-sided body measuring 71 cm (28 in) in dia; onboard data recorder and radio; chemical batteries with 1-wk lifetime.		1110 (689.7)	990 (615.2)	106	29.7	Second Japanese orbital success. While test satellite did not achieve planned elliptical orbit, final near-circular orbit was satisfactory. Still in orbit.

Launch Date	Name, Country, International Designation, Vehicle	Payload Data	Apogee in Kilometers (and st mi)	Perigee in Kilometers (and st mi)	Period in Minutes	Inclination in Degrees	Remarks
Feb. 17	DOD Spacecraft (United States) 1971–12A Thor-Burner II	Total weight: Not available. Objective: Develop space flight techniques and technology. Payload: Not available.	832 (517)	767.7 (477)	100.8	98.8	Four spacecraft launched with single booster. Still in orbit.
	and						
	Calsphere 3 1971–12C	Total weight: 9 kg (2 lbs). Objective: Provide target for radar calibration and evaluate surface material erosion and drag effects vis-a-vis inert gold surface. Payload: 25-cm (10-in) sphere with gold surface.	834 (518.2)	764 (474.7)	100.8	98.8	Still in orbit.
	and						
	Calsphere 4 1971–12D	Total weight: 9 kg (2 lbs). Objective: Provide target for radar calibration and evaluate surface material erosion and drag effects vis-a-vis inert gold surface. Payload: 25-cm (10-in) sphere with gold surface.	833 (517.6)	762 (473.5)	100.8	98.8	Still in orbit.
	and						
	Calsphere 5 1971–12E	Total weight: 9 kg (2 lbs). Objective: Provide target for radar calibration and evaluate surface material erosion and drag effects vis-a-vis inert aluminum surface. Payload: 25-cm (10-in) sphere with aluminum surface.	833 (517.6)	772 (479.7)	100.8	98.8	Still in orbit.
Feb. 17	Cosmos 395 (U.S.S.R.) 1971–13A Not available	Total weight: Not available. Objective: "Continuation of Cosmos scientific satellite series." Payload: Not available.	545 (338.7)	529 (328.7)	95.3	74.0	Still in orbit.
Feb. 18	Cosmos 396 (U.S.S.R.) 1971–14A Not available	Total weight: Not available. Objective: "Continuation of Cosmos scientific satellite series." Payload: Not available.	268 (166.5)	189 (117.4)	89.0	65.4	Reentered 3/3/71.

Date	Name/Designation	Details	Col4	Col5	Col6	Remarks
Feb. 25	Cosmos 397 (U.S.S.R.) 1971-15A Not available	Total weight: Not available. Objective: "Continuation of Cosmos scientific satellite series." Payload: Not available.	2241 (1392.5)	584 (362.9)	113.8	65.8	Cosmos 397 maneuvered into orbital path of Cosmos 394 (launched 2/9/71), intercepted and inspected satellite on fast pass, and exploded into 77 objects. Still in orbit.
Feb. 26	Cosmos 398 (U.S.S.R.) 1971-16A Not available	Total weight: Not available. Objective: "Continuation of Cosmos scientific satellite series." Payload: Not available.	232 (144.2)	191 (118.7)	89.0	51.6	Still in orbit.
Mar. 3	Cosmos 399 (U.S.S.R.) 1971-17A Not available	Total weight: Not available. Objective: "Continuation of Cosmos scientific satellite series." Payload: Not available.	438 (272.2)	196 (121.8)	90.8	64.9	Reentered 3/17/71.
Mar. 3	Mao 2 (People's Republic of China) 1971-18A Not available	Total weight: 220.4 kg (486 lbs). Objective: Perform scientific experiments. Payload: Transmits at frequencies of 20.009 and 19.995 megacycles.	1815 (1127.8)	266 (165.3)	105.9	69.9	Second orbital success for Mainland China. Launched from Shuang-ch'eng space facility, Kansu Province, northwestern China. Satellite transmitted data nearly 3 wks. Still in orbit.
Mar. 13	Explorer 43 (United States) 1971-19A Thor-Delta M-6	Total weight: 288 kg (635 lbs). Objective: Investigate, during period of decreasing solar activity, through several solar rotations, nature of interplanetary-magnetospheric interaction including characteristic features of solar wind, interplanetary fields and sector structure, and modulation effects on cosmic rays. Payload: 16-sided, 180-rm (71-in)-high, drum-shaped structure, consisting of aluminum honeycomb nr shield panels and 3 bands of solar panels (16 panels per band) mounted on honeycomb aluminum platform; experiment and instrument modules mounted in instrument compartment in top side of honeycomb shelf; 2 diametrically opposed experiment booms, each 3.5 m (11.5 ft) long, and 2 attitude control system booms, each 1.5 m (5 ft) long, spaced 90° from experiment booms, appended to spacecraft exterior and deployed incrementally	206 049 (128 032)	241 (149.8)	6012	28.6	Explorer 43 was launched into highly elliptical orbit by thrust-augmented improved Thor-Delta. First of series of second-generation Explorer spacecraft; largest and most advanced in series. Exceeded all mission objectives, with 11 of 12 scientific experiments operating successfully. Still in orbit.

Launch Date	Name, Country, International Designation, Vehicle	Payload Data	Apogee in Kilometers (and st mi)	Perigee in Kilometers (and st mi)	Period in Minutes	Inclination in Degrees	Remarks
		after launch to total length of 122 m (400 ft) tip to tip along both X and Y spacecraft axes and 6.1 m (20 ft) tip to tip along spin axes. Spacecraft fitted with 8 equally spaced sv antennas (turnstile type) which extend from spacecraft top. Spacecraft spin-stabilized and containing 12 scientific experiments, onboard experimental computer, and silver cadmium battery.					
Mar. 19	Cosmos 400 (U.S.S.R.) 1971–20A Not available	Total weight: Not available. Objective: "Continuation of Cosmos scientific satellite series." Payload: Not available.	1005 (624.5)	983 (610.8)	104.9	65.8	Cosmos 400 was employed as target for maneuverable interceptor satellite Cosmos 404 (launched 4/4/71), which made inspection during slow pass. Still in orbit.
Mar. 21	DOD Spacecraft (United States) 1971–21A Titan IIIB-Agena	Total weight: Not available. Objective: Develop space flight techniques and technology. Payload: Not available.	39 266.4 (24 399)	328.3 (204)	701.8	63.2	Still in orbit.
Mar. 24	DOD Spacecraft (United States) 1971–22A Thorad-Agena	Total weight: Not available. Objective: Develop space flight techniques and technology. Payload: Not available.	235 (146)	172.2 (107)	88.5	81.5	Reentered 4/12/71.
Mar. 27	Cosmos 401 (U.S.S.R.) 1971–23A Not available	Total weight: Not available. Objective: "Continuation of Cosmos scientific satellite series." Payload: Not available.	291 (180.8)	186 (115.6)	89.2	72.8	Reentered 4/9/71.
Mar. 31*	Isis 2 (United States–Canada) 1971–24A Thor-Delta E	Total weight: 264 kg (582 lbs). Objective: Inject spacecraft into near-circular earth orbit which will permit study of topside of ionosphere above electron peak of F region; continue and extend cooperative Canadian-U.S. program of ionospheric studies ini-	1423 (884.2)	1355 (842)	113.5	88.2	Fourth and final mission in joint U.S.-Canadian space program of ionospheric studies. Launched successfully into near-circular earth orbit by thrust-augmented improved Thor-Delta. All spacecraft systems performing satisfactorily; all 12

Date	Name/Designation	Payload/Objective	Weight kg (lb)	Apogee/Perigee km (mi)	Period (min)	Inclination (°)	Remarks
		tiated by *Alouette 1*, by combining sounder data with correlative direct measurements for a time sufficient to cover latitudinal and diurnal variations during period of declining solar activity. Payload: 8-sided oblate spheroid with 127 cm (50 in) dia, 122 cm (48 in) high; spacecraft frame of central 41-cm (16-in)-dia thrust tube which supports 8 radial ribs; 16 solar cell panels and 8 equatorial panels mounted on ribs; 12 ionospheric experiments, telemetry and tracking transmitters, command receiver and decoder, PCM equipment, tape recorder, housekeeping and power electronics, and attitude and spin-control systems; appendages including 8 telemetry antennas, 2 38-cm (15-in) probes, and beacon antenna 36-cm (14-in) in dia; when in orbit, 4 sounding antennas extended to form orthogonal dipoles 73.2 m (240 ft) and 18.7 m (61.5 ft) tip to tip.					scientific instruments operational. Exceeded all mission objectives. Still in orbit.
Apr. 1	*Cosmos 402* (U.S.S.R.) 1971-25A Not available	Total weight: Not available. Objective: "Continuation of Cosmos scientific satellite series." Payload: Not available.		1035 (643.2) / 948 (589.1)	104.9	64.9	Still in orbit.
Apr. 2	*Cosmos 403* (U.S.S.R.) 1971-26A Not available	Total weight: Not available. Objective: "Continuation of Cosmos scientific satellite series." Payload: Not available.		213 (132.4) / 205 (127.4)	88.6	81.3	Reentered 4/14/71.
Apr. 4	*Cosmos 404* (U.S.S.R.) 1971-27A Not available	Total weight: Not available. Objective: "Continuation of Cosmos scientific satellite series." Payload: Not available.		1009 (627) / 817 (507.7)	103.1	65.7	*Cosmos 404* maneuvered into orbital path of *Cosmos 400* (launched 3/19/71) and intercepted and inspected satellite on slow pass. Reentered 4/4/71.
Apr. 7	*Cosmos 405* (U.S.S.R.) 1971-28A Not available	Total weight: Not available. Objective: "Continuation of Cosmos scientific satellite series." Payload: Not available.		681 (423.2) / 674 (418.8)	98.2	81.2	Still in orbit.

Launch Date	Name, Country, International Designation, Vehicle	Payload Data	Apogee in Kilometers (and st mi)	Perigee in Kilometers (and st mi)	Period in Minutes	Inclination in Degrees	Remarks
Apr. 14	Cosmos 406 (U.S.S.R.) 1971-29A Not available	Total weight: Not available. Objective: "Continuation of Cosmos scientific satellite series." Payload: Not available.	222 (137.9)	199 (123.7)	88.6	81.3	Reentered 4/24/71.
Apr. 15	Tournesol (France) 1971-30A Diamant-B	Total weight: 90 kg (198 lbs). Objective: Determine probable presence of hydrogen at several points of celestial sphere by measuring Lyman-alpha radiations originating from sun, from geocorona, and from stars and nebulae located in direction opposite to sun. Payload: Cylindrical spacecraft 75 cm high x 70 cm in dia (29.6 in high x 27.6 in in dia); 4 solar panels laterally deployed after orbit for total extension of 250 cm (98.4 in); 5 scientific experiments; cold gas jet stabilization.	696 (432.5)	457 (284)	96.2	46.3	Tournesol (D-2A); seventh French satellite, launched from Kourou, French Guiana. Transmitting on command on 136.631 mhz. Still in orbit.
Apr. 17	Meteor 8 (U.S.S.R.) 1971-31A Not available	Total weight: Not available. Objective: Photograph cloud cover and snow and study atmospheric thermal energy radiated by earth. Payload: Cylindrical body with 2 large solar paddles attached; 3 data collection systems, TV cameras, infrared sensors, and actinometric scanners; 3-axis attitude-control system.	633 (393.3)	609 (378.4)	97.1	81.2	Meteor 8 meteorological satellite still in orbit.
Apr. 19	Salyut 1 (U.S.S.R.) 1971-32A Not available	Total weight: 18 427 kg (40 625 lbs). Objective: Perfect elements of design and onboard systems and conduct scientific research and experiments in space. Payload: 20-m (65.6-ft)-long cylindrical module with 4-m (13.1-ft) maximum dia; consists of several compartments, 3 pressurized; first is transfer compartment about 2-m (6.6-ft) dia ex-	269 (167.2)	256 (159.1)	89.7	51.5	Mission of Salyut 1, world's first manned space laboratory, consisted of 2 phases. First phase was 4/24 joint flight with Soyuz 10, containing Cosmonauts Vladimir A. Shatalov, Aleksey S. Yeliseyev, and Nikolay N. Rukavishnikov. Soyuz 10 linked up with Salyut 1 for 5½ hrs, undocked, and reentered 4/25. Second phase began with 6/7 dock-

Date	Name, source, designation, launch vehicle	Weight (kg, lb), description, objective, payload	Orbital data: Perigee km (mi)	Apogee km (mi)	Period, min	Inclination, deg	Remarks
Apr. 22	DOD Spacecraft (United States) 1971-33A Titan IIIB-Agena	panding to 3 m (9.8 ft); main habitable compartment, 4-m (13.1-ft) dia; 3rd pressurized compartment contains control and communications equipment, power supply, and life support system; 4th compartment not pressurized, about 2-m (6.6-ft) dia, contains engine installations and associated control equipment; internally, about 100 cu m (3530 cu ft) of space in pressurized compartments; also contains chemical batteries, oxygen and water reserves, and scientific instruments; externally, 2 double sets of solar-cell panels, placed at opposite ends, extend winglike from small compartments. Total weight: Not available. Objective: Develop space flight techniques and technology. Payload: Not available.	400.7 (249)	130.4 (81)	89.8	110.9	ing with *Soyuz 11*, containing Cosmonauts Georgy T. Dobrovolsky, Vladislav N. Volkov, and Viktor I. Patsayev. Three-man crew transferred to *Salyut 1* for 23-day stay. While on board, crew carried out comprehensive scientific and technical studies and medico-biological experiments; also conducted TV transmissions. During reentry 6/30 after longest manned space flight (near 24 days), 3 cosmonauts died as result of sudden pressure loss in *Soyuz* capsule. *Salyut 1* reentered 10/11, after spending nearly 6 mos in earth orbit. Reentered 5/13/71.
Apr. 23**	*Soyuz 10* (U.S.S.R.) 1971-34A Not available	Total weight: 6577 kg (14 500 lbs). Objective: Conduct joint experiments with *Salyut 1*; make comprehensive check of onboard systems; test manual and automatic control systems; check out spacecraft orientation and stabilization in different flight conditions; conduct medical-biological research on influence of space flight factors on human organism. Payload: 3-unit spacecraft about 9.5 m (31 ft) long with 2 crew cabins totaling about 8.9 cu m (315 cu ft) of interior space; TV cameras; 2 large winglike solar panels; docking equipment.	224 (139.2)	200 (124.3)	88.2	51.3	*Soyuz 10* — carrying Cosmonauts Vladimir A. Shatalov, Aleksey S. Yeliseyev, and Nikolay N. Rukavishnikov—docked 4/24 with unmanned *Salyut 1* for 5½ hrs. *Soyuz 10* automatically closed within 180 m (590 ft) of target vehicle and approach and docking were effected manually. Rigid mechanical link-up achieved. No crew transfer. After separation, *Soyuz 10* circled *Salyut 1* for 1 hr, taking pictures. *Soyuz 10* reentered and landed near Karaganda, Kazakhstan, 4/25. Total flight time 47 hrs 46 min.
Apr. 23	*Cosmos 407* (U.S.S.R.) 1971-35A Not available	Total weight: Not available. Objective: "Continuation of Cosmos scientific satellite series." Payload: Not available.	818 (508.3)	791 (491.5)	100.9	74.0	Still in orbit.
Apr. 24	*San Marco 3* (Italy) 1971-36A Scout	Total weight: 171.5 kg (378 lbs). Objective: Investigate earth's equatorial atmosphere in terms of neutral density, composition, and temperature	723 (449.3)	222 (137.9)	94.1	3.2	Third spacecraft in joint Italian–U.S. cooperative space program. Launched successfully by NASA Scout vehicle from Indian Ocean

Launch Date	Name, Country, International Designation, Vehicle	Payload Data	Apogee in Kilometers (and st mi)	Perigee in Kilometers (and st mi)	Period in Minutes	Inclination in Degrees	Remarks
		and its response to diurnal or sporadic changes in atmospheric heat input. Payload: 71-cm (28-in)-dia sphere with 4 canted 48-cm (19-in) monopole antennas for telemetry and command; spacecraft structure integral part of air-density balance; drag balance configuration consisting of light external shell connected through elastic elements of air-drag measuring balance to heavier internal structure (drum) of spacecraft; solar cells on periphery of inner structure; 3 experiments; thermal control system; 2 nickel-cadmium batteries; command and telemetry systems.					platform off coast of Kenya, Africa, by Italian crew. Scientific instruments performed as designed and produced excellent data. Exceeded expected lifetime. Reentered 11/28/71.
Apr. 24	Cosmos 408 (U.S.S.R.) 1971-37A Not available	Total weight: Not available. Objective: "Continuation of Cosmos scientific satellite series." Payload: Not available.	1510 (983.3)	200 (124.3)	102	81.8	Reentered 12/29/71.
Apr. 28	Cosmos 409 (U.S.S.R.) 1971-38A Not available	Total weight: Not available. Objective: "Continuation of Cosmos scientific satellite series." Payload: Not available.	1228 (763)	1172 (728.3)	109.4	74.0	Still in orbit.
May 5	DOD Spacecraft (United States) 1971-39A Titan IIIC	Total weight: Not available. Objective: Develop space flight techniques and technology. Payload: Not available.	35 787 (22 237) After apogee motor firing, 35 841.7 (22 271)	295 (183.3) 35 653.4 (22 154)	631 1434	26.4 .9	Still in orbit.
May 6	Cosmos 410 (U.S.S.R.) 1971-40A Not available	Total weight: Not available. Objective: "Continuation of Cosmos scientific satellite series." Payload: Not available.	288 (179)	203 (126.1)	89.2	65	Reentered 5/18/71.

ASTRONAUTICS AND AERONAUTICS, 1971

Date	Satellite	Description					Remarks
May 7	Cosmos 411 (U.S.S.R.) 1971–41A	Total weight: Not available. Objective: "Continuation of scientific satellite series." Payload: Not available.	Cosmos	1493 (927.7)	1317 (818.3)	113.8	74	Eight satellites launched with single booster. Still in orbit.
	Not available and						
	Cosmos 412 1971–41B	Total weight: Not available. Objective: "Continuation of scientific satellite series." Payload: Not available.	Cosmos	1536 (954.4)	1482 (920.9)	116.1	74	Still in orbit.
	and						
	Cosmos 413 1971–41C	Total weight: Not available. Objective: "Continuation of scientific satellite series." Payload: Not available.	Cosmos	1508 (937)	1476 (917.2)	115.7	74	Still in orbit.
	and						
	Cosmos 414 1971–41D	Total weight: Not available. Objective: "Continuation of scientific satellite series." Payload: Not available.	Cosmos	1495 (928.9)	1428 (887.3)	115.1	74	Still in orbit.
	and						
	Cosmos 415 1971–41E	Total weight: Not available. Objective: "Continuation of scientific satellite series." Payload: Not available.	Cosmos	1501 (932.7)	1453 (902.9)	115.4	74	Still in orbit.
	and						
	Cosmos 416 1971–41F	Total weight: Not available. Objective: "Continuation of scientific satellite series." Payload: Not available.	Cosmos	1493 (927.7)	1373 (853.1)	114.4	74	Still in orbit.
	and						
	Cosmos 417 1971–41G	Total weight: Not available. Objective: "Continuation of scientific satellite series." Payload: Not available.	Cosmos	1494 (928.3)	1345 (835.7)	114.1	74	Still in orbit.
	and						
	Cosmos 418 1971–41H	Total weight: Not available. Objective: "Continuation of scientific satellite series." Payload: Not available.	Cosmos	1494 (928.3)	1401 (870.5)	114.8	74	Still in orbit.

Launch Date	Name, Country, International Designation, Vehicle	Payload Data	Apogee in Kilometers (and st mi)	Perigee in Kilometers (and st mi)	Period in Minutes	Inclination in Degrees	Remarks
May 10	Cosmos 419 (U.S.S.R.) 1971-42A Not available	Total weight: Not available. Objective: "Continuation of Cosmos scientific satellite series." Payload: Not available.	340 (211.3)	203 (126.1)	87.5	51.5	Reentered 5/12/71.
May 18	Cosmos 420 (U.S.S.R.) 1971-43A Not available	Total weight: Not available. Objective: "Continuation of Cosmos scientific satellite series." Payload: Not available.	248 (154.1)	200 (124.3)	88.9	51.7	Reentered 5/29/71.
May 19	Cosmos 421 (U.S.S.R.) 1971-44A Not available	Total weight: Not available. Objective: "Continuation of Cosmos scientific satellite series." Payload: Not available.	464 (288.3)	273 (169.6)	91.9	70.9	Reentered 11/8/71.
May 19	Mars 2 (U.S.S.R.) 1971-45A Not available	Total weight: 4650 kg (10 250 lbs). Objective: Conduct complex research of Mars and its atmosphere and study characteristics of solar plasma, cosmic rays, and radiation along Mars route. Payload: Composed of 2 basic units, orbital station and descent stage; orbital station contains instrument compartment, motor assembly tank unit, correction motor, solar panels, antenna feeder unit, and temperature regulation radiators; descent stage contains instrument and parachute container in shape of ring, aerodynamic deceleration cone, and coupling frame; spacecraft instruments include IR radiometer, photometer and multichannel ultraviolet photometer, photoradiometric complex, radiotelescope, and 2 TV camera systems.	In Areocentric orbit				Mars 2, first of 2 spacecraft in U.S.S.R. 1971-72 Mars mission, entered Martian orbit 11/27/71 after flight of 470 million km (292 million mi). Ejected capsule bearing Soviet coat of arms 1380 km (858 mi) above Mars on direct trajectory to planet's surface. Became first man-made object to reach surface. Along with Mars 3, launched 5/28, explored surface relief of Mars from eccentric orbit, took photos, measured temperature, pressure, density and chemical composition of planet's atmosphere; discovered "hot spot" on Mars surface 293–298K (20°–25°C; 68°–77°F) higher than surrounding area; discovered surface pressure at measured areas to be within 400–800 newtons per sq m (4–8 millibars); determined boundary of Mars atmosphere to be 80–110 km (50–68 mi). By 3/1/72, Mars 2 had completed 127 orbits of Mars. Still in orbit.

Date	Name	Payload	Weight kg (lbs)	Period min.	Incl. deg.	Remarks
May 22	Cosmos 422 (U.S.S.R.) 1971-46A Not available	Total weight: Not available. Objective: "Continuation of Cosmos scientific satellite series." Payload: Not available.	1011 (628.2)	986 (612.7)	105	74	Still in orbit.
May 27	Cosmos 423 (U.S.S.R.) 1971-47A Not available	Total weight: Not available. Objective: "Continuation of Cosmos scientific satellite series." Payload: Not available.	489 (303.9)	279 (173.4)	91.9	71	Reentered 11/26/71.
May 28	Cosmos 424 (U.S.S.R.) 1971-48A Not available	Total weight: Not available. Objective: "Continuation of Cosmos scientific satellite series." Payload: Not available.	287 (178.3)	198 (123)	88.9	65.3	Reentered 6/10/71.
May 28	Mars 3 (U.S.S.R.) 1971-49A Not available	Total weight: 4650 kg (10 250 lbs). Objective: Conduct complex research of Mars and its atmosphere and study characteristics of solar plasma, cosmic rays, and radiation along Mars route. Payload: Composed of 2 basic units, orbital station and descent stage; orbital station contains instrument compartment, motor assembly tank unit, correction motor, solar panels, antenna feeder unit, and temperature regulation radiators; descent stage contains instrument and parachute container in shape of ring, aerodynamic deceleration cone, and coupling frame; spacecraft instruments include IR radiometer, photometer and multichannel ultraviolet photometer, photoradiometric complex, radiotelescope, 2 TV camera systems, and French "Stereo" instrument which registered sun's shortwave radiation.	In Areocentric orbit				Mars 3, second of 2 spacecraft in U.S.S.R. 1971-72 Mars mission, entered Martian orbit 12/2/71. On same day ejected capsule which parachuted to softlanding in Mars southern hemisphere between Electris and Phaetonis in area 45° S latitude and 158°W longitude. Capsule sent back video signals for 20 sec before being shut down by adverse surface conditions. Mars 3 continued in elliptical orbit of planet. Along with Mars 2, launched 5/19, explored surface relief of Mars from eccentric orbit, took photos, measured temperature, pressure, density and chemical composition of planet's atmosphere; discovered "hot spot" on Mars surface 293–298K (20°–25°C; 68°–77°F) higher than surrounding area; discovered surface pressure at measured areas to be within 400–800 newtons per sq m (4–8 millibars); determined boundary of Mars atmosphere to be 80–110 km (50–68 mi). By 3/1/72, Mars 3 had completed 7 orbits of Mars. Still in orbit.
May 29	Cosmos 425 (U.S.S.R.) 1971-50A Not available	Total weight: Not available. Objective: "Continuation of Cosmos scientific satellite series." Payload: Not available.	550 (341.8)	508 (315.7)	95.1	74	Still in orbit.
May 30	Mariner 9 (United States) 1971-51A Atlas-Centaur	Total weight: 1000 kg (2200 lbs) at launch. Objective: To select orbit which will permit, during primary orbital opera-	In Areocentric orbit				Mariner 9 inserted into elliptical orbit around Mars 11/13/71 after 400-million-km (248-million-mi) flight; became first spacecraft to

Launch Date	Name, Country, International Designation, Vehicle	Payload Data	Apogee in Kilometers (and st mi)	Perigee in Kilometers (and st mi)	Period in Minutes	Inclination in Degrees	Remarks
		tional lifetime of 90 days, viewing of about 70% of planet's surface with wide-angle imaging camera at resolution of about 1 km (0.6 mi) per TV line; study dynamic characteristics and time-variable features of Mars from Martian orbit.					orbit another planet in solar system. First photos from orbit shown on live TV, but detail limited by surface dust storm which began in Sept. As storm cleared, photos revealed bulge at Mars equator, huge volcano with 500-km (300-mi) base, 100-km (60-mi)-long channel with numerous side branchings, and water vapor above south polar cap. Spacecraft took first closeup pictures of Martian moons Deimos and Phobos. Mars atmosphere primarily carbon dioxide but also contained atomic hydrogen and atomic oxygen. Many volcanic piles appeared geologically young and suggested Mars was geochemically young planet. By 5/31/72 spacecraft had completed 400th revolution of Mars, mapped about 85 percent of planet, and obtained Mars gravity field map. Primary mission objectives completed and adjudged success. Still in orbit.
June 4	Cosmos 426 (U.S.S.R.) 1971-52A Not available	Payload: Basic octagonal structure 138.4 cm (54.5 in) high with 4 solar panels attached, each 214.6 cm long and 90.2 cm wide (84.5 x 35.5 in). Spacecraft measures 2.4 m (8 ft) from separation plane to top of low-gain antenna and span, 6.9 m (22 ft, 7.5 in) with solar panels extended. Carries 2 sets of attitude control jets, consisting of 6 jets each, mounted on ends of solar panels; high-gain antenna 101.6 cm (40 in) in dia; low-gain antenna 10.2 cm (4 in), extended 144.8 cm (57 in) from top of octagonal structure; medium-gain antenna mounted on solar panel outrigger; Canopus star tracker on upper ring structure of octagon; 6 scientific experiments including TV, uv spectrometer, IR radiometer, and IR interferometer spectrometer; large propulsion subsystem for orbital insertion; nickel-cadmium battery with 600-w-hr capacity. Total weight: Not available. Objective: "Continuation of Cosmos scientific satellite series." Payload: Not available.	1996 (1240.3)	389 (241.7)	109.2	74	Still in orbit.
June 6	Soyuz 11 (U.S.S.R.) 1971-53A Not available	Total weight: 6577 kg (14 500 lbs). Objective: Conduct joint experiments with Salyut 1, make comprehensive check of onboard systems, test manual	237 (147.3)	163 (101.3)	88.4	51.5	Soyuz 11 — carrying Cosmonauts Georgy T. Dobrovolsky, Vladislav N. Volkov, and Viktor I. Patsayev— docked with unmanned Salyut 1

ASTRONAUTICS AND AERONAUTICS, 1971

	and automatic control systems, check out spacecraft orientation and stabilization in different flight conditions, conduct medical-biological research on influence of space flight factors on human organism. Payload: 3-unit spacecraft about 9.5 m (31 ft) long with 2 crew cabins totaling about 8.9 cu m (315 cu ft) of interior space; TV cameras; 2 large winglike solar panels; docking equipment.				6/7. Crew transferred and Salyut-Soyuz station became first manned orbiting laboratory in space. During flight crew conducted experiments, made astronomical observations, transmitted live TV, reared tadpoles, grew vegetables, and took photographs. Flight broke 17-day 17-hr flight record of *Soyuz 9* June 2–19, 1970. On 6/29 crew transferred to *Soyuz 11* and undocked from *Salyut 1*. Crew reported they were in good health and said all systems functioned normally. At 1:35 am Moscow time 6/30 (6:35 pm EDT 6/29) spacecraft's braking engine fired. At end of engine firing communication with *Soyuz 11* crew ceased. After proper automatic landing, rescue helicopter team found *Soyuz 11* crew dead in capsule. Crew died when imperfect seal of hatch between command module and work compartment permitted air supply to evacuate following retrofire. Total flight time 23 days 18 hrs 23 min.
June 8	DOD Spacecraft (United States) 1971-54A Thor-Burner II	Total weight: Not available. Objective: Test infrared celestial mapping sensor system. Payload: Not available.	579.4 (360)	544 (338)	95.8	90.2	Still in orbit.
June 11	*Cosmos 427* (U.S.S.R.) 1971-55A Not available	Total weight: Not available. Objective: "Continuation of Cosmos scientific satellite series." Payload: Not available.	301 (187)	207 (128.6)	89.7	72.9	Reentered 6/23/71.
June 15	DOD Spacecraft (United States) 1971-56A Titan IIID	Total weight: Not available. Objective: Develop space flight techniques and technology. Payload: Not available.	289.7 (180)	178.6 (111)	89.1	96.3	First use of new Titan IIID booster. Reentered 8/6/71.
June 24	*Cosmos 428* (U.S.S.R.) 1971-57A Not available	Total weight: Not available. Objective: "Continuation of Cosmos scientific satellite series." Payload: Not available.	248 (154.1)	204 (126.8)	89	51.7	Reentered 7/6/71.

Launch Date	Name, Country, International Designation, Vehicle	Payload Data	Apogee in Kilometers (and st mi)	Perigee in Kilometers (and st mi)	Period in Minutes	Inclination in Degrees	Remarks
July 8	*Solrad 10* (*Explorer 44*) (United States) 1971-58A Scout	Total weight: 115 kg (253.5 lbs). Objective: Place satellite into orbit that will enable satellite to monitor sun's x-ray and ultraviolet emissions in order to understand solar physical processes better and to improve prediction techniques of solar activity and ionospheric disturbances. Payload: Twelve-sided cylinder, 58 cm high with 76-cm dia (23 in high x 30-in dia); 4 symmetrically placed 18- x 53-cm (7- x 21-in) solar-cell panels deployed after 3rd-stage burnout; spin-stabilized; 15 scientific instruments; 2 redundant telemetry transmitters.	630.3 (391.7)	436.3 (271.1)	95.3	51	Third of series of cooperative NRL-NASA missions. All spacecraft operations functioning normally; all 15 scientific instruments operating satisfactorily and returning data. Still in orbit.
July 16	*Meteor 9* (U.S.S.R.) 1971-59A Not available	Total weight: Not available. Objective: Photograph cloud cover and snow and study atmospheric thermal energy reflected and radiated by earth. Payload: Cylindrical body with 2 large solar paddles attached; 3 data-collection systems, TV cameras, infrared sensors, and actinometric scanners; 3-axis attitude control system.	642 (398.9)	615 (382.1)	97.2	81.1	*Meteor 9* meteorological satellite still in orbit.
July 16	DOD Spacecraft (United States) 1971-60A Thorad-Agena	Total weight: Not available. Objective: Develop space flight techniques and technology. Payload: Not available.	506.9 (315)	487.6 (303)	94.5	75	Still in orbit.
July 20	*Cosmos 429* (U.S.S.R.) 1971-61A Not available	Total weight: Not available. Objective: "Continuation of Cosmos scientific satellite series." Payload: Not available.	256 (159.1)	177 (110)	88.8	51.7	Reentered 8/2/71.
July 23	*Cosmos 430* (U.S.S.R.) 1971-62A Not available	Total weight: Not available. Objective: "Continuation of Cosmos scientific satellite series." Payload: Not available.	265 (164.7)	187 (116.2)	89	65.4	Reentered 8/5/71.

Date	Vehicle	Payload/Objective	Orbital data	
July 26	Apollo 15 (United States) 1971-63A Saturn V	Total weight: 140 310 kg (309 330 lbs) at initial earth orbit insertion. Objective: Perform selenological inspection, survey, and sampling of materials and surface features in preselected area of Hadley-Apennine region; emplace and activate surface experiments; evaluate capability of Apollo equipment to provide extended lunar surface stay time, increased EVA operations, and surface mobility; conduct inflight experiments and photographic tasks from lunar orbit. Payload: 34.7-m (114-ft)-long S-IVB/IU/spacecraft LM adapter/command and service modules/lunar module/lunar roving vehicle.	171.4 (106.5) Lunar orbit, 314.8 (195.6)	169.4 (105.3) 107.5 (66.8)	Apollo 15, 4th successful manned lunar landing mission, carried Astronauts David R. Scott, Alfred M. Worden, and James B. Irwin. After translunar injection and CSM separation from S-IVB, LM was extracted from S-IVB. Two midcourse corrections during coast. Spacecraft entered lunar orbit 7/29 at 4:06 pm EDT. LM undocked, initiated descent, and touched down about 600 m (656 yds) northwest of planned target of Hadley-Apennine region at 6:16 pm EDT 7/30. Crew performed 3 EVAs. During first EVA (6 hrs 33 min), Astronauts Scott and Irwin rode lunar roving vehicle for first time, deployed ALSEP, and collected lunar samples. Rover used during second EVA (7 hrs 12 min). Third EVA shortened (4 hrs 50 min) but crew explored Hadley Rille, using Rover. Total astronaut EVA time 18 hrs 35 min. Total weight of lunar rocks 77 kg (170 lbs). LM liftoff from moon at 1:11 pm EDT 8/2 covered with live TV from Rover camera stationed nearby. Total lunar stay time 66 hrs 55 min. After CSM-LM docking and crew transfer, LM ascent stage deorbited and impacted near landing site. Impact recorded by Apollo 12, Apollo 14, and Apollo 15 seismometers. Scientific subsatellite launched into lunar orbit by CSM. Transearth injection initiated and Astronaut Worden left CSM for inflight EVA to retrieve film from SM camera. One midcourse correction required during transearth coast. CM separated from SM and splashed down in Pacific 4:47 pm EDT 8/7. Total flight time 12 days 7 hrs 11 min 53 sec.
	Apollo 15 Subsatellite 1971-63D	Total weight: 36 kg (80 lbs). Objective: Pursue extensive, continuing lunar scientific studies from lunar orbit.	Lunar orbit, 141.3 (87.8)	102 (63.4)	First subsatellite launched from lunar orbit. Stored in service module SIM bay (scientific instrument module), subsatellite was

Launch Date	Name, Country, International Designation, Vehicle	Payload Data	Apogee in Kilometers (and st mi)	Perigee in Kilometers (and st mi)	Period in Minutes	Inclination in Degrees	Remarks
		Payload: Hexagonal satellite 79 cm (31 in) long and 36 cm (14 in) in dia; three 1½ m (5 ft) booms deploy after launch; carries 3 experiments, S-band transponder, particle shadows/boundary layer experiment, and magnetometer experiment; solar-cell array and nickel-cadmium battery.					spring-ejected into lunar orbit 8/4 and began scientific studies of moon. Still in orbit.
July 28	Molniya 1-18 (U.S.S.R.) 1971-64A Not available	Total weight: Not available. Objective: Ensure operation of long-distance telephone, telegraph, and radio communications and also transmission of U.S.S.R. central TV programs to points of Orbita network in far north, Siberia, Far East, and Central Asia. Payload: Satellite contains TV and radio transmission equipment, orientation and orbit-correction system, and electric power supply.	39 340 (24 444.7)	995 (618.3)	707.2	65.3	Still in orbit.
July 30	Cosmos 431 (U.S.S.R.) 1971-65A Not available	Total weight: Not available. Objective: "Continuation of Cosmos scientific satellite series." Payload: Not available.	284 (176.5)	165 (102.5)	89	51.7	Reentered 8/11/71.
Aug. 5	Cosmos 432 (U.S.S.R.) 1971-66A Not available	Total weight: Not available. Objective: "Continuation of Cosmos scientific satellite series." Payload: Not available.	252 (156.6)	200 (124.3)	89.1	51.7	Reentered 8/18/71.
Aug. 6	Ov 1-20 (United States) 1971-67A Atlas and	Total weight: Not available. Objective: Study intensities and energy distribution of protons trapped in earth's magnetic field; study effects of space plasma on experimental antennas and uv solar radiation intensity variations in wavelength and time. Payload: Spacecraft carries energetic proton analyzer and particle energy and flux thermal detector.	1932.8 (1201)	136.8 (85)	105.9	92	Nine satellites launched with single Atlas booster. Ov 1-20 reentered 8/28/71.

ASTRONAUTICS AND AERONAUTICS, 1971

Ov 1-21 1971–67B	Total weight: Not available. Objective: Measure ion densities, composition and temperature. Payload: Spacecraft carries velocity mass spectrometer; atmospheric composition sensor; ELF/VLF antenna effects transceiver, operating on 7 frequencies and 7 volts, in 14 narrow bands and 1 broad band; 2 18.3 m (60 ft) dipole antennas.	917.3 (570)	788.6 (490)	101.9	87.6	Still in orbit.
and						
Cannonball 2 1971–67C	Total weight: 363 kg (800 lbs). Objective: Measure atmospheric density. Payload: 66-cm (26-in) dia hollow brass sphere 2.5 cm (1 in) thick; contains accelerometer, telemetry, tape recorder, batteries and radar beacon; solar cells on outer surface.	1794.4 (1115)	130.4 (81)	104.2	92	Ejected from Ov 1-20. Densest satellite ever launched. Reentered 1/31/72.
and						
Musketball 1971–67D	Total weight: 61 kg (135 lbs). Objective: Measure atmospheric density. Payload: 32-cm (12.4-in)-dia sphere contains C-band transponder, electronics, power supply, and ion density and composition sensors.	653.4 (406)	130.4 (81)	92.3	87.6	Ejected from Ov 1-21. Reentered 9/19/71.
and						
AVL 802 1971–67E	Total weight: Not available. Objective: Gather aerodynamic data for use in design and development of orbiting vehicles. Payload: 0.6-m (2-ft) rigid aluminum sphere.	917.3 (570)	791.8 (492)	101.9	87.6	Ejected from Ov 1-21. Still in orbit.
and						
AVL 802 1971–67F	Total weight: Not available. Objective: Gather aerodynamic data for use in design and development of orbiting vehicles. Payload: 2.1-m (7-ft) inflatable sphere of aluminum foil and mylar.	917.3 (570)	762.8 (474)	101.6	87.5	Ejected from Ov 1-21. Still in orbit.
and						

Launch Date	Name, Country, International Designation, Vehicle	Payload Data	Apogee in Kilo- meters (and st mi)	Perigee in Kilo- meters (and st mi)	Period in Minutes	Incli- nation in Degrees	Remarks
	AVL 802 1971-67G	Total weight: Not available. Objective: Gather aerodynamic data for use in design and development of orbiting vehicles. Payload: 2.1-m (7-ft) inflatable sphere of wire mesh and subliming plastic skin.	917.3 (570)	777.3 (483)	101.8	87.8	Ejected from Ov 1-21. Still in orbit.
	and						
	AVL 802 1971-67H	Total weight: Not available. Objective: Gather aerodynamic data for use in design and development of orbiting vehicles. Payload: 2.1-m (7-ft) inflatable grid sphere of wire mesh and subliming plastic skin.	917.3 (570)	777.3 (483)	101.8	87.6	Ejected from Ov 1-21. Still in orbit.
	and						
	RTD 701 1971-67J	Total weight: Not available. Objective: Provide radar calibration target. Payload: 112-cm (44-in) rigid sphere of aluminum.	915.7 (569)	774.1 (481)	101.7	87.6	Ejected from Ov 1-21. Still in orbit.
Aug. 8	Cosmos 433 (U.S.S.R.) 1971-68A Not available	Total weight: Not available. Objective: "Continuation of Cosmos scientific satellite series." Payload: Not available.	299 (185.8)	112 (69.6)	88.6	49.4	Reentered 8/9/71.
Aug. 12	Cosmos 434 (U.S.S.R.) 1971-69A Not available	Total weight: Not available. Objective: "Continuation of Cosmos scientific satellite series." Payload: Not available.	11 798 (7331)	187 (116.2)	228.2	51.5	Still in orbit.
Aug. 12	DOD Spacecraft (United States) 1971-70A Titan IIIB–Agena	Total weight: Not available. Objective: Develop space flight tech- niques and technology. Payload: Not available.	402.3 (250)	133.6 (83)	89.8	110.9	Reentered 9/3/71.

Date	Name	Payload/Objective				Remarks
Aug. 16	*Eole* (France–United States) 1971-71A Scout	Total weight: 85 kg (187 lbs). Objective: Place spacecraft into appropriate earth orbit to obtain meteorological data from balloons; analyze meteorological data acquired from constant density surface balloons for study of characteristics and movements of air masses. Payload: Octagonal prism 58 cm (23 in) long x 71 cm (28 in) dia; contains 8 61 cm (24 in) long x 27 cm (11 in) wide rectangular solar panels which extend from satellite body; 10-m (33-ft)-long gravity gradient boom with 3-kg (6.6-lb) tip mass deployed after launch for stabilization; 400 mhz antenna resembling truncated cone transmits and receives data; 4- 136 mhz turnstile antennas; 5920 n-on-p solar cells charge 5 amp-hr silver cadmium battery.	906.1 (563.6)	677.8 (421.2)	100.7	50.2	Second cooperative France–U.S. project. French-built Cooperative Applications Satellite launched successfully by NASA. All spacecraft operations functioned normally and monitored air temperature and pressure from 500 balloons released daily from 3 sites in Argentina. Mission considered success. Still in orbit.
Aug. 27	*Cosmos 435* (U.S.S.R.) 1971-72A Not available	Total weight: Not available. Objective: "Continuation of Cosmos scientific satellite series." Payload: Not available.	478 (297)	272 (169)	92	70.9	Reentered 1/28/72.
Sept. 2	*Luna 18* (U.S.S.R.) 1971-73A Not available	Total weight: Not available. Objective: Carry out further scientific research of moon and near-moon space. Payload: Not available.	Lunar orbit. 100 (62.1)		119	35	*Luna 18* unmanned lunar probe launched successfully. Entered lunar orbit 9/7. Made 54 revolutions of moon before attempting landing. Spacecraft reached moon surface 9/11 near Sea of Fertility, 3°34′N latitude and 56°30′E longitude. All communications stopped upon landing; spacecraft assumed to have crashlanded.
Sept. 7	*Cosmos 436* (U.S.S.R.) 1971-74A Not available	Total weight: Not available. Objective: "Continuation of Cosmos scientific satellite series." Payload: Not available.	542 (336.8)	510 (316.9)	95.1	74	Still in orbit.
Sept. 10	*Cosmos 437* (U.S.S.R.) 1971-75A Not available	Total weight: Not available. Objective: "Continuation of Cosmos scientific satellite series." Payload: Not available.	545 (338.7)	520 (323.1)	95.2	74	Still in orbit.

Launch Date	Name, Country, International Designation, Vehicle	Payload Data	Apogee in Kilometers (and st mi)	Perigee in Kilometers (and st mi)	Period in Minutes	Inclination in Degrees	Remarks
Sept. 10	DOD Spacecraft (United States) 1971–76A Thorad-Agena and	Total weight: Not available. Objective: Develop space flight techniques and technology. Payload: Not available.	220.5 (137)	169 (105)	88.3	74.9	Dual payload launched with single booster. 76A reentered 10/5/71.
	DOD Spacecraft 1971–76B	Total weight: Not available. Objective: Develop space flight techniques and technology. Payload: Not available.	506.9 (315)	487.6 (303)	94.5	75	Still in orbit.
Sept. 14	Cosmos 438 (U.S.S.R.) 1971–77A Not available	Total weight: Not available. Objective: "Continuation of Cosmos scientific satellite series." Payload: Not available.	273 (169.6)	181 (112.5)	89	65.3	Reentered 9/27/71.
Sept. 21	Cosmos 439 (U.S.S.R.) 1971–78A Not available	Total weight: Not available. Objective: "Continuation of Cosmos scientific satellite series." Payload: Not available.	278 (172.7)	208 (129.3)	89.3	65.4	Reentered 10/2/71.
Sept. 24	Cosmos 440 (U.S.S.R.) 1971–79A Not available	Total weight: Not available. Objective: "Continuation of Cosmos scientific satellite series." Payload: Not available.	789 (490.3)	271 (168.4)	95.2	70.9	Still in orbit.
Sept. 28	Shinsei (Japan) 1971–80A Mu-4S-3	Total weight: 66 kg (145 lbs). Objective: Study cosmic rays and solar electric waves. Payload 26-sided body, 75 cm (30 in) in dia, covered with 5184 silicone solar cells; "gravity-turn" device employed for guidance and control; solar radio receivers, cosmic ray detectors, and ionospheric probes.	3892 (2418.4)	619 (384.6)	133	31.2	Japan successfully launched its third satellite, Shinsei ("New Star"). Was Japan's first scientific satellite. Functioned perfectly and relayed scientific information. Still in orbit.
Sept. 28	Cosmos 441 (U.S.S.R.)	Total weight: Not available. Objective: "Continuation of Cosmos	266 (165.3)	207 (128.6)	89.2	65	Reentered 10/10/71.

Date	Name / Designation	Description	Weight (kg/lb)	Orbit params	Inclination	Remarks
	1971-81A Not available	scientific satellite series." Payload: Not available.				
Sept. 28	*Luna 19* (U.S.S.R.) 1971-82A Not available	Total weight: Not available. Objective: Conduct scientific investigation of moon and near-lunar space from lunar orbit. Payload: Spacecraft contains optical-mechanical TV camera, instruments to measure moon's gravitation field, propulsion and control system, and scientific experiment.	Lunar orbit, 140 (87)	122	40.6	*Luna 19* unmanned lunar probe launched successfully. Entered lunar orbit 10/3. All systems operating normally; spacecraft conducting geophysical research of moon's gravitational field and relaying photographs of lunar surface. As of 3/10/72, spacecraft had made 1810 orbits of moon. Still in orbit.
Sept. 29	*Oso 7* (United States) 1971-83A DSV-3N and	Total weight: 635 kg (1400 lbs). Objective: Obtain high resolution data from solar corona in particular spectral bands in EUV and in visible regions during one solar rotation. Payload: Main body is 9-sided wheel 142 cm (56 in) in dia, made of aluminum honeycomb material; rectangular sail, mounted on top of wheel, carries solar-cell array and pointed experiments; overall height 200 cm (80 in); spin-stabilized. Six scientific experiments contain 2 x-ray spectroheliographs, EUV spectroheliograph, x-ray polarimeter, white light coronograph, EUV coronograph, celestial x-ray telescope, gamma-ray spectrometer, cosmic ray telescope, and solar x-ray telescope.	575.7 (357.7)	329.1 (204.5)	93.6	33.1	*Oso 7* placed into noncircular orbit by 2-stage long-tank, thrust-augmented, Thor-Delta anomaly, but pitch angle corrected by ground controllers and all spacecraft operations functioning normally. *Tetr 3* launched as secondary payload. *Oso 7* made first x-ray observations from spacecraft of beginning solar flare and of solar streamers. Obtained first observations of corona in white light and extreme ultraviolet. Also discovered sun has polar caps cooler than other regions. Still in orbit.
	Tetr 3 1971-83B	Total weight: 20.4 kg (45 lbs). Objective: Test Manned Space Flight Network and train MSFN personnel. Payload: Carries S-band transponder mounted in rear section of Delta 2nd stage; magnetically stabilized.	571 (354.8)	391 (243)	94.3	33	Launched as secondary payload. Still in orbit.
Sept. 29	*Cosmos 442* (U.S.S.R.) 1971-84A Not available	Total weight: Not available. Objective: "Continuation of Cosmos scientific satellite series." Payload: Not available.	272 (169)	199 (123.7)	89.7	72.8	Reentered 10/12/71.
Oct. 7	*Cosmos 443* (U.S.S.R.) 1971-85A Not available	Total weight: Not available. Objective: "Continuation of Cosmos scientific satellite series." Payload: Not available.	297 (184.6)	201 (124.9)	89.4	65.4	Reentered 10/19/71.

Launch Date	Name, Country, International Designation, Vehicle	Payload Data	Apogee in Kilometers (and st mi)	Perigee in Kilometers (and st mi)	Period in Minutes	Inclination in Degrees	Remarks
Oct. 13	Cosmos 444 (U.S.S.R.) 1971–86A Not available	Total weight: Not available. Objective: "Continuation of scientific satellite series." Payload: Not available. Cosmos	1510 (938.3)	1323 (822.1)	114.1	74	Eight satellites launched with single booster. Still in orbit.
and	Cosmos 445 1971–86B	Total weight: Not available. Objective: "Continuation of scientific satellite series." Payload: Not available. Cosmos	1513 (940.1)	1352 (840.1)	114.4	74	Still in orbit.
and	Cosmos 446 1971–86C	Total weight: Not available. Objective: "Continuation of scientific satellite series." Payload: Not available. Cosmos	1513 (940.1)	1383 (859.4)	114.8	74	Still in orbit.
and	Cosmos 447 1971–86D	Total weight: Not available. Objective: "Continuation of scientific satellite series." Payload: Not available. Cosmos	1516 (942)	1412 (877.4)	115.1	74	Still in orbit.
and	Cosmos 448 1971–86E	Total weight: Not available. Objective: "Continuation of scientific satellite series." Payload: Not available. Cosmos	1518 (943.2)	1442 (896)	115.5	74	Still in orbit.
and	Cosmos 449 1971–86F	Total weight: Not available. Objective: "Continuation of scientific satellite series." Payload: Not available. Cosmos	1542 (958.2)	1485 (922.7)	116.2	74	Still in orbit.

and						
	Cosmos 450 1971-86G		Total weight: Not available. Objective: "Continuation of Cosmos scientific satellite series." Payload: Not available.	1531 (951.3)	1464 (909.7)	115.9	74	Still in orbit.
and						
	Cosmos 451 1971-86H		Total weight: Not available. Objective: "Continuation of Cosmos scientific satellite series." Payload: Not available.	1575 (978.7)	1490 (925.8)	116.6	74	Still in orbit.
Oct. 14	DOD Spacecraft (United States) 1971-87A Thor-Burner II	Total weight: Not available. Objective: Develop space flight techniques and technology. Payload: Not available.	878.7 (546)	795 (494)	101.6	98.9	Still in orbit.
Oct. 14	*Cosmos 452* (U.S.S.R.) 1971-88A Not available	Total weight: Not available. Objective: "Continuation of Cosmos scientific satellite series." Payload: Not available.	280 (174)	175 (108.7)	89	64.9	Reentered 10/27/71.
Oct. 17	*Sesp 1971-2* (United States) 1971-89A Thorad-Agena	Total weight: Not available. Objective: Test advanced development payloads; test closed loop radiator/pump cooler and gather celestial background radiation pointed toward tracking ballistic missiles during midcourse of flight; verify mechanics and long-term power generation of flexible rolled-up solar cell array (FRUSA); test for secure command and control system; study energetic particles interaction with ionosphere. Payload: Basic Agena stage carries 1.7-m-wide x 4.9-m-long (5.5- x 16-ft) solar-cell panels rolled up like window shades, which roll out on parallel extension arms stored in ribbon form; arms extend as stiff tubes to total length of 9.8 m (32 ft); also earth reflecting ionospheric sounder (ERIS); 34 000 solar cells supply 500 w power.	801.5 (498)	774.1 (481)	100.5	92.7	*Sesp 1971-2* (Space Experiment Support Program) spacecraft operating normally. Abundant power provided by new FRUSA (flexible rolled-up solar array) panels may triple spacecraft's 6-month lifetime. Still in orbit.

ASTRONAUTICS AND AERONAUTICS, 1971

Launch Date	Name, Country, International Designation, Vehicle	Payload Data	Apogee in Kilometers (and st mi)	Perigee in Kilometers (and st mi)	Period in Minutes	Inclination in Degrees	Remarks
Oct. 19	Cosmos 453 (U.S.S.R.) 1971-90A Not available	Total weight: Not available. Objective: "Continuation of Cosmos scientific satellite series." Payload: Not available.	492 (305.7)	270 (167.8)	92.1	70.9	Reentered 3/19/72.
Oct. 21	Debris of ITOS-B (United States) 1971-91A Thor-Delta N-6	Total weight: 306 kg (675 lbs). Objective: Place satellite in sun-synchronous orbit having local equator-crossing time between 3:00 pm and 3:20 pm, conduct in-orbit evaluation and check-out and, upon completion, turn over to NOAA for daytime and nighttime cloud-cover observations in both direct readout and stored modes of operation. Payload: 102- x 125- x 102-cm (40- x 49- x 40-in) rectangular, box-shaped spacecraft with 3-panel solar array; 3-axis stabilized, earth-oriented satellite carries 2 advanced vidicon camera subsystems (AVCS) and 2 automatic picture transmission (APT) camera subsystems for daytime coverage and 2 scanning radiometer (SR) subsystems (infrared) for nighttime coverage; solar panels measure 0.9 x 1.5 m (3 x 5 ft) and total 4.5 sq m (48 sq ft); spacecraft measures 4.3 m (14 ft) with panels deployed; thermal control system; 4 antennas.	Crashed during preliminary orbit				ITOS-B, second operational spacecraft of Improved Tiros Operational System (ITOS) series, launched by 2-stage long-tank thrust-augmented Thor-Delta. During coast, 2nd-stage nitrogen gas leak caused tumbling. Upon reignition, spacecraft pointed in wrong direction and plunged into uninhabited area above Arctic Circle 90 min after launch.
Oct. 23	DOD Spacecraft (United States) 1971-92A Titan IIIB-Agena	Total weight: Not available. Objective: Develop space flight techniques and technology. Payload: Not available.	392.7 (244)	133.6 (83)	89.7	110.9	Reentered 11/17/71.
Oct. 28	Prospero (United Kingdom) 1971-93A Black Arrow	Total weight: 65.8 kg (145 lbs). Objective: Test equipment for future satellites and conduct scientific experiment to measure incidence of micrometeoroids. Payload: Spherical spacecraft 1.1 m	1540 (956.9)	552 (343)	106.4	82	U.K. successfully launched technology satellite Prospero from Woomera Test Range, Australia. U.K. became sixth nation to launch its own satellite with its own launch vehicle. Still in orbit.

ASTRONAUTICS AND AERONAUTICS, 1971

Date	Name	Description	Orbit (km/mi)	Apogee (km/mi)	Period (min)	Inclination	Remarks
Nov. 2	*Cosmos 454* (U.S.S.R.) 1971–94A Not available	(44 in) in dia and 0.7 m (28 in) long without boom or paddles deployed; spin-stabilized; telemetry system; light-weight silicone solar cells. Total weight: Not available. Objective: "Continuation of Cosmos scientific satellite series." Payload: Not available.	346 (215)	204 (126.8)	90	65.4	Reentered 11/16/71.
Nov. 2*	*Dscs 2–1* (United States) 1971–95A Titan IIIC	Total weight: 544 kg (1200 lbs). Objective: Provide communications channels for automatic voice, automatic digital, and secure voice communications networks of Defense Dept. Payload: Cylindrical 2.7-m-dia x 4-m-high (9. x 13-ft) spacecraft; carries single frequency conversion X-band repeaters with bandwidth of 410 mhz to carry up to 1300 duplex voice channels or 1 million data bits per sec; 8 solar-array curved panels generate 520 w.	35 816 (22 255)	35 807.9 (22 250)	1436.1	2.6	Two improved DSCS (Defense Satellite Communications System) satellites launched by single Titan IIIC booster. First of series of six. Both spacecraft experiencing serious dynamic instability in orbit as result of mass shift that occurred when antennas were deployed. Lifetime may be shortened. Still in orbit.
	and						
	Dscs 2–2 1971–95B	Total weight: 544 kg (1200 lbs). Objective: Provide communications channels for automatic voice, automatic digital, and secure voice communications networks of Defense Dept. Payload: Cylindrical 2.7-m-dia x 4-m-high (9. x 13-ft) spacecraft; carries single frequency conversion X-band repeaters with bandwidth of 410 mhz to carry up to 1300 duplex voice channels or 1 million data bits per sec; 8 solar-array curved panels generate 520 w.	35 816 (22 255)	35 807.9 (22 250)	1436.1	2.6	Still in orbit.
Nov. 15	*Explorer 45* (Italy–U.S.) 1971–96A Scout	Total weight: 50 kg (110 lbs). Objective: Measure characteristics and formation of earth's ring current and development of main-phase magnetic storms, relation between magnetic storms, aurora, and acceleration of	27 034.5 (16 798.5)	220.8 (137.2)	469	3.6	Fourth spacecraft in joint Italian-U.S. cooperative space program. Launched successfully from Italian platform in Indian Ocean off coast of Kenya, Africa, by Italian crew. Most compact and most advanced

399

Launch Date	Name, Country, International Designation, Vehicle	Payload Data	Apogee in Kilometers (and st mi)	Perigee in Kilometers (and st mi)	Period in Minutes	Inclination in Degrees	Remarks
		particles within inner magnetosphere and relative importance of various diffusion mechanisms and populating radiation zones at 4- to 5-earth-radii elliptical equatorial orbit. Payload: 26-sided polyhedral satellite approximating 63.5-cm (25-in)-dia sphere; 76.2-cm (30-in) boom along spin axis supporting 3-axis fluxgate magnetometer; search coil magnetometers packaged on two 61-cm (24-in)-long radial booms; electric-field spheres located on two 2.7 m (9 ft) booms; total length with booms deployed 5.8 m (19 ft); spacecraft body covered with solar cells; spin-stabilized; silver cadmium battery; 4 antennas; 7 scientific and 3 engineering experiments.					satellite in NASA Explorer series. New data-handling system permits ground experimenter to make in-flight changes in onboard stored programs. Still in orbit.
Nov. 17	Cosmos 455 (U.S.S.R.) 1971-97A Not available	Total weight: Not available. Objective: "Continuation of Cosmos scientific satellite series." Payload: Not available.	487 (302.6)	271 (168.4)	92.1	70.9	Reentered 4/9/72.
Nov. 19	Cosmos 456 (U.S.S.R.) 1971-98A Not available	Total weight: Not available. Objective: "Continuation of Cosmos scientific satellite series." Payload: Not available.	287 (178.3)	175 (108.7)	89.1	72.8	Reentered 12/2/71.
Nov. 20	Cosmos 457 (U.S.S.R.) 1971-99A Not available	Total weight: Not available. Objective: "Continuation of Cosmos scientific satellite series." Payload: Not available.	1221 (758.7)	1184 (735.7)	109.4	74	Still in orbit.
Nov. 24	Molniya II-1 (U.S.S.R.) 1971-100A Not available	Total weight: Not available. Objective: Continue operation of long-range telephone and telegraph radio communications system within Soviet Union and transmission of U.S.S.R. central TV programs to stations in	39 553 (24 577.1)	516 (320.6)	712	65	New communications satellite launched successfully as part of Soviet satellite communications system. Still in orbit.

Date	Name	Details					Remarks
		Orbits and participating international networks. Payload: Not available.					
Nov. 29	Cosmos 458 (U.S.S.R.) 1971-101A Not available	Total weight: Not available. Objective: "Continuation of Cosmos scientific satellite series." Payload: Not available.	497 (308.8)	271 (168.4)	92.2	70	Reentered 4/20/72.
Nov. 29	Cosmos 459 (U.S.S.R.) 1971-102A Not available	Total weight: Not available. Objective: "Continuation of Cosmos scientific satellite series." Payload: Not available.	286 (177.7)	199 (123.7)	89.4	65	Cosmos 459 was employed as target for maneuverable interceptor satellite Cosmos 462 (launched 12/3/71), which made inspection during fast pass. Reentered 12/27/71.
Nov. 30	Cosmos 460 (U.S.S.R.) 1971-103A Not available	Total weight: Not available. Objective: "Continuation of Cosmos scientific satellite series." Payload: Not available.	539 (334.9)	518 (321.9)	95.2	74	Still in orbit.
Dec. 2	Intercosmos 5 (U.S.S.R.) 1971-104A Not available	Total weight: Not available. Objective: Continue investigation of radiation and dynamics of corpuscular flux in near-earth space and investigation of nature and spectrum of low-frequency electromagnetic oscillations in natural plasma. Payload: Carries instruments recording protons, electrons, neutrons, and alfa-particles.	1055 (655.6)	197 (122.4)	97.2	48.4	Intercosmos 5 carried scientific instrumentation from Soviet Union and Czechoslovakia. Still in orbit.
Dec. 2	Cosmos 461 (U.S.S.R.) 1971-105A Not available	Total weight: Not available. Objective: "Continuation of Cosmos scientific satellite series." Payload: Not available.	508 (315.7)	468 (303.2)	94.5	69.2	Still in orbit.
Dec. 3	Cosmos 462 (U.S.S.R.) 1971-106A Not available	Total weight: Not available. Objective: "Continuation of Cosmos scientific satellite series." Payload: Not available.	1782 (1107.3)	227 (141.1)	105.2	65.7	Cosmos 462 was maneuvered into orbital path of Cosmos 459 (launched 11/29/71), intercepted and inspected satellite on fast pass, and exploded. Still in orbit.
Dec. 6	Cosmos 463 (U.S.S.R.) 1971-107A Not available	Total weight: Not available. Objective: "Continuation of Cosmos scientific satellite series." Payload: Not available.	285 (177.1)	204 (126.8)	89.2	64.9	Changes in orbital parameters and early reentry indicated Cosmos 463 may have been launched as maneuvering observation satellite to watch Indo-Pakistan war. Perigee was lowered to 188 km (116.8 mi) on

Launch Date	Name, Country, International Designation, Vehicle	Payload Data	Apogee in Kilometers (and st mi)	Perigee in Kilometers (and st mi)	Period in Minutes	Inclination in Degrees	Remarks
							16th revolution; revolutions 30 and 46 crossed East Pakistan 12/8 and 12/29 before apogee was raised again. Satellite reentered 12/11/71.
Dec. 10	Cosmos 464 (U.S.S.R.) 1971–108A Not available	Total weight: Not available. Objective: "Continuation of Cosmos scientific satellite series." Payload: Not available.	311 (193.3)	180 (111.9)	89.4	72	Changes in orbital parameters and early reentry indicated Cosmos 464 may have been launched as maneuverable observation satellite to watch Indo-Pakistan war. Apogee and perigee were lowered on 46th revolution after spacecraft drifted into position, for 3 daylight passes over East Pakistan before 12/16/71 reentry.
Dec. 11	Ariel 4 (U.S.–U.K.) 1971–109A Scout	Total weight: 99.6 kg (220 lbs). Objective: Investigate interactions between electromagnetic waves, plasma, and energetic particles in upper ionosphere. Payload: Cylinder with conical top section, 76 cm in dia x 91 cm long (30 x 36 in); 4 booms extending after launch and increasing overall dia to 3.4 m (11 ft); deployable (dipole) aerial extending after launch to approximately 13.7 m (45 ft) tip to tip; 4 short antennas affixed to conical top; solar cell arrays mounted on spacecraft body and booms, totaling 8064 solar cells; nickel-cadmium battery; data-handling system consisting of high-speed encoder, low-speed encoder and tape recorders, programmer, and command pulse conditioner; 5 experiments.	587 (364.8)	472 (293.3)	95.2	83	Fourth U.S.–U.K. cooperative satellite launched successfully. Orbital path of U.K.-built spacecraft more elliptical than desired, but all experiments expected to fulfill objectives. Still in orbit.
Dec. 14	DOD Spacecraft (United States) 1971–110A Thorad-Agena	Total weight: Not available. Objective: Develop space flight techniques and technology. Payload: Not available.	1000 (621.4)	982 (610.2)	104.9	69.9	Four satellites launched with single Thorad-Agena vehicle. Still in orbit.

ASTRONAUTICS AND AERONAUTICS, 1971

and non Spacecraft 1971–110C	Total weight: Not available. Objective: Develop space flight techniques and technology. Payload: Not available.	998 (620.1)	983 (610.8)	104.9	69.9	Still in orbit.
and non Spacecraft 1971–110D	Total weight: Not available. Objective: Develop space flight techniques and technology. Payload: Not available.	997 (619.5)	981 (609.6)	104.8	70	Still in orbit.
and non Spacecraft 1971–110E	Total weight: Not available. Objective: Develop space flight techniques and technology. Payload: Not available.	997 (619.5)	981 (609.6)	104.8	70	Still in orbit.
Dec. 15	*Cosmos 465* (U.S.S.R.) 1971–111A Not available	Total weight: Not available. Objective: "Continuation of Cosmos scientific satellite series." Payload: Not available.	1011 (628.2)	969 (602.1)	104.8	74	Still in orbit.
Dec. 16	*Cosmos 466* (U.S.S.R.) 1971–112A Not available	Total weight: Not available. Objective: "Continuation of Cosmos scientific satellite series." Payload: Not available.	375 (233)	175 (108.7)	90	64	Reentered 12/27/71.
Dec. 17	*Cosmos 467* (U.S.S.R.) 1971–113A Not available	Total weight: Not available. Objective: "Continuation of Cosmos scientific satellite series." Payload: Not available.	465 (288.9)	266 (165.3)	91.8	71	Reentered 4/18/72.
Dec. 17	*Cosmos 468* (U.S.S.R.) 1971–114A Not available	Total weight: Not available. Objective: "Continuation of Cosmos scientific satellite series." Payload: Not available.	808 (502.1)	786 (488.4)	100.7	74	Still in orbit.
Dec. 19*	*Intelsat-IV F–3* (United States) 1971–116A Atlas-Centaur	Total weight: 1412 kg (3090 lbs) at launch; 700 kg (1544 lbs) after apogee motor fire. Objective: Place satellite and apogee	36 052.9 (22 402.2)	35 745.5 (22 211.2)	1442	0.47	Launched by NASA into good transfer orbit for ComSatCorp. Apogee motor fired 12/20/71 stationing satellite in synchronous orbit over

403

Launch Date	Name, Country, International Designation, Vehicle	Payload Data	Apogee in Kilometers (and st mi)	Perigee in Kilometers (and st mi)	Period in Minutes	Inclination in Degrees	Remarks
		motor into proper transfer orbit; provide telemetry and backup calculations through transfer orbit so satellite can be injected into synchronous orbit for commercial communications. Payload: 528-cm x 238-cm-dia (208-in x 93.7-in-dia) cylindrical satellite capable of carrying 3000–9000 telephone circuits simultaneously or 12 color TV channels or a combination; spin-stabilized; 12 communications repeaters (transponders); 6 antennas (2 transmit horns, 2 receive horns, and 2 steerable 127-cm (50-in) dish spot-beam antennas; 42 240 solar cells.					Atlantic Ocean at 19.5° west longitude to service North America, South America, and Western Europe. Second satellite in improved Intelsat IV series. Spacecraft operating normally; began operational service 2/72. Still in orbit.
Dec. 20**	*Molniya 1-19* (U.S.S.R.) 1971-115A Not available	Total weight: Not available. Objective: Continue operation of long-range telephone, telegraph, and radio communications system and transmission of U.S.S.R. central TV programs to stations in Orbita network in areas of extreme north, Siberia, Far East and Central Asia. Payload: Satellite contains TV and radio transmission equipment, orientation and orbit correction system, and electric power supply.	39 146 (24 324.2)	493 (306.3)	703.3	65.4	Still in orbit.
Dec. 25	*Cosmos 469* (U.S.S.R.) 1971-117A Not available	Total weight: Not available. Objective: "Continuation of Cosmos scientific satellite series." Payload: Not available.	262 (162.8)	252 (156.6)	89.6	64.9	Still in orbit.
Dec. 27	*Cosmos 470* (U.S.S.R.) 1971-118A Not available	Total weight: Not available. Objective: "Continuation of Cosmos scientific satellite series." Payload: Not available.	257 (159.7)	190 (118.1)	88.9	65.4	Reentered 1/6/72.

Dec. 27	Oreol (U.S.S.R.-France) 1971-119A Not available	Total weight: Not available. Objective: Investigate physical phenomena in earth's upper atmosphere at high latitudes and study nature of aurora borealis. Payload: Carries 3-axis magnetometer, telemetry system, telecontrol system, and apparatus for study of physical phenomena.	2477 (1539.2)	400 (248.6)	114.6	73.9	Oreol joint Soviet-French experiment in Arcade program launched successfully. Still in orbit.
Dec. 29	Meteor 10 (U.S.S.R.) 1971-120A Not available	Total weight: Not available. Objective: Photograph cloud cover and snow and study atmospheric thermal energy radiated by earth. Payload: Cylindrical body with 2 large solar paddles attached; 3 data collection systems, TV cameras, infrared sensors, and actinometric scanners; 3-axis attitude control system.	890 (553)	876 (544.3)	102.6	81.2	Meteor 10 meteorological satellite still in orbit.

* Local time at site; 1 day later by Greenwich time.
** Local time at site; 1 day earlier by Greenwich time.

Appendix B

CHRONOLOGY OF MAJOR NASA LAUNCHES, 1971

This chronology of major NASA launches in 1971 is intended to provide an accurate and ready historical reference, compiling and verifying information previously scattered in several sources. It includes launches of all rocket vehicles larger than sounding rockets launched either by NASA or under "NASA direction" (e.g., in 1971 NASA provided vehicles and launch facilities and launched ComSatCorp's two Intelsat IV satellites and NATO's *Natosat 2*, as well as *Isis 2* in a joint U.S.-Canadian program, *Eole* in a cooperative U.S.-France program, and *Ariel 4* in a cooperative U.S.–U.K. program; also, under a NASA and University of Rome agreement, an Italian team launched a NASA satellite, *Explorer 45*, from Italy's San Marco platform off the coast of Kenya, on NASA's Scout booster; an Italian-built spacecraft, *San Marco 3*, also was launched by the same team from the same site using a NASA Scout).

An attempt has been made to classify performance of both the launch vehicle and the payload and to summarize total results in terms of primary mission. Three categories have been used for evaluating vehicle performance and mission results—successful (S), partially successful (P), and unsuccessful (U). A fourth category, unknown (Unk), has been added for payloads when vehicle malfunctions did not give the payload a chance to exercise its main experiments. These divisions are necessarily arbitrary; many of the results cannot be neatly categorized. Also they ignore the fact that a great deal is learned from missions that may have been classified as unsuccessful.

Date of launch is referenced to local time at the launch site. Open sources were used, verified when in doubt with the project offices in NASA Headquarters and with NASA Centers. For further information on each item, see Appendix A of this volume and the entries in the main chronology as referenced in the index. The information was compiled in May 1972 by Leonard C. Bruno of the Science and Technology Division of the Library of Congress.

Date	Name (NASA Code)	General Mission	Launch Vehicle (Site)	Performance Vehicle	Performance Payload	Performance Mission	Remarks
Jan. 25*	*Intelsat-IV F-2*	Operational communications satellite	Atlas-Centaur (ETR)	S	S	S	Launched into elliptical orbit by NASA for ComSatCorp; on 1/26/70 ComSatCorp fired apogee motor to circularize synchronous orbit and put satellite over Atlantic. First in improved Intelsat IV series and first launched by Atlas-Centaur. Operating normally.
Jan. 31	Apollo 14 (AS-509, CSM-110, LM-8)	Manned lunar landing Apollo flight	Saturn V (KSC)	S	S	S	Fourth manned lunar landing mission; third successful landing mission. Astronauts Alan B. Shepard, Jr., commander; Stuart A. Roosa, CM pilot; Edgar D. Mitchell, LM pilot. Launched from KSC at 4:03 pm EST 1/31. Touched down at planned Fra Mauro site at 4:17 am EST 2/5. Astronauts Shepard and Mitchell both performed two EVAs totaling 9 hrs 24 min. Deployed ALSEP and used mobile equipment transporter (MET) to obtain 43 kg (94.8 lbs) of lunar samples. Took photographs and terrain descriptions. Crew nearly reached Cone Crater but steep slopes and late timeline necessitated return to LM. LM lifted off from lunar surface at 142:25 GET 2/6. CM landed in Pacific near U.S.S. *New Orleans* at 4:05 pm EST 2/9. Total mission time 9 days 2 min.
Feb. 2*	*Natosat 2* (NATO-B)	Non-NASA mission; military communications satellite	Thor-Delta DSV-3M (ETR)	S	S	S	Launched by NASA for NATO into elliptical transfer orbit. Apogee kick motor placed spacecraft in synchronous orbit 2/4/70. Still in orbit at 26° west longitude.
Mar. 13	Explorer 43 (IMP-I)	Interplanetary Monitoring Platform	Thor-Delta M-6 (ETR)	S	S	S	First of series of second-generation Explorer spacecraft; largest and most advanced in series. Injected into highly elliptical orbit. Exceeded all mission objectives, with 11 of 12 scientific experiments operating successfully.
Mar. 31*	*Isis 2* (ISIS-B)	Scientific satellite, ionospheric physics	Thor-Delta E (WTR)	S	S	S	Fourth and final mission in joint U.S.-Canadian space program of ionospheric studies. Launched into near-circular orbit. All 12 scientific instruments performed satisfactorily; exceeded all mission objectives.
Apr. 24	*San Marco 3* (San Marco-C)	Scientific satellite, earth atmosphere	Scout (Kenya)	S	S	S	Third spacecraft in joint U.S.-Italian cooperative space program. Launched by NASA Scout from Indian Ocean platform off coast of Kenya, Africa, by Italian crew. Three experiments performed satisfactorily; exceeded expected lifetime. Reentered 11/28/71.
May 8	*Mariner 8* (Mariner-H)	Scientific interplanetary probe	Atlas-Centaur (ETR)	U	Unk	U	Mariner 8 Mars probe first of 2 spacecraft in 1971 Mariner-Mars mission. Liftoff normal at 9:11 pm EDT but anomalies occurred with Centaur engine start. Centaur stage oscillated in pitch, tumbled out of control, and reentered atmosphere approximately 1500 km (900 mi) down range and 400 km (250 mi) north of Puerto Rico.

Date	Name	Purpose	Vehicle			Results
May 30	*Mariner 9* (Mariner-I)	Scientific interplanetary probe	Atlas-Centaur (ETR)	S	S	First spacecraft to orbit another planet in solar system. Took first close up photos of Martian moons Deimos and Phobos; mapped about 85 percent of planet and obtained Mars gravity field map. Photos from successful mission revealed much new information about Mars surface.
June 20	PAET	Planetary atmosphere experiments test	Scout (WS)	S	S	Investigated means of determining structure and composition of unknown planetary atmosphere. PAET blunt cone achieved planned trajectory; reached 6500 m-per-sec (21 327 fps) entry vehicle velocity at time of separation, and impacted northeast of Bermuda.
July 8	*Solrad 10* (*Explorer 44*) (Solrad-C)	Scientific satellite, solar radiation	Scout (WS)	S	S	Third of series of cooperative NRL-NASA missions. All spacecraft operations functioned normally. All 15 scientific instruments operated satisfactorily and returned good data.
July 26	*Apollo 15* (AS-510, CSM-112, LM-10)	Manned lunar landing Apollo flight	Saturn V (KSC)	S	S	Fifth manned lunar landing mission; fourth successful landing mission. Astronauts David R. Scott, commander; Alfred M. Worden, CM pilot; James B. Irwin, LM pilot. Launched from KSC at 9:34 am EDT 7/26. Landed on moon's Hadley-Apennine region at 6:16 pm EDT 7/30. Astronauts Scott and Irwin performed three EVAS totaling 18 hrs 35 min. Crew deployed ALSEP, obtained 77 kg (170 lbs) of lunar samples, took photos, explored Hadley Rille, and drove lunar roving vehicle for first time. LM lifted off moon at 1:11 pm EDT 8/2. CM landed in Pacific near U.S.S. *Okinawa* at 4:47 pm EDT 8/7. Total mission time 12 days 7 hrs 11 min 53 sec.
and	*Apollo 15 Subsatellite*	Scientific satellite, lunar orbit studies		S	S	First subsatellite launched from lunar orbit. Stored in Service module SIM (scientific instrument module) bay, subsatellite was spring-ejected into lunar orbit and began scientific studies of moon. Still in orbit.
Aug. 16	*Eole* (CAS-1)	Cooperative applications satellite, meteorology	Scout (WS)	S	S	Second cooperative France-U.S. project. French-built satellite launched by NASA. Monitored air temperature and pressure from 500 balloons released from three sites in Argentina.
Sept. 20	Barium Ion Cloud (BIC)	Study electric and magnetic fields in magnetosphere	Scout (WS)	S	S	Cooperative W. German-U.S. effort. Barium released at planned altitude, latitude, and longitude; 12 000-km (7500-mi)-long cloud recorded and observed by scientific instruments.
Sept. 29	*Oso 7* (OSO-H)	Scientific satellite, solar physics	Thor-Delta DSV-3N (ETR)	P	S	*Oso 7* placed into noncircular orbit by launch vehicle anomaly, but pitch angle corrected by ground controllers. Made first x-ray observations from spacecraft of beginning solar flare and solar "streamers." Obtained first observations of corona in white light and extreme ultraviolet.
and						

Date	Name (NASA Code)	General Mission	Launch Vehicle (Site)	Performance			Remarks
				Vehicle	Payload	Mission	
	Tetr 3 (TETR-D)	Test and training satellite			S	S	Secondary payload on *Oso 7* launch. Employed for test and calibration of equipment and crew training for Manned Space Flight Network (MSFN).
Oct. 21	ITOS-B	Operational meteorological satellite	Thor-Delta DSV-N-6 (WTR)	U	Unk	U	Second operational spacecraft of Improved Tiros Operational System (ITOS) series. Flight appeared normal through first burn. During coast, 2nd-stage nitrogen gas leak caused tumbling. Upon reignition, spacecraft was pointing in wrong direction and plunged into uninhabited area above Arctic Circle 90 min after launch.
Nov. 15	Explorer 45 (SSS-A)	Scientific satellite magnetosphere	Scout (Kenya)	S	S	S	Fourth spacecraft in joint Italian-U.S. cooperative space program. Launched into highly elliptical orbit from Italian platform in Indian Ocean off coast of Kenya, Africa, by Italian crew. Most compact and most advanced satellite in NASA Explorer series.
Dec. 11	Ariel 4 (UK-4)	Scientific satellite, ionospheric physics	Scout (WTR)	S	S	S	Fourth U.S.-U.K. cooperative satellite. Orbital path of U.K.-built spacecraft more elliptical than desired, but all experiments expected to fulfill objectives.
Dec. 19*	Intelsat-IV F-3	Commercial communications satellite	Atlas-Centaur (ETR)	S	S	S	Launched by NASA into elliptical orbit. Transferred by ComSatCorp into synchronous orbit over Atlantic at 19.5° west longitude. Began operational service 2/72.

*Time at launch site; one day later by Greenwich time.

Appendix C

CHRONOLOGY OF MANNED SPACE FLIGHT, 1971

This chronology contains basic information on all manned space flights during 1971 and, taken with Appendix C to the 1965, 1966, 1968, 1969, and 1970 volumes of this publication, provides a summary record of manned exploration of the space environment through 1971. The information was compiled by Leonard C. Bruno of the Science and Technology Division of the Library of Congress.

The year 1971 saw a moderate increase in manned space flight activity. Contrasting with 1970, in which one U.S. Apollo flight and one Soviet Soyuz flight placed only 5 men in space, 1971 saw two Apollo and two Soyuz flights place a total of 12 men in space.

After last year's *Apollo 13* mission failure, the 1971 success of *Apollo 14* demonstrated again the viability of the Apollo system. Two of its astronauts spent nearly 9½ hours exploring the moon's surface and returned with 43 kg (95 lbs) of lunar material. These accomplishments were virtually doubled six months later when *Apollo 15's* lunar crew spent more than 18½ hours exploring the moon's surface and employed the lunar roving vehicle for the first time. The crew returned to earth with about 77 kg (170 lbs) of lunar samples.

The Soviet three-man *Soyuz 10* docked with *Salyut 1*, a huge unmanned space station launched four days earlier. Some difficulties were encountered, for no crew transfer was effected and *Soyuz 10* returned to earth 14 minutes short of a two-day flight. *Soyuz 11* initially met with better luck. It docked with *Salyut 1* and transferred its three-man crew to inhabit the station. After conducting many successful experiments and setting a new world record for the longest manned space flight, the three Soviet cosmonauts returned to *Soyuz 11* but died just before reentry when an imperfect hatch seal vented their air supply.

By the end of 1971, the United States had conducted a total of 25 manned space flights—2 suborbital, 16 in earth orbit, 3 in lunar orbit, and 4 lunar landings—with a total of 30 different crewmen. Of the 30 American astronauts, 10 had participated in two flights each, 5 had flown three times, and 1 had flown four times. The Soviet Union had conducted 18 manned flights, all in earth orbit, with 25 cosmonauts. Three had participated in two flights each and two had flown three times. Cumulative totals for manned spacecraft hours in flight had reached 2958 hours 45 minutes for the United States and 2097 hours 16 minutes for the Soviet Union. Cumulative total man-hours in space were 7796 hours 24 minutes for the United States and 4403 hours 12 minutes for the U.S.S.R.

Data on U.S. flights are the latest available to date within NASA, although minor details are subject to modification as data are refined. Major aspects of all U.S. manned flights remain subject to direct observation by interested citizens of the world, with a significant portion of recent missions seen live on worldwide color television.

ASTRONAUTICS AND AERONAUTICS, 1971

Date Launched	Date Recovered	Designation (NASA Code)	Crew	Weight in Kilograms (and in lbs)	Revolutions	Maximum Distance from Earth in Kilometers (and at mi)	Duration	Remarks
Jan. 31	Feb. 9	*Apollo 14* (AS-509, CSM-110, LM-8)	Alan B. Shepard, Jr. Stuart A. Roosa Edgar D. Mitchell	137 268.8 (302 626)	34 (of moon)	386 437 (240 121)	216 hrs 2 min	Fourth manned lunar landing mission; third successful one. Spacecraft launched by Saturn V. After insertion into translunar trajectory and CSM separation, difficulties were encountered in CSM-LM docking. Successful hard docking accomplished on sixth attempt. S-IVB evasive maneuver accomplished; stage was set on lunar impact trajectory and impacted moon 1:01 am EST 2/4/70. During translunar coast, only one midcourse correction made; live color TV transmitted. Spacecraft entered lunar orbit 2/4. LM undocked, initiated descent, and touched down in Fra Mauro 4:17 am EST 2/5. Shepard and Mitchell performed 2 EVAS. During first EVA 2/5 (4 hrs 49 min) Apollo lunar surface experiments package (ALSEP) and laser-ranging retroreflector deployed; about 19.5 kg (43 lbs) of lunar samples collected. Live color TV transmitted. Crew reentered LM. During second EVA (4 hrs 35 min) crew employed mobile equipment transporter (MET) to carry tools, cameras, and portable magnetometer during geology traverse. Furthermost point of traverse established short of rim of Cone Crater. Live color TV transmitted. Crew returned to LM with 23.5 kg (51.8 lbs) of lunar samples for total of 43 kg (94.8 lbs). Total astronaut EVA time 9 hrs 24 min. LM liftoff from moon at 142:25 CST 2/6. Total lunar stay time 33 hrs 31 min. After CSM-LM docking and crew transfer, LM ascent stage deorbited and impacted near landing site. Impact recorded by *Apollo 12* and *Apollo 14* ALSEP. Transearth injection initiated and one midcourse correction required during coast. CM separated from SM and splashed down in mid-Pacific 4:05 pm EST 2/9. Recovery by U.S.S. *New Orleans*.
Apr. 23**	Apr. 25**	*Soyuz 10*	Vladimir A. Shatalov Aleksey S. Yeliseyev Nikolay N. Rukavishnikov	6577 (14 500)	32	224 (139.2)	47 hrs 46 min	*Soyuz 10* launched from Baykonur 4:54 am local time 4/23/71 into same orbital plane as unmanned *Salyut 1* station (launched 4/19/71). Automatic rendezvous brought *Soyuz 10* within 180 m (590 ft) of Salyut target. Cosmonaut Shatalov then manually effected

412

Date	Mission	Crew	Weight kg (lb)	Perigee/Apogee km (mi)	Duration	Remarks
June 6	June 30	Soyuz 11	Georgy T. Dobrovolsky Vladislav N. Volkov Viktor I. Patseyev	6577 (14 500)	237 (147.3)	385	570 hrs 23 min	approach and actual docking. Rigid mechanical link-up achieved; no crew transfer. Two spacecraft remained docked for 5½ hrs. After separation, Soyuz 10 circled Salyut 1 taking pictures. Soyuz 10 reentered and landed near Karaganda, Kazakhstan, at 4:40 am Baykonur time 4/25. Total flight time 47 hrs 46 min. Difficulty in rendezvous and docking, lack of crew transfer, and early return to earth indicate mission not altogether successful. Soyuz 11 launched from Baykonur 9:55 am local time 6/6/71. Automatic rendezvous brought Soyuz 11 within 100 m (328 ft) of Salyut 1 space station. Cosmonaut Dobrovolsky manually effected approach and actual docking 26 hrs 50 min after launch. Patseyev and Volkov transferred to Salyut and prepared station to function as habitated craft; Dobrovolsky transferred. Salyut-Soyuz station became first manned orbiting laboratory in space. During flight, crew conducted experiments, made astronomical observations, transmitted live TV, reared tadpoles, grew vegetables, and took photographs. Flight broke 17-day 17-hr flight record of Soyuz 9 6/2–19/70. On 6/29 crew transferred to Soyuz 11 and undocked from Salyut 1. Crew reported they were in good health and said all systems functioned normally. At 1:35 am Moscow time 6/30 (6:35 pm EDT 6/29) spacecraft's braking engine fired before reentry. At end of firing communication with Soyuz 11 crew ceased. After proper automatic landing, rescue helicopter team found Soyuz 11 crew dead in capsule. Crew died of pulmonary embolism when imperfect seal of hatch between command module and work compartment permitted air supply to evacuate following retrofire.
July 26	Aug. 7	Apollo 15 (AS-510, CSM-112, LM-10)	David R. Scott Alfred M. Worden James B. Irwin	140 309.7 (309 330)	404 779 (251 518)	74 (of moon)	295 hrs 12 min	Fifth manned lunar landing mission; fourth successful one. Spacecraft launched by Saturn V booster. After insertion into lunar trajectory and CSM separation, LM was extracted from S-IVB. S-IVB evasive maneuver accomplished; stage set on lunar impact trajectory and impacted moon 4:58 pm EDT 7/29/71. Two midcourse corrections during coast; live color TV transmitted. Spacecraft entered lunar orbit 7/29 at 4:06 pm EDT. LM undocked, initiated

Date		Designation (NASA Code)	Crew	Weight in Kilograms (and in lbs)	Revolutions	Maximum Distance from Earth in Kilometers (and st mi)	Duration	Remarks
Launched	Recovered							
								descent, and touched down in Hadley-Apennine region 6:16 pm EDT 7/30. Scott and Irwin performed 3 EVAS. During first EVA 7/31 (6 hrs 33 min), astronauts drove lunar roving vehicle (LRV) for first time, deployed Apollo lunar surface experiments package (ALSEP), and collected lunar samples. LRV used during second EVA 8/1 (7 hrs 12 min), and crew completed heat flow experiment, collected core sample, and stopped at Spur Crater, Dune Crater, and Hadley Plains. Third EVA 8/2 shortened (4 hrs 50 min), but crew explored Hadley Rille, Scarp Crater, and Rim Crater using Rover. Live color TV transmitted during all 3 EVAS. Total weight of collected lunar samples 77 kg (170 lbs). Total astronaut EVA time 18 hrs 35 min. LM liftoff from moon at 1:11 pm EDT 8/2 covered with live TV from Rover camera stationed nearby. Total lunar stay time 66 hrs 55 min. After CSM-LM docking and crew transfer, LM ascent stage deorbited and impacted near landing site. Impact recorded by *Apollo 12*, *Apollo 14*, and *Apollo 15* seismometers. Scientific subsatellite launched into lunar orbit by CSM. Transearth injection initiated and Astronaut Worden left CSM for 38 min 12 sec inflight EVA to retrieve film from SM camera. One midcourse correction required during transearth coast. CM separated from SM and splashed down in Pacific 4:47 pm EDT 8/7. Recovery by U.S.S. *Okinawa*.

** Local time at site; 1 day earlier by Greenwich time.

Appendix D

ABBREVIATIONS OF REFERENCES

Listed here are abbreviations for sources cited in the text. The list does not include all sources provided in the chronology, for some of the references cited are not abbreviated; only references that appear in abbreviated form are listed below. Abbreviations used in the chronology entries themselves are cross-referenced in the Index.

A&A	AIAA's magazine, *Astronautics & Aeronautics*
A&A 1971	NASA's *Astronautics and Aeronautics, 1971* [this publication]
ABC	American Broadcasting Company
AEC Release	Atomic Energy Commission News Release
Aero Daily	*Aerospace Daily* newsletter
Aero Med	*Aerospace Medicine* magazine
AF Mag	Air Force Association's *Air Force Magazine*
AFFTC Release	Air Force Flight Test Center News Release
AFHF *Newsletter*	*Air Force Historical Foundation Newsletter*
AFJ	*Armed Forces Journal* magazine
AFNS Release	Air Force News Service Release
AFOSR Release	Air Force Office of Scientific Research News Release
AFRPL Release	Air Force Rocket Propulsion Laboratory News Release
AFSC *Newsreview*	Air Force Systems Command's *Newsreview*
AFSC Release	Air Force Systems Command News Release
AF&SD	*Air Force and Space Digest* magazine
AFSSD Release	Air Force Space Systems Division News Release
AIA Release	Aerospace Industries Association News Release
AIAA *Facts*	American Institute of Aeronautics and Astronautics' *Facts*
AIAA *News*	American Institute of Aeronautics and Astronautics' *News*
AIAA Release	American Institute of Aeronautics and Astronautics News Release
AIP *Newsletter*	*American Institute of Physics Newsletter*
Amer Av	*American Aviation* magazine (formerly *Aerospace Technology*)
AP	Association Press news service
ARC *Astrogram*	NASA Ames Research Center's *Astrogram*
Atlanta J&C	*Atlanta Journal and Constitution* newspaper
Av Daily	*Aviation Daily* newsletter
Av Wk	*Aviation Week & Space Technology* magazine
B News	*Birmingham News* newspaper
B Sun	*Baltimore Sun* newspaper
Battelle *Sci Pol Rev*	Battelle Memorial Institute's *Science Policy Reviews*
Bull Atomic Sci	*Bulletin of the Atomic Scientists*
Bus Wk	*Business Week* magazine
C Daily News	*Chicago Daily News* newspaper
C Trib	*Chicago Tribune* newspaper
Can Press	Canadian Press news service
CBS	Columbia Broadcasting System
C&E News	*Chemical & Engineering News* magazine

Cl Press	*Cleveland Press* newspaper
Columbia J Rev	*Columbia Journalism Review* Magazine
ComSatCorp Release	Communications Satellite Corporation News Release
CQ	*Congressional Quarterly*
CR	*Congressional Record*
CSM	*Christian Science Monitor* newspaper
CTNS	Chicago Tribune News Service
DASA Release	Defense Atomic Support Agency News Release
D News	*Detroit News* newspaper
D Post	*Denver Post* newspaper
DJ	Dow Jones news service
DOC PIO	Department of Commerce Public Information Office
DOD Release	Department of Defense News Release
DOT Release	Department of Transportation News Release
EOP Release	Executive Office of the President News Release
FAA Release	Federal Aviation Administration News Release
FBIS-Sov	Foreign Broadcast Information Service, Soviet number
FonF	*Facts on File*
FRC Release	NASA Flight Research Center News Release
FRC X–Press	NASA Flight Research Center's *FRC X–Press*
GE Forum	*General Electric Forum* magazine
Goddard News	NASA Goddard Space Flight Center's *Goddard News*
GSFC Release	NASA Goddard Space Flight Center News Release
GSFC *SSR*	NASA Goddard Space Flight Center's *Satellite Situation Report*
GT&E Release	General Telephone & Electronics News Release
H Chron	*Houston Chronicle* newspaper
H Post	*Houston Post* newspaper
JA	*Journal of Aircraft* magazine
JPL *Lab-Oratory*	Jet Propulsion Laboratory's *Lab-Oratory*
JPL Release	Jet Propulsion Laboratory News Release
JPRS	Department of Commerce Joint Publications Research Service
JSR	American Institute of Aeronautics and Astronautics' *Journal of Spacecraft and Rockets* magazine
KC Star	*Kansas City Star* newspaper
KC Times	*Kansas City Times* newspaper
KSC Release	NASA John F. Kennedy Space Center News Release
LA *Her-Exam*	Los Angeles *Herald-Examiner* newspaper
LA Times	*Los Angeles Times* newspaper
Langley Researcher	NASA Langley Research Center's *Langley Researcher*
LaRC Release	NASA Langley Research Center News Release
LATNS	Los Angeles Times News Service
LC *Info Bull*	Library of Congress *Information Bulletin*
LeRC Release	NASA Lewis Research Center News Release
Lewis News	NASA Lewis Research Center's *Lewis News*
M Her	*Miami Herald* newspaper
M News	*Miami News* newspaper
M Trib	*Minneapolis Tribune* newspaper
Marshall Star	NASA George C. Marshall Space Flight Center's *Marshall Star*
MJ	*Milwaukee Journal* newspaper
MSC Release	NASA Manned Spacecraft Center News Release
MSC *Roundup*	NASA Manned Spacecraft Center's *Space News Roundup*
MSFC Release	NASA George C. Marshall Space Flight Center News Release
N Hav Reg	*New Haven Register* newspaper
N News	*Newark News* newspaper
N Va Sun	*Northern Virginia Sun* newspaper
NAA *News*	National Aeronautic Association *News*
NAC Release	National Aviation Club News Release
NAE Release	National Academy of Engineering News Release
NANA	North American Newspaper Alliance
NAS Release	National Academy of Sciences News Release

NAS–NRC Release	National Academy of Sciences–National Research Council News Release
NAS–NRC–NAE *News Rpt*	National Academy of Sciences–National Research Council–National Academy of Engineering *News Report*
NASA Ann	NASA Announcement
NASA Hist Off	NASA Historical Office
NASA Hq *WB*	NASA Headquarters *Weekly Bulletin*
NASA Int Aff	NASA Office of International Affairs
NASA *Lar*, X/8	NASA *Legislative Activities Report*, Vol. X, No. 8
NASA Proj Off	NASA Project Office (for project reported)
NASA Release	NASA Headquarters News Release
NASA Rpt SRL	NASA Report of Sounding Rocket Launching
NASA SP–4014	NASA Special Publication #4014
NASC Release	National Aeronautics and Space Council News Release
Natl Obs	*National Observer* magazine
NBC	National Broadcasting Company
NGS Release	National Geographic Society News Release
NMI	NASA Management Instruction
NN	NASA Notice
NOAA Release	National Oceanic and Atmospheric Administration News Release
NR *News*	North American Rockwell Corp. *News*
NR Release	North American Rockwell Corp. News Release
NR *Skywriter*	North American Rockwell Corp. *Skywriter*
NSC *News Letter*	National Space Club *News Letter*
NSC Release	National Space Club News Release
NSF *Highlights*	National Science Foundation's *Science Resources Studies Highlights*
NSF Release	National Science Foundation News Release
NY News	*New York Daily News* newspaper
NYT, 5:4	*New York Times* newspaper, section 5 page 4
NYTNS	New York Times News Service
O Sen	*Orlando Sentinel* newspaper
Oakland Trib	*Oakland Tribune* newspaper
Omaha W–H	*Omaha World-Herald* newspaper
OST Release	Office of Science and Technology News Release
P *Bull*	Philadelphia *Evening* and *Sunday Bulletin* newspaper
P *Inq*	*Philadelphia Inquirer* newspaper
PAO	Public Affairs Office
PD	National Archives and Records Service's *Weekly Compilation of Presidential Documents*
PIO	Public Information Office
PMR *Missile*	USN Pacific Missile Range's *Missile*
PMR Release	USN Pacific Missile Range News Release
Pres Rpt 72	*Aeronautics and Space Report of the President; 1971 Activities*, March 1972
SBD	*Space Business Daily* newsletter
SAO Release	Smithsonian Astrophysical Observatory News Release
Sci Amer	*Scientific American* magazine
SciServ	Science Service news service
SD	*Space Digest* magazine
SD Union	*San Diego Union* newspaper
SF	*Spaceflight* magazine
SF Chron	*San Francisco Chronicle* newspaper
SF Exam	*San Francisco Examiner* newspaper
Sov Aero	*Soviet Aerospace* newsletter
Sov Rpt	Center for Foreign Technology's *Soviet Report* (translations)
SP	*Space Propulsion* newsletter
Spaceport News	NASA John F. Kennedy Space Center's *Spaceport News*
Spacewarn	IUWDS World Data Center A for Rockets and Satellites' *Spacewarn Bulletin*
SR	*Saturday Review* magazine

SR list	NASA compendium of sounding rocket launches
SSN	*Soviet Sciences in the News*, publication of Electro-Optical Systems, Inc.
St Louis G–D	*St. Louis Globe-Democrat* newspaper
St Louis P–D	*St. Louis Post-Dispatch* newspaper
Testimony	Congressional testimony, prepared statement
Text	Prepared report or speech text
Transcript	Official transcript of news conference or Congressional hearing
UPI	United Press International news service
USGS Release	U.S. Geological Survey News Release
US News	*U.S. News & World Report* magazine
W News	*Washington Daily News* newspaper
W Post	*Washington Post* newspaper
W Star	Washington *Evening Star/Sunday Star* newspaper
WH Release	White House News Release
WJT	*World Journal Tribune* newspaper
WS Release	NASA Wallops Station News Release
WSJ	*Wall Street Journal* newspaper

INDEX AND LIST OF ABBREVIATIONS AND ACRONYMS

A

A-300B (European consortium airbus), 308-309
AAAS. See American Assn. for the Advancement of Science.
AAS. See American Astronautical Society.
ABC. See American Broadcasting Co.
Abelson, Dr. Philip H., 241, 331
ABM. See Antiballistic missile system.
Accident
  aircraft, 277, 288, 305
    C-5, 271, 282, 284, 302-303
    Concorde, 48
    F-14, 11, 46-47, 112, 141
    F-111, 108
    YF-12, 174
  helicopter, 13, 14, 290
  lunar landing trainer, 22-23
  spacecraft, 357
    Soyuz, 156, 180-181, 185, 186, 191-192, 196-197, 277, 283, 299-300
    cause, 187, 192
Achernar (star), 149
Ad Hoc Tropical Task Force, 303
Adak, Alaska, 326
Advanced Research Projects Agency (ARPA), 23, 300
AE. See Atmosphere Explorer.
AEC. See Atomic Energy Commission.
AEC-NASA Space Nuclear Propulsion Office, 52, 96, 167, 250, 323
Aeritalia S. P. A., 315
Aero Spacelines, Inc., 242
Aerobee (sounding rocket)
  150
    micrometeorite study, 54
    solar astronomy, 161
    stellar data, 13
    ultraviolet astronomy, 14, 23, 119
    upper-atmosphere data, 117
    x-ray astronomy, 132
  170,
    airglow experiment, 14, 41, 209
    solar astronomy, 228
    stellar data, 227, 300
    test and support mission, 292
    ultraviolet astronomy, 60, 327
    x-ray astronomy, 79, 119, 226, 296
  170B, 190
  350, 52, 159, 174
Aerojet-General Corp., 57, 346, 354
Aerojet-General Liquid Rocket Co., 60

Aeronautical Radio, Inc., 260
Aeronautical Research Associates, 15
Aeronautical Research Assn. of Princeton, Inc., 290
Aeronautical Satellite Meeting, Second, 165
Aeronautics (see also Federal Aviation Administration), 37, 52, 101
  aerospace industry. See Aerospace industry.
  air pollution. See Air pollution.
  aircraft. See Aircraft.
  anniversary, 196
  awards and honors, 15, 75, 101, 115, 131, 135, 139, 263, 266, 298-299, 322, 326, 354
  benefits, 71
  cooperation, 39, 53, 70, 122, 132, 138, 165, 168, 170, 171, 174, 187, 188, 218, 224, 242, 272, 278, 286, 290, 295
  employment, 44, 81, 94, 125, 170, 295, 309, 349
  exhibit, 147, 150, 196, 198, 214, 248, 301
  facilities, 195, 365
  funds, 20, 53, 94, 221
  general aviation. See General aviation.
  history, 234, 261, 282, 301
  meeting, 122, 266, 307
  military (see also U.S. Air Force, aircraft), 37, 70, 165, 174, 267, 272, 278, 303, 312, 349
  NASA program. See National Aeronautics and Space Administration, programs.
  noise abatement. See Noise, aircraft.
  research (see also Aircraft, research; Lifting body; and TF-8A, X-15, YF-12, etc.), 23, 71, 73, 163, 171, 189, 193, 287, 393
  space technology use in, 73, 120, 151, 187, 219, 256, 297
  statistics, 105, 134, 371-372
Aeronautics and Space Report of the President, 37
Aeros (scientific satellite), 356
Aerospace Corp., 1, 361
Aerospace Industries Assn. of America, Inc. (AIA), 116, 125
Aerospace industry, 79, 82, 213, 232, 346
  employment, 3, 21, 41, 76, 81, 85, 94, 118, 125, 130, 141, 170, 295, 309, 333, 360, 368

419

financial problems, 124, 132, 216, 359–360
NASA role, 280, 370
union contract, 343
Aerospace Mechanism Symposium, Sixth, 254
Aerospace Medical Assn., 114
Aerospace Research Pilot School, 11, 186
Aerospace Safety Research and Data Institute (LaRC), 362
Aerospace Sciences Meeting, Ninth annual AIAA, 15
Aérospatiale, 266, 351
AFCRL. See Air Force Cambridge Research Laboratories.
AFETR. See Eastern Test Range.
Africa, 329
AFSC. See Air Force Systems Command.
Agena (booster upper stage), 11, 79, 80, 107, 195, 227, 254, 289, 296, 350
Agency for International Development (AID), 120, 369
Agnew, Vice President Spiro T., 24–25, 31, 135, 139, 264, 346
Agreement. See International cooperation; International cooperation, space; and Treaty.
Agriculture, Dept. of, 58, 98, 120, 132, 168, 286, 333, 351
AIA. See Aerospace Industries Assn. of America, Inc.
AIA Aerospace Research Center, 116
AIAA. See American Institute of Aeronautics and Astronautics.
AIAA Space Shuttle Development, Testing, and Operations Conference, 75
AID. See Agency for International Development.
Aiken, William S., Jr., 272
Air-cushion railroad, 82
Air-cushion research vehicle, 182
Air Defense Command, 166, 209
Air Force Academy, 150, 266
Air Force Assn., 262, 276
Air Force Cambridge Research Laboratories (AFCRL), 153, 199
Air Force Museum, 247, 248
Air Force One (Presidential aircraft), 212, 326–327
Air Force Science Advisory Board, 57
Air Force Systems Command (AFSC), 8, 12, 94, 95, 259, 267, 346
Aeronautical Systems Div., 300, 336
Air Force Two (Presidential aircraft), 212
Air France, 308
l'Air Liquide, 312
Air pollution, 2–3, 70, 122, 150, 231, 242, 251, 290, 322, 331
Air Products and Chemicals, Inc., 89
Air Quality Advisory Board, 290
Air traffic control
automation, 9, 80, 121, 187
contract, 85, 187
FAA regulation, 107, 139, 193, 277
funds, 40
NASA role, 121, 187
satellite use, 3, 219, 362
Air University, 354
Air Yugoslavia, 305
Airborne Warning and Control System (AWACS), 21
Airbus Industries, 242
Aircraft (see also individual aircraft, such as C–5A, F–14, F–111, X–15), 56, 239
accident, 11, 13, 46, 48, 108, 112, 141, 174, 271, 277, 282, 284, 288, 290, 302, 305
air pollution. See Air pollution, aircraft.
amphibious, 288, 330
antisubmarine, 12, 17, 21, 38
award, 101, 131, 135, 139, 326
bomber, 21, 70, 87, 101, 119, 122, 224, 239, 248, 255, 260, 262, 263, 272, 279, 284, 298, 303, 309, 314, 337, 344, 348, 358
cargo, 141, 242, 260, 288
carrier, 219, 224, 286, 303
close-support, 21, 70, 80
collision study, 10, 107, 219–220
contract dispute, 2, 36, 38, 147, 294
cost, 54, 152, 210, 246, 252, 260, 351
exhibit, 147, 214, 248
fighter, 46, 70, 87, 101, 140, 152, 170, 171, 210, 212, 239, 246, 248, 255, 260, 262, 264, 271, 300, 323, 329, 341, 344, 345, 352, 360
flying boat, 197, 330
foreign, 12, 140, 171, 211, 214, 224, 248, 260, 264, 286, 312, 329, 346, 360
general-aviation, 56, 276
helicopter, 13, 21, 30, 74, 80, 147, 248, 290, 294, 295, 351, 357, 365
hijacking. See Hijacking of aircraft.
industry, 36, 44, 54, 60, 94, 132, 232, 294, 346, 370
instruments, 12
interceptor, 101
noise. See Noise, aircraft.
pilotless, 1
private, 147, 226, 310, 329
race, 282
reconnaissance, 94–95, 101, 120, 168, 281, 329
record, 12, 17, 38, 226, 310, 330, 366
research (see also Aeronautics, research; and TF–8A, X–15, X–24, YF–12, etc.), 53–54, 70–71, 104, 120, 133, 138, 165, 224, 239, 245, 256–257, 278, 303, 307, 337, 338, 356, 370
safety, 122, 262, 276, 305, 314
sensor bearing, 120, 231, 286
sonic boom. See Sonic boom.
statistics, 54, 277
STOL, 19, 21, 22, 39, 71, 97, 122, 127, 173, 187, 218, 233, 267, 280, 298, 300, 315, 337, 338, 363, 365, 369
supercritical wing. See Wing, aircraft, supercritical.

supersonic. See Supersonic transport, Concorde, F–14A, F–111, Tu–144, C–15, YF–12, etc.
tanker, 129
traffic, 271
traffic control. See Air traffic control.
training, 101, 271
transport (see also Supersonic transport), 101, 109, 129, 308–309
  jet, 47, 88, 124, 130, 132, 141, 147, 174, 192, 211, 212, 216, 220, 278, 301, 312, 332, 370
  v./STOL, 18, 65, 329, 365
  VTOL, 135, 262, 298, 300
Airglow, 14, 41, 108, 130, 131, 159, 209
Airline Pilots Assn., 305, 334
Airlines (see also Air pollution; Air traffic control; Airports; Noise, aircraft; and Supersonic transport)
  aircraft, 44, 54, 124, 211, 220, 301, 307, 314
  fares, 36
  finances, 54, 359–360
  hijacking. See Hijacking of aircraft.
  safety, 305
  services, 129, 305
  statistics, 104
  supersonic transport, 84–85, 88, 109, 129, 137, 148, 157, 211, 226, 232, 346
Airlock module (AM) (Skylab component), 183, 284, 288, 323, 355
Airports (see also Air pollution; Air traffic control; Noise, aircraft; and individual airports such as Washington National Airport), 312, 328
  environmental problems, 49, 248, 271, 320
  facilities, 334, 370
  offshore, 137, 152, 182, 282
  regulation, 194
  runway research, 278
  safety, 276–277, 305
  statistics, 134, 370
  supersonic transport (SST), 314
  v./STOLport, 287
  weapon detection devices, 133
AirResearch Manufacturing Co., 147
Airways, 80
Akazawa, Shoichi, 82
Aksenov, Ye. P., 311
Alabama A&M Univ., 128
Alabama Regional Council of Governments, 128
Alabama Space and Rocket Center, 182
Alaska, 96, 101, 271, 370
Alaska, Univ. of, 76, 79, 272
Albert, Col. John G. (USAF), 75
Aldrin, Col. Edwin E., Jr. (USAF), 11, 63, 111, 135, 186, 198, 214, 232, 242
Aleksandrov, L. A., 217
Aleutian Islands, 300, 310
Alexander, Dr. Calvin, 102
Alexandria, Va., 176
Alfven, Dr. Hannes O., 281, 332
Alibrando, Alfred P., 200

Allegheny Airlines, 128
Allen, Joseph, 222
Allen, William M., 131, 139, 354
Aller, Robert O., 267
Alouette 1 (Canadian satellite), 88, 106
Alouette 2, 88
Alper, Dr. Marshall E., 55
Alps, 320
ALSEP. See Apollo lunar surface experiments package.
Altimeter, 11
Altman, Dr. Manfred, 129
Altus AFB, Okla., 271, 282, 284, 303
AM. See Airlock module.
Amchitka Island, 300, 310, 326
Amelia Earhart Lives, 243
American Airlines, 19, 211, 220, 260, 298, 326
American Assn. for the Advancement of Science (AAAS), 1, 102, 177, 187, 358
American Astronautical Society (AAS), 68, 89, 103, 178, 239
American Board of Surgery, 245
American Broadcasting Co. (ABC), 88
American Business Women's Assn., 314
American Chemical Society, 256
American College of Surgeons, 289
American Federation of Labor (AFL), 131
American Geophysical Union, 345
American Institute of Aeronautics and Astronautics (AIAA), 282
  Aerospace Sciences Meeting, 15
  Annual Meeting and Technical Display, 8th, 297
  awards, 15, 75, 233, 298, 325
  Board of Governors conference, 136
  Electric Power Systems Committee, 244
  National Capitol Section, 280, 326
  officers, 15
  Offshore Airport Center Planning Conference, 282
  Propulsion Joint Specialist Conference, Seventh, 161
  Space Shuttle Development, Testing, and Operations Conference, 75
  The Supersonic Transport: A Factual Basis for Decision, 60
  Symposium on Uranium Plasmas: Research and Applications, 2nd, 320
American Institute of Physics (API), 265
American Interplanetary Society, 282
American Museum–Hayden Planetarium, 233, 279
American Newspaper Publishers Assn., 107
American Nuclear Society, 63, 294
American Ordnance Assn., 18
American Philosophical Society, 314
American Physical Society, 36, 113
American Revolution Bicentennial Commission, 212
American Rocket Society, 13, 282
American Science and Engineering, Inc., 89, 113, 278, 286

American Society of Mechanical Engineers, 84
American Society of Newspaper Editors, 103
American Telephone & Telegraph Co. (AT&T), 170, 292
Ames, Capt. Lionel E., Jr. (USN), 124, 246
Ames Research Center (ARC) (NASA), 39, 64, 251, 262, 340
 air pollution research, 170
 Airborne Science Office, 219
 aircraft research, 70, 271, 335
 biomedical research, 167, 199, 210, 253, 256, 257, 261, 369
 employment, 238
 Exobiology Branch, 322
 ILLIAC IV computer, 23
 meeting, 253, 254, 301, 352
 micrometeorite research, 54
 oil spills measurement, 81
 personnel, 216, 297, 314, 330, 356
 sun, rotation of, 350
 water vapor electrolysis system, 235, 308
Amino acids, 167, 249, 256, 302, 322, 369
An-10 (U.S.S.R. airliner), 140
An-24 (U.S.S.R. airliner), 312
Anacostia River, 129
Analysis of Apollo 10 Photography and Visual Observations (NASA SP-232), 193
Anatuvuk Pass, Alaska, 271
Anchorage, Alaska, 271
Anders, L/C William A. (USAF, R.), 68
Anderson, Charles A., 291
Anderson, Sen. Clinton P., 19, 41, 44, 132, 265
Anderson, Rep. Glenn M., 141
Anderson, Jack, 124
Andoeya, Norway, 12
Andover, Me., 69, 292
Andreyanov, V., 162
Andromeda (galaxy), 1
Angra do Heroisimo, The Azores, 350
Anik (Canadian comsat), 128
Animal experiments, space, 127, 182, 235
Anniversary
 AEC, 215
 aeronautics, 165
 American Institute of Physics, 265
 atomic battery, 326
 computer, 218
 Kepler, Johannes, 359
 manned space flight, 100, 122, 142, 152, 167, 197, 198
 NASA, 32, 122, 196, 197, 198, 215
 nuclear power, space, 180
 radio, 347
 satellite, 98
 Smithsonian Institution, 266
 U.N., 318
 USN, 276
 U.S.S.R., 100, 152, 167
Antarctic, 304

Antenna, 69, 74, 95, 102, 231, 277, 279–280, 292, 308, 350
Antiballistic missile (ABM), 4, 43, 77, 137, 152, 194, 284, 347, 361
Apennine Mountains (moon), 145, 183, 227
API. See American Institute of Physics.
Applications Technology Satellite. See *Ats 1*, etc.
Apollo (program), 212, 230, 284, 328
 achievements, 44, 47, 76, 113, 183, 197, 208, 217, 223, 312, 342, 361, 368
 anniversary, 196, 198, 326
 astronaut. See Astronaut.
 award, 57, 70, 179, 279
 criticism, 133
 funds, 17, 20–21, 161, 173, 229, 324, 342
 history, 30, 50, 215
 launch
  Apollo 14 (AS-509), 24–30
  Apollo 15 (AS-510), 202–208
 management, 70
 news coverage, 215–216
 plans, 4, 21–22, 35, 61, 63–64, 68, 86–87, 89, 98, 108, 141, 142–145, 154–156, 215–216, 302, 305, 312, 324
 policy, 38, 50, 201, 258
 press comment (see also Apollo missions), 17, 161, 197, 223–224, 229
Apollo (spacecraft)
 anomalies, 255
 command and service module. See Command and service module.
 command module. See Command module.
 contamination, 56
 design, 17, 43–44, 86–87
 equipment, 55–56, 86–87, 247, 252
 exhibit, 353
 heat shield, 335
 instrumentation, 35, 86–87, 223
 launch
  Apollo 14 (AS-509), 24–30
  Apollo 15 (AS-510), 202–208
 launch vehicle. See Saturn.
 life support system, 156, 196, 205
 lunar module. See Lunar module.
 lunar roving vehicle. See Lunar roving vehicle.
 record, 30, 35, 207–208
 safety methods, 215
 scientific instrument module, 202–203, 206–207, 223
 service module. See Service module.
 test, 90, 268
Apollo Day, 199
Apollo 8 mission, 239, 264
Apollo 10 mission, 193, 255
Apollo 11 mission,
 commemorative dollar, 243, 308
 commemorative stamp, 232–233
 crew, 239
 light flashes, 64
 lunar samples, 5–7, 12–13, 16, 23, 30, 39, 45, 46, 83, 115, 236
 exchange of, 159, 190, 369

Of a Fire on the Moon (book about), 4
press comment, 232-233
quarantine, 115
Robert Hutchings Goddard—Father of the Space Age (book carried), 214
significance, 47, 76
viewers, 43, 80
Apollo 11 Mission Report (NASA SP-238), 130
Apollo 12 (spacecraft), 152
Apollo 12 mission, 57, 69, 345
  atomic battery, 326
  ion detector, 288
  lightning strike, 199
  lunar samples, 5-7, 12-13, 16, 23, 30, 39, 45, 46, 83, 115, 236, 257
    exchange of, 159, 190, 369
  quarantine, 115
  results, 44, 51, 255
  seismometer, 203, 207, 236
  significance, 47, 76
Apollo 13 mission, 30, 57, 59, 69, 75, 146, 244, 266
Apollo 14 (spacecraft), 4, 35, 43-44, 46, 56
Apollo 14 mission, 24-30, 80, 122, 208, 361, 368
  anomalies, 24, 25, 30, 35, 43, 255
  atomic battery, 326
  award, 57, 59, 106
  Bible carried, 214
  biological aspects, 5, 14, 17, 30, 35, 44, 63, 261
  computer, 25, 30
  Congress, report to, 61
  cost, 209
  docking, 25, 29, 30, 35, 38
  exhibit, 153
  experiments, 27, 69
    ALSEP, 27-30
    cold-cathode gauge, 64
    extrasensory perception test, 46, 170
    laser, 27
    magnetometer, 27, 51
    seismometer, 25-27, 44, 63, 64, 113, 203, 207, 236, 245
    solar wind, 27
    suprathermal ion detector, 288
    thumper shots, 27
  extravehicular activity, 27, 29, 41, 47, 48, 51, 86
  foreign reaction, 35, 43
  golf shot, 29, 106
  launch, 24
  lunar landing, 25
  lunar samples, 24, 27-30, 41, 45, 59, 98, 236, 265, 365
  medallions carried, 247
  Nixon, President Richard M., messages and statements, 24, 31, 39, 42, 59
  photographs, 27, 29, 30, 47, 48, 59
  preparations, 4, 10, 17
  press comment, 17, 35, 38, 40, 43, 44, 47
  press conference, 4, 35, 41, 59, 170
  quarantine, 5, 30, 115
  records, 30
  results, 30, 42, 44, 47, 51, 60, 183, 212, 365, 368
  splashdown, 30
  TV broadcasts, 25, 30
  U.S.S.R., reaction to, 35, 43
Apollo 14 Preliminary Science Report (NASA SP-272), 183
Apollo 15 (spacecraft), 96, 98, 110
Apollo 15 mission, 44, 61, 128, 139, 201-208, 253, 271, 272, 313
  anomalies, 203, 207, 255
  atomic battery, 326
  awards, 239, 279, 346
  biological aspects, 188, 196-197, 208, 222, 225, 228, 261
  celebration, 222, 238, 239, 256, 266
    seismometer, 142-143, 202, 207, 223, 236, 245
    solar wind spectrometer, 144
    suprathermal ion detector, 144
  extravehicular activity, 141, 201-202, 204-205, 207, 217, 222, 225, 227, 228, 229, 255
  foreign reaction, 215, 216, 222, 225, 226
  launch, 201
  launch schedule, 108
  lunar landing, 202, 203, 212
  lunar liftoff, 205, 217
  lunar roving vehicle, 24, 51, 55, 108, 110, 115, 120, 128, 141, 142, 144, 202, 204-206, 213, 215, 227, 256, 365
  lunar samples,
    core samples, 227, 238, 260
    rock samples, 142, 205, 206, 217, 223, 225, 226, 228, 229, 230, 253, 308, 365
  commemorative stamp, 215
  Congress, report to, 253
  cost, 209
  countdown, 193, 194, 200
  criticism, 219
  docking, 202, 206
  exhaust plume measurement, 199
  experiments, 86-87, 202, 223, 225
    ALSEP, 142, 144, 205
    cold cathode gauge, 142, 144
    heat flow, 142, 144, 205, 245, 291
    laser, 315
    magnetometer, 142, 144, 207
  NASA film, 227, 255, 298
  New York Times Special Features agreement, 200, 211, 336
  Nixon, President Richard M., messages and statements, 201, 208, 212, 213, 221-222, 253, 279
  photographs, 35, 142, 205, 206, 207, 223, 226
  preparations for, 96, 110, 115, 156, 193, 194
  press comment, 213, 215-216, 217-218, 222, 223-224, 225-226, 230
  press conference, 120, 141, 142-145, 156, 217, 227, 228, 245, 312
  press coverage, 215, 361
  quarantine, 188, 222
  records, 207-208

rendezvous, 206
results, 207–208, 213, 217, 218, 222, 223, 226–227, 245, 365, 368
splashdown, 207, 221, 268
subsatellite, 87, 207, 225
test, 200
TV broadcasts, 197, 202, 204, 206, 237, 264
U.S.S.R. reaction, 222
Apollo 16 mission, 44, 69, 216, 251, 324, 342
  atomic battery, 326
  crew, 61
  extravehicular activity, 228, 289, 293
  launch schedule, 293, 356, 368
  launch vehicle, 254
  lunar landing site, 166
  lunar roving vehicle, 128, 245, 267, 323
  medical aspects, 289
  preparations for, 349
  space flight data analysis, 35
Apollo 17 mission, 44, 61, 209, 216, 324, 342
  crew, 228–229, 234, 236, 312
  experiments, 21, 98, 289, 326
  launch schedule, 21, 356, 368
  photographs, 35
  press comment, 236
  press conference, 234
  subsatellite, 87
Apollo 18 mission, 20, 21, 247, 305
Apollo 19 mission, 20, 22, 247
Apollo Lunar Landing Commemorative Trust Fund, 63
Apollo lunar surface experiments package (ALSEP), 27–30, 142, 144, 205, 209
Applications Technology Satellite (ATS), 61
Apollo Telescope Mount (ATM) (Skylab component), 8, 11, 183, 252, 284, 286, 323, 361
APS. See Auxiliary propulsion system.
Aquanaut, 262, 292
Arbator, Georgy, 129
ARC. See Ames Research Center.
Arcas (sounding rocket), 54, 63, 64, 218, 222, 225, 230
  Boosted Arcas I, 50, 63
  Boosted Arcas II, 22, 226, 285
Arecibo (Puerto Rico) Observatory, 178, 231, 344
Arenosillo, Spain, 112
Ares (Greek mythological god), 331
Argentina, 230
Ariel 1 (U.K. satellite), 348
Ariel 2, 348
Ariel 3, 348
Ariel 4, 348, 369
Arizona, 281
Arkansas, 351
Arking, Dr. Albert, 50
Arlington County (Va.) Board, 165–166
Armenia, U.S.S.R., 369
Armstrong, Neil A.
  aircraft materials, lecture on, 71
  Apollo 11 mission, 111, 162, 198
  awards and honors, 121, 135, 209, 242
  resignation, 239, 243
Armstrong, Neil, Airport (proposed), 209
Army Ballistic Missile Agency (ABMA), 5
Army Corps of Engineers, 122
Arnold Engineering Development Center, 60, 276
Arnold, Dr. James R., 291
ARPA. See Advanced Research Projects Agency.
ARTS III. See Automated radar terminal system.
Arveson, John C., 81, 261
Ash, M. E., 310
Assateague National Seashore Park, 105
An Assessment of Industrial Activity in the Field of Biomedical Engineering, 322
Assn. of Bay Area Governments, 170
Asteroid, 282, 310, 327, 332, 334
Astrachan, Anthony, 111
Astris (West German booster), 310
Astronaut (see also Cosmonaut; Extravehicular activity; Space biology), 35, 96, 191, 279, 341, 342
  accident, 13, 290
  anniversary, 122, 198
  Apollo mission. See Apollo missions.
  appointment, 11, 47, 176, 253
  astronaut-astronomer, 359
  awards and honors, 57, 59, 65, 70, 72, 78, 106, 121, 239, 240, 242, 256, 260, 262–263, 266, 346–347
  celebrations for, 222, 239, 240, 256, 266
  colleges, visits to, 68
  commemorative coin, 308
  commemorative stamp, 215
  Congress, report to, 61, 253
  crew assignment, 61, 108
  damage suit, 10, 131
  European tour, 125, 150, 152, 311, 312, 323
  former, 292, 301, 309, 330
  hazards, 63, 99, 115, 163, 181, 185, 261, 325
  memorial, 227, 264, 301, 307, 320
  memorial ceremony, 199
  Nixon, President Richard M., messages and statements, 24, 31, 39, 42, 197, 201, 208, 212, 221, 253
  physiology, 14, 17, 44, 197, 225, 228, 256, 261, 276, 277, 289
  press conference, 4, 41, 59, 141, 227, 312
  promotion, 240
  public image of, 4–5, 32
  quarantine, 5, 30, 35, 188, 222
  record, 35, 208
  religion, 49, 65
  resignation, 52, 166, 239, 243
  retirement, 186, 345
  scientist-astronaut, 87, 90, 133, 138, 201, 217, 229, 230, 312
  Skylab, 10, 87, 132, 133, 288, 346
  space center visits, 271, 272
  space rescue, 64, 65
  syndicated article agreement, 200, 230

training, 115, 189
tribute to, 42, 76, 253
U.N. visit, 239
White House dinner and visit, 59, 256
Astronaut-astronomer,
Astronaut Information: American and Soviet (Revised) (71-204 SP), 248
Astronautics Engineer Award, 70
Astronomy (see also individual observatories, planets, probes, sounding rockets, and satellites, such as Asteroid; Comet; Extraterrestrial life; Galaxy; *Mariner 9*; Moon; *Oao 2*; Pulsar; Quasar; Radioastronomy; Star; Telescope), 40, 161, 253, 281, 310, 342, 347
  award, 15, 17, 102
  gamma ray, 181
  infrared, 1, 54, 219, 321, 358–359
  international cooperation, 10, 13, 16, 88–89, 159, 190, 216–217, 293, 321, 340, 355, 360, 369
  meeting, 6, 178, 248–249, 291, 334, 344, 369
  NASA program, 2–3, 9, 20–21, 21–22, 57, 67, 69, 74–75, 90, 94–95, 96–97, 98, 103, 121, 127, 148, 174, 178–179, 186, 189, 211, 219, 231, 240, 334, 345, 356, 365–366, 368
  solar, 56, 67, 68, 73, 83, 113, 121, 161, 167, 237, 240, 252, 268, 269, 277, 287, 302, 333, 335, 339, 349, 350
  stellar, 13, 36, 38, 60, 113, 194, 226, 239, 299, 300
  ultraviolet. See Ultraviolet.
  U.K., 356
  U.S.S.R., 101, 190, 218, 357, 362
  x-ray. See X-ray.
AT&T. See American Telephone & Telegraph Co.
Ataka and Co., 84–85, 138
Atlanta, Ga., 130, 320
Atlantic City, N.J., 289
Atlantic Ocean, 3, 13, 37, 103, 137, 165, 180, 219, 303, 310, 349, 354, 357, 362
Atlantic Tropical Experiment, 303
Atlas (booster), 220
  Atlas-Agena (booster), 341, 352
  Atlas-Centaur (booster), 14, 126–127, 136, 146, 163, 164, 168, 244, 275, 342, 354
Atlas (missile), 199
Atlas, geographical, 11
ATM. See Apollo Telescope Mount.
Atmosphere
  stellar, 168
  upper (earth), 3, 5, 8, 9, 78, 107, 109, 168, 171, 187, 192, 213, 220, 236, 324, 333, 343, 370
  weight, 309
Atmosphere Explorer C (AE–C) satellite, 81
Atmosphere Explorer D, 81
Atmosphere Explorer E, 81
Atomic Energy Commission (AEC) (see also AEC–NASA Space Nuclear Office; and NERVA and SNAP programs), 219, 324
  anniversary, 215
  Brookhaven National Laboratory, 113
  budget, 20–21, 58, 198–199, 363
  cooperation, 180, 197, 250
  criticism, 303
  laser, 288
  nuclear reactor, 180, 209
  nuclear rocket engine, 52
  Oak Ridge National Laboratory, 99
  organization, 346
  personnel, 65, 194, 199, 213, 222, 233, 253, 290
  policy, 294
  reorganization, proposed, 69, 82, 153–154
  Space Nuclear Systems Div., 250
  undersea nuclear test, 300, 303, 310, 326
  water recovery and waste management system, 95–96
Atomic Industrial Forum, 294
Ats 1 (Applications Technology Satellite), 271
Ats 3, 239, 252
Ats 4, 163, 275
Ats 5, 362
ATS–F, 22, 161, 293, 362
ATS–G, 22
Atsugi Naval Air Station, Japan, 12
Aubiniere, Gen. R. (France), 352
Auburn, Mass., 17
Aurora, 12, 47, 76, 79, 89
Australia, 15, 102, 220, 226, 358
Automated radar terminal system (ARTS III), 9
Automation, 295
Auxiliary propulsion system (APS), 175
Avco Corp., 335
Aviation Hall of Fame, 354
Aviation/Space Writers Assn., 54, 121, 147, 351
AVL 802 (satellite), 221
AWACS. See Airborne Warning and Control System.
Awards, 106, 262, 307
  civic, 239, 256
  foreign, 180, 191, 320
  Government, 17, 38–39, 49, 138, 165, 322
  NASA. See National Aeronautics and Space Administration.
  institutions, 79, 84, 266, 318, 327
  military, 72, 121, 150, 266
  society
    achievement, 60, 242, 314
    aeronautics, 15, 101, 131, 135, 139, 263, 266, 298, 326–327
    astronautics, 13, 14, 70, 85, 179, 266–267, 298
    science, 102, 314
AX (close-support aircraft), 21, 70, 80
Axford, William I., 15

B

B-1 (advanced strategic bomber), 70, 121, 122, 255, 263, 272, 284, 298, 303, 309, 344
B-29 (bomber), 239
B-47 (bomber), 239, 354
B-52 (Stratofortress), 42, 56, 87, 142, 303, 337, 354
B-57 (NASA research aircraft), 247
B-70 (supersonic bomber), 248
Babakin, Dr. Georgy N., 218
BAC. See British Aircraft Corp.
Back contamination, 115
Backfire (U.S.S.R. supersonic strategic bomber), 248, 260
Bahamas, 2
Bailey, J. Martyn, 242
Baker, Miss (space monkey), 182
Bal Harbour, Fla., 294
Baldeschwieler, John D., 197
Baldone, Latvia, 287
Baldwin, Mrs. Betty J., 314
Ball Brothers Research Corp., 8
Balloon, 121, 127, 167, 236, 250, 254
Baltimore, Md., 281
Bank of America, 162
Bankers Trust Co., 162
Barbados, 3
Barber, Marvin R., 307
Barium-ion-cloud (BIC) probe, 145, 251, 259, 357, 370
Barking Sands, Kauai, Hawaii, 54, 63, 64
Barstow, Calif., 108
Barth, Dr. Charles A., 148, 195
Bartlett, Gov. Dewey H., 2
Baryon particle, 36
Batavia, Ill., 181
BATEAM. See Biomedical Technology Transfer Team.
Baton Rouge, La., 89
Battelle Memorial Institute, 86
Baum, Dr. William A., 174
Baur, Dr. Paul, Jr., 111
Baykonur, U.S.S.R., 8, 66, 115, 135
 launch
  Cosmos, 7, 61, 93, 124, 128, 133, 174, 197, 212, 219, 222, 227, 343, 352, 358
  Luna 18, 246
  Luna 19, 267
  Mars 2, 134
  Mars 3, 147
  Salyut 1, 105, 107
  Soyuz 10, 107
  Soyuz 11, 154, 226
Baylor Univ., 193
 School of Medicine, 138
Bazhenov, Georgey N., 264
Beattie, Donald A., 144
Bechtel Corp., 307
Beck, L/Cdr Preston E. (USN, Ret.), 244
Beechcraft 99 (transport aircraft), 128
Beechcraft Baron (light aircraft), 226
Beeler, De E., 115, 255
Beggs, James M., 88
Bell, Rep. Alphonzo, 38
Belgium, 291, 310
Belgrade, Yugoslavia, 305
Bell Aerospace Co. See Textron, Inc.
Bell Helicopter Co. See Textron, Inc.
Bell Telephone Laboratories, 78, 299
Bellcom, Inc., 41
Bellis, B/G Benjamin N. (USAF), 212
Belson, Judge James A., 114
Bendix Corp., 278
 Aerospace Systems Div., 278
 Instrument and Life Support Div., 12
Beregovoy, Georgy T., 16
Beresford, Spencer M., 264
Berkner, Lloyd V., Award, 179
Berlin, West, 142
Bermuda, 46
Berry Aviation Co., 77
Berry, Dr. Charles A., 114
 Apollo 14 mission, 14, 44, 261
 Apollo 15 mission, 225, 228
 appointment, 245
 Soyuz 11 mission, 180
 U.S.-U.S.S.R. Joint Space Biology and Medicine Working Group, 284, 359
Berry, Doyle G., 77
Berry, Ronald L., 298
Beryllium, 70
Bethpage, N.Y., 276
Bhandari, Dr. Narenda, 6
Bible (carried on Apollo 14 mission), 214
BIC. See Barium-ion-cloud probe.
"Big Bird" (USAF reconnaissance satellite), 328
Bikle, Paul F., 114
Binsack, J. H., 64
Biomedical Technology Transfer Team (BATEAM), 235
Biosatellite, 21
Biostack (cosmic radiation experiment), 325–326
Birds, migration of, experiment, 160
Bisplinghoff, Dr. Raymond L., 170
Black Arrow (U.K. booster), 211, 300
Black Brant (sounding rocket), 63
Black Brant IIIB, 248
Black Brant VB, 54
Black Brant VC, 199
Black hole (space phenomenon), 239
Black Sea, 262
Blackhawk (helicopter), 357
Blagonravov, Dr. Anatoly A., 9, 234
Blamont, Jaques E., 93
Blankenship, L. Vaughn, 201, 258
Blue Streak (U.K. booster 1st stage), 310
Bochum Observatory (West Germany), 81, 324
Bodenschatz, Dr. Manfred, 142
Boeing 707 (jet passenger transport), 130, 192, 212, 326–327, 350, 354
Boeing 727 (jet passenger transport), 278
Boeing 747 (jet passenger transport), 139, 354
Boeing 767 (subsonic jet transport), 192

Boeing Co., 1, 79
  award, 131, 139, 279, 299, 354
  Commercial Airplane Group, 101
  contract, 8, 24, 71, 115, 198, 255, 290
  employment, 82
  jet passenger transport. See Boeing 707, 727, 747, and 767.
  lunar roving vehicle, 24, 51, 245, 323
  Mariner Venus-Mercury spacecraft, 115
  missile, 8, 262
  personnel, 57, 131
  Space Center, 68
  STOL aircraft, 290, 315
  supersonic transport, 56, 81, 82, 129, 135, 139, 192
  Vertol Div., 188–189
Boileau, O. C., 68
Bolam, Mrs. Irene, 243
Bolane, Rep. Edward P., 222
Bombay, India, 6
Bonn, West Germany, 277
Boosted Arcas. See Arcas.
Borman, Col. Frank (USAF, Ret.), 194
Boston, Mass., 102, 209, 308
Boston Museum of Science, 327
Boulder, Colo., 302
Bower, Robert E., 191
Brandt, Dr. John C., 36
Bras d'Or (Canadian hydrofoil warship), 312
Brauer, Dr. Richard D., 18
Brayton cycle, 147
Brazil, 329
Bremen, West Germany, 249
Brennwald, Louis H., 216
Bretigny, France, 150, 254
Brevard County, Fla., 65
  Circuit Court, 10
Brezhnev, Leonid I., 102, 299
Bridgeport, Conn., 309
Bridgeport, Univ. of, 309
A Briefe and True Report of the New Found Land of Virginia, 97
Bristol Aircraft Corp., 326
British Aircraft Corp. (BAC), 37, 48, 267, 351
British Airports Authority, 326
British Astronomical Society, 202
British Interplanetary Society, 85
Broglio, Prof. Luigi, 262
Brooke, Sen. Edward W., 124
Brookhaven National Laboratory, 113
Brooklyn Polytechnic Institute, 17
Brooks AFB, Tex., 298
Brooks, Dr. Harvey, 277
Brooksbank, William A., 334
Brougham, Trevor, 226
Brown, B. Porter, 84
Brown, Gen. George S. (USAF), 259
Brown Engineering Co., 158, 180
Brown Univ., 151
Browne, Dudley E., 307
Browne, Secor D., 348–349
Broyhill, Rep. Joel T., 209
Brussels, Belgium, 48, 168, 260
Buccaneer (U.K. strike aircraft), 224

Buckley, Sen. James L., 24
Budinger, Dr. Thomas F., 63
Bulgaria, 237, 249, 263, 319
Bundy, William P., 358
Burch, Dean, 60
Burroughs Corp., 23
Bush, George, Ambassador to U.N., 70, 120
Butler, Charles F., 312
Byurakan Astrophysical Observatory, U.S.S.R., 248

C

C–5 (military cargo transport aircraft), 282, 284, 302–303
C–5A, 335
  accident, 271, 307
  contract, 2, 36, 294
  cost, 260, 359
  modification, 348, 359
  Paris Air Show exhibit, 147
C–8 (Buffalo)) (jet STOL aircraft), 187
C–47 (research transport aircraft, 120
C–130 (Hercules) (NASA research transport aircraft), 104, 138–139, 171, 370
C–130B, 120
C–141 (research aircraft), 278
CAB. See Civil Aeronautics Board.
Cabell, Gen. Charles P. (USAF, Ret.), 145
Cabell, Rep. Earle, 38
Caetano, Prime Minister Marcello (Portugal), 349–350
Cafferty, Michael, 107
Cairns, Robert W., 170
California, 81, 112, 360, 370
California Institute of Technology (Cal Tech), 1, 7, 13, 102, 151, 170, 235, 245, 279, 296
California State Air Resources Board, 170
California State College, 299
California Statewide Air Pollution Research Center, 170
California, Univ. of, 6, 15, 23, 109, 281, 327
  Berkeley, 1, 36, 63, 102, 145, 242, 248, 299
  Los Angeles (UCLA), 18, 212, 231
  San Diego, 261, 269
Calisphere 3 (calibration satellite), 48
Caliphere 4, 48
Calisphere 5, 48
Calverton, N.Y., 112, 141
Calvin, Prof. Melvin, 36
Cambridge, Mass., 313
Cambridge, U.K., 348
Cambridge Univ., 324
Camera, 27, 35, 38, 56, 85, 128
  underwater, 131
Cameron, Col. Lyle W. (USAF), 267
Camp, Rep. John N. Happy, 38
Canada, 311, 313–314
  aircraft, 280–281
  award, 102

cooperation, 3, 15, 63, 88, 96, 104, 106, 220, 305, 369
Defence Ministry, 312
Dept. of Communications, 106
satellite, 88, 105, 369
Canadian Centre of Inland Waters, 104
Canadian Domestic Communications System, 128
Canberra V-8 (U.K. bomber), 224
Caneel Bay, Virgin Islands, 24, 31
Cannikin (underground nuclear test), 300, 303, 311, 326
Cannon AFB, N. Mex., 314
Cannon Ball 2 (USAF satellite), 220–221
Cannon, Sen. Howard W., 19, 213, 263, 292
Cannon, Dr. Robert H., Jr., 296
Canopus (star), 149, 306
Cape Kennedy, Fla., 65, 217
Carbon, 39
Carbon monoxide, 299
CARD. See Civil Aviation Research and Development (CARD) Policy Study.
Cardiovascular medicine, 235
Career Service Award, 60
CARETS. See Central Atlantic Regional Ecological Test Site Project.
Carey General Contractors, Inc., 195
Cargo, David F., 42
Caribbean Sea, 171
Carlos, Prince Juan, 25
Carnegie Institution of Washington, 1, 18
Carolinas, 2
Carpenter, Cdr. M. Scott (USN, Ret.), 292
Carroll, Thomas E., 13
CAS. See Cooperative Applications Satellite.
Case Western Reserve Univ., 151, 181
Cassiopeia (constellation), 1
Castel Gondolfo, Italy, 215
CATS. See Computer-aided telemetry system.
Causse, J. P., 352
Cayenne, French Guiana, 248
CBS. See Columbia Broadcasting System.
Centaur (booster upper stage), 14, 96, 98, 148, 354
Centaur Quality and Workmanship Review Board, 168, 243–244
Centaure (French sounding rocket), 83, 235, 259, 370
Centaurus X-3 (pulsar), 113, 131
Centerville, Iowa, 213
Central Atlantic Regional Test Site (CARETS) project, 281
Central Intelligence Agency (CIA), 117, 124
Centre National d'Etudes Spatiales (CNES), 83, 230, 235, 254, 291, 357
Century Development Corp., 166
Cernan, Capt. Eugene A. (USN), 13, 14, 228, 234
Cerro Tololo Inter-American Observatory, 178
Cessna 401 (light aircraft), 81
Cessna Aircraft Co., 1

CETI. See Communication with Extraterrestrial Intelligence.
CH-54B (helicopter), 365
Chafee, Secretary of the Navy John H., 72, 240
Chaffee, L/Cdr Roger B. (USN), 10, 357
Chalkyitsik, Alaska, 271
Chang, Sherwood, 39
Chanute, Octave, Award, 75
Chapel of the Astronauts, Inc. (proposed), 264, 301, 307, 320
Charyk, Dr. Joseph V., 60
Chatham, George N., 56
Chauncey, Dr. David L., 102
Chavkin, Jerold M., 65
Chelson, Paul O., 179
Chernomor (U.S.S.R. oceanographic laboratory submarine), 262
Chernykh, N., 357
Cherry, George W., 233, 349
Chesapeake Bay, 97
Cheyenne (helicopter), 80, 294
Chicago, Ill., 1, 9, 46, 65, 220, 236, 262, 280
Chicago, Univ. of, 117, 240
Chicom 1 (People's Republic of China satellite), 53, 61
Chicom 2, 61, 368
Chien Wei-chang, 111
Chile, 178, 329, 330
Chiles, Sen. Lawton, 145
China, Communist. See People's Republic of China.
China Lake, Calif., 115
Chincoteague Coast Guard Station, 105
Chincoteague National Wildlife Refuge, 105
Chinghua Univ., 111
Chou En-lai, 84
Chouteau, Dr. Jacques, 311–312
Chrysler Corp., 99
Church of Good Shepherd, Maitland, Fla., 5
Chrysler Corp., 278
Churchill Research Range, Canada, 5, 9, 14, 50, 63, 89, 174, 187, 192, 246
CIA. See Central Intelligence Agency.
Cincinnati, Univ. of, 239
Citizen's Advisory Committee on Environmental Quality, 194
Citizens Bank and Trust Co. of Maryland, 53
Civil Aeronautics Board (CAB), 124, 288, 348–349
Civil Aviation. See Aeronautics.
Civil Aviation Research and Development (CARD) Policy Study, 122–123, 218, 296, 369
CL-215 (Air Tanker), 129
Clark AFB, Philippines, 352
Clark, Adm. John E. (USN), 159
Clark Univ., 214
Clarke, Arthur C., 280
Clarke, Gilmore D., 330
Clauser, Francis H., 170
Clements, Henry E., 197

Cleveland, Ohio, 2, 18, 106, 292
Cloud study, 105, 112, 213, 251, 257
CM. See Command module.
CNES. See Centre National d'Etudes Spatiales.
Cochran, Jacqueline, 354
Cocoa Beach, Fla., 122, 309
Cokeley, Ralph C., 47
Cold cathode gauge experiment, 144
College of Eastern Utah, 57
College Park, Md., 69
Collier, Robert J., Trophy, 131
Collins, Col. Michael (USAF, R.), 52, 101, 135, 195, 242, 301, 309, 330
Collision avoidance, aircraft, 10, 107, 219–220
Colorado Aviation Historical Society, 322
Colorado Springs, Colo., 266
Colorado, Univ. of, 2, 93, 108, 148, 195
Columbia Astrophysics Laboratory, 52
Columbia Broadcasting System (CBS), 88, 307, 327
Columbia, Mo., 78
Columbia Univ., 144, 245, 302, 322, 345
Columbine III (presidential aircraft), 248
Columbus, Christopher, 31, 243
Columbus Day, 243
Coma (galaxy), 132
Comecon. See Council of Mutual Economic Assistance.
Comet, 93, 334, 358
Command module (CM)
  Apollo 9, 286
  Apollo 14 (*Kitty Hawk*), 24–30, 40, 247
  Apollo 15 (*Endeavor*), 96, 141, 201–202, 207–208, 268, 313
  Apollo 16, 326
  Skylab, 64, 258
  Soyuz, 152
Command and service module (CSM), 153, 233
  Apollo 14 (*Kitty Hawk*) (CSM-110), 24–30, 35, 365
  Apollo 15 (*Endeavor*) (CSM-112), 201–202, 206–207, 211, 229, 365
  Skylab, 164, 183
Command Pilot Astronaut Badge (USAF), 72
Commerce, Dept. of, 212
Committee for Nuclear Responsibility, 300
Committee for the Future, 219
Committee on Space Research (COSPAR), 166–167
Common Market, 58
Communication with Extraterrestrial Intelligence (CETI) conference, 248
Communications satellite (see also individual satellites: *Echo I, Intelsat I, Intersat-IV F-2, Molniya I–18,* etc.)
  agreement, 106, 138, 219, 252, 272, 319
  anniversary, 98
  award, 298
  contract, 128, 173
  cooperation, international, 17, 18, 58, 106, 128, 138, 219–220, 252, 265, 271, 272, **283**, 319, 352, **356**

earth station, 238, 283
FCC regulations, 65, 71, 74, 88, 176, 292
launch
  Dscs 2–1, 306
  Dscs 2–2, 306
  Intelsat-IV F–2, 14
  Intelsat-IV F–3, 354
  Molniya I–18, 210
  Molniya I–19, 355
  Molniya II–1, 330
  Natosat 2 (*Nato 2*), 36
  military, 36, 79, 94, 306, 313, 336, 366, 369
  NASA program, 21–22, 41, 106, 128, 173, 356
  use, 83, 98, 117, 197, 219, 252, 265, 271, 272, 283, 341, 357, 362
  U.S. policy, 3, 17, 48
Communications Satellite Corp. (ComSat Corp), 294
  contract, 138
  Early Bird. See *Intelsat I.*
  earth station, 60, 69, 82, 292
  FCC regulation, 176
  INTELSAT, 138, 237
  Intelsat I (*Early Bird*), 14
  Intelsat-III series, 14, 69
  Intelsat-IV series, 14, 294
  Intelsat-IV F–2, 14, 82, 294, 341, 354, 356–357, 365
  Intelsat-IV F–3, 342, 352, 354, 365
  Intelsat-IV F–4, 356
  Intelsat-IV F–5, 356
  Intelsat-IV F–6, 356
  revenues, 50, 69, 104
  satellite program, 60, 69, 294, 356
  services, 82
Community Development, Dept. of (proposed), 81
Comparison of Military Research and Development Expenditures of the United States and the Soviet Union, 201
Computer, 127
  anniversary, 218
  automated transit system, 47
  ILLIAC IV, 23
  industry use, 309
  medical use, 106–107
  NASA, 32, 98, 135, 162, 174, 309
    Apollo 14, 25, 30
    contract, 187, 264
    program exchange, 231
  pollution research use, 103, 288
  U.S.S.R., 122, 309
  weapon detection system, 212
Computer-aided telemetry system (CATS), 179
Computer Software Management and Information Center (COSMIC), 81, 232
Comstock, G. M., 83
Concept verification testing (CVT), 334
Concorde (Anglo-French supersonic transport), 126, 127, 148, 214, 232, 242, 266, 280, 332, 349–350, 351
  cost, 73
  environmental problems, 150

flights, 47, 125, 248, 249
Paris Air Show exhibit, 147, 214
sales price, 157, 346
Cone (lunar crater), 27, 41, 48
Conference on Remote Sensing of Chesapeake Bay, 97
Congress, 260, 296
  antiballistic missile system, 4
  atomic energy,
  business enterprises emergency loan guarantee, 254
  C-5 cargo transport, 302
Defense Dept. of, 87, 152, 286
  Government Accounting Office report to, 1
  Joint Committee on Atomic Energy, 52, 300
  Joint Committee on the Environment and Technology, 77
  Joint DOT-NASA Civil Aviation Research and Development Policy Study Report, 122–123
  Kennedy, President John F., message to, 142, 215
  metric system, 212
  national security, 303
  Presidential aircraft request, 326–327
  President's messages
    Aeronautics and Space Report of the President, 37
    budget, 20–21
    business enterprises emergency loan guarantee, 124, 201
    economic program, 253
    foreign policy, 55
    hijacking of aircraft, 36
    marine resources and engineering development, 99
  National Science Board, 180
  Smithsonian Institution, 301
  space program, 14, 60, 64, 162, 344
  supersonic transport (SST), 365–366
Congress, House of Representatives, 130, 137, 142, 165, 178, 195, 198, 209, 211, 212, 222
  astronauts report to, 61, 253
  bills defeated, 78, 79, 344
  bills introduced, 42, 44, 63, 94, 142, 173, 209, 221, 301
  bills passed, 152, 157, 175, 210, 211, 212, 320, 323
  Committee of the Whole, 130
  Committee on Appropriations, 74, 173
    Subcommittee on Department of Interior and Related Agencies, 112
    Subcommittee on HUD-Space-Science, 80, 172
  Committee on Armed Services, 68, 163
    Subcommittee on Investigations, 112, 312
  Committee on Banking and Currency, 63
    Subcommittee on Consumer Affairs, 247
  Committee on Foreign Affairs, 280
    Subcommittee on Arms Control, 194
  Committee on Interstate and Foreign Commerce, 209
  Committee on Science and Astronautics, 42, 94, 280, 307, 315, 347, 350
    Advisory Panel, 318
    International Science Panel, 18
    membership, 38, 153
    NASA budget, 107
      testimony, 61–62, 63, 66–67, 69, 70–71, 75–76
    Subcommittee on International Cooperation in Science and Space, 134
    Subcommittee on Manned Space Flight, 264
    Subcommittee on NASA Oversight, 163–164, 244, 275
  Committee on Ways and Means, 280
  resolutions passed, 38
  Select Committee on Small Business, Subcommittee on Small Business Problems in Small Towns and Urban Areas, 166
Congress, Senate, 85, 308, 320, 349
  astronauts report to, 61, 253
  bills defeated, 80–81
  bills introduced, 17, 44, 69, 71–72, 76–77, 82, 124, 134, 197
  bills passed, 77, 78, 135, 178, 180, 198, 210, 216, 280, 284
  Committee on Aeronautical and Space Sciences, 132, 265
    committee members, 24, 38
    Fletcher, James C., nomination, 68
    NASA budget, 41, 44, 157
      testimony, 52–53, 77, 85–86, 93–94, 96–97
  Committee on Appropriations, 131, 195, 325
    Subcommittee on Appropriations for Dept. of Interior and Related Agencies, 79
    Subcommittee on HUD-Space-Science, 80
  Committee on Armed Services, 124, 219, 346
  Committee on Banking and Currency, 158, 194
  Committee on Commerce, Subcommittee on Oceans and Atmosphere, 292–293
  Committee on Foreign Relations, 327
    Subcommittee on Arms Control, 195
  Committee on Interior and Insular Affairs, 17
  Committee on Nutrition, 333
  Committee on Rules and Administration, 80
  conventions ratified, 252
  military procurement, 255, 266, 271
  NASA budget, 171, 178, 198–199, 209, 216
  nominations approved and confirmed, 69, 221, 257, 341
  nominations submitted to, 11, 45, 57, 152, 199, 254, 312

President's messages, 103
resolutions passed, 19, 24, 170
space shuttle, 177–178, 218
supersonic transport (SST), 80–81, 137, 138, 211
Connally, Secretary of the Treasury John B., 124
Constellation (transport aircraft),
Contamination, space, 56
Contract (see also under agencies, such as NASA, USAF).
  cost-plus-award-fee, 49, 71, 80, 81, 98, 115, 126, 131, 135, 175, 191, 193, 211
  cost-plus-fixed-fee, 19, 49, 105, 147, 171, 188, 255, 279, 344, 361
  fixed-price, 132, 171, 186, 195, 328
  fixed-price-incentive-fee, 341
  incentive, 69, 153, 286
  level-of-effort, 245, 356
  study, 85, 101, 132, 174, 190, 276, 330, 345–346
Contraves, A. G., 305
Convair 990 (*Galileo*) (jet research aircraft), 219, 251, 257
Convention for the Suppression of Unlawful Seizure of Aircraft, 252
Convention on International Liability for Damage Caused by Space Objects, 313
Conversion Research, Education and Assistance Act of 1971, 286
Cook, Don, 190
Cooke, H. Lester, 289
Cooperative Applications Satellite C (CAS–C), 106
Copernicus (lunar crater), 39
Coralie (French booster), 310
Corfield, Aerospace Minister Frederick (U.K.), 211
Corn Blight Watch Experiment, 98, 134, 168, 286
Cornell Aeronautical Laboratory, Inc., 131, 194, 290
Cornell Univ., 5, 102, 160, 194, 237, 249, 318, 344
  Laboratory for Planetary Studies, 331
Corp. for Public Broadcasting (CPB), 161
Corps of Engineers, 351–352
Cortright, Edgar M., 179, 191, 192
COSMIC. See Computer Software Management and Information Center.
Cosmic ray, 6–7, 63, 73, 83, 99, 148
Cosmonaut, 175, 247
  anniversary, 152, 167
  awards and honors, 180, 266
  death, 156, 180–181, 185, 186, 188, 191, 192, 196–197, 242
  memorials to, 200, 227, 239
  Nixon, President Richard M., message, 226
  Salyut 1–Soyuz 10 mission, 119, 283
  Salyut 1–Soyuz 11 mission, 154–156, 177, 180, 181, 283, 366
  Soyuz 9 mission, 167
  Soyuz 10 mission, 107, 110
  Soyuz 11 mission, 154

space cooperation, 100, 238, 340
training, 16
Cosmonaut Yuri Gagarin (U.S.S.R. tracking ship), 350
Cosmonautics (U.S.S.R. space encyclopedia), 91
COSMOS (consortium), 18
Cosmos 379 (U.S.S.R. satellite), 249
Cosmos 382, 249
Cosmos 390, 7
Cosmos 391, 9
Cosmos 392, 11
Cosmos 393, 15
Cosmos 394, 42, 66, 296
Cosmos 395, 49
Cosmos 396, 49
Cosmos 397, 54, 66, 296
Cosmos 398, 56, 249
Cosmos 399, 61
Cosmos 400, 78, 116, 296
Cosmos 401, 84
Cosmos 402, 93
Cosmos 403, 94
Cosmos 404, 96, 116, 296
Cosmos 405, 98
Cosmos 406, 102
Cosmos 407, 108
Cosmos 408, 110
Cosmos 409, 115
Cosmos 410, 124
Cosmos 411, 125
Cosmos 412, 125
Cosmos 413, 125
Cosmos 414, 125
Cosmos 415, 125
Cosmos 416, 125
Cosmos 417, 125
Cosmos 418, 125
Cosmos 419, 128
Cosmos 420, 133
Cosmos 421, 135
Cosmos 422, 139
Cosmos 423, 146
Cosmos 424, 148
Cosmos 425, 148
Cosmos 426, 153
Cosmos 427, 160
Cosmos 428, 174
Cosmos 429, 197
Cosmos 430, 200
Cosmos 431, 212
Cosmos 432, 219
Cosmos 433, 222
Cosmos 434, 227, 249
Cosmos 435, 240
Cosmos 436, 250
Cosmos 437, 254
Cosmos 438, 255
Cosmos 439, 262
Cosmos 440, 265
Cosmos 441, 268
Cosmos 442, 270
Cosmos 443, 280
Cosmos 444, 284
Cosmos 445, 284
Cosmos 446, 284

Cosmos 447, 284
Cosmos 448, 284
Cosmos 449, 284
Cosmos 450, 284
Cosmos 451, 284
Cosmos 452, 287
Cosmos 453, 291
Cosmos 454, 306
Cosmos 455, 323
Cosmos 456, 325
Cosmos 457, 327
Cosmos 458, 333
Cosmos 459, 333
Cosmos 460, 334
Cosmos 461, 337
Cosmos 462, 339
Cosmos 463, 343
Cosmos 464, 347-348
Cosmos 465, 351
Cosmos 466, 352
Cosmos 467, 353
Cosmos 468, 353
Cosmos 469, 358
Cosmos 470, 359
COSPAR. See Committee on Space Research.
Cotter, Rep. William R., 38, 240, 353-354
Coughlin, Rep. R. Lawrence, 38
Council for Advancement of Science Writing Seminar on Science and Public Policy, 53
Council for Scientific and Industrial Research (CSIR), South Africa, 267
Council of Mutual Economic Assistance (Comecon), 110
Transport Commission, 79
Counterbudget—A Blueprint for Changing National Priorities, 1971-1976, 50
Court Line Aviation, 301
Courten, Dr. Henry C., 167
CPB. See Corp. for Public Broadcasting.
Crab Nebula, 131
Crawford, David L., 36
Crater, 27, 39, 41, 48, 203
Crimea, U.S.S.R., 190
Crimean Astrophysics Observatory, 315
Cronkite, Walter, 327
Crozaz, G., 83
CSM. See Command and service module.
CTOL (conventional take off and landing aircraft), 130
Cuba, 215, 238, 312-313, 319
Cunningham, R. Walter, 166
Curtis, Sen. Carl T., 24
CVT: concept vertical testing.
Cyclic adenyl acid, 282
Cygnus (constellation), 89
Cygnus X-1 (pulsar), 131
Czechoslovakia, 237, 319, 337

D

Daley, Mayor Richard J., 256
Dana, William H., 56, 200, 225, 239, 265, 319, 337, 352

Daniel and Florence Guggenheim Foundation, 13
Daniel and Florence Guggenheim International Astronautical Award, 262
Daniel and Florence Guggenheim Space Theater, 279
Daniel Guggenheim Fund for the Promotion of Aeronautics, Inc., 13
Dankhoff, Walter F., 296
Dannenberg, Konrad K., 334
d'Arrest (comet), 358
Dartmouth College, 151
Darwin, Australia, 226
David, Dr. Edward E., Jr., 18, 63, 94, 119, 363
Davis, L/C Benjamin O., Jr. (USAF, Ret.), 193
Davis, Rep. John W., 38
Davis, Rep. Mendel J., 153
Davis, Dr. Raymond, Jr., 113
Dayton, Ohio, 248
Daytona Beach, Fla., 283
DC-8 (jet transport), 130
DC-10 (jet transport), 211, 220
Dean, James H., 289
Debré, Defense Minister Michel (France), 168
The Decision To Go to the Moon, 201, 258
Deckert, Wallace H., 70
Deep Space Network (DSN), 69, 313, 318
Defence Research Board (DRB) (Canada), 88
Defense, Dept. of (DOD) (see also U.S. Air Force, U.S. Army, and U.S. Navy, 39, 117, 124, 146, 246, 286
aircraft. See Aircraft.
award, 57
budget, 20-21, 37, 58, 124, 180, 212, 286, 322, 323, 325, 329, 344, 354, 364
communications satellite system, 74, 95, 153, 306, 366
computer programs, 23
contract, 8, 52, 310
cooperation, 123
NASA, 69, 75, 87, 224, 368
cooperation, international, 36-37
criticism, 78
MAST, Project, 351
missile program, 4, 8-9, 21, 43, 52, 125, 347
nuclear tests, underground, 300
Pentagon papers, 320
personnel, 49-50, 246, 294, 350
procurement, 360
R&D, 40, 118, 363
space program, 95, 306, 310, 366, 368
weapon systems, 78, 288
Defense Intelligence Agency, 95
Defense Navigation Satellite System (DNSS), 95
Defense Satellite Communications System (DSCS), 95, 306, 366
DeFlorez Training Award, 298

Deimos (Martian moon), 314, 331, 332, 340, 342, 365
Delaware, 97
Delta (booster upper stage) (see also Thor-Delta), 37, 164, 186, 269, 360–361
Delta Launch Vehicle System Review Board, 300
Dementyev, Pavel, 140, 148
Denmark, 96, 341
Denver, Colo., 281, 322, 346, 351, 353, 355
Denver mint, 243
Denver, Univ. of, 259
Descartes (lunar landing site), 29, 166, 323
De Simone, Herbert F., 45
Dewhirst, Dr. David, 324
DGLR. See German Society for Aviation and Astronautics.
Dial-a-plane system, 121
Diamant-B (French booster), 103, 342
di Carroba, R., 352
Dicke, Dr. Robert H., 18
Dickeson, Bob., 226
Dickson, Paul, 336
Dickstein, Morris, 5
Digital Equipment Corp., 16
Dijon, France, 150
Dillon, Mrs. Betty Crites, 312
Diritis, M., 357
Disarmament, 55, 137, 148, 189, 190, 194, 195, 218, 252, 264–265, 266
Distinguished Public Service Medal (NASA), 57
Distinguished Service Medal (NASA), 57, 59, 72, 302, 346–347
Distinguished Service Medal (USAF), 72, 266
Distinguished Service Medal (USN), 72
District of Columbia, 99
Dixon, Mrs. Jeane L., 189
DNSS. See Defense Navigation Satellite System.
Dobrovolsky, Georgy T., 154, 180, 181, 186, 192, 200, 226, 283
Dobrovolsky, Maria, 226
Dobrovolsky, Natasha, 226
Dobrynin, Ambassador Anatoly R. (U.S.S.R.), 70, 191, 194, 242
Docking
  Apollo 14, 25, 29, 30, 35, 38
  Apollo 15, 203
  multiple docking adapter, 284, 323, 346, 353, 355
  Soyuz 10–Salyut 1, 108, 110, 283
  Soyuz 11–Salyut 1, 154, 158, 191, 283, 366
  trainer, 16, 77
  U.S.–U.S.S.R. cooperation, 117, 177, 244, 249, 325, 332, 334, 340–341, 347, 353, 369
DOD. See Defense, Dept. of.
Domestic Action Council, 351
Donaldson, Dr. Coleman D., 15, 290
Donlan, Charles J., 75, 302

Donnelly, John P., 200, 230, 337, 353
Donner, Frederic G., 152
Doppler Geodetic Satellite Program, 95
DOT. See Transportation, Dept. of.
Doub, William O., 221, 290
Douglas Aircraft Co., 338
DOVAP (ground transmitter), 187
Dowling College, 167
Downing, Rep. Thomas N., 38, 152
Drake, Dr. Frank D., 249, 344
Draley, Eugene C., 85
Draper, Dr. Charles Stark, 261
Draper, William, 315
DRB. See Defence Research Board (Canada).
Drug addiction, 215
Dryden, Dr. Hugh L., 106, 112–113
Dryden, Hugh L., Memorial Fellowship, 70
Dryden, Hugh L., Memorial Fund, 112
Dryden, Hugh, Research Lecture, 15
DSCS. See Defense Satellite Communications System.
Dscs 2–1, 306, 313
Dscs 2–2, 306, 313
DSN. See Deep Space Network.
Duboshin, G. N., 311
DuBridge, Dr. Lee A., 16
Dudley Observatory (Albany, N.Y.), 107, 108, 335
Duffy, B/G Robert A. (USAF), 178
Duke, L/C Charles M., Jr. (USAF), 61, 323
Dulles International Airport, Va., 212, 278
Dune Crater (moon), 205
Dunkelman, Lawrence, 335
Dupik, Dr. Vladimir, 177
Dupree, Dr. A. Hunter, 151
Durham, N.C., 170
Dutourd, Jean, 216
Dynalectron Corp., Land-Air Div., 13

E

Earhart, Amelia, 243
Earth
  atmosphere. See Atmosphere, upper.
  crater, 39, 41, 48
  density, 310
  Earth Awareness Foundation, 137
  earthquake, 64
  energy resources, 154
  environmental problems (see also Air pollution; Noise, aircraft; Pollution), 10, 17, 20, 42, 49, 96, 147, 160, 162, 172, 178, 191, 215, 368
  Environmental Science: Challenge for the Seventies, 180
  magnetic field, 104
  marine research and development, 99, 104
  oceans. See Oceanography.
  origin of, 345
  photographs, 286

resources measurement (see also Earth Resources Technology Satellite), 10–11, 33, 44, 67, 97, 119–120, 132–133, 133–134, 138–139, 139–140, 174–175, 224, 324
This Island Earth (NASA SP–250), 48
Earth Awareness Foundation, 137
Earth Observations Aircraft Program, 104, 139, 171
Earth Resources Aircraft Program (ERAP), 120
Earth resources experiment package (EREP), 10, 107, 173, 175
Earth Resources Laboratory, 19, 49
Earth Resources Technology Satellite (ERTS), 19, 22, 61, 67, 87, 95, 97, 119–120, 135, 173, 198, 281, 356, 369, 370
Eastern Airlines, Inc., 54, 124, 133, 165
Eastern Test Range (ETR) (see also Cape Kennedy and Kennedy Space Center).
  contract, 98, 310
  launch
    Atlas-Agena, 341
    Atlas-Centaur, 14
    Dscs 2–1, 306, 313
    Dscs 2–2, 306, 313
    Explorer 43 (IMP–I), 72
    failure, 126–127
    High Energy Astronomy Observatory (HEAO), 329
    Intelsat-IV F–2, 14
    Mariner 8 (Mariner-H), 126–127
    Mariner 9 (Mariner-I), 148
    Nato 2 (NATO–B), 36
    Oso 7 (OSO–H), 269
    Pioneer-F, 322
    unidentified satellite, 122, 341
Easton, Richard A., 179
Echo 1 (comsat), 157
Eckert, Dr. John P., 218
Eclipse, lunar, 44
Eclipse, solar, 68, 167
Economic Affairs, Dept. of (proposed), 81
EDP Technology, Inc., 131, 194
Edwards AFB, Calif., 11, 108, 131, 174, 186, 255, 256
Effelsberg, W. Germany, 277
Eggers, Dr. Alfred J., 61
Eglin AFB., Fla., Climatic Laboratory, 258
Egypt, 124, 360
Eisenhower, President Dwight D., 59, 243, 248, 265, 308
El Centro, Calif., 247, 264
ELDO. See European Launcher Development Organization.
Electric field, 74, 78, 79
Electric propulsion, 244
Electris (Mars), 337
Electronics Research Center (ERC), 362
Ellington AFB, Tex., 22, 207, 222
Embolism, 187
Emergency Conversion Loan Act, 141

Emergency loan guarantee, 194, 201, 212, 216
Emergency Loan Guarantee Board, 254
Emme, Dr. Eugene M., 234, 265
Emory Univ., 221
Employment of New Ph.D.'s and Postdoctorals in 1971 (NRC study), 252
Encke (comet), 358
Endeavor (Apollo 15 command and service module). See Command and service module.
Energy absorber, 251, 259
Energy sources, 129, 154
Eneyev, Timur M., 127
Engine (see also individual engines, such as F–1, H–1)
  aircraft, 271, 302, 307
    CTOL, 130
    jet, 38, 44, 88, 124, 125, 173–174, 212, 221
    noise, 61, 83, 130, 241, 247, 251, 290, 295–296
    quiet, 83, 96, 130, 241
    STOL, 97
    supersonic transport, 61, 126, 174
    VTOL, 300, 329
  electric, 244
  nuclear (see also NERVA), 235, 250, 320
  rocket, 96, 136, 158
    cooperation, 87
    space shuttle, 80, 87, 98, 160, 161, 193, 218, 233, 240, 245, 276, 320, 356
    testing
      liquid, 87
      hydrogen, 87
      hydrogen/oxygen, 161
Engineers, 107, 119, 308
  employment, 77, 104–105, 117–118
Engle, Capt. Joseph H., (USAF), 234
Engler, Martin, Jr., 242
Enrollment Increase in Science and Mathematics in Public Secondary Schools, 1948–49 to 1969–70 (NSF 71–30), 288
Environmental Data Service, 219
Environmental program, U.S., 42
Environmental Protection Agency (EPA), 104, 290, 351
Environmental Science: Challenge for the Seventies, 180
Environmental Science Services Administration (ESSA), 19
Eole (CAS–1) (French Cooperative Applications Satellite), 230, 250, 254, 357, 369
Eole Data Interpretation Group, 231
EPA. See Environmental Protection Agency.
epndb: effective perceived noise in decibels.
ERAP. See Earth Resources Aircraft Program.
ERC. See Electronics Research Center.
EREP. See Earth resources experiment package.

Ericson, David B., 302
ERNO-Raumfahrttechnik GmbH, 3, 19
Eros (asteroid), 310
EROWS. See Expendable remote-operating weather station.
ERTS. See Earth Resources Technology Satellite.
ERTS-A, 97, 135, 173
ERTS-B, 135, 173
ESC. See European Space Conference.
Esch, Rep. Marvin L., 38
ESP. See Extrasensory perception.
ESRANGE (European Sounding Rocket Range), Kiruna, Sweden, 213
ESRO. See European Space Research Organization.
ESRO-4 (scientific satellite), 356
ESSA. See Environmental Science Services Administration.
"Estimated Academic R&D Direct Price Trends 50 Percent Higher Over Decade 1961-71," 305
Etam, W. Va., 292
ETR. See Eastern Test Range.
Europa I (ELDO booster), 328, 352
Europa II, 310, 327-328, 352
Europa III, 297, 352
European Broadcasting Union, 23
European Launcher Development Organization (ELDO), 278, 291, 297, 310, 347, 352
European Organization for Nuclear Research, 19
European Space Conference (ESC), 16, 45, 48, 75, 165, 335, 340, 369
European Space Research Organization (ESRO), 19, 177, 220, 265, 297, 308, 347, 356
European Space Symposium, Eleventh, 142
Eutrophication, 104
EVA. See Extravehicular activity.
Evans, Albert J., 272
Evans, L/Cdr Ronald E. (USN), 228-229, 234
Exceptional Scientific Achievement Medal (NASA), 255, 279, 302
Exceptional Service Medal (NASA), 188, 255, 279, 302, 314
Exhibit, 152, 153, 345
Expendable remote-operating weather station (EROWS), 295
Explorer (spacecraft), 186
Explorer 1 (satellite), 5
Explorer 30 (*Solrad 8*) (Solar Radiation Satellite), 190
Explorer 35 (Interplanetary Monitoring Satellite (IMP-E), 64
Explorer 37 (*Solrad 9*), 190
Explorer 41, 73
Explorer 42 (*Uhuru*) (Small Astronomy Satellite), 89, 113, 131, 319
Explorer 43 (IMP-I), 240, 365
Explorer 44 (*Solrad 10*), 189-190, 240, 365
Explorer 45 (Small Scientific Satellite), 319, 365, 369
Extrasensory perception (ESP), 46, 170
Extraterrestrial life, 167, 248, 338, 345, 368, 370
Extravehicular activity (EVA), 153, 328
  Apollo 14, 25-26, 41, 47, 48, 51
  Apollo 15, 115, 141, 142, 156, 201 203-207, 217, 221, 225, 227, 228, 229
  Apollo 16, 61, 289, 293, 323
  Skylab, 288
Eyewitness to Space, 289
Eyles, Donald E., 57

F

F-1 (rocket engine), 1, 158
F-1-m (U.S.S.R. launch vehicle), 362
F-1-r (U.S.S.R. launch vehicle), 362
F-4 (Phantom (fighter aircraft), 124, 246, 286, 303
F-5D (fighter aircraft), 239
F-9 (People's Republic of China supersonic fighter aircraft), 140
F-14 (Tom Cat) (supersonic fighter aircraft), 46, 70, 112, 124, 141, 152, 210, 211, 212, 246, 252, 255, 271, 272, 284, 286-287, 323, 365
F-14A, 11, 165, 298
F-14B, 210, 211, 212
F-15 (fighter aircraft, 70, 170, 212, 255, 272, 298, 303, 344
F-15 (ELDO launcher), 352
F-90 (fighter aircraft), 115
F-100 (supersonic fighter), 239
F-101 (supersonic fighter), 239
F-102 (supersonic fighter), 239
F-104 (fighter aircraft), 239
F-111 (supersonic fighter aircraft), 87, 108, 116, 131, 165, 212, 262, 272, 300, 341, 344, 356
F-111A, 14
F-111C, 353
F-111D, 314
F-147 (fighter aircraft), 219
FAA. See Federal Aviation Administration.
FAA Statistical Handbook of Aviation, 327
FAI. See International Aeronautical Federation.
Fairbanks, Alaska, 47, 76, 77, 310
Fairchild Hiller Corp., 74, 85, 98, 115
Fairchild Industries, Inc., 115, 300, 308
Fairchild, Sherman M., 85
Falcon (*Apollo 15* LM). See Lunar module.
Faller, Dr. James E., 144
Fanale, Dr. Fraser P., 12
Farley, Clare F., 197
FB-111 (supersonic fighter), 262, 279
FBI. See Federal Bureau of Investigation.
FCC. See Federal Communications Commission.

Federal Aviation Administration (FAA)
  air traffic control, 85, 107, 121, 139, 193, 219–220, 260
  aircraft certification, 218–219
  airports, 134, 165, 182, 194, 305, 320, 334
  airways, 80
  award, 165
  collision prevention, 10, 107, 193
  contract, 85, 182, 198, 320
  cooperation, 3, 71, 122, 187, 218–219, 219–220, 255, 271, 297
  General Aviation Accident Prevention Program, 277
  landing system, 305, 334
  noise, aircraft, 83, 198
  organization, 65, 82
  personnel, 2, 239, 266, 323
  satellite communications systems, 3, 165, 255, 297
  statistics, 104–105, 134, 255, 277, 327
  transport, supersonic (see also Supersonic transport), 332, 351
Federal Bureau of Investigation (FBI), 266
Federal Communications Commission (FCC), 60, 65, 69, 71, 74, 176, 293
Federal Electric Corp., 158
Federal Funds for Academic Science, Fiscal Year 1969 (NSF 71-7), 58
Federal Funds for Research, Development, and Other Scientific Activities, 251
Federal R&D Expenditures Related to Budget Functions, 1960–72, 175
Federal Railway Administration, 323
Federal Reserve System, 308
Federal Support to Universities and Colleges, Fiscal Year 1970, 161
Federal Trade Commission (FTC), 247
Federal Women's Award, 38
Fédération Aéronautique Internationale (FAI). See International Aeronautical Federation.
Federation of American Scientists, 124
Fedoseyev, Anatoly, 147, 168, 171, 176, 190, 329
Feld, Bernard T., 273
Feoktistov, Konstantin P., 111, 119
Fiat, 3
Final Report of the Ad Hoc Supersonic Transport Review Committee of the Office of Science and Technology, March 30, 1969, 232
Fine Arts Commission, 309, 323, 330
Finerman, Dr. Aaron, 192
Fink, Daniel J., 133
Finland, 341
Flandro, Dr. Gary A., 85
Fleischer, R. L., 83
Fletcher, Dr. James C., 139, 298
  aeronautics, 280, 307
  Apollo program, 324
  appointment, 49–50, 57, 59, 64, 68, 69, 72, 114, 368
  awards and honors presented by, 279, 302
  budget, 57, 172, 280, 324
  international cooperation, space, 275
  Langley Research Center visit, 152
  Lewis Research Center visit, 246–247
  Lunar module (LM) presentation to National Air and Space Museum, 195
  Manned Spacecraft Center visit, 146–147
  Mariner 9 (Mariner-I), 146
  noise prevention, aircraft, 307
  Paine, Dr. Thomas O., portrait unveiling, 314–315
  press conference, 57, 127, 246
  space program, national, 57, 127, 172, 256, 263–264, 272, 275, 280
  space shuttle, 172, 218, 272, 275, 280
  space station, 172
  technology utilization, 160, 172, 272–273, 340
  U.S.S.R. space program, 191–192
Flight Achievement Award (AAS), 179
Flight Research Center (FRC) (NASA)
  aircraft electronic control system, 73
  contract, 345
  employment, 238
  lifting-body flight
    M2-F3, 42, 200, 225, 239, 337
    X-24A, 11, 39, 49, 65, 85, 142, 210, 351
  personnel, 339
  space shuttle pilot study, 308
  supercritical wing. See Wing, aircraft, supercritical.
Flight Safety Foundation, 314
Florida, 2, 301
Florida, Legislature, 145
Florida, Univ. of, 18, 288, 292
Flowers, Rep. Walter, 38
Flugge-Lotz, Dr. Irmgard, 298
Flying Spot (x-ray unit), 278
FOBS. See Fractional orbiting bombardment system.
Food and Agricultural Organization, 104, 171
For the Benefit of All Mankind: A Survey of the Practical Returns from Space Investment (House report), 350–351
Ford Foundation, 177
Ford, Rep. Gerald R., 131
Ford Motor Co., 309
Ft. Greely, Alaska, 222, 225, 230
Fort Pierce, Fla., 24
Fort Worth, Tex., 212
47J Sioux (helicopter), 248
Foster, Charles R., 295
Foster, Dr. John S., Jr., 95, 300
Foundation for Research on the Nature of Man, 170
Fox, Dr. Leo, 44
Fox Main, Dew Line Station, Alaska, 74, 78, 79
Fra Mauro (lunar crater), 24, 27, 47, 64, 98, 167, 265

Fractional orbiting bombardment system (FOBS), 303, 362
France, 242, 252
 air-cushion railroad, 82
 aircraft (see also Concorde), 224, 346
 Apollo 14 mission, reaction, 43
 Apollo 15 mission, reaction, 216
 award, 191
 booster, 310
 Concorde (U.K.-France supersonic transport), 48, 147, 214, 248, 249, 280, 332, 346
 international cooperation, 50, 287, 299
 international cooperation, space, 16, 18, 57–58, 96–97, 148, 230, 235, 259, 291, 297, 299, 310, 330, 358, 359, 370
 launch, satellite, 103, 230, 342, 368
 missile program, 224
 National Assembly, 73
 satellite, 57–58, 230, 342, 357, 365
 space program, 58, 230, 234, 254, 311–312, 330, 358, 362
 UV panorama (S–138) experiment, 38
Frankfurt, Univ. of, 325
Franklin Institute, 359
Franklin Mint, 247
Franklin Mint Collectors Society, 247
Franz Josef Archipelago, 304
FRC. See Flight Research Center.
Freedom 7 (spacecraft), 122
Freedom 7 mission, 42
Freeman, Dr. John W., Jr., 288, 291, 293
Freitag, Capt. Robert F. (USN, Ret.), 12
Frey, Rep. Louis, Jr., 38
Friedheim, Jerry W., 146, 152, 211, 212
Friedman, Dr. Herbert, 16, 73
Froehlke, Secretary of the Army Robert F., 186, 294
Frog experiment, 127
Frontier Airlines, 128
Frost, Dr. James D., 193
FRUSA (flexible rolled-up solar array), 8
Frutkin, Arnold W., 133, 198
FTC. See Federal Trade Commission.
Fulton, Rep. James G., 38, 280, 282
Fuqua, Rep. Don., 38

G

Gabor, Dr. Dennis, 307, 309
Gabriel, David S., 167, 243, 323
Gagarin, Yuri, Gold Medal, 266
Gagarin, Col. Yuri A. (U.S.S.R.), 16, 100, 152, 167
Galaxy, 1, 201, 253, 279, 315
Galilei, Galileo, 97
Gambrell, Sen. David H., 37
Gamma ray, 181
Gangler, James J., 291
GAO. See General Accounting Office.
GARP. See Global Atmospheric Research Program.
Garrett Corp., 147
Gast, Dr. Paul W., 7, 45, 190, 226, 228, 245

GATE (GARP Atlantic Tropical Experiment), 180
Gates Learjet Corp., Aircraft Div., 44
Gatland, Kenneth, 115
Gault, Donald E., 39
Gavin, James J., Jr., 57
Gazenko, Prof. Oleg G., 36, 235
GCA Corp., 236, 251, 257
GE. See General Electric Co.
Gebben, Vernon D., 106
Gehrig, James J., 265
Gemini (program), 324
Gemini (spacecraft), 286
Gemini 5 mission, 240
Gemini 7 mission, 153
Gemini 8 mission, 240
Gemini 11 mission, 240, 345
General Accounting Office (GAO), 1, 78, 201, 218, 233, 246, 252, 294, 353, 356
General aviation, 105, 122, 276–277
General Dynamics Corp.
 Aerospace Div., 262, 300
 award, 17
 CL–215 Air Tanker, 129
 contract, 98, 101, 341
 Convair Aerospace Div., 337
 Convair Div., 101, 127, 136, 150
 F–111, 165, 212, 262, 300, 341, 344
 research and applications module (RAM), 3, 337
General Electric Co. (GE), 83, 133
 Aerospace Group, 195
 Aircraft Engine Group, 83, 241
 contract, 82, 174
 GE Houston Operations, 171
 quiet jet engine, 83, 130, 241, 365
 R&D Center, 99
 research and applications module (RAM), 3
 space shuttle, 195
 supersonic transport engine, 82, 135
General Motors Corp., 24
General Services Administration (GSA), 357
Geneva, Switzerland, 19, 23, 88, 132, 173, 177, 182, 199, 250, 257
Gentry, Maj. Jerauld R. (USAF), 42
George Washington Univ., 176, 258, 342
Georgia Institute of Technology, 13
Georgia, Univ. of, 81, 232
German Society for Aviation and Astronautics (Deutsche Gesellschaft fuer Luft- und Raumfahrt E. V.) (DGLR), 142
Germantown, Md., 115
Germany, East, 237, 287, 319
Germany, West, 81, 84–85, 312
 aircraft, 246
 Apollo 14 mission, reaction to, 43
 computer program, 122
 cooperation, 287
 cooperation, space, 19, 58, 96, 145, 259, 310, 352, 356, 357, 370
 R&D, 362
 space program, 3, 16, 325
 space shuttle, 291

Gervais, Joseph, 243
GET: ground elapsed time
Getlein, Frank, 289
Giacobini-Zinner comet, 358
Giaconni, Dr. Riccardo, 89, 113
Gilbert, B/G William W. (USAF), 193
Gillilland, Whitney, 348–349
Gilruth, Dr. Robert R., 14, 84, 171, 177, 181, 305, 325, 332, 334
Glenn, Col. John H., Jr. (USMC, Ret.), 198
Glennan, Dr. T. Keith, 253, 257
Glider research, 320
Global Associates, 191
Global Atmospheric Research Program (GARP), 103, 178, 303–304
  Atlantic Tropical Experiment (GATE), 180
  Data Management Panel, 303
Glushko, Petrovich, 91
Goals and Means in the Conquest of Space, 183, 363
Goat experiment, 311
Goddard Institute for Space Studies (GSFC), 32, 45, 50, 358–359
Goddard, Mrs. Esther, 69, 214, 347
Goddard Memorial Dinner, 69
Goddard Memorial Lecture Award, 13
Goddard Memorial Trophy, 69
Goddard, Dr. Robert H., 13, 17, 234, 347
Goddard, Robert H., Library, 214
Goddard Space Flight Center (GSFC) (NASA), 250, 335
  Apollo 15 tracking, 207
  astronomical research, 35–36, 49–50, 102, 339
  award, 302
  contract, 134–135, 356
  cooperation, 102–103
  employment, 238
  Laser Data Systems Branch, 154
  meeting, 129, 146, 297
  patent, 269
  position-location and aircraft communications experiment (PLACE), 362
  satellite monitoring
    Explorer 43 (IMP–I), 72, 240
    Explorer 45 (SSS–A), 319
    ITOS–B, 294
    Oso 7, 269
  Scientific Colloquium, 154
  sounding rocket experiments (see also Sounding rockets)
    airglow, 108
    atmospheric data, 2, 5, 8, 78, 174–175, 187
    electric field, 78, 79
    energetic particles and fields, 14
    grenade experiment, 259
    ionosphere, 22, 31, 115–116
    magnetic field, 78, 79
    micrometeoroid collection, 108
    ozone, 54, 63, 64–65, 74–75, 222, 225, 230
    polar cap absorption, 246
    stellar astronomy, 13, 119, 323
  test and support mission, 292
  x-ray studies, 226
Goland, Martin, 15, 75
Gold, Dr. Thomas, 5, 237
Goldhaber, Dr. Gerson, 36
Goldich, S. S., 83
Goldstein, Richard M., 335
Goldstone, Calif., tracking antenna, 69, 102, 313, 318
Goldwater, Sen. Barry M., 24, 79
Goldwater, Rep. Barry M., Jr., 38
Golf Magazine, All-America Award, 106
Golovine Award, 85
Gordon, Capt. Richard J., Jr. (USN, Ret.), 345
Gossick, M/G Lee V. (USAF), 193
Gotlieb, Alan, 106
Government-Industry System Safety Conference, 146
Government Printing Office (GPO), 327
Grants, N. Mex., 39
Gravel, Sen. Mike, 272
Greeley, Ronald, 39
Green, Constance McLaughlin, 75
Greenville, S.C., 140
Griffin, Sen. Robert P., 134
Grissom, L/C Virgil I. (USAF), 10, 199, 357
Grissom, Mrs. Virgil I., 10, 131
Griswold, Solicitor Gen. Erwin, 49
Grobecker, Dr. Alan J., 323
Gromyko, Foreign Minister Andrey A. (U.S.S.R.), 153, 252, 272
Grumman Corp., 290, 294, 295
  Grumman Aerospace Corp., 167, 191
  aircraft, 128, 219, 288, 290, 338
    F–14, 46, 112, 141, 152, 163
    F–14A, 11, 112
  award, 15, 57, 279
  contract, 253, 288, 294, 338
  space shuttle, 281
GSA. See General Services Administration.
GSFC. See Goddard Space Flight Center.
Gudkov, Oleg, 211
Guerney, Sen. Edward J., 134
Guggenheim, Daniel, 13
Guggenheim, Daniel, Medal, 326
Guggenheim, Daniel and Florence, International Astronautics Award, 262
Guggenheim, Harry F., 13, 534
Gulf of Mexico, 200
Gum Nebula, 36
Guppy (cargo aircraft), 242
Gurovsky, Dr. N. N., 284, 359
Gvishiani, Dzherman M., 287
GX3+1 (x-ray source), 356

H

H–1 (rocket engine), 158
Hadley-Apennine region (lunar landing site), 202, 203, 254
Hadley Delta (lunar mountain), 120, 223
Hadley Plains (moon), 203

Hadley Rille (lunar landing site), 153, 144, 183, 219, 226, 291
Hafele, Dr. Joseph C., 296
Hage, George H., 299
Hagen, Dr. John P., 135
Haggerty, James J., 58, 90
Haglund, Howard H., 47, 55
The Hague, Netherlands, 103
Haise, Fred W., Jr., 61, 75, 171, 266
Halaby, Najeeb E., 109, 148, 257, 297
Haley Astronautical Award, 75
Hall, Albert C., 294
Hall, Asaph, 331
Hall, Floyd D., 54
Hall, R. Cargill, 261
Halley's comet, 358
Ham (chimpanzee), 32
Hanbury, Una, 106
Hamilton, Jeffrey T., 192, 301
Hammarskjold, Knut, 177
Handler, Dr. Philip, 106, 287
Hanna, Rep. Richard T., 38, 222
Hansen, Grant L., 186, 276
Hansen, James E., 50
Haramura, Hiroshi, 83
Harbert, Mary Ann, 352
Hardsite Defense, 43
Hardy, Dr. James P., 90
Hargis, Dr. William J., Jr., 292
Harmon International Aviator's Trophy, 135
Harnett, Daniel J., 76, 354
Harr, Dr. Karl G., Jr., 125
Harrier (U.K. v/STOL aircraft), 80, 300, 329
Harriot, Thomas, 97
Harris, Louis, survey, 220
Harrison, Dr. Donald C., 235, 301
Hart, H. R., Jr., 83
Hartzog, George B., 112
Harvard Univ., 18, 277
Harvest Moon (proposed lunar mission), 219
Harvey Mudd College, 152
Haskin, Dr. Larry A., 279
HATS. See Huntsville Assn. of Technical Writers.
Haughton, Daniel J., 2, 36, 161, 221, 294
Hawaii, 98, 222
Hawaii, Univ. of, 161, 345
Hawker Siddeley Group Ltd., 3
  Hawker Siddeley Aviation, Ltd., 329
  Hawker Siddeley Dynamics Ltd., 19
Hawkins, Howard R., 71
Hayes International Corp., 158
Health, Education, and Welfare (HEW), Dept. of, 38, 58, 161
HEAO. See High Energy Astronomy Observatory.
Heat flow experiment, 142, 144, 205, 245, 291, 294, 295
Heat shield, 335
Hebert, Rep. F. Edward, 112, 163
Hechler, Rep. Ken, 38
Heckscher, John L., 199
Hedrick, Ira G., 15

Heffner, Dr. Hubert, 197
Helicopter, 295
  accident, 13, 14, 290
  astronaut pickup, 30
  comsat communication, 74
  military, 101, 294, 351, 357, 365
  record, 365
  U.S.S.R., 147
Helicopter Corp. of America, 330
Helio Aircraft Corp., 300
Helms, Richard, 124
Helsinki, Finland, 148, 189, 252
Hempstead, N.Y., 233
Henize, Dr. Karl, 359
HEOS-A2 (Highly Eccentric Orbit Satellite), 356
Heppenheimer, T. A., 315
Hercules (constellation), 93
Hercules, Inc., 170
Hero of Socialist Labor, 218
Hero of the Soviet Union, 180
Herzberg, Gerhard, 307
Heuss Island, Arctic Ocean, 304
HEW, See Health, Education, and Welfare (HEW), Dept. of.
Hewlett Packard Co., 350
Hibbs, Dr. Albert R., 47
Hickam AFB, Hawaii, 207, 219
High Energy Astronomy Observatory (HEAO), 67, 85, 189, 276
Highly Eccentric Orbit Satellite. See HEOS-A2.
High Speed Test Center, 182
Hijacking of aircraft
  Congress, 103
  ICAO consideration of, 55
  Nixon, President Richard M., 36, 54, 103
  protective measures against, 133, 212–213, 278
  statistics, 157
  treaty, 55, 103
  U.S., 36, 55, 103, 157
Hill, Louis W., Space Transportation Award, 298, 323
Hillier, James, 170
Hills, Dr. H. Kent, 144, 288, 293
Himmel, Dr. Seymour C., 349
Hines, William, 192
Hiroshima, Japan, 224
Hirsch, Richard, 9
Hirten, John E., 107
History Manuscript Award (AIAA), 75
History of Astronautics Symposium, Fifth, 261
Hitachi Central Research Laboratory, 231–232
Hitt, Patricia R., 38
Hodgson, Secretary of Labor James D., 94
Hoff, Dr. Nicholas J., 15
Hoffman, Gen. H., 352
Hoffman, Dr. William F., 358
Hofstein, Steven R., 360
Hogan, Joseph M., 49
Hogarth, Laurence T., 161
Holloway, Frederic A. L., 170

Holmes, Jay E., 58
Holmfeld, John D., 126
Holmquest, Dr. Donald, 138
Holography, 142, 307
Homestake Gold Mine, South Dakota, 113
Honeywell, Inc., Aerospace Div., 264
Honolulu, Hawaii, 349
Hood, Robert H., 231
Hope, Bob, 70
Hord, Dr. Charles W., 195
Horowitz, Dr. Norman H., 90
Horttor, Richard L., 335
Hotz, Robert B., 119
Houghton, Dr. John, 242
Housing and Urban Development (HUD), Dept. of, 3, 135, 198, 209, 211, 216, 226
 funding,
Houston Intercontinental Airport, 278
Houston, Tex., 138, 166, 288, 290
 Apollo 14 mission, 25, 30
 Apollo 15 mission, 203, 222, 225–226, 302, 335
 Lunar Receiving Laboratory, 30, 226
 meeting, 5, 114, 312, 369
Houston, Univ. of, 50, 63
Hovertrain, 348
Hubbard, Dr. Jerry S., 90
Hubbel, Charles H., 282
HUD. See Housing and Urban Development, Dept. of.
Huddle, Franklin P., 56
Hudson (bomber aircraft), 115
Hughes Aircraft Co., 298, 312
Hughes, Howard, 44, 197
Hughes Tool Co., 197
Human Resources, Dept. of (proposed), 81
Humphrey, Sen. Hubert H., 162, 195, 266, 358
Humphreys, Dr. James W., Jr., 12, 245
Hungary, 43, 237, 242–243
Hunn, Spencer S., 3
Huntington Beach, Calif., 38
Huntington Institute of Applied Medical Research, 112
Huntsville, Ala., 5, 142, 279, 321
Huntsville Assn. of Technical Writers (HATS), 321
Hurricane Camille, 172
Husain, Dr. Liaquat, 257, 260, 265
Hyde, Floyd H., 3
Hydrofoil craft, 18
Hydrofoil warship, 312
Hydrogen, 60, 89, 93, 175, 359
Hydrogen bomb, 260, 273
Hydrogen gas cloud, 2
Hydroxyl, 194
Hypo, Project, 104

I

IAA. See International Academy of Astronautics.
IAEA. See International Atomic Energy Agency.
IAF. See International Aeronautical Federation and International Astronautical Federation.
IATA. See International Air Transport Assn.
IBM. See International Business Machines Corp.
ICAO. See International Civil Aviation Organization.
Icarus (asteroid), 310
ICBM. See Intercontinental ballistic missile.
IEEE. See Institute of Electrical and Electronic Engineers.
IFR: instrument flight rules.
Il–18 (U.S.S.R. airliner), 12, 17, 38, 140
Il–62 (U.S.S.R. airliner), 140
ILLIAC IV (computer), 23
Illinois, Univ. of, 236
ILS: instrument landing system.
IMP. See Interplanetary Monitoring Platform.
IMP–E (Interplanetary Monitoring Platform), 64
IMP–H, 356
Implantable Biotelemetry Systems (NASA SP–5094), 193
Improved Tiros Operational Satellite. See *Itos 1*, etc.
Independence (presidential aircraft), 248
India, 19, 105, 111, 115, 235, 264, 282
Indian Ocean, 109, 150, 319–320
Indian River, Fla., 13, 290
Indiana, 199
Industrial R&D Spending, 1970 (NSF 71–39), 352
Industrial Research Magazine award, 262
Information retrieval, 32, 81
Informatics Tisco, Inc., 69
Ingersoll-Rand Co., 78
Institute of Electrical and Electronic Engineers (IEEE), 308
Institute of Planners of California, 107
Institute of Rehabilitation Medicine, 189
Institution of Mechanical Engineers (London), 84
INTELSAT. See International Telecommunications Satellite Consortium.
Intelsat (communications satellite series), 198
Intelsat I (Early Bird), 14, 98
Intelsat III series, 14, 69
Intelsat IV series, 14, 294
Intelsat-IV F–2, 14, 82, 341, 354, 356–357, 365
Intelsat-IV F–3, 342, 352, 354, 365
Intelsat-IV F–4, 356
Intelsat-IV F–5, 356
Intelsat-IV F–6, 356
Interagency Aircraft Noise Abatement Program, 295
Interagency Committee on Back Contamination, 115
Intercontinental ballistic missile (ICMB), 4, 68, 77, 84, 103, 107, 117, 135, 146, 150, 248, 264–265, 303, 357–358, 362

Intercosmos 5 (U.S.S.R. satellite), 337, 366
Interdisciplinary research relevant to problems of our society (IRPOS), 178
Interferometry, 102
Interior, Dept. of, 81, 82, 351–352
International Academy of Astronautics (IAA), 261, 277
International Aeronautical Federation (IAF), 262
   awards, 262–263
International Aerospace Hall of Fame, 242
International Air Safety Seminar, 295
International Air Transport Assn. (IATA), 177
International Assn. of Fire Fighters, 23
International Astronautical Federation (IAF), 260–261
International Atomic Energy Agency (IAEA), 253, 257
International Business Machines Corp. (IBM), 78
   Federal Systems Div., 85
International City Management Assn., 177
International Civil Aviation Organization (ICAO), 55, 312, 327, 349
International Conference on Peaceful Uses of Atomic Energy, 250
International Congress of the History of Science, Thirteenth, 234
International cooperation (see also Disarmament and Treaty)
   astronomy, 102, 190, 248–249
   computer programs, 230–232
   ecology, 17
   natural resources, 164
   nuclear power, 50, 197, 250
   science and technology, 17, 55, 177, 299
   systems analysis, 287
International cooperation, space (see also European Launcher Development Organization; European Space Research Organization; Global Atmospherics Research Program; International Telecommunications Satellite Consortium), 15–16, 17, 37, 44, 45, 55, 69–70, 77, 101, 119–120, 127, 275, 311
International Cooperation in Outer Space: A Symposium, 132
   lunar surface samples, 6, 10, 13, 16, 58, 159, 190, 308, 369
   military, 36–37, 41, 369
   satellite, 15, 18, 57–58, 219–220
      communications. See Communications satellite, cooperation.
   earth resources, 10, 119, 132, 198, 369
   Europe, 15–16, 19, 57–58, 297, 310, 319
   NASA-
      -Canada, 88, 356, 369
      -Europe, 75
      -France, 230, 357, 369
      -Germany, West, 356
      -Italy, 109, 369
      -NATO, 36–37, 41, 369
      -U.K., 348, 369
   U.S.S.R.-Czechoslovakia, 337
      -France, 358–359
   sounding rocket. See Sounding rocket, international programs.
   space research, 16
      Europe, 16, 141–142, 237, 347, 352
      France-U.S.S.R., 147, 235, 330, 358, 359
      U.S.-Australia, 15
         -Canada, 15–16, 96
         -Denmark, 96
         -Europe, 15, 16, 335
         -France, 96
         -Germany, West, 96, 145, 259, 325, 370
         -Japan, 15–16
         -Sweden, 96
         -U.K., 96
      U.S.S.R., 7, 10, 16, 20, 43, 68, 77, 100, 133, 139, 159, 171, 176–177, 185, 191, 197, 217, 238, 243, 247, 249, 275, 283, 291, 293–294, 321, 325, 330, 332, 334, 337, 340, 343, 353, 358, 369
   space shuttle, 3, 16, 45, 291
   space station, 3, 45
   space tug, 278
   Statements by Presidents of the United States on International Cooperation in Space—A Chronology: October 1957–August 1971 (Sen. Doc. 9240), 265
   tracking
      France-U.S.S.R., 235
International Coordinating Center of Marine Exploration in U.S.S.R., 110
International Decade of Ocean Exploration, 99
International Galabert Prize, 191
International Institute for Strategic Studies, 247–248, 360
International Institute of Applied Systems Analysis (proposed), 287
International Orbital Laboratory Symposium, 261
International Solar Energy Society, 129
International Symposium on Basic Environmental Problems of Man in Space, Fourth, 277
International Symposium on Space Technology, Ninth, 133
International Telecommunications Satellite Consortium (INTELSAT), 14, 45, 48, 102, 138, 242–243, 265
   charter, 237
International Telecommunications Union, 319
International Union of Geodesy and Geophysics, 241–242
International Workshop on Earth Resources Survey, 119, 369–370
Interplanetary Monitoring Platform (IMP), 64, 356
Intersputnik (international space com-

munications organization), 319
Intravehicular activity (IVA), 328
Ionosphere, 19, 22, 31–32, 88, 111, 115, 176, 193, 218, 236–237, 369
Iorilla, Anthony J., 298
Iowa, Univ. of, 348
Iran, 313
Ireland, 43, 220
IRPOS. See Interdisciplinary research relevant to problems of our society.
Irwin, Col. James B. (USAF)
  Apollo 15 mission
    celebrations for, 222, 238
    Congress, report to, 254
    flight, 202–208
    lunar rock samples, 226
    medical aspects, 188, 225, 228, 289
    preparations for, 141, 188
    press comment, 222
    press conference, 120–121, 227
    report, 223
    splashdown, 207
    syndicated article agreement, 200–201
  Apollo 17 mission, 228–229
  awards and honors, 239, 260, 262–263, 266, 346–347
  goodwill tour, 311
  space center visits, 271, 272
Is There an R&D Gap? (report), 124
Isayev, Aleksey, 176
Isis 1 (International Satellite for Ionospheric Studies), 88
Isis 2 (ISIS–B), 88, 369
Israel, 124, 279, 329
Israel, David R., 2
Italy, 3, 43, 291, 310, 315, 333, 355
  launch, satellite, 109–110, 319, 365, 368, 369
Itek Corp., 330
Itos 1 (Tiros-M) (Improved Tiros Operational Satellite), 295
ITOS–B, 294, 295, 300
ITOS–C, 356
ITOS–D, 356
ITOS–E, 356
IVA. See Intravehicular activity.

J

J–2 (rocket engine), 60, 158
Jackson, Sen. Henry M., 85, 103, 329
Jackson, Roy P., 39, 96, 130, 296, 337
Jaffe, Leonard, 179, 211, 340
Jaffe, Dr. Leonard D., 57
Jamaica, 104, 138, 171, 370
Jamaica Bay and Kennedy Airport: A Multidisciplinary Environmental Study, 49
Jamaica Geological Survey, 104, 171
James, Lee B., 116, 240
Jane's All the World's Aircraft, 296
Japan, 243
  international cooperation, 231–232
  international cooperation, space, 15, 111
  launch, satellite, 48, 161, 200, 268, 368
  science and technology, 122, 291, 363
  space objects liability treaty, 313–314
  space program, 9, 220, 305
  supersonic transport, 82, 84–85, 138
Japanese Space Development Committee, 161
Jastrow, Dr. Robert, 32, 45
Javelin (sounding rocket), 176
Javits, Sen. Jacob K., 171, 177
Jeffries, John, Award, 298
Jeppesen and Co., 322
Jeppesen, Elrey B., 322
Jet Propulsion Laboratory (JPL) (Cal Tech), 191–192, 211, 353, 355
  astronomical research, 102
  award, 85, 314
  computer, 179
  contract, 47, 175, 211, 344
  Deep Space Network, 318
  Goldstone Tracking Station, 102, 313, 318
  Jupiter (planet) research, 178
  Mariner program, 175
  Mariner 8 (Mariner-H) failure investigation, 126–127, 136
  Mariner 9 (Mariner-I), 148, 301–302, 306, 313, 321, 331, 339, 353
  Mars (planet) research, 90
  patent, 335
  personnel, 46–47, 55, 159–160, 191–192, 261
  Pioneer spacecraft, 178, 322
  Surveyor 3 camera, 57
  Thermoelectric Outer Planet Spacecraft (TOPS), 179
  transit system, 47, 55
  Viking, Project, 211, 344
JetStar (jet transport), 115
Jha, Ambassador Lakshmi Kant (India), 70
Johns Hopkins Univ., 10, 14, 132, 159
Johnson, Dr. Clarence L., 115
Johnson, David S., 217
Johnson, E. G., 261
Johnson, Dr. Francis S., 144
Johnson, Dr. Fred M., 299
Johnson, Harry W., 44
Johnson, Dr. Howard W., 325
Johnson, Col. Leonard W., Jr. (USAF), 352
Johnson, President Lyndon B., 162, 265, 282
Johnson, U. Alexis, 335
Johnson, Vincent L., 136, 164, 302
Johnson-Sea-Link (submersible research vessel), 24
Johnsson, Olof, 170
Johnston, Richard S., 298
Joint Dot–NASA Civil Aviation Research and Development Policy Study Report, 122
Joint Working Group (NASA-Soviet Academy of Sciences), 243
Jones, R/A Don A. (USN), 12
Jones, James C., 254

Jones, Lawrence C., 331
Jordan, Sen. B. Everett, 197
JPL. See Jet Propulsion Laboratory.
JT3D (turbofan engine), 130, 198
JT8D (turbofan engine), 198
Jupiter (missile), 182
Jupiter (planet)
  atmosphere, 179
  exploration
    spacecraft, 22, 74, 179, 322, 327, 367
    unmanned, 22, 67, 148, 322, 327, 356
  gravity, 178
  magnetic field, 178
  moon, 97
  photographs, 174
  radiation belt, 178, 322
  rotation, 178

K

Kahurin, L. I., 167
Kansas City, Mo., 328
Kaplan, Dr. Ian R., 39
Kappa 9-M (Japanese rocket), 9
Kapryan, Walter J., 57
Kapustin-Yar, U.S.S.R., 337
Karaganda, Kazakhstan, U.S.S.R., 108
Karth, Rep. Joseph E., 38, 280
Katys, G., 174
Katyusha (U.S.S.R. rocket), 261
Kayten, Gerald G., 233
Kazakhstan, U.S.S.R., 135
Keathley, William H., 251
Keldysh, Prof. Mstislav V., 10, 89, 94, 100, 140, 145, 217, 268, 284, 311
Kelly, K. K., 148
Kelly, Dr. Marvin J., 78
Kendall, Dr. Henry W., 183
Kennecott Copper Corp., 78
Kennedy, Sen. Edward M., 17, 76, 166, 286, 308
Kennedy, President John F., 142, 162, 201, 215, 265, 282
Kennedy, John F., International Airport, 2, 49, 139
Kennedy Space Center (KSC) (NASA), 126, 301, 320, 328
  anniversary, 180
  Apollo/Saturn. See Apollo missions.
  astronauts at, 5, 13, 17, 120, 188–189, 271
  award, 57
  computer, 287
  contract, 71, 278, 279, 292
  economy program, 287
  employment, 65
  exhibit, 122, 353
  facilities, 361
  launch operations (see also Launch Complex 17, 34, 39 and 40; and Apollo missions), 7, 24, 30, 46, 95, 98, 110, 128, 160, 200–201, 245, 254–255, 347
  lightning study, 201
  meeting, 162, 269
  personnel, 57, 122, 357
  press conference, 120
  space shuttle, 2, 145
  Visitor Information Center, 264
  visits to, 343
    Agnew, Vice President Spiro T., 31
Kenney, Dr. V. Paul, 114
Kent, Wash., 55, 68, 128
Kentucky, 167, 256
Kenya, Africa, 109, 323, 369
Kepler, Johannes, 97, 233, 359
Keweenaw Rocket Launch Site (KRLS), 22, 31
Key Biscayne, Fla., 74
Khanina, F., 357
Kharkov, Univ. of, 325
King, Gov. Bruce, 42
King, John W., 357
King, Robert E., 49
KingCobra (helicopter), 357
Kingston, Ontario, 10
Kirilenko, Andrey P., 186
Kitt Peak National Observatory, 36, 40, 178, 299
Kitty Hawk (aircraft), 248
Kitty Hawk (*Apollo 14* command and service module). See Command and service module.
Klass, Joseph, 243
Klass, Philip J., 243
Klein, Milton, 52, 71, 96, 250, 323
KLM Royal Dutch Airlines, 88
Knight of St. Patrick Award, 78
Komarov, Vol. Vladimir M. (U.S.S.R.), 266
Komarov, V. M., Diploma, 266
Korn, Steve, 345
Kourou Space Center, French Guiana, 64, 74, 78, 83, 103, 235, 259, 310, 342, 358, 370
Kozlov, Mikhail V., 264
Kraemar, Robert S., 322, 358
Kraft, Christopher C., Jr., 23, 96, 302, 328
Kraft, Joseph, 129
Kramer, James J., 251
Kranz, Eugene F., 57
Kranzberg, Dr. Melvin, 151
Kremlin Wall, Moscow, 186
Krier, Gary, 114, 279, 287
KRLS. See Keweenaw Rocket Launch Site.
Kronauer, M/G Clifford J. (USAF), 193
Kushiro, Ikuo, 83
Kuznetsov NK-144 (U.S.S.R. turbofan engine), 126
Kvenvolden, Keith A., 39
Kwajalein Missile Range, 125

L

L-1011 (TriStar) (jet transport), 36, 38, 44, 47, 88, 124, 132, 147, 173–174, 190, 216, 221, 301
Labor, Dept. of, 141
  Manpower Administration, 119

LaGarde, Jean Bernard, 75
Laird, Secretary of Defense Melvin R., 68, 107, 124, 264, 286, 351
Lajes Field, The Azores, 351
Lake Pontchartrain, La., 200
Land-Air Div., Dynalectron Corp., 131
Lane, A. L., 148
Langley Gold Medal for Aerodromics, 266
Langley Research Center (LaRC) (NASA)
  award, 179, 302
  brake, diagonal, studies, 51, 278
  camera, underwater, 104
  conference on aircraft safety and operating problems, 122
  contract, 49
  cooperation, 104, 210
  8-Foot Tunnels Branch, 299
  employment, 238
  energy absorber, frangible-tube, 259
  exhibit, 152
  flight research vehicle, 233
  meeting, 231
  patent, 259
  personnel, 107, 186, 191
  Scout launch vehicle, 319
  Space systems Div., 299
  Visitor Information Center, 152
  V/STOL Wind Tunnel, 365
Langley, Dr. Samuel P., 266
Langseth, Dr. Marcus E., 144, 245
Lapp, Dr. Ralph, 325
LaRC. See Langley Research Center.
Large Space Telescope (LST), 67, 330
Las Vegas, Nev., 63, 196
Laser, 88, 288
  lunar experiments, 27, 63, 144, 278, 299, 315
Latham, Dr. Gary V., 7, 64, 113, 142–143, 236, 245, 291
Latvian Academy of Sciences Radioastrophysical Observatory, 287
Latvian State Univ., 357
Launch Complex 36, 30, 126
Launch Complex 37, 30, 357
Launch Complex 39, 202, 349
Launch platform, floating, 262
Lawless, James G., 39
Lawrence, Lovell, Jr., 13
Le Bourget Airport, France, 147, 168
Lebedev, P. N., Physics Institute, 315
Lecky-Thompson, Squadron Leader Leslie, 135
Lederberg, Dr. Joshua, 91
Lederer, Jerome F., 295, 314
Lee, Capt. Chester M. (USN), 35, 156
Lefevre, Theo, 45, 48, 335
Lehigh Univ., 239
Leonov, L/C Aleksey A. (U.S.S.R.), 16
Leopold, King of Belgium, 260
LeRC. See Lewis Research Center.
Lerner, Lawrence, 137
Lessing, Lawrence P., 89
Lewis, Dr. George W., 196
Lewis Research Center (LeRC) (NASA), 238, 298, 320, 362
  Aerospace Environment Branch, 56

air-pollution study, 2
anniversary, 196
Atlas-Centaur, 127, 136, 150
award, 262, 302
Awareness Program, 329
Centaur launch failure investigation, 127, 136
computer, 106–107
contract, 96, 98, 147
cooperation, 292
Mariner 9, 150
noise abatement, 295, 296
personnel, 349
Quiet Engine Project, 83, 96, 130, 241, 246–247
research, 56, 72, 262, 292, 320
solar cell, 12
Library of Congress
  Congressional Research Service, 248
  Science Policy Research Div., 7, 56, 368
Lick Observatory, Calif., 1, 299
Life, origin of, 167
Life support system, 96, 105, 156, 181, 193, 196, 206, 235, 308, 311–312
Lifting body
  M2-F2, 42
  M2-F3, 42, 56, 200, 225, 237, 265, 319, 337, 352, 365
  X-24A, 11, 39, 49, 65, 85, 130, 142, 153, 210, 351, 356, 365
Lightning, 199, 201
Lightning Strikes to In-Flight Missiles program, 199
Lilienthal, Cdr. Donald H. (USN), 12, 17, 41
Lillie, Dr. Charles, 93
Lindbergh, B/G Charles A. (USAF, Ret.), 13
Lindner, Clarence H., 170
Lindsay, John C., Memorial Award, 154
Lindsay, Dr. John F., 238
Lindsay, Mayor John V., 239
Liquid hydrogen, 60, 89, 219, 331
Liquid oxygen, 60, 219, 297
Lithicon, 360
Littlewood, William, 326
Littlewood, William, Memorial Lecture, 326
Litton Industries, 182
  Environetics Div., 137
  Guidance and Control Systems Div., 175
LLTV. See Lunar landing training vehicle.
LM. See Lunar module.
Lockheed Aircraft Corp., 65, 307
  AH-56A (Cheyenne) (helicopter), 294, 357
  award, 115
  C-5A (cargo transport), 2, 36, 147, 260, 271, 294, 307, 335, 348, 358
  contract, 2, 36, 52, 147, 186
  financial problems, 44–45, 124, 132, 147, 161, 173–174, 194, 201, 212, 216, 254, 294
  L-1011 (TriStar jet transport), 36,

38, 44, 47, 88, 124, 132, 147, 173–174, 190, 216, 221, 301
  patent, 171
  Poseidon missile, 52
  sounding rocket experiment, 174
  STOL aircraft, 290
  T-33 (jet trainer), 115
  U-2 (high-altitude aircraft), 95, 115
Lockheed Aircraft Service Co., 126
Lockheed Electronics Co., 19, 49, 190
Lockheed Missiles & Space Co., 12, 195, 281, 299, 305, 335
Lockheed-Georgia Co., 101, 338, 348
Loehwing, David A., 359–360
Loftin, Laurence K., Jr., 186, 191
Logsdon, Dr. John M., 50, 201, 258
Logunov, Anatoly, 168
LOI: lunar orbit insertion.
Lomask, Milton, 75
London Airport, U.K., 220
London, U.K., 45, 48, 84, 85, 126, 300, 311, 312, 314, 329, 360
Long Beach, Calif., 211, 343
Long Beach, Calif., Harbor Commission, 44
Long Beach, N.Y., 137
Long Island Lighting Company, 295
Long, Elgin L., 310, 329
Long, James E., 178
Loosbrock, John F., 214
Los Angeles, Calif., 80, 220, 257, 309
Los Angeles Air Pollution Control District, 170
Los Angeles Chamber of Commerce, 275
Los Angeles Harbor Commission, 197
Losey, Robert M., Award, 15
Louis, Victor, 187
Louisiana, 351–352
Louisiana State Univ., 351–352
Lousma, Maj. Jack R. (USMC), 68
Love, Eugene S., 299
Lovell, Capt. James A., Jr. (USN), 14, 75, 137, 266, 327
Low, Dr. George M., 306, 368
  aeronautical research, 52–53, 54
  aerospace unemployment, 85–86, 87
  Apollo 14 mission, 43, 44, 80
  awards by, 57
  backup capability in space missions, 163–164
  cooperation, 52–53, 54
  international cooperation, space, 15, 16, 76–77, 105–106, 216–217
  Mariner H, 136
  NASA budget, 21, 22, 52–53, 61, 62, 80, 85, 86, 93, 163–164
  Nuclear engine for rocket vehicle application (NERVA), 52
  press conference, 11–12, 136, 181
  quarantine of astronauts, 115
  Skylab program, 163–164, 181, 183
  Soyuz 11 mission, 180–181
  space program, 21–22, 23, 51, 61, 62, 75–76, 80, 93, 121, 181
  space shuttle, 22, 80, 93
  This Island Earth (NASA SP-250), 48
  U.S.–U.S.S.R. space cooperation, 10, 16, 76–77, 88–89, 216, 283–284
Lowell Observatory, Ariz., 174
LPM. See Lunar portable magnetometer.
LRL. See Lunar Receiving Laboratory.
LRV. See Lunar roving vehicle.
LSI. See Lunar Science Institute.
LST. See Large Space Telescope.
LTTAT. See Thor-Delta, long-tank thrust-augmented.
LTV Aerospace Corp., 344, 345
Lubbock (Tex.) Regional Airport, 278
Lucas, Dr. William R., 46, 300–301
Lucerne, Switzerland, 266
Ludwig, Dr. George H., 358
Lukas, Dr. Richard C., 298
Lukasic, Dr. Stephen J., 300
Luna 16 (U.S.S.R. lunar probe), 4, 6, 145, 152, 159, 190, 369
Luna 17, 4, 9, 41, 66, 98, 188, 217, 220, 231, 246, 256, 272, 306, 322, 368
Luna 18, 246, 250, 259, 268
Luna 19, 267
Lunar and Planetary Laboratory (Tucson, Ariz.), 40
Lunar landing training vehicle (LLTV), 22–23
Lunar module (LM), 55–56, 57, 171, 233, 244, 294
  Apollo 9, 286
  Apollo 11, 130–131, 198
  Apollo 12, 130, 288
  Apollo 13, 244
  Apollo 14 (*Antares*), 24–30, 35, 40, 47, 130, 365
  Apollo 15 (*Falcon*), 98, 108, 110, 120, 199, 201–203, 213, 216–218, 229, 365
  Apollo 16, 293
  Apollo 17, 228
  Apollo 18, 305
  contract, 102–103, 255
Lunar Orbiter (program), 212, 253, 294
Lunar portable magnetometer (LPM), 27
Lunar roving vehicle (LRV) (Rover) (see also *Lunokhod 1*), 42
  Apollo 15, 24, 51, 55–56, 108, 110, 115, 120, 128, 141, 142, 144, 201, 203, 213, 215, 245, 289, 308, 353
  lunar excursion, 204–206, 213, 226, 254, 365
  Apollo 16, 128, 245, 267, 289, 323
  Apollo 17, 289
  cost, 24
Lunar Sample Preliminary Examination Team, 239
Lunar Science Conference, Second, 5–7, 369
Lunar Science Institute, 291, 335, 342, 369
Lundin, Bruce T., 2, 167, 168, 196, 302
Lundquist, B/G Gustav E. (USAF, Ret.), 2, 308
Lunney, Glynn S., 57
Lunokhod 1 (U.S.S.R. lunar surface explorer), 145, 152, 246, 315, 368

ceases operations, 272, 378
exploration, 4, 8, 9, 41, 66, 73, 88, 133, 139, 168, 188, 196, 217, 221, 231, 246, 256
press comment, 20, 43
results, 9, 59, 100, 209, 220, 278, 302, 322
Lyon, b/g Herbert A. (usaf), 193

M

M2-F2 (lifting body), 42
M2-F3, 42, 56, 200, 225, 239, 265, 319, 337, 352, 365
M31 (Andromeda) (galaxy), 121, 253
M32 (galaxy), 121, 253
M82 (galaxy), 194
McClellan AFB, Calif., 14-15
McClintock, Barbara, 18
McConnell, Dr. Dudley G., 199
McConnell, Joseph H., 69
McCreery, Frank E., 307
McCurdy, Richard C., 75-76, 172
McDade, Rep. Joseph M., 222
McDivitt, b/g James A. (usaf), 57
MacDonald, J. Ross, 170
MacDonald Observatory, Tex., 64
McDonnell Douglas Corp., 3, 170, 186, 195, 211, 220, 281, 298, 299
  McDonnell Douglas Aircraft Co., 344
  McDonnell Douglas Aeronautics Co., 12, 38, 265, 276, 288, 355
McElroy, Dr. William D., 61, 196, 320
McGannon, Dr. William J., 292
McGovern, Sen. George S., 71-72, 122, 263
McIntosh, Patrick S., 302
McKay, Kenneth G., 170
McMurtry, Thomas C., 65-66, 77, 94, 101, 141, 145, 233, 240, 252, 256, 270, 279, 346, 349
McNamara, Robert S., 320
Madras, India, 282
Madrid, Spain, 69, 165, 219
Maffei 1 (galaxy), 1
Maffei 2, 1
Maffei, Paolo, 1
Magellan, Ferdinand, 310
Magellanic Premium, 314
Magnesium tetrabenzporphine, 299
Magnetic field, 78, 79, 104, 308, 341, 354
Magnetoglow, 23
Magnetometer, 27, 51, 133, 144, 219, 285-286, 293
Magnetosphere, 23, 357
Magnetron, 190
Magnuson, Sen. Warren G., 19
Magruder, William M., 75, 84-85, 138, 253
Mahoney, David J., 212
Mailer, Norman, 4-5, 32
Maine, Univ. of, 363
Maitland, Fla., 5
Making of an Ex-astronaut, The, 90-91
Management Study of NASA Acquisition Processes, 178

Manchester, Univ. of (U.K.), 64
Manke, John A., 11, 49, 85, 142, 153
Manned space flight (see also Apollo program; *Apollo 8* through *Apollo 17* missions; Gemini missions; Astronaut; Cosmonaut; *Salyut 1*; *Salyut 1-Soyuz 10*; *Salyut 1-Soyuz 11*; Skylab; Soyuz missions; Space biology; and Space station), 1, 4, 43, 217, 325
achievements, 30, 42, 47, 48, 207, 208, 223, 254, 312, 362
advantage, 59, 218, 224, 315
anniversary, 100, 122, 197, 198
award, 84, 298
cooperation (see also Docking), 110-111, 185, 216, 238, 248, 325, 340-341, 343
eva. See Extravehicular activity.
funding, 20-21, 57, 64, 253, 368
hazards, 59, 62, 115, 181, 185, 225, 228, 255, 289
lunar landing. See Moon landing, manned.
military aspects, 158, 191
policy and plans
  U.S., 10, 13, 62, 96, 176, 178, 181, 216, 249, 256, 280, 283, 312, 328-356, 368
  U.S.S.R., 61, 105, 110, 111, 154-155, 158, 188, 191-192, 249, 268
press comment, 215-216
record, 30
safety aspects, 163-164
Manned Space Flight Network (msfn), 27, 69, 269
Manned Space Program Accident/Incident Summaries, 90
Manned Spacecraft Center (msc) (nasa), 8, 14, 38-39, 111, 133, 139, 181, 286, 305
aircraft electronic control system, 73
Apollo spacecraft, 30-31, 208, 328
Apollo 13 mission, 57
Apollo 14 mission, 35, 44, 48, 57, 64
Apollo 15 mission, 208, 215, 217, 225, 226, 228, 234, 279
Astronaut Office, 254
astronauts at, 4, 225, 234
award, 57, 84, 279
contract, 49, 102-103, 105, 106, 131, 132, 171, 177, 186, 190, 195, 255, 320, 328
Earth and Lunar Sciences Div., 7
Earth Observations Aircraft program, 138-139, 171
earth resources experiment package (erep), 107
employees, 238
Lunar Receiving Laboratory. See Lunar Receiving Laboratory.
lunar roving vehicle, 267
Lunar Science Institute. See Lunar Science Institute.
meeting, 243, 291

patent, 251
personnel, 113–114, 137, 176, 245, 247, 357
Planetary and Earth Sciences Div., 45, 245
press conference, 4, 35, 228, 234
Skylab, 171, 258, 276, 284
sounding rocket experiments, 130, 131, 135, 171
space environment simulation chamber, 307
space shuttle, 105, 132, 160, 195, 320, 328
Tito, President Josip Broz, visit, 286
Mao Tse-tung, 72
Maran, Dr. Stephen P., 36
Marconi Space and Defence Systems, 19
Marietta, Ga., 253–254
Marine Resources and Engineering Development Act, 99
Marine Science Affairs: Annual Report of the President to the Congress on Marine Resources and Engineering Development, 99
Mariner 6 (Mars probe), 69, 148
Mariner 7, 69, 148
Mariner 8 (Mariner-H), 126–127, 136, 146, 148, 149, 275, 365
Mariner 9 (Mariner-I), 127, 139, 164, 166, 246, 275, 353
  control, 153, 179, 199, 263, 293, 306, 318, 321, 361
  cooperation, 340, 355, 360, 369
  cost, 330
  launch, 146, 148–150
  Launch Readiness Review, 146
  photographs, 313, 314, 315, 321, 327, 332, 340, 342, 347, 353, 365
  press comment, 314, 324, 327, 330, 332–333
  press conference, 301, 339
  results, 314, 318, 324, 337–338, 339–340, 364
  test, 136
Mariner Venus-Mercury (spacecraft), 75, 115, 175
Mariner-Venus 1967, Final Project Report (NASA SP-190), 364
Mark, Dr. Hans M., 298
Marov, M., 338
Mars (planet) (see also *Mariner 6 through 9*, *Mars 1 through 3*, and Viking Project)
  atmosphere, 195–196, 246, 337–338, 350, 359
  dust storm, 301–302, 321, 332–333, 339, 355, 361, 365
  exploration, 45, 162, 191, 318, 324, 345, 351, 354–355
  international cooperation, 174–175, 293–294, 321, 340, 355, 360
  manned, 22, 51, 320–321
  plans, 22, 146, 147
  softlanding, 337
  spacecraft, 62, 264
  life on, 90, 338, 345
  mapping, 246, 339, 361
  moon, 310, 331, 340, 342, 364
  photographs, 246, 321, 339, 353, 355, 360
  poles, 321, 360
  radiation, 219, 338
  spectrometer data, 195, 340
  surface, 246, 321, 337
  temperature, 359
  volcanism, 340, 353
  weather, 174
Mars 1 (U.S.S.R. Mars probe), 134
Mars 2, 134, 147, 157, 209, 293, 318, 324, 330, 332, 338, 340, 341, 351, 353, 354, 355, 357, 360, 368, 369
Mars 3, 147, 157, 209, 293, 318, 324, 330, 334, 337, 340, 341, 347, 351, 353, 354, 355, 359, 360, 368, 454
Marseilles, Univ. of, 311–312
Marshall Space Flight Center (MSFC) (NASA)
  Apollo Telescope Mount, 9, 11, 252, 284
  astronaut visits, 272
  commemorative space stamps, 215
  concept verification testing, 334
  contract, 99, 106, 158–159, 174, 175, 180, 190, 245–246, 286, 330, 345–346, 356
  cooperation, 348
  employment, 238
  launch vehicle. See Saturn.
  Life Scientists Program, 210
  lunar roving vehicle (LRV), 68
  meeting, 42, 142, 278, 337
  Neutral Buoyancy Space Simulator, 287
  offshore oil and gas operations study, 348
  personnel, 46, 98, 116, 240
  remote sensing techniques, 128
  Skylab, 152, 175, 265, 284, 323, 353
  space shuttle, 3, 60, 106, 160, 175, 193, 245–246, 345–346, 356
  space station, 12, 192–193
  space tug, 224, 278
  Stratoscope II (balloon borne astronomical telescope), 121
Marsik, Stanley J., 72, 262
Martin, B/G Abner B. (USAF), 193
Martin, John L., Jr., 54
Martin Marietta Corp., 107, 175, 211, 281, 294, 310, 346, 351, 353
Maryland, 97, 224
Maryland, Univ. of, 102, 299, 322, 356
Mascon (mass concentration of gravitational pull), 109, 143
Masefield, Peter G., 326
Maser, 89, 154
Mason-Rust Co., 80
Massachusetts Institute of Technology (MIT), 13, 64, 247, 248, 261
  award, 57, 279
  experiment, 102, 119, 183, 289, 310, 369
  funding, 248, 325

meeting, 126
seismic research, 345
Massachusetts, Univ. of, 239
MAST. See Military assistance to safety and traffic.
Masursky, Dr. Harold, 339-340
Materials technology, 23, 71, 132, 195, 279, 320, 335
Mathematics, 288
Mathews, Charles W., 340
MATRA (S. A. Engins MATRA, France), 3, 19
Mattingly, L/CDR Thomas K., II (USN), 61, 171
Mauchly, Dr. John W., 218
Maverick (missile), 312
Max Planck Institute for Radio Astronomy, 277
Maxwell AFB, Ala., 354
May, Dr. Charles E., 72
MCC: midcourse correction.
MCI Lockheed Satellite Corp., 65
MDA. See Multiple docking adapter.
MDS. See Modulation demodulation subsystem.
Mead, Merrill H., 330
Meaney, George, 131
Mechanics and Control of Flight Award (AIAA), 233
Mecom, John W., Jr., 345
Medaris, M/G John B. (USA, Ret.), 5
Medvedev, Zhores, 210
Melbourne, Univ. of, 181
Members of Congress for Peace Through Law, 122
Memorandum of Understanding, 88, 231, 297
Mental telepathy, 46, 170
Mercer, Robert D., 335
Mercury (planet), 22, 167, 350
Mercury (spacecraft), 122, 286
Merrill, Capt. Henry T. (Dick), 266
MESA. See Modularized equipment stowage assembly.
MESH (consortium), 19
Messerschmitt-Boelkow-Blohm, 3
MET. See Mobile equipment transporter.
Meteor (U.S.S.R. meteorological satellite), 366
Meteor 7, 11
Meteor 8, 105
Meteor 9, 195
Meteor 10, 360
Meteorite, 9, 39, 167, 256, 315, 350, 369
Meteoroid detection, 40-41
Meteorological satellite (see also individual satellites, such as *Ats 3, Ats 5, Essa 9, Itos 1, Meteor 7,* and *Nimbus 4*)
cooperation, 133-134
Global Atmospheric Research Program, 103, 178, 180
ITOS program, 294-295, 356
Nimbus program, 22, 151, 220, 329, 356
Synchronous Meteorological Satellite, 21, 22, 356
Tiros program, 294-295
U.S.S.R., 4, 11, 105, 195, 360, 366
Meteorology, 3, 10, 62, 70, 103, 180, 219, 230, 303-304, 327
Metric America (SP-345), 212
Metropolitan Construction Co., 195
Meudon Observatory, France, 194
Mexico City, Mexico, 295
Mi-12 (U.S.S.R. helicopter), 147
Miami, Univ. of, 242
Michigan, Univ. of, 13, 22, 31, 41, 119-120, 209, 234, 236, 315, 331, 369-370
Michoud Assembly Facility (MSFC), 80, 200, 254-255, 291, 348
Mickelwait, Lowell P., 81
Micrometeorite, 54
Micrometeoroid, 108
Microwave Communications, Inc., 65
Midwest Research Institute, 328
MIG-15 (U.S.S.R. fighter aircraft), 248
MIG-19, 140
MIG-23 (Foxbat), 124, 260, 264
MIG-25, 260
MIG-27, 260
Military assistance to safety and traffic (MAST), 351
Military Balance 1971-1972, The, 247-248
Military Construction Authorization Act, 344
Milky Way (galaxy), 1, 36, 89, 113, 299
Millard, John P., 81
Miller, Rep. George P., 38, 42, 94, 301, 347
Miller, William H., 141
Millionschikov, M. D., 182
Minalov, J. D., 64
Miner, Dr. Ellis D., 318
Minimech (self-deploying arm), 269
Minnesota, 83
Minnesota, Univ. of, 279
Minow, Newton N., 320
Minuteman (missile), 117, 193, 199, 224
Minuteman III, 21
Mirabelle (particle accelerator), 168, 299
Mirage IV-A (French supersonic bomber), 224
MIRV. See Multiple independently targetable reentry vehicle.
Missile, 182, 193, 319
air-to-surface (ASM), 312
antiballistic (ABM), 4, 43, 77-78, 137, 152, 194, 284, 311, 347, 360-361
contract, 8, 52
foreign
France, 224
People's Republic of China, 68, 119, 122, 150, 224, 248, 297, 320, 352, 357
United Kingdom, 224
U.S.S.R., 84, 85, 104, 107, 117, 135, 137, 146, 152, 194, 218, 224, 248, 362
guidance, 112

intercontinental ballistic (ICBM), 4, 21, 68, 77, 84, 104, 107, 117, 135, 146, 150, 152, 194, 199, 224, 248, 264, 303, 357, 362
  limitation of. See Strategic Arms Limitation Talks.
  multiple independently targetable reentry vehicle (MIRV), 265, 364
  NATO, 224–225
  short-range-attack (SRAM), 8, 262
  statistics, 103, 107, 224
  submarine-launched (SLM), 19, 21, 52, 103, 224, 283, 286, 309, 335
  test, 4, 78, 91, 95, 125, 135, 150, 199, 265, 310–311, 346, 361
  tracking, 150, 176, 357–358
Mississippi, 351–352
Mississippi State Univ., 351–352
Mississippi Test Facility (MTF) (NASA), 60, 137, 174, 191, 341, 351–352
  Central Instrument Facility, 158
  Earth Resources Laboratory, 19, 49
Missouri, Univ. of, College of Engineering at Columbia, 78
MIT. See Massachusetts Institute of Technology.
Mitchell, Capt. Edgar D. (USN)
  Apollo 14 mission
    Bible carried, 214
    Congress, report to, 61
    docking, 25, 27
    extrasensory perception test, 46, 170
    extravehicular activity, 25, 27, 41, 47, 48, 51
    flight, 24–30
    medical aspects, 5
    Nixon, President Richard M., 24, 31, 42
    preparations for, 5, 10
    press conference, 4, 41, 59, 170
    quarantine, 5, 30
    splashdown, 30
  Apollo 16 mission, 61
  awards and honors, 72
  European tour, 150, 152
  promotion, 47
  tribute to, 41
Mitchell, Dr. James K., 144
Mitre Corp., 1
MMH. See Monomethyl hydrazine.
Mobile equipment transporter (MET), 27, 30
Mobile quarantine facility (MQF), 30
Modularized equipment stowage assembly (MESA), 27, 204
Modulation demodulation subsystem (MDS), 344
Moffet, Dr. Alan T., 102
Molniya I (U.S.S.R. comsat series), 156, 238
Molniya I–18, 210
Molniya I–19, 355, 366
Molniya II–1, 330, 366
Mondale, Sen. Walter F., 145, 177–178, 180
Mongolia, 319

Monomethyl hydrazine (MMH), 268
Monopole, 6
Montegani, Francis J., 83
Montoya, Sen. Joseph M., 69
Moon (see also Apollo missions, Lunar Science Institute, Lunar Receiving Laboratory, Sea of Rains, etc.), 120–121, 234, 313
  age, 257, 260, 265–266, 306–307, 365
  Apollo lunar surface experiment package (ALSEP), 27, 142, 206–207
  atmosphere, 223
  atomic battery, 326
  base, 47, 51, 249
  cold-cathode gauge experiment, 64, 142, 144, 223
  colonization, 312
  communications, 207
  crater, 27, 39, 41, 48, 204, 205, 306–307
  dust, 27
  eclipse, 44
  electrical properties experiment, 287–290
  exploration, 281, 293
  equipment, 27, 47, 120–121, 204–206, 238–239, 365
  international cooperation, 6, 247–248
  manned, 4, 25–30, 40, 41, 47, 48, 51, 59–60, 61, 86–87, 108, 120–121, 130–131, 141, 183, 196–197, 202–208, 211–212, 219, 222–223, 223–224, 237–238, 247, 297, 302, 305, 324, 342, 365
  unmanned, 4, 6, 8, 9, 10, 20, 43, 73, 98, 100, 196, 220, 221, 237–238, 246, 250, 256, 259, 267–268, 278, 306, 366–368
  heat flow experiment, 142, 144, 205, 245, 291
  landing
    manned, 27, 39, 40, 69–70, 76, 80
    anniversary, 326
    commemorative coin, 243, 308
    commemorative stamp, 215, 232
    implications, 208, 217–218, 368
    mementoes, 152
    U.S.S.R. plans, 259, 268, 331
    unmanned, U.S.S.R. (see also *Lunokhod I*), 246
  landing site, 30, 166, 202–203, 213, 217, 219, 254
  laser retroflector experiment, 27, 64, 144, 278, 299, 315
  life on, 6
  lunar law treaty, 153–154
  lunar roving vehicle. See Lunar roving vehicle and *Lunokhod I*.
  Lunar Science Conference, 5–7, 369
  magnetic field, 51, 64, 144
  magnetometer, lunar surface, 51, 142, 144, 293
  mapping, 141, 293–294, 305
  mountains, 223, 230
  navigation, 207
  origin, 109, 228, 245

photographs, 29, 30, 35, 141, 220, 223, 225, 226–227, 254, 268, 278
probe
   Luna 18, 246
   Luna 19, 267
quake, 64, 113, 142–143, 236, 288, 291
radioactivity, 291
rilles, 7, 205
seismic experiment, 7, 25, 27, 44, 63, 64, 113, 142–143, 203, 207, 236, 245, 288
shape, 223
solar wind experiment, 27
suprathermal ion detector experiment, 142, 288, 291
surface, 7, 41, 223, 229, 306–307, 342, 365
   sample, 147, 167, 238, 250,308, 369
      Apollo 11, 5–7, 12–13, 14, 23, 30, 39–40, 45–46, 83, 115, 190, 236
      Apollo 12, 5–7, 12–13, 16, 23, 30, 39–40, 45–46, 83, 115, 190, 236, 257
      Apollo 14, 24, 25–30, 41, 45–46, 59, 98, 115, 236, 265
      Apollo 15, 142, 203, 207–208, 217, 225–226, 228–229, 245, 253, 257, 365
      core sample, 228, 238, 260
      exchange of, 10, 13, 16, 88, 159, 190, 293, 369
      irradiation, 111
      Luna 16, 6, 16, 190
      storage, 251
   temperature, 220, 245, 291
   volcanic activity, 212, 230, 291
   water on, 6, 288, 289–290, 291, 293, 297
Moran, Capt. Walter P., 298
Morgan, Brig. Gen. Thomas W. (USAF), 36
Morgantown, W. Va., 47
Morgenthau, Dr. Hans J., 117
Morison, R/A Samuel E. (USN, Ret.), 310
Moritz, Bernard, 302
Morrison, L. V., 356
Morton, Dr. Louis, 151
Moscow, 6–7, 19–20, 73, 122, 162, 168, 187, 200, 250, 268, 299, 321, 325, 340, 357
   communications, 171, 265, 271, 272
   International Congress of the History of Science, Thirteenth, 234
   international space cooperation meetings, 16, 159, 216–217, 243, 284, 293, 332, 334, 340–341, 343–344, 353
   International Union of Geodesy and Geophysics meeting, 241
   Intersputnik agreement signing, 319
   Mutual Economic Assistance meeting, 79
   Nixon, President Richard M., visit, 290
   Soviet Academy of Sciences meeting, 145
   Soyuz 11 cosmonaut burial ceremonies, 185–186
   treaty signing, 45, 313–314

Vostok 1 anniversary celebration, 100
Mosher, Rep. Charles A., 38, 222
Mossinghoff, Gerald J., 231
Motorola, Inc., Government Electronics Div., 171
Mt. Hadley (lunar mountain), 223
Mount Palomar Observatory, Calif., 1
Moutsoulas, M. D., 64
Moynahan, Dr. Daniel P., 16–17
MQF. See Mobile quarantine facility.
MR-12 (U.S.S.R. rocket), 235
MSC. See Manned Spacecraft Center.
MSFC. See Marshall Space Flight Center.
MSFN. See Manned Space Flight Network.
MTF. See Mississippi Test Facility.
MTS (Meteoroid Technology Satellite), 356
Mu-4S (Japanese booster), 48, 163–164
Mueller, Dr. George E., 17, 138
Mukden, Manchuria, 140
Muller, Paul M., 314
Multiple docking adapter (MDA) (Skylab component), 175, 183, 281, 284, 323, 346, 353, 355
Multiple independently targetable reentry vehicle (MIRV), 266, 364
Murchison meteorite, 167, 256
Murphy, James T., 46, 266
Murphy, Rep. Morgan F., 38
Murray meteorite, 167, 256
Museum of New Mexico, 42–43
Museum of Science and Industry, 262
Musketball (USAF satellite), 221, 366
Muskie, Sen. Edmund S., 82, 195
Myers, Dale D., 57, 63, 86, 93, 96, 133, 139–140, 160, 192, 224, 240, 299, 326, 328

N

NAA. See National Aeronautic Assn.
NAC. See National Aviation Club.
NACA. See National Advisory Committee for Aeronautics.
NAE. See National Academy of Engineering.
NAS. See National Academy of Sciences.
NAS–NRC Space Science Board, 67
NASA. See National Aeronautics and Space Administration.
NASA Aerospace Advisory Panel, 42
NASA Central Planning Office, 161
NASA Contractor Equal Employment Opportunity Program, 49
NASA–DOT Office of Noise Abatement, 295
NASA Historical Advisory Committee, 151–152
NASA Interplanetary Monitoring Platform (IMP) Management Team, 46
NASA–NOAA Space Meteorology Team, 70
NASA Office of Advanced Research and Technology (OART), 14, 70, 130, 140, 169, 232, 233, 239, 272, 349

NASA Office of Applications (OA), 340, 369
NASA Office of Industry Affairs and Technology Utilization, 199
NASA Office of Legislative Affairs, 231
NASA Office of Manned Space Flight (OMSF), 12, 30–31, 84, 160, 188, 208, 267, 291, 302, 314, 334
NASA Office of Space Science (OSS), 340
NASA Office of Space Science and Applications (OSSA), 67, 73, 93–94, 136, 150, 190, 270, 295, 302, 319, 340
NASA Office of Technology Utilization, 82
NASA Office of Tracking and Data Acquisition (OTDA), 30–31, 69, 208
NASA Pasadena Office, 238
NASA Performance Evaluation Board, 153
NASA Project Designation Committee, 337
NASA/RECON (Remote Control), 81
NASA Safety Office, 90
NASA Scientific and Technical Information Facility, 69
NASA Source Selection Board, 353–354
NASA Space Program Advisory Council, 106
NASA Space Shuttle Technology Conference, 75
NASA Space Systems Advisory Committee, 188
NASA/STIMS (Scientific and Technical Information Modular System), 81
NASC. See National Aeronautics and Space Council.
Nash, Douglas, B., 12–13
Nassau County, N.Y., 295
National Academy of Engineering (NAE), 49, 115, 170, 306
  Aeronautics and Space Engineering Board, 39
  Committee on Interplay on Engineering with Biology Medicine, 322
  Founders Medal, 115
National Academy of Sciences (NAS), 187–188, 210, 265
  Jamaica Bay and Kennedy Airport: A Multidisciplinary Environmental Study, 49
  meeting, 166–167, 248–249, 299
  Plan for U.S. Data Management in the Global Atmospheric Research Program, 303
  Plans for U.S. Participation in the GARP Atlantic Tropical Experiment, 303
  Space Science Board, 67
  United States Science Program, 167
National Advisory Committee for Aeronautics (NACA), 49–50, 176
  Gas Dynamics Laboratory, 290
National Advisory Committee on Oceans and Atmosphere, 292
National Aeronautic Assn. (NAA), 263
National Aeronautics and Space Administration (NAS( ) (see also NASA Centers, programs, probes, satellites, and related headings, such as Ames Research Center, Apollo program, *Mariner 9*; *Intelsat-IV F-2*; and Space program, national), 9, 306
accident, 22, 290
accomplishments. See Space program, national.
agreement. See International cooperation; space; and Treaty.
anniversary, 32, 180, 196, 198, 326
astronaut. See Astronaut.
awards and honors, 17, 38, 57, 60, 69, 72, 75, 78, 84, 106, 121, 135, 138, 154, 179, 233, 239, 242, 262–263, 266–267, 298–299, 318
  Distinguished Service Medal, 59, 302, 364–365
  Exceptional Scientific Achievement Medal, 255, 279, 302
  Exceptional Service Medal, 255, 279, 302, 314
  Group Achievement Award, 279, 302
budget, FY 1972, 14, 20–21, 22, 32, 37, 38, 64, 116, 119, 121, 161, 280, 326, 328, 368, 369
  bills introduced, 42, 44
  bills passed, 152, 175, 180, 198–199, 210, 211, 216
  bill signed, 221, 226
  House consideration, 175
    appropriations, 80, 172, 173, 175, 209, 211
    authorization, 61–62, 63, 65, 69, 70–71, 75–76, 107, 152, 198, 209
  Senate consideration, 209
    appropriations, 172, 180, 198, 209, 216
    authorization, 52–53, 76, 85–88, 93, 96, 171, 178, 180, 198, 210
contract, 221, 322
  aeronautics, 101, 160, 187, 189, 290, 338, 365
  communication system, 128, 344
  computer services, 190
  engine, 1, 96, 345–346
  facilities, 69, 195
  history, 292
  instrumentation, 50, 158
  launch vehicle, 99, 153, 158
  life support system, 180
  liquid hydrogen, 89
  lunar roving vehicle, 24
  nuclear power, 147, 276
  space equipment, 175, 328
  space shuttle, 60, 101, 105, 106, 132, 165, 186, 193, 218, 233, 240, 245–246, 272, 276, 281, 298, 320, 328, 345–346, 353–354, 356, 361, 368–369
    controversy, 218, 246, 353–354, 356
  space station, 174
  spacecraft, 2, 81, 115, 135, 153, 211, 294–295, 328
  study, 85, 101, 132, 189–190, 276, 345–346, 362

support services, 19, 71, 80, 98, 107, 158, 173, 190, 191, 279
telescope, 286, 330
cooperation, 2, 93-94, 97, 98, 102, 104, 105, 130, 160-161, 170, 253
  AEC, 52-53, 96, 99, 116
  AFCRL, 199
  Agriculture, Dept. of, 98, 132, 168, 286
  ARPA, 23
  DOD, 87, 93, 123
  DOT, 39, 53, 70, 122-123, 218, 247, 264, 290, 295, 369
  FAA, 70, 218, 271, 278, 290
  Geological Survey, 224
  HEW, 161
  HUD, 369
  NSF, 369
  Smithsonian Institution, 97
  USA, 70, 188, 233, 272
  USAF (see also X-24A), 70, 87, 165, 174, 218-219, 225, 272, 278, 290, 328, 365
  USN, 70, 301
cooperation, international. See International cooperation, space; Sounding rocket, international programs.
criticism, 1, 32, 84, 133, 145-146, 211, 218, 219, 277
employment, 22, 61, 65, 76, 85-86, 238, 368
exhibit, 147, 152, 353
facilities, 2, 8, 19, 20, 32, 42, 71, 80, 94, 112, 126, 135, 152, 158, 191, 195, 209, 218, 220, 226, 247, 251, 291-292, 343, 357, 365
history, 9, 31, 50, 151, 292
launch
  Apollo 14 (AS-509), 24
  Apollo 15 (AS-510), 201
  failure, 4, 126-127, 136, 146, 275, 294, 300, 365
  postponed, 24, 342, 352
  probe
    barium-ion-cloud project, 259
    Mariner 9 (Mariner-I), 148
    PAET (Planetary Atmosphere Experiments Test), 168-169, 240
  satellite
    Ariel 4 (UK-4), 348
    Eole 1 (CAS-1), 230-231
    Explorer 43 (IMP-I), 72
    Explorer 45 (SSS-A), 319
    Intelsat-IV F-2, 14
    Intelsat-IF F-3, 354
    Isis 2 (ISIS-B), 88
    Natosat 2 (*Nato 2*; NATO-B), 36-37
    Oso 7 (OSO-H), 269
    San Marco 3 (San Marco-C), 109-110
    Solrad 10 (*Explorer 44*) (Solrad-C), 189-190
  schedule for 1972, 356
  sounding rocket
    Aerobee 150, 13, 14, 54, 119, 161, 171
    Aerobee 170, 14, 41, 60, 79, 119, 209, 226, 228, 292, 296, 300, 327
    Aerobee 170B, 190-191
    Aerobee 350, 52, 159, 174
    Arcas, 54, 63, 64-65, 218, 222, 225, 230
      Boosted Arcas I, 50, 63
      Boosted Arcas II, 22, 226, 285
    Black Brant IIIB, 248
    Black Brant VB, 54
    Black Brant VC, 199
    Javelin, 176
    Nike-Apache, 14, 19, 22, 31-32, 68, 73, 105, 107, 108, 111, 115, 213, 234, 236, 246, 251, 257, 343
    Nike-Cajun, 3, 5, 8, 64, 68, 74, 78, 112, 130, 131, 187, 192, 234, 236, 259, 285, 343
    Nike-Tomahawk, 12, 47, 74, 77, 78, 79, 89, 193, 323
    Viper Dart, 234-235, 236
    Stratoscope II (balloon-borne telescope), 253
lawsuit, 65
management, 41, 69
manpower. See Employment.
meeting, 5-7, 12, 53, 63, 75, 97, 98, 105, 119-120, 142, 146, 173, 252, 254, 278
organization, 12, 44, 49, 69, 82, 98, 106, 224, 233, 242, 251, 272, 295-296, 340, 369
patents, 171, 251, 259, 305, 308
personnel, 101, 106, 112, 137, 252-253, 257, 290, 314-315, 368
  appointment, 12, 44, 46, 55, 56-57, 59, 61, 68, 69, 72, 79, 84, 98, 114, 116, 139, 159, 161, 167-168, 176, 186, 188, 191, 192, 194, 196, 197, 200, 216, 230, 231, 233, 240, 245, 251, 255, 267, 295-296, 323, 330, 357, 368
  death, 78, 85, 145, 176, 188
  promotion, 36, 47, 240
  resignation, 11, 51, 52, 166, 177, 192, 239, 243, 247, 368
  retirement, 114, 116, 186, 240, 314, 345
procurement, 62
programs, 22, 365-370
  aeronautics (see also Wing, aircraft, supercritical), 21, 23, 37, 39, 53-54, 56, 66-67, 70-71, 73, 77, 82, 94, 96-97, 101, 119-120, 121, 122-124, 127-128, 130, 165, 172, 173, 174, 187, 188-189, 195, 218, 221, 233, 262, 269, 272, 278, 290, 307-308, 335, 337, 356, 365, 369
  astronomy, 2, 8, 9, 21, 22, 57, 67, 69, 74-75, 89, 95, 97, 98, 103, 121, 127, 148, 174, 178-179, 186, 189, 211, 219, 231, 240, 334, 344, 356, 365, 368
  communications, 22, 41, 106, 128, 173, 356

computer, 23, 32, 81, 98–99, 106–107, 179, 187, 231–232, 264, 309
earth resources, 10–11, 21, 48, 61, 67, 87, 95, 96, 97, 119–120, 135, 170, 173, 175, 198, 221, 224, 281, 324, 340, 356, 369–370
manned space flight (See also Apollo program)
   achievements, 20, 30, 42, 47, 48, 207–208, 221–222, 223, 253, 361–362, 368
   criticism, 20, 43, 133, 236
   funding, 20–21, 57, 64, 254, 368
   international cooperation (see also Docking), 110, 185, 216, 238, 247, 325, 340–341, 343
   plans, 10, 21–22, 61, 88–89, 96, 176, 216, 328, 356, 368
   policy, 12, 59, 61–62, 115, 127, 181, 215, 230, 256, 280–281, 283
meteorology, 22, 62, 103, 134, 168, 240, 252, 294, 329, 356
nuclear propulsion, 21, 22, 41, 52–53, 71, 89, 93, 115, 220, 226, 235, 250, 276, 368
sounding rocket, 366, 370
space biology, 21, 115, 127, 132, 138, 189, 235, 245, 253, 256, 261, 284, 288, 289, 325, 359, 369
space rescue, 23, 64, 68–69, 140, 173, 198–199
space science, 20–21, 22, 88–89, 93, 167, 369
space shuttle. See Space shuttle.
space station, 14, 38–39, 61, 84, 95, 127, 164, 176, 183, 249, 267, 290–291, 302, 324, 328, 361–362
   contract, 174
   crew preparation, selection, and training, 38, 87, 106, 132, 173, 193–194, 284, 342
   criticism, 145–146
   design, 3, 12, 71, 174, 183
   equipment, 265, 283, 283, 288, 353, 355
   experiments, 10–11, 38, 87, 107, 132–133, 173, 175, 183, 193–194, 342, 359
   funds, 20–21, 93, 157, 173, 198–199, 221, 368
   international cooperation, 3, 10–11, 38, 45, 173, 278
   meeting, 42, 346
   plans, 68–69, 89, 127–128, 164, 172, 192–193, 198, 229, 256, 290–291, 324, 334, 342
   test, 11, 64, 99, 132, 355
space tug, 224, 278, 340
technology utilization. See Technology utilization, space.
tracking and data acquisition (see also Tracking), 21, 69, 97, 173, 267, 369–370
test
   aircraft, 83, 87, 233, 262, 278, 284
   concept verification testing (CVT) system, 334

lifting body, 42, 200, 225, 239, 265, 317, 337, 352, 365
lunar module, 98, 200
lunar roving vehicle, 24, 51, 55–56
parachute, 247, 264, 268
remote sensing, 97, 368–369
space shuttle, 60
space station, 11, 152, 252, 258, 276, 355
spacecraft, 93, 96, 193, 327, 369
universities, 68, 85, 210
National Aeronautics and Space Council (NASC), 9, 45, 68, 123, 175
National Air and Space Museum, 52, 79–80, 101, 130, 135, 195, 301, 309, 323, 329
National Atlas of the United States of America, 11
National Aviation Club (NAC), 326
National Broadcasting Co. (NBC), 88
National Capital Planning Commission, 301, 309
National Center for Atmospheric Research (NCAR), 121, 178, 231
National Civil Service League, 60
National Commission on Peacetime Transition (proposed), 72
National Endowment for the Humanities, 97
National Gallery of Art, 289
National Geographic Society, 310
National League of Cities, 177
National Medal of Science, 17–18, 138
National Moon Walk Day, 134, 170, 197–198
National Museum of Natural History, 79–80
National Oceanic and Atmospheric Administration (NOAA)
   award, 279
   budget, 99, 349
   contract, 302
   cooperation, 119–120, 231, 351
   personnel, 12, 19, 217
   satellite, 356
     launch, 294
   sounding rocket experiment, 176, 234, 236
   Space Environmental Laboratory, 302
National Park Service, 17, 112
National Press Club, 238
National Radio Astronomy Observatory, 40, 102, 178
National Religious Broadcasters, 18
National Research Council (NRC), 252
   United States Space Science Program, 167
   U.S. Committee for the Global Atmospheric Research Program, 303
National Research Council of Canada, 63, 307
National Rocket Club, 157
National Science Board, 180, 320
National Science Foundation (NSF), 197
   contract, 368–369
   cooperation, 231, 369
   energy sources, 154

Enrollment Increases in Science and Mathematics in Public Secondary Schools, 1948–49 to 1969–70, 288
Estimated Academic R&D Direct Price Trends 50 Percent Higher Over Decade 1961–71, 305
Federal Funds for Academic Science, Fiscal Year 1969, 58
Federal Funds for Research, Development, and Other Scientific Activities, 251
Federal R&D Expenditures Related to Budget Functions, 1960–72, 175
Federal Support to Universities and Colleges, Fiscal Year 1970, 161
funds, 20, 40, 58, 157, 198, 209, 216, 226
grants, 221
industrial R&D, 13, 117–118, 352
Industrial R&D Spending, 1970, 352
interdisciplinary conference, 359
NASA report on R&D to, 116
organization, 61, 69
personnel, 61, 103, 170, 196, 301, 320, 341
Recent Trends in Enrollment and Manpower Resources in Graduate Science Education, 1969–70, 146
Research and Development in Industry, 1969: Funds, 1969; Scientists & Engineers, January 1970, 117–118
Research and Development in Local Governments, Fiscal Years 1968 & 1969, 33
Research Applied to National Needs (RANN), 178
Scientific Activities of Independent Nonprofit Institutions 1970, 58
scientists and engineers, unemployed, 76, 117–118, 188
Unemployed Rates for Scientists, Spring 1971, 188
National Science Foundation Act of 1950, 197
National Science Teachers Assn. (NSTA), 268
National security, 32–33, 37, 86, 99, 136–137, 150, 263, 275, 281, 303, 360
National Security Industrial Assn., 263
National Society of Engineers, 166
National Space Club (NSC), 15, 69–70, 106, 113, 324, 347
Press Award, 69–70
National Technology Support: A Study of Research and Development Trends and Their Implications (AIA study), 116
National Transportation Award, 139
National Transportation Safety Board, 82
National Urban Coalition, 50
National Weather Service, 219
National Zoo, 32
NATO. See North Atlantic Treaty Organization
Natosat 2 (NATO-B) (military comsat), 36–37, 41, 369

Natural Resources, Dept. of (proposed), 81–82, 154
Naugle, Dr. John E., 66, 93–94, 133, 164, 217, 300–301, 359
Naval Air and Material Center, 244
Naval Air Missile Test Center, Calif., 276
Naval Air Systems Command, 190, 329
Naval Observatory, 296, 324
Naval Oceanographic Office, 119–120
Naval Research Laboratory (NRL), 8, 16, 23, 54, 132, 189, 226, 240, 269, 277–278, 349
Naval Warfare Panel, 57
Navigation satellite, 95
NBC. See National Broadcasting Co.
NCAR. See National Center for Atmospheric Research.
Nebraska, 128
Nefyuduv, Yuri G., 167
Nelson, Sen. Gaylord, 253
Nepoklonov, Boris, 306
Neptune (planet), 22, 179, 324
NERVA. See Nuclear energy for rocket vehicle application.
Netherlands, 291, 310
Neupert, Dr. Werner M., 339
New Cities Research and Experimentation Administration (proposed), 286
New Hampshire, Univ. of, 269
New Jersey Dept. of Transportation, 287
New Mexico, 42
New Mexico State Univ., 313, 339
New Orleans (aircraft), 248
New Orleans International Airport, 312
New Orleans, La., 60, 77, 345
New Technology Opportunities Program, 303, 344
New York Academy of Sciences, 167
New York, N.Y., 2, 279, 282, 350
  air services and traffic, 19, 126, 182, 257, 288
  Apollo 14 mission, reaction, 43
  astronauts in, 65, 238, 239
  communications via satellite, 271
  meeting, 15, 36
New York State Court of Appeals, 131
New York Times Special Features agreement, 200, 211
New York Univ. Medical Center, 189
Newell, Dr. Homer E., 108, 126, 133
Newport News, Va. 157
NGC 7662 (Planetary Nebula), 253
NGC-253 (galaxy), 194
Nicks, Oran W., 48, 302
Nierenberg, Dr. William A., 292
Nike-Apache (sounding rocket), 226
  airglow, 108
  cloud study, 105, 251, 257
  energetic particles and fields, 14
  ionosphere study, 19, 22, 31–32, 111, 115
  micrometeoroid collection, 108
  pilot density probe, 343
  polar cap absorption data, 246
  solar eclipse data, 68
  upper-atmosphere data 108, 213, 234

Nike-Cajun (sounding rocket)
  airglow, 130, 131
  cloud study, 112
  cooperative experiment, 64–65, 74–75, 112, 370
  equatorial atmosphere data, 83
  grenade experiment, 3, 5, 8, 68, 174, 187, 192, 259
  ionosphere study, 285–286
  solar eclipse comparison data, 68
  upper-atmosphere data, 3, 5, 8, 78, 174, 187, 192, 234, 236, 343
Nike-Tomahawk (sounding rocket)
  auroral study, 12, 47, 77, 89
  electric-field study, 74–75, 78, 79
  ionosphere study, 193
  stellar astronomy study, 323
Nikitrine, Ignatiy A. See Fedoseyev, Anatoly.
Nikolayev, L/C Andrian G. (U.S.S.R.), 16, 100, 152, 167, 266
Nikolayeva-Tereshkova, Valentina, 16
Nikon-F (underwater camera), 104
Nimbus (meteorological satellite), 22, 329
Nimbus 4, 63, 64–65, 151, 220, 222, 225, 230
Nimbus-E (R&D satellite), 356
Nix Olympica (Mars), 321
Nixon, President Richard M., 3, 186, 292
  Aeronautics and Space Report, 37
  Apollo 11 lunar samples, 42–43
  Apollo 14 mission, messages and statements, 24, 31, 39, 42, 59
  Apollo 15 mission, messages and statements, 201, 208, 212, 213, 221–222, 253, 279
  appointments and nominations by, 11, 13, 16–17, 19, 45, 49–50, 56–57, 59, 132, 152, 193, 194, 197, 252, 254, 290, 292, 294, 312, 320, 348–349, 368
  astronauts
    promotion, 47
    report to, 67
    resignation, 52
    White House dinner, 256
  awards by, 59, 138
  bills signed, 221, 226, 354
  budget, 20, 38, 40
  Columbus Day, 243
  conference on employment, 63
  Defense, Dept. of, 354
  emergency loan guarantee, 124, 201, 216
  environmental program, 42
  Environmental Science: Challenges for the Seventies, National Science Board Report, 180
  Fulton, Rep. James G., death of, 282
  hijacking of aircraft, 36, 55
  international cooperation, space, 37, 55, 138, 166, 265
  messages to Congress, 37, 55, 74, 81–82, 99, 103, 154, 180
  NASA appropriations, 20–21, 221, 226

National Moon Walk Day, 197–198
New Technology Opportunities Program, 303
nuclear war deterrent agreement, 264–265
nuclear warhead test, 300, 303
patent policy statement, 238–239
presidential aircraft, 212, 326–327
public relations activities order, 37
reorganization plans, 81–82
reports transmitted to Congress, 37, 54–55, 76, 180
resignations accepted by, 196, 199, 348
SALT agreement, 137, 264–265
science, 40, 180, 251
Seabed Arms Control Treaty, 45
Soyuz 11 cosmonauts death, message to U.S.S.R., 180–181
space program, national, 37, 55, 59, 76, 161, 263–264, 265
space shuttle, 297
supersonic transport, 37, 81, 130, 349–350, 351
Technology Mobilization and Reemployment Program, 94
United Nations Day, 191
USAF Museum dedication statement, 247
U.S.S.R. message to on *Apollo 14* mission, 43
World Law Day, 166
NK-144 (turbofan engine), 126
NOAA. See National Oceanic and Atmospheric Administration.
Noaa 1 (National Oceanic and Atmospheric Administration meteorological satellite), 294
Nobel Conference, 281–282
Nobel Prize, 89
  in Chemistry, 307
  in Medicine, 282
  in Physics, 307, 309
NOISE (aircraft noise organization), 137–138
Noise, aircraft, 137, 138, 271
  contract, 198, 365
  NASA–DOT Office of Noise Abatement, 295
  NASA research, 241, 242, 251, 295, 307, 365
  supersonic transport, 2, 150, 307, 351
Nolan, James P., Jr., 151–152
NORAD. See North American Air Defense Command.
Norfolk, Va., 349
Norite, 7
North American Air Defense Command (NORAD), 61
North American Aviation, Inc., 10
North American Rockwell Corp. (NR)
  aircraft, 290, 298, 309, 344
  award, 279
  contract, 1, 178, 186, 190, 343
  Rocketdyne Div., 60, 158, 175, 193, 218, 233, 240, 245, 353, 356

Space Div., 153, 171, 195
space shuttle, 37, 281
North Atlantic Treaty Organization (NATO), 36–37, 41, 168, 224, 248, 365, 369
North Pole, 310, 329
Northeast Corridor, 308
Northern Illinois Univ., 83
Northrop Corp., 1
  Northrop Aircraft Corp., 216
  Northrop Services, Inc., 158–159
Northwest Environmental Technology Laboratories, Inc., 320
Norway, 12, 341
Notre Dame Univ., 114
NR. See North American Rockwell Corp.
NRC. See National Research Council.
NRL. See Naval Research Laboratory.
NSC. See National Space Club.
NSF. See National Science Foundation.
NSTA. See National Science Teachers Assn.
Nuclear engine for rocket vehicle application (NERVA)
  funds, 21, 41, 94, 195, 198, 226, 292
  NASA program, 22, 52, 71, 116, 251, 276, 368
Nuclear power, 83, 100, 154, 196, 199, 213, 250
Nuclear propulsion, 41, 52, 96–97, 221, 235, 250, 320–321
Nuclear submarine, 21
Nuclear test, 84, 310–311, 320, 324–325, 326, 331
Nuclear Test Ban Treaty, 162
  underground, 250, 300
Nuclear weapons, 218, 224, 247–248, 252, 265, 266
Nygren, R/A Harley D. (USN), 12

O

OA. See NASA Office of Applications.
Oak Ridge National Laboratory, 99, 102
Oao 2 (Orbiting Astronomical Observatory), 93, 239, 246
OAO-B, 163, 276
OAO-C, 356
OART. See NASA Office of Advanced Research and Technology.
Obata, Gyo, 301, 309, 323, 329
Oberth, Hermann, 234
Obregon, Mauricio, 310
O'Brian, Hugh, Youth Foundation, 162
Ocean of Storms (Oceanus Procellarum) (moon), 7, 12–13, 236, 291
Oceanography
  international aspects, 110, 112–113
  record, 262
  research, 24, 119–120, 262
  treaty, 45, 47, 75–76
  U.S. program, 99, 292–293, 348, 349
Odessa, U.S.S.R., 200, 350
O'Donnell, John J., 305, 334

OECD. See Organization for Economic Cooperation and Development.
Of a Fire on the Moon, 4
Office of Management and Budget (OMB), 199, 260, 297, 326
Office of Naval Research (ONR), 244
Office of Science and Technology (OST) (President's), 232
Office of Technology Assessment (proposed), 197
Office of Technology Transfer (proposed), 173
Offshore Airport Center Planning Conference, 282
Offshore oil study, 348
Ofo (Orbiting Frog Otolith satellite), 40–41, 127
OGO. See Orbiting Geophysical Observatory.
Ogo V (Orbiting Geophysical Observatory), 2, 93
O'Hair, Mrs. Madalyn Murray, 49, 65
O'Hare International Airport, 9
Ohio Historical Society, 198
Ohsumi (Japanese satellite), 48, 268
Oklahoma, 2, 288
Oklahoma City, Okla., 2
O'Leary, Dr. Brian T., 90–91
Olsen, Kenneth H., 16–17
Olympus (turbojet enginge), 126
OMB. See Office of Management and Budget.
OMSF. See NASA Office of Manned Space Flight.
On the Moon with Apollo 15, 183
ONR. See Office of Naval Research.
OPTAC. See Optical aimpoint guidance system.
Optical aimpoint guidance (OPTAC) system, 346
Optical plasmatron, 323
Optical Society of America, 142
Orbita network (U.S.S.R. comsat system), 238, 330
Orbital Workshop (Skylab component), 132, 152, 164, 183, 284, 323
Orbiting Geophysical Observatory (OGO), 2
Orbiting Solar Observatory (OSO) (see also *Oso 1*, etc.), 2, 21, 67
Order of Lenin, 163
Order of Leopold, 260
Oreol (Soviet-French satellite), 359, 366
Organization for Economic Cooperation and Development (OECD), 201
Orgueil meteorite, 256
Orion (constellation), 299
Orion (NASA barge), 152, 284
Orion Nebula, 121
Orlando, Fla., 131
Ormsby, Robert B., 348
OSO. See Orbiting Solar Observatory.
Oso 1, 269
Oso 2, 269
Oso 3, 269
Oso 4, 269

Oso 5, 269
Oso 6, 269
Oso 7 (OSO-H), 269, 277, 302, 333, 349, 361, 365
OSO-C, 269
OSO-H, 2
OSS. See NASA Office of Space Science.
OSSA. See NASA Office of Space Science and Applications.
OST. See Office of Science and Technology (President's).
O'Sullivan, William J., 157
OTDA. See NASA Office of Tracking and Data Acquisition.
O'Toole, Thomas, 24, 47, 112, 133, 223, 337, 349
Outer Planets Mission Steering Group, 96
Ov 1-20 (Orbiting Vehicle Research Satellite), 220, 366
Ov 1-21, 220, 366
OV-10A (Bronco) (research aircraft), 301
Oxford, Ohio, 242
Oxford Univ., 242
Oximeter, 199
Ozone, 54, 63, 64–65, 74–75, 222, 225, 230

P

P-3A (turboprop research aircraft), 120
P-3C (antisubmarine patrol aircraft), 12, 17, 38, 41
P-38 (fighter aircraft), 115
Pacific Missile Range (PMR), 152, 276
Pacific Ocean, 165, 296–297
   Apollo missions, 30, 108, 349
   communications satellite, 3, 74, 98, 176, 177, 219–220, 313
   missile interception tests, 4, 77–78, 125
   oil spills, 81
   Salyut 1, 282–283
Packard, David M., 158, 350
PAET (Planetary Atmosphere Experiment Test), 168–169, 240
Pago Pago, 30
Paine, Dr. Thomas O., 56–57, 314–315, 368
Pakistan, 140
Palestine, Tex., 121
Palmdale, Calif., 47
Palmer Observatory, 311
Palos, Spain, 31
Pan American World Airways, Inc., 109, 148, 257, 297–298, 330
Panofsky, Dr. Wolfgang K. H., 183
Parachute test, 247, 264, 268
Parin, Dr. Vasily V., 163
Paris, France, 148, 287, 291, 347, 351, 352
Paris International Air Show, 147, 150, 152, 168, 214, 249
Particle accelerator, 36, 50, 168, 281–282, 299
Particle and field experiment, 248

Patent, 171, 238–239, 251, 259, 268, 305, 308, 335, 360
Patrick AFB, Fla., 13, 272
Patsayev, Viktor I., 154, 180, 186, 192, 200, 283
Patuxent Naval Air Station, Md., 12
Paul, Dr. Rodman W., 151–152
Paul VI, Pope, 43, 215, 232
Pearce, J. B., 148
Pecora, Dr. William T., 120
Peebles, Ohio, 241
Pegasus 11 (VTOL aircraft engine), 300, 329
Peking, People's Republic of China, 111, 257, 305
Pelly, Rep. Thomas M., 38
Pendray, G. Edward, Award, 15
Pendray, Mrs. Leatrice M., 282
Pennsylvania State Univ., 56
Pennsylvania, Univ. Of, 129, 187, 218
Pensacola, Fla., 182
Pentagon, 4, 240
Pentagon papers (background report on U.S. in Vietnam war). 320
Penzias, Dr. Arno A., 299
People mover transit system, 47, 55
People's Republic of China, 306, 352, 361
   aircraft, 140
   disarmament, 252
   launch, satellite, 61, 365, 368
   missile program, 68, 119, 122, 150, 224, 248, 296–297, 319–320, 357–358
   nuclear test, 84, 324–325
   satellite, 75, 81
   space program, 53, 72, 111, 368
   technology, 273
Pepin, Dr. Robert O., 279
Perel'man, R. G., 183, 363
Perry, Geoffrey E., 249
Perry, Robert L., 346
Perseus (constellation), 1
Peters, Harry E., 154
Petrone, Dr. Rocco A., 70, 215, 323
Petrov, Dr. Boris N., 101, 171, 188, 217, 325, 334
Pettis, Rep. Jerry L., 146
Phaetonis (Mars), 337
Philadelphia, Pa., 314
Philipp, Dr. Warren H., 72, 262
Philippines, 219–220, 352
Philips, Ronald J., 177, 192
Phillips, Nizer, Benjamin, Krim, and Ballou, 79
Phillips, L/C Samuel C. (USAF), 101–102, 188, 266
Phinney, Dr. William, 228
Phobos (Martian moon), 310, 331, 340, 342, 365
Phoenicis Lacus (moon), 353
Phoenix, Ariz., 75, 281
Photography (see also Moon, photographs), 35–36, 72
Pickering, Dr. William H., 55, 159, 192, 313, 314
Pin Main, Alaska, 74
Pinkel, I. Irving, 362–363

Pioneer (program), 21, 69
Pioneer-F (interplanetary probe), 74, 250, 322, 327, 369
Pioneer-G (spacecraft), 322
Piper Aztec (light aircraft), 220
Piper Navajo (light aircraft), 310, 329
PLACE. See Position-location and aircraft communications experiment.
Planetary Explorer mission, 186
Planetoid, 357
Plans for U.S. Participation in the GARP Atlantic Tropical Experiment, 303
Plasma, 320, 323
Playboy International Writers Convocation, 280
Plesetsk, U.S.S.R., 8, 249
  Cosmos launch, 9, 49, 66, 98, 102, 108, 125, 139, 146, 148, 153, 200, 240, 250, 254, 255, 262, 265, 270, 291, 306, 323, 325, 327, 333, 334, 337, 347, 351, 353, 359
  Meteor 8 launch, 105
  Meteor 9 launch, 195
  Meteor 10 launch, 360
  Molniya I-19 launch, 355
  Molniya II-1, 330
  Oreol launch, 259
Plight of Soviet Science Today, 210
PLSS. See Portable life support system.
Plutarch, 227
Pluto (planet), 22, 179, 310
Plutonium 244, 102, 250
PMR. See Pacific Missile Range.
pndb: perceived noise in decibels.
Pobedonostsev, Yu. A., 261
Podgorny, President Nikolay V. (U.S.S.R.), 43, 180–181, 222
Pogue, Maj. William R. (USAF), 68
Point Barrow, Alaska, 3, 5, 174, 187, 192, 288, 343
Point Mugu, Calif., 152, 276
Polaire (French satellite), 342
Poland, 43, 287, 319
Polar cap absorption, 196, 246
Polaris (missile), 125, 176
Pollack, Dr. James B., 340
Pollock, Howard W., 19
Pollution control, 32, 42, 49
Pomeroy, Dr. John H., 251
Pompidou, President Georges (France), 125, 299, 349–350
Ponnamperuma, Dr. Cyril A., 39, 167, 256, 322, 356
Popovich, Pavel R., 152
Port of New York Authority, 49
Portable life support system (PLSS), 156, 205
Porter, Daniel R., 198
Porter, Frank C., 82
Porter, Sylvia, 17
Portugal, 219–220
Poseidon (missile), 52, 224, 309
Poseidon (nuclear submarine), 21
Position-location and aircraft communications experiment (PLACE), 361–362
Post, Wiley, 288

Postal Service, U.S., 232
Potomac River, 288
Powell, Maj. Cecil (USAF), 37, 65, 130
Powers, Francis Gary, 329
Pratt & Whitney Div. See United Aircraft Corp.
Present, Stuart M., 22–23
President's Science Advisory Committee, 16–17, 49–50
Press, Dr. Frank, 345
Press comment
  AAAS convention, 1
  aerospace industry, 309
  antiballistic missile (ABM), 4
  Apollo program, 161, 197, 223–224, 229
  Apollo 11 mission, 232–233
  Apollo 14 mission, 17, 35, 38, 40, 42, 43, 44, 47
  Apollo 15 mission, 213, 215–216, 217–218, 222, 223–224, 225–226, 230
  Apollo 17 mission, 236
  atomic bomb test, 311
  computer, invention of, 218
  emergency industrial loan guarantee, 194
  engineers, 104–105
  F-14 (fighter aircraft), 271, 287
  Fedoseyev, Anatoly, 176
  Fletcher, Dr. James C., appointment of, 64, 72
  international cooperation, space, 13, 19–20, 247–248, 249–250, 347
  Kennedy Space Center economy drive, 287
  Luna 18 mission, 259
  lunar exploration, 4, 223–224
  McDonnell Douglas Corp., 170
  Mariner 9, 314, 324, 327, 330, 332–333
  Mars 2, 324, 330
  Mars 3, 324, 330, 347
  military procurement bill, 284
  moon, water on, 293
  NASA, 139
    public information policy, 84
    National Air and Space Museum, 329–330
    New Technologies Opportunity Program, 344–345
  New York Times Special Features agreement, 211
  noise, aircraft, 242
  nuclear warhead test, 303
  nuclear weapons deterrent agreement, 266, 271
  People's Republic of China satellite, 72
  Pioneer-F probe, 74
  Rowe, Henry S., resignation, 320
  SALT agreement, 189
  Salyut 1–Soyuz 10 mission, 110–111, 156–157, 158
  Schlesinger, Dr. James R., appointment, 213
  scientist-astronaut, 230
  Seabed Arms Control Treaty, 47
  Seaborg, Dr. Glenn T., retirement, 213
  Soyuz 11 accident, 185

space program, national, 32, 74, 223–224
space shuttle, 14
supersonic transport, 78, 79, 85, 119, 129–130, 139
technology, 331–332
technology utilization, 140
Toro (asteroid), 332–333
U.S.S.R. space program, 24, 101, 111, 119
Venus 7, 20
Press conference
  aerospace industry employment, 125–126
  aircraft, military, 121
  Apollo program, 215
  Apollo 14 mission, 4, 35, 41, 59
    extrasensory perception tests, 170
  Apollo 15 mission, 120, 141, 142–144, 156, 217, 227, 228, 245, 312
    lunar samples, 238, 257–258
  Apollo 17 mission, 234
  C–5A contract, 2
  comet, 93
  Concorde (supersonic transport), 351
  earth resources technology satellite, 96
  F–14B (jet fighter), 212
  international cooperation, space, 191
  L–1011 (TriStar) (jet transport), 221
  launch vehicles, U.S., sale of, 48
  Lewis Research Center, 246–247
  lunar rock sample storage, 251
  lunar roving vehicle, 120
  manned space flight, 312
  Mariner 8, 136
  Mariner 9, 301–302, 321
  military expenditures, 350
  National Air and Space Museum, 301
  nuclear test ban, 250
  People's Republic of China
    nuclear test, 84
    satellite, 111
  Pioneer program, 322
  plutonium, 102
  Salyut 1–Soyuz 10 mission, 110, 119, 191–192
  space program, national, 119, 121, 127, 138, 146
  space shuttle, 272
  space station, 249–250
  supersonic transport, 2, 127, 148, 150
  Toro (quasi-moon), 281–282
  Tu–144, 264
  unmanned spacecraft exploration, 57
  U.S.–U.S.S.R. space docking, 177
  U.S.S.R.
    computer program, 121
    missile program, 146, 286
    space program, 329
Price, Dr. P. Buford, 6
Price, Rep. Robert D., 38
Princeton Electronics, Inc., 360
Princeton, N.J., 282
Princeton Univ., 18
Priorities for Space Research, 1971–1980 (NAS–NRC Space Science Board report), 67

Pritchard, Wilbur L., 299
Probe (see also individual probes, such as *Luna 18*, *Luna 19*, Mariner 8, *Mariner 9*, *Mars 2*, *Mars 3*, and *Venus 7*)
  international cooperation, 145, 259
  interplanetary, 15, 20, 22, 67, 115, 175, 179
  Jupiter, 22, 67, 74, 322, 327, 356, 369
  lunar. See *Luna 16*, *Luna 17*, *Luna 18*, *Luna 19*, *Surveyor 3*, and *Zond 2*.
  Mars, 22, 146, 147–148, 191, 293, 318, 321, 324, 337, 345, 351, 354–355, 360
  Mercury, 22, 175
  Neptune, 22, 179
  Pluto, 22, 179
  Saturn, 22, 121
  sun, 237, 252
  Uranus, 22
  U.S.S.R., 4, 9, 15, 20, 134, 147–148, 209, 237, 246, 267–268, 293
  Venus, 15, 22, 67, 115, 175, 186, 333
Professor Zubov (U.S.S.R. ship), 358
Propulsion, 95, 98
Prospero (X–3) (U.K. technology satellite), 300, 368
Proton, 19
Proton (U.S.S.R. booster), 246, 249
Proxmire, Sen. William 73, 81, 145, 177, 252, 260, 271, 284
Public Service Award (NASA), 279
Pudenz, Dr. Robert H., 112
Pueblo, Colo., 182
Puerto Rico, 71, 133, 134
Pulsar, 36, 67, 74, 89, 131
Pulver, W. A., 307
Purdue Univ., 98
Pyrimidines, 167

Q

Q-fan engine, 363
Quarantine, 5, 30, 115, 188, 222
Quasar, 67, 74, 89, 102, 190
Queens Univ., 10
QUESTOL. See Quiet, experimental, short takeoff and landing transport aircraft.
Quiet Engine Project, 83, 96, 130, 241
Quiet, experimental, short takeoff and landing (QUESTOL) transport aircraft, 218, 337, 338, 356, 365

R

Radar, 9, 48, 64, 222, 231, 260, 267, 310, 347
Radar-cross-section (RCS) test vehicle, 261
Radiation (see also Ultraviolet and X-ray), 9
  cosmic, 63, 197, 209, 219, 325
  effects, 235, 241, 242, 261, 325
  gamma, 181
  measurement, 197, 209, 325
  solar, 189

Radioactivity, 311
Radioastronomy, 89, 277, 279, 287
Radioelectronics, 162
Radiotelescope, 249, 344, 346
Rae, Dr. John B., 151
Ragsdale, Robert G., 320
Rakovsky, Mikhail Y., 122
Raleigh, Sir Walter, 97
RAM. See Research and Applications Modules.
Ramo-Woolridge Corp., 57
RAND Corp., 145, 320, 336, 346
R&D. See Research and development.
Rangel, Rep. Charles B., 38
RANN. See Research Applied to National Needs program.
Rapp, Rita M., 38
Rathbun Dam, 213
Raushenbakh, Boris V., 119
RB–57F (reconnaissance research aircraft), 120, 168
RB–211 (jet engine), 38, 44, 88, 124, 221
RCA, 71, 81, 349
  RCA Alaska Communications, Inc., 71
  RCA Global Communications, Inc., 71
  RCA Service Co., 174
Rea, Dr. Donald G., 178
Recent Trends in Enrollment and Manpower Resources in Graduate Science Education, 1969–70 (NSF report); 146
Rechtin, Eberhardt, 363
Reconnaissance satellite, 243, 296, 328
Record
  aircraft, 12, 17, 38, 41, 220, 226, 330
  spacecraft, 30, 35, 207
  underwater, 314
Redstone (booster), 122
Redstone (missile), 362
Redstone Arsenal, Ala., 5, 279
Redstone Arsenal Army Airfield, Ala., 253
Rees, Dr. Eberhard F. M., 68, 196, 224, 242, 344
Rehovoth, Israel, 18
Reitsch, Hanna, 320
Relay telemetry subsystem (RTS), 343
Relativity, theory of, 296
Reliability, 164
Remote sensing (see also Earth resources experiment package; Earth Resources Technology Satellite), 96, 97, 104, 119, 128, 132, 139, 198, 286, 318, 333, 369
Rendezvous, 68
  Apollo 14 mission, 30
  Soyuz 10–Salyut 1 mission, 108, 110, 121
  U.S.–U.S.S.R. cooperation plans, 77, 243, 249, 325, 369
Research and applications modules (RAM), 3, 101, 192, 330, 334, 337, 341
Research and development (R&D), 46, 178, 276, 320, 334, 363
  aeronautics (see also Aeronautics, research), 123, 271, 369
  cost, 116, 305
  criticism, 291
  Federal support, 13, 20, 33, 89, 116, 117, 248, 251, 303, 336, 368–369
  funds, 13, 20, 33, 89, 116, 117, 175, 251, 331
    AEC, 332
    DOD, 124, 332
    NASA, 20, 42, 93, 116, 175, 251, 332, 336, 368
    NOAA, 99
    science, 20, 33, 40, 89, 251
    U.S.S.R., 86, 124, 201
Research and Development in Industry, 1969: Funds, 1969; Scientists & Engineers, January 1970 (NSF 71–18), 117–118
Research and Development in Local Governments, Fiscal Years 1968 & 1969 (NSF 71–6), 33
Research and Technology Operating Plan (RTOP) Summary, 46, 334
Research Applied to National Needs (RANN program), 178
Research, Inc., 106
Research Triangle Institute, 200
Resolute Bay, Canada, 248
Reusable nuclear stage (RNS), 276
Review of Recent Launch Failures (House report), 275
Reynolds, Clarke G., 363
RF: radio frequency.
RFP: request for proposals.
Rice Univ., 47, 144, 288, 291
Rickenbacker, Edward V., 165
Rieger, Siegfried H., 298
Riga, Latvia, 110
Riga (planetoid), 357
Rim Crater (moon), 205
Rio de Janeiro, Brazil, 248, 249
RL–10 (rocket engine), 96
RM (Radiation/Meteoroid satellite), 40–41
RNS. See Reusable nuclear stage.
Robert Hutchings Goddard—Father of the Space Age, 214
Robertson, R. E., 261
Rockefeller Univ., 265
Rockwell Standard Corp., 10
Rockwell Standard Co.,
Roe, Rep. Robert A., 38
Rogers, Will, 288
Rogers, Secretary of State William P., 17, 272
Rohini (Indian rocket), 282
Rohr Corp., 307
Rolla, Mo., 296
Rolls-Royce Ltd., 38, 44, 88, 124, 173, 221, 267, 300
Romania, 243, 319
Rome, Italy, 257, 323
Rome, Univ. of, 262, 319
Roosa, L/C Stuart A. (USAF)
  Apollo 14 mission
    Congress, report to, 61
    docking, 25, 29
    flight, 24–30

lunar photography, 27
medical aspects, 5, 30, 44, 261
Nixon, President Richard M., 24, 31, 42, 43
preparations for, 4, 10
press conference, 4, 41, 59
quarantine, 5, 30
splashdown, 30, 42
Apollo 16 mission, 61
awards and honors, 72
European tour, 150, 152
promotion, 47
tribute to, 42
Rosamond Dry Lake, Calif., 265
Rose, Tokyo, 243
Rosen, George, 363
Rosenberg, Morris D., 343
Rosman, N.C., 272
Ross, Miles, 271
Rotating-cylinder-flap system, 301
Rothrock, Dr. Addison M., 176
Rotor, 188–189
Roush, Rep. J. Edward, 173, 222
Rover. See Lunar roving vehicle.
Rowen, Henry S., 320
Royal Astronomical Society, 202
Royal Greenwich Observatory, 327, 356
RTD 701 (calibration target satellite), 221
RTOP. See Research and Technology Operating Plan Summary.
Rukavishnikov, Nikolay, N., 107, 283
Rumford Premium, 102
Rusk, Dr. Howard A., 189
Russell, Dr. Archibald E., 326
The Russian Space Bluff, 331
Ryan, Gen. John D. (USAF), 11

S

S–IC. See Saturn V (booster), stage, 1st.
S–II. See Saturn V, stage, 2nd.
S–IVB. See Saturn V, stage, 3rd.
S–210 (Japanese rocket), 9
S. A. Engins MATRA, France, 3, 19
SAAB Aktieboloag, 3, 19
Sabin, Dr. Albert E., 18
SAC. See Strategic Air Command.
SAE. See Society of Automotive Engineers.
Safeguard (antiballistic missile system), 4, 43, 77, 78, 125, 311, 361
Sagan, Dr. Carl E., 248–249, 318, 331
Sagittarius (constellation), 93
St. Louis, Mo., 288
St. Moritz, Switzerland, 99
St. Vincent Charity Hospital (Cleveland), 106
Salon-de-Provence, France, 150
Salpeter, Edwin E., 90
SALT. See Strategic Arms Limitation Talks.
Salt Lake City, Utah, 161
Salyut 1 (U.S.S.R. scientific space station), 105, 108, 110, 112, 122, 154–155, 162–163, 177, 198, 229, 282, 366–367

Salyut 1–Soyuz 10 mission, 108, 110, 112, 119, 121, 282
Salyut 1–Soyuz 11 mission, 156, 158, 177, 191, 200, 277, 282
Salyut Crater (moon), 203
Samoa, 30
SAMSO. See Space and Missile Systems Organization.
San Antonio, Tex., 325
San Clemente, Calif., 243
San Diego, Calif., 207, 281–282, 291
San Francisco Bay, Calif., 281
San Francisco, Calif., 32, 80, 307
San Francisco International Airport, 310
San Marco (Italian launch site), Indian Ocean, 319
San Marco 1 (Italian satellite), 109–110
San Marco 2, 109–110
San Marco 3 (San Marco-C), 109–110, 333, 355, 368, 469
Sandage, Dr. Allan R., 18
Sands Point, N.Y., 13
Santa Clara, Univ. of, 254
Santa Fe, N. Mex., 42–43
SAO. See Smithsonian Astrophysical Observatory.
Saphier, Lerner, Schindler-Environetics, 182
Sapp, Capt. Earle W. (USN), 349
Sarnoff, David, 349
SAS. See Small Astronomy Satellite.
SAS–B (Small Astronomy Satellite), 356
Sashin, O., 220
Satellite Early Warning System (SEWS), 319
Saturn (booster), 24–25, 57, 201, 276
Saturn IB, 64, 99, 164, 328
S–IB stage, 291–292
Saturn V, 60, 110, 245
contract, 153, 353–354
cost, 209
launch, 217
AS–509, 24, 30
AS–510, 202, 207
program, 57, 328–329
stage
1st (S–IC–511), 254–255
2nd (S–II), 153
3rd (S–IVB), 25, 202–203, 365
Saturn (planet), 22, 121
Sawyer, Paul, 79, 84
Scarp Crater (moon), 205
Schaefer, Hermann J., 241
Schaeffer, Dr. Oliver A., 265
Schaibley, John R., 188
Scheer, Julian W., 51, 200, 232
Scherer, Lee R., 159, 255
Schilling Trophy, 262–263
Schlesinger, Dr. James R., 199, 213, 222, 233, 294, 300, 311, 346
Schmickrath, Bernard A., 178
Schmitt, Dr. Harrison H., 200–201, 217, 229, 230, 234, 236, 312
Schneider, William C., 358
Schneiderman, Dan, 321
Schultz, M/G Kenneth W. (USAF), 193

SCI Systems, Inc., 193
sci Electronics, Inc., 193
Science (see also National Academy of Sciences and National Science Foundation), 5, 58, 76, 234
  award, 17, 102, 138, 282, 307, 311
  benefits, 88, 146
  criticism, 147, 277, 283, 291, 303
  education, 288
  Government support of, 20, 40, 58, 88, 99, 108, 116, 187, 216, 260, 363
  human needs, 89, 108, 363
  international cooperation, 177
  national policy and goals, 17, 20, 51, 54–55, 69, 81, 89, 167, 187, 296, 363
  President's Science Advisory Committee, 16, 49
  U.S.S.R., 82, 100, 110, 140, 163, 183, 190, 210, 262, 291, 309, 311, 315, 323
Science and Technology, Dept. of (proposed), 69
Science Information Exchange, 82
Science Research Council (SRC) (U.K.), 348, 356
Scientific Activities of Independent Nonprofit Institutions 1970 (NSF 71-9), 58
Scientific Balloon Flight Station, Tex., 121
Scientific instrument module (SIM), 202, 207, 223
Scientist-astronaut, 87, 90, 133, 138, 200–201, 217, 228–229, 230, 234, 236, 312
Scientists, 1, 248
  employment, 58, 76–77, 104–105, 118, 119, 188, 252
  immigrant, 107
  NASA programs, 96, 210
  reorientation, 221
  Union of Concerned Scientists, 209
Scorpius (constellation), 93
Scott AFB, Ill., 132
Scott, Col. David R. (USAF)
  Apollo 15 mission, 114, 223, 226, 257–258, 298
  celebrations for, 222, 236, 237
  Congress, report to, 253–254
  extravehicular activity, 120–121, 203–204
  flight, 202–208
  Kennedy Space Center visit, 271
  Marshall Space Center visit, 272
  medical aspects, 118, 207, 222, 225, 228, 261, 289
  New York *Times* article, 200–201, 228
  Nixon, President Richard M., 201, 208, 221–222
  preparations for, 114
  press conference, 120–121, 227
  tribute to, 121–122
  U.N. visit, 239
  Apollo 17 mission, 229
  awards and honors, 239, 260, 262, 266, 346–347
  European visit, 311
Scott, Sen. Hugh, 191

Scott, Miss Sheila, 220
Scout (booster), 110, 164, 168, 189, 230, 259, 319, 348
Scripps Institution of Oceanography, 292
Sea Grant Program, 99
Sea of Fertility (Mare Fecunditatis) (moon), 7, 190, 246
Sea of Rains (Mare Imbrium) (moon), 4, 10, 66, 115, 133, 143, 167, 168, 196, 223, 227, 278, 291
Sea of Serenity (Mare Serenitatis) (moon), 143
Sea of Tranquility (Mare Tranquillitatis) (moon), 6, 13, 121
Sea Sciences Corp., 292, 293
Seabed Arms Control Treaty, 45, 47, 55
Seaborg, Dr. Glenn T., 65, 194, 199, 213, 215, 219, 233, 250, 253, 257
Sealab Project, 292
Seamans, Secretary of the Air Force, Dr. Robert C., Jr., 72, 87, 267
Seattle-Tacoma International Airport, 278
Seattle, Wash., 57, 63, 81, 166, 333
Secret Sentries in Space, 243
Seed experiment, space, 235
Seiberling, Rep. John F., Jr., 38
Seismometer experiment, lunar, 291
  Apollo 12, 7, 25, 63, 113, 203, 207, 236, 245
  Apollo 14, 25, 30, 44, 63, 64, 13, 203, 207, 236, 245
  Apollo 15, 143, 203, 207, 222, 236, 245
Seitz, Dr. Frederick, 112, 113, 265
Self-test and repair (STAR) computer, 179
Serenitatis, Mare (Sea of Serenity), 143
Sergievsky, Boris, 330
Serphukov accelerator, U.S.S.R., 168, 182, 299
Servan-Schreiber, Jean Jacques, 73
Service module (SM), 30, 141, 207
SES-100B (Surface Effect Ship), 200
Sesp 1971-2 (Space Experiments Support Program satellite), 289
SEVA. See Stand-up extravehicular activity.
Sevastyanov, Vitaly I., 152, 167, 266
SEWS. See Satellite Early Warning System.
Seyfert Galaxy, 73–74, 89
Shaffer, John H., 2, 65, 107, 194, 212, 237, 287, 266, 332, 334, 351
Shapiro, Dr. Irwin I., 102, 310
Shapley, Willis H., 60, 267
Sharp, Mitchell W., 250
Shatalov, Vladimir A., 16, 107, 283
Shea, Joseph F., 102
Sheldon, Dr. Charles S., II, 7, 268, 362
Shemya, Amchitka Island, 326
Shepard, R/A Alan B., Jr. (USN), 247
  Apollo 14 mission, 35
  Congress, report to, 60–61
  docking, 25, 29–30
  extravehicular activity, 25–26, 27–28, 41, 46–47, 48, 51
  flight, 24–30
  golf shot, 28–29, 106
  medical aspects, 5, 30, 44, 261

Nixon, President Richard M., 24, 31, 42, 43
 preparations for, 4, 10, 22-23
 press conference, 4, 41, 59-60
 quarantine, 5, 30
 significance, 61
 splashdown, 30, 42
appointment, 176, 254
awards and honors, 72, 106, 242, 266
European tour, 150, 152
Freedom 7 mission, 122
promotion, 240
tribute to, 42
Shepherd, James T., 116
Shields, Jack E., 133
Shinsei (New Star) (Japanese satellite), 268, 368
Shklovsky, Dr. Yosif S., 248-249
Short-range attack missile (SRAM), 8, 262
Shuang-ch'eng spaceport, People's Republic of China, 53, 61
Shuttle Technology Program—Fiscal Year 1972, 263
Siberia, 210, 243, 288
Significant NASA Inventions, Available for Licensing in Foreign Countries, 129
Sikorsky, Aircraft Div., United Aircraft Corp., 150, 357
Sikorsky, Igor I., 150
Silver, Dr. Leon T., 245, 257-258, 278
SIM. See Scientific instrument module.
Simmons, Dr. Gene M., 183, 247
Sinkiang Province, People's Republic of China, 150
Sirius (star), 306
Sjoberg, Sigurd A., 57
Sjogren, William L., 314
Skully, Richard P., 237
Skylab (spacecraft) (see also Airlock module; Apollo Telescope Mount; Multiple docking adapter; Orbital Workshop), 42, 152, 172, 183, 265
Skylab program, 44, 56, 172, 176, 328-329
 contract, 99, 106, 171, 175, 278, 279, 286
 crew preparation, selection, and training, 86-87, 132, 133, 284, 288, 346
 design, 64, 183
 experiments, 10-11, 38, 87, 106-107, 173, 175, 183, 193-194, 268-269, 342, 358-359, 361
 funds, 20-21, 94, 157, 173, 198-199, 221, 368
 hardware, 64, 265, 281, 284, 323, 346, 353, 355, 368-369
 international cooperation, 10, 38, 44, 173
 medical aspects, 167, 193-194, 256, 276, 342
 NASA program, 10-11, 39, 62-63, 84, 86-87, 94-95, 127, 164, 172, 267, 291, 302, 324, 361, 363-364, 368
 safety aspects, 42
 space rescue program, 64, 68-69, 163-164, 173, 198-199
 test, 11, 64 132, 152, 252, 258, 276, 353, 355
Skylab Student Education Conference, 268-269
Skylab Student Project, 268-269
Slater, John C., 18
Slayton, Donald K., 39, 176, 181
Small Astronomy Satellite (SAS), 67
 Explorer 42 (Uhuru), 89, 113, 131, 319
 SAS-B, 356
Small Scientific Satellite, 319, 365, 369
Small Business Administration, 334
Smelt, Dr. Ronald, 102
Smith, Dr. Bradford A., 313, 339
Smith, Delbert E., 117
Smith, Sen. Margaret Chase, 24, 68
Smith, W. B., 310
Smithsonian Astrophysical Observatory (SAO) (Cambridge, Mass.), 7, 40, 313
Smithsonian Institution, 24, 52, 82, 97, 101, 130, 135, 153, 195, 301, 309
Smog research, 170
SMS. See Synchronous Meteorological Satellite.
SMS-A (Synchronous Meteorological Satellite), 356
Smyth, Robert K., 141
SNAP-3A (radioisotope thermoelectric generator), 180
SNAP-19, 250
SNAP-27, 250
SNECMA (Société Nationale d'Étude et de Construction de Moteurs d'Aviation), 267
Snow, Edgar, 84
Snow, Dr. Joel A., 301
SNSO. See AEC-NASA Space Nuclear Systems Office.
Snyder, Dr. Conway W., 144
Society of Aerospace Material & Process Engineers, 279
Society of American Test Pilots, 259
Society of Automotive Engineers (SAE), 130, 161, 326
Society of Exploration Geophysicists, 312
Society of Sigma Xi, 283
Sodium experiment, 251, 256
Sokolsky, V. N., 234, 261
Solar cell, 4, 12, 369
Solar corona, 277-278
Solar energy, 129, 154
Solar flare, 56, 190, 277, 302, 304, 349, 365
Solar neutrino, 113
Solar radiation satellite. See *Solrad 8, 9, and 10.*
Solar wind, 67, 73, 93, 341, 354
 experiment, 27, 144
Solrad 8 (Explorer 30) (solar radiation satellite), 190
Solrad 9 (Explorer 37), 190
Solrad 10 (Explorer 44) (Solrad-C), 189-190, 240, 365
Somnium, 233-234
Sonett, Dr. Charles P., 5, 64, 350
Sonic boom, 78, 112, 122, 351

Sophia, Princess (Spain), 24–25
Sorells, A. R., 90
Sortie can, 334, 340
Sounding rocket (see also individual sounding rockets: Aerobee, Arcas, Black Brant VB, Centaure, Javelin, Nike-Apache, Nike-Cajun, Nike-Tomahawk; Vertikal-2)
  international programs, 236, 252, 369
  NASA-Canada, 8, 9, 14, 50, 63, 89, 174, 192, 246, 248
  -France, 64, 74, 78, 82, 259, 370
  -India, 111, 115
  -India-Japan, 111
  -U.S.S.R., 10, 133–134
  U.S.S.R.-France, 359
Sours, Wayne P., 269
South Africa, 267
South Carolina Lions Clubs, 140
South Dakota, 128
South Pole, 310, 329
Southwest Research Institute, 15
Soviet Academy of Medical Sciences, 163
Soviet Academy of Sciences, 73, 94, 100, 119, 127, 140, 145, 163, 181, 182, 218, 268, 277, 357
  Council for International Cooperation in Space Research, 217
  honorary membership, 65, 194
  Intercosmos Council, 171
  international cooperation, 249
  international cooperation, space, 10, 16, 77, 88, 133, 217, 242, 283, 293, 340, 347, 353, 359, 369
  lunar rock samples, exchange of, 88, 159, 190
Soviet Council of Ministers, 200
Soviet Party Congress, 94
Soviet Space Programs, 1966–70, 132
The Soviet SST, 126
Soviet State Committee for Nuclear Energy, 219
Soviet State Planning Committee, 121
Soyuz (program), 284
Soyuz (U.S.S.R. spacecraft), 16, 152, 177
Soyuz 1 mission, 266
Soyuz 9 mission, 155–156, 169
Soyuz 10 mission, 107, 112, 119, 121, 283
Soyuz 11 mission, 154–156, 180–181, 185, 186, 187, 226, 283, 299, 366
Space and Missile Systems Organization (SAMSO), 102, 178, 188, 266, 290, 310, 336
Space biology, 99, 325
  animal experiments, 127, 182, 235
  award, 163, 298, 325
  back contamination, 115
  environment, effects of, 167, 225, 228, 256, 261, 277
  extraterrestrial life, 167, 248–249, 322, 337–338, 345, 369
  international cooperation, 11, 36, 284, 325, 359
  International Symposium on Basic Environmental Problems of Man in Space, Fourth, 277
  life support system, 95–96, 105, 156, 181, 193–194, 196–197, 208, 231, 235, 244, 308, 311–312
  lunar soil experiment, 6, 111
  medical benefits, 183, 189, 235, 292, 368
  meeting, 253
  NASA program, 138, 189, 210, 235, 245, 284
    Apollo 14 mission, 5, 14, 17, 30, 35, 44, 63, 261
    Apollo 15 mission, 188, 193, 196–197, 207, 223, 225, 228, 289
    Skylab, 132, 193–194, 288
  nutrition, 177, 276, 289
  radiation effects, 63, 235, 261, 325–326
  Soyuz 11 accident cause, 187
  U.S.S.R. program, 176–177
  weightlessness, effects of, 127, 132, 163, 181, 342
Space debris, 166, 173, 181–187, 209
Space Electronics Corp., 57
Space Flight Award (AAS), 179
Space General Corp., 57
Space law, 153
Space, military use of, 77
  communications, 37, 41, 74, 80, 87, 95, 306, 336, 365
  NATO, 36, 41, 365
  navigation, 94
  reconnaissance, 124, 328
  space shuttle, 75, 87, 95, 328
  U.S. (see also unidentified satellite), 74, 87, 95, 124, 306, 328, 336
  U.S.S.R., 7, 24, 115, 158, 172, 176, 191–192, 296, 338, 362
Space, peaceful use of, 10, 16, 91, 101
Space program, national (see also individual programs, such as Apollo program; and National Aeronautics and Space Administration, budget), 67, 162, 172, 243, 275–276, 282, 283, 328–329, 342–343, 365–368
  achievements, 7–8, 51, 167, 179, 225, 253, 364–369
    manned space flight, 30, 42, 47, 48, 208, 209, 223–224, 253–254, 361, 364, 368
  Aeronautics and Space Report of the President, Transmitted to the Congress, January 1971, 37
  anniversary, 142, 180
  benefits. See Space results.
  budget, 7–8, 14, 20–21, 31, 32, 38, 57, 62, 112, 119, 121, 220, 221, 292, 363–364, 368
  cost, 50–51, 62
  criticism, 20, 32, 43, 133, 145, 146, 211, 229, 236
  employment, 61–62, 75–76, 172, 179, 368
  international aspects (see also International cooperation, space), 7–8, 10–11, 15–16, 17, 18, 37, 45, 54–55, 60–61, 68, 96, 133, 138–139, 159, 165, 185, 191, 247–248, 249–250, 263, 265, 324

management, 69–70, 72, 106
military aspects, 7–8
Nixon, President Richard M., 42, 54–55, 59, 161, 263–264, 265
objectives, 67, 105, 117, 215, 263–264, 272–273, 280–281, 343–344
policy, 7–8, 32–33, 51, 67, 121, 127–128, 145, 146, 163–164, 165, 181, 183, 199, 200, 212, 213, 267, 324
post-Apollo, 21–22, 45, 51, 57–58, 61–62, 67, 74, 89, 93, 96–97, 105, 139, 145, 263, 324, 328–329
press comment, 32, 74, 223–224
press conference, 21–22, 119, 122, 127, 139
significance, 31, 42, 59, 126, 225, 281
Space Program Advisory Council, 105
U.S.S.R. vs. U.S. See Space race.
Space race, 7, 16, 24, 25, 62, 91, 111–112, 121, 158, 185, 223, 229, 329, 330, 331, 366
Space rescue, 22, 64, 68–69, 140, 164, 173, 198
Space results (see also Earth; Mars; Moon; Venus; individual probes, satellites, and sounding rockets), 43, 47, 51, 93, 126, 136–137, 167, 172, 215–216, 217–218, 221–229, 298, 312, 321, 368
  agriculture, 37, 98, 368, 369
  aircraft, 70, 74
  astronomy, 67, 93, 113, 223, 239, 245, 257, 265–266, 281, 314, 315, 318, 321, 332–333, 342, 349, 358–359, 368
  communications, 37, 126, 281, 368
  computer, 32, 98–99, 106–107, 121–122, 162
  earth sciences, 19, 32–33, 44, 48, 67, 119–120, 128, 137, 171, 174–175, 198, 230, 253, 261, 281, 309, 324, 369–370
  economic benefits, 51, 166, 172
  education, 37, 117, 243
  engineering, 140, 259, 308, 309, 368–369
  environmental sciences, 37, 67, 162, 231, 320–321, 322, 324, 348, 368
  For the Benefit of all Mankind: A Survey of the Practical Returns from Space Investment, 350–351
  international relations, 44
  materials technology, 23, 72, 74, 350–351
  medicine, 98–99, 106–107, 188, 189, 193–194, 199–200, 235, 253, 256, 261, 271–272, 292, 368
  meteorology, 37, 281, 303, 327
  military, 70, 75, 87, 95
  navigation, 37
  oceanography, 120, 348
  science and technology, 315
  social problems, 342
  transportation, 37, 251, 257, 259
Space Science Award (AIAA), 15
Space shuttle, 44, 58, 62, 95, 139–140, 213, 229
  contract, 60, 101, 105, 106, 132, 165, 187, 193, 218, 233, 240, 245–246, 272, 276, 281, 298, 320, 328, 345, 353–354, 356
  cooperation, 87, 93, 244, 257, 276
  cost, 24, 62, 94, 127–128, 145, 172, 182–183, 257, 315
  criticism, 145–146
  design, 60, 64, 80, 87, 132, 161, 165, 171, 218, 272, 281, 352, 356, 362–363, 365
  development site, 145, 281
  engine, 60, 87, 160, 193, 232, 240, 245–246, 345, 354, 356
  funds for, 14, 20, 22, 37, 50–51, 80, 87, 94, 116, 129, 152, 157, 177–178, 198–199, 221, 226, 229, 368
  international cooperation, 3, 16, 37, 45, 278, 291, 340–341, 369–370
  launch and landing sites, 2, 281
  meeting, 42, 75, 98, 279, 291, 352
  military application, 87, 280, 328
  NASA program, 50–51, 62–63, 93, 127–128, 129, 139–140, 145, 160, 164, 165, 182–183, 192–193, 213, 263, 264, 275–276, 280, 297, 324, 330, 334
  patent, 171
  press comment, 14
  safety, 42
  shuttle pilot study, 308
  Shuttle Technology Program—Fiscal Year 1972, 263
  tests, 60, 87
  test facilities, 8, 60
Space Shuttle Aerothermodynamics Technology Conference, 352
Space Shuttle Technology Conference, 98
Space Shuttle Technologies, FY 1971 Programs, 129
Space station (see also Salyut 1 and Skylab), 44, 58, 249
  contract, 174
  criticism, 145
  design, 3–4, 12, 175, 183
  international cooperation, 3, 10, 38, 45, 173, 278, 333
  military, 229
  NASA program, 10–11, 12, 14, 50, 88, 127–128, 139–140, 145, 172, 174–175, 192, 228–229, 334
  use, 10–11, 90, 342
  U.S.S.R., 61, 100, 101, 107–108, 110–111, 115, 154–155, 163, 188, 189–190, 197, 282–283, 365–366
Space tug, 224, 278, 340
Spacecraft (see also individual spacecraft, such as Apollo, Explorer, Luna, Mariner, Skylab)
  accident. See Accident, spacecraft.
  airlock, 284
  command and service module. See Command and service module.
  command module. See Command module.
  computer, 179
  contamination, 56

debris, 313–314
design, 15, 64, 68, 80
electrical systems, 244
equipment, 106, 183, 207, 254, 264, 265, 284, 302, 306, 323, 328, 330, 344
exhibit, 152, 153
instrumentation, 8, 10–11, 14, 35–36, 38, 174–175, 236–237, 240, 277, 278, 321, 355
international cooperation, 105
life support system, 95–96, 99, 156, 181, 193–194, 196–197, 235, 308
lunar module. See Lunar module.
materials technology, 87, 195, 279
multiple docking adapter, 284, 323, 346, 353, 355
propulsion. See Engine; NERVA; Nuclear propulsion; and individual launch vehicles, such as Saturn.
research and applications module (RAM), 3, 101, 192–193, 330, 334, 337, 340
safety, 42, 64, 68–69, 178, 261, 362–363
scientific instrument module, 203, 207, 223
service module. See Service module.
test, 86–87, 132, 152, 232, 258, 264, 327
Spacesuit, 153, 156, 196, 223, 286, 288
Spacetrack system, 290–291
Spain, 43, 112
Spartan (missile), 4, 311, 326, 346
Spectrograph, 8
Spectroheliograph, 8
Spectrometer, 10–11, 144, 148, 175, 195–196, 220–221, 265–266, 318
Speer, Dr. F. A., 98
Sperry, Lawrence, Award, 298
Sperry Rand Corp., 159
Spin-scan camera, 15
Spinar (celestial object), 73
Spirit of '76, The (Presidential aircraft), 212, 326–327, 350
Spitz, Dr. Armand N., 103
Spitz Laboratories, 103
Spivak, Jonathan, 182–183
Spring Mill State Park, Ind., 199
Sprint (missile), 4, 43, 77–78, 125, 361
Spruce Goose (flying boat), 45, 197
SPS. See Service propulsion system.
Spur Crater (moon), 205
Sputnik (U.S.S.R. satellite), 145
Sputnik 1, 24, 59, 113, 209
Sputnik 2, 24
SR–71 (reconnaissance aircraft), 115
SRAM. See Short-range attack missile.
SRC. See Science Research Council (U.K.).
Sriharikota Island, India, 282
SS–6 (Sapwood) (U.S.S.R. ICBM), 362
SS–9 (U.S.S.R. ICBM), 84, 117, 135, 152, 194, 362
SS–11 (U.S.S.R. ICBM), 146
SST. See Supersonic transport.

Staats, Comptroller General Elmer B., 218
Stafford, Maj. Thomas P. (USAF), 176, 186
Stand-up extravehicular activity (SEVA), 203–204
Standard Oil Co. of New Jersey, 170
Stanford Linear Accelerator Center, 36
Stanford Research Institute, 291
Stanford Univ., 13, 15, 98, 183, 197, 235, 248, 298
  Stanford Univ. Medical Center Cardiology Dept., 301
Star, 13, 36, 38, 60, 113, 226, 239, 350
STAR. See Self-test and repair computer.
STAR (Satellites for Telecommunications, Applications, and Research) (consortium), 19
Star City, U.S.S.R., 16
State, Dept. of, 45, 48, 77, 102, 120, 237, 272, 358, 369–370
State Univ. of New York
  Buffalo, 201
  Stony Brook, 192, 257, 260, 265
Statements by Presidents of the United States on International Cooperation in Space—A Chronology: October 1957–August 1971 (Sen. Doc. 92–40), 265
Stecher, Theodore P., 36
Stein, Andrew J., 2
Steinbacher, Dr. Robert H., 301, 318
Stennis, Sen. John C., 19, 351
Stereo-1 (U.S.S.R.–France Mars experiment), 354
Stevens, Peter, 347
Stevenson, M/G John D. (USAF, Ret.), 12
Stever, Dr. H. Guyford, 320, 341
Steward Observatory, 40
Stockholm, Sweden, 282, 329
Stokes, Mayor Carl B., 2
STOL (short takeoff and landing) aircraft, 298, 363
  Canada, 147, 280
  contract, 19, 187, 300, 338, 365
  cooperative research, 39, 71, 218, 267, 290, 315
  Italy, 315
  NASA program, 21, 22, 39, 71, 97, 127, 171, 187, 218, 233, 290, 301, 308, 337, 338, 365, 369
STOLAND (advanced STOL avionics system), 187
STOLport, 19
Stoner, George H., 57
Strasser, Gabor, 363
Strategic Air Command (SAC), 279
Strategic Arms Limitation Talks (SALT), 55, 74, 148, 189, 190, 194, 195, 252, 264–265, 266
Stratoscope II (balloon-borne telescope), 121, 253
Strughold, Dr. Hubertus, 298, 325
Stubbs, Dr. Peter, 57

A Study of NACA and NASA Published Information of Pertinence in the Design of Light Aircraft, 56
Stuhlinger, Dr. Ernst, 334
Submarine, missile-carrying, 224, 283, 286
Subsatellite, 87, 207, 225, 365
Suffolk County, N.Y., 295
Sukhanovsky, A., 315
Sukhoi Su-11 (U.S.S.R. fighter aircraft), 360
Sullivan, Rep. Leonor K., 247
Sullivan, Walter S., 223, 287, 342
Sun, 113, 223, 257, 350
   satellite data, 269, 335, 349
   sounding rocket data, 228, 252
Sunspots, 56, 304
Suomi, Verner E., 15
Super Guppy (NASA cargo aircraft), 281, 288, 355
Super nova, 36
Superconstellation (transport aircraft), 101
Supercritical wing. See Wing, aircraft.
Supersonic transport (SST) (see also Concorde and Tu-144), 119, 192
   cost, 73, 82, 84, 157, 346, 351
   criticism, 73
   design and development, 60–61, 90, 264, 323, 326
   economic aspects, 56, 82, 109, 130, 131
   Final Report of the Ad Hoc Supersonic Transport Review committee of the Office of Science and Technology, March 30, 1969, 232
   foreign, 82, 85, 147, 214, 280–281, 332
   funds, 21, 74, 78, 80–81, 82, 98, 129–130, 131, 135, 137, 138, 139, 211, 232, 366
   hazards, 60–61, 150, 240–241, 242
   international aspects, 84–85, 109
   Nixon, President Richard M., 37, 81, 82, 130, 349–350, 351
   press comment, 78, 79, 85, 119, 129–130, 139
   press conference, 48, 138, 148, 351
   sonic boom, 351
The Supersonic Transport: (Study), 56
The Supersonic Transport: A Factual Basis for Decision, 60
Suprathermal ion detector experiment, 142, 288, 291
Surface propulsion system (SPS), 203
Surveyor (program), 47, 212
Surveyor 3 (lunar probe), 57, 83
Sutherland, Dr. Earl W., Jr., 282
Sutter, Dr. John F., 257, 265
Swan experiment, 10
Swann, Dr. Gordon A., 245, 279
Swearingen Aircraft Co., 309
Sweden, 3, 19, 96, 198, 213, 313, 341, 345, 346
Swenson, Dr. Loyd S., Jr., 234
Swigert, John L., Jr., 68, 75, 179, 266
Swiss Transport Museum, 267
Switzerland, 50, 305
Symington, Rep. James W., 38

Symington, Sen. Stuart, 19
Symphonie (West German comsat), 352
Symposium on Clean Room Technology in Surgery Suites, 139
Symposium on Uranium Plasmas: Research and Applications, 2nd, 320
Synchronous Meteorological Satellite (SMS), 21–22, 356
Synthesis (radio telescope), 279
Syromyatnikov, V. S., 254
Szent-Gyorgi, Dr. Albert, 296

T

T-33 (jet trainer), 115, 308
Tacsat 1 (comsat), 74
TACT. See Transonic Aircraft Technology Program.
Tactical Air Command, 262
Takizawa, Ataru, 138–139
Talkeetna, Alaska, 69
Tango-Kilo (satellite reconnaissance intelligence), 124
Tansei (MST-1) (Japanese satellite), 48, 161–162, 200, 368
Tanum, Sweden, 341
Task Force on Science and Technology, 320
Tata Institute of Fundamental Research, 6
Taurus (constellation), 350
Taylor, Dr. Gerald R., 6
Taylor, John W. R., 296
TD-1 (ESRO scientific satellite), 356
TDRSS. See Tracking and data relay satellite system.
Teague, Rep. Olin E., 38, 63, 142, 222
Technology, 4, 16, 89, 146–147, 283
   benefits, 20, 126, 216, 272
   exhibit, 99, 301
   funding, 20, 109, 116, 118, 331–332
   international aspects, 17, 55
   policy, 50, 55, 81, 86, 109, 116, 213, 248, 263–264, 275, 290–291, 303, 331–332, 344, 362
   U.S.S.R., 311
Technology, Inc., 276
Technology Mobilization and Reemployment Program, 94, 119, 141–142
Technology unitization, space, 53, 76, 81, 129, 283, 350–351
   agriculture, 10–11, 368
   communications, 18, 37, 368
   earth resources, 10–11, 325
   ecology, 160, 321
   economic problems, 160, 283
   education, 37, 306
   engineering, 311, 334–335
   environmental problems, 96, 307, 324–325, 368
   health, 306
   materials technology, 250, 259
   medicine, 98–99, 180, 199–200, 235, 250, 253, 292, 301, 322, 350–351, 368

meteorology, 37
oceanography, 10–11, 104
Office of Technology Transfer (proposed), 173
pollution control, 322
press comment, 140
safety, 146
science, 350
social problems, 160, 301, 303, 306, 342
transportation, 37, 160, 309
urban planning, 281
Tektite I, Project, 263
Tele-Communications, Inc., 74
Telesat Canada, 128
Telesat-A (Canadian comsat), 128, 356
Telescope (see also Radiotelescope), 54, 67, 121, 253
Telescope, space (see also Apollo Telescope Mount), 67, 330
Television, 52, 83, 186, 237–238
  Apollo 14, 10, 23, 25–30, 40, 41, 43, 65
  Apollo 15, 197, 203, 206–207, 213, 215–216
  Apollo 16, 267
  educational, 161, 264
  Lunokhod 1, 8, 220
  Mariner 9, 318
  Soyuz 10, 108
  Soyuz 11, 156
  via satellite, 23, 71, 88, 98, 128, 161, 210, 330, 354, 355
Teller, Dr. Edward, 152
Tennessee Technological Univ., 298–299
Tennessee, Univ. of, Space Institute, 240
Tereshkova, Valentina. See Nikolayeva-Tereshkova, Valentina.
Terhune, L/G Charles H. (USAF, Ret.), 159–160
TERLS. See Thumba Equatorial Rocket Launching Station.
Test Vehicle 3 (TV-3), 135
Teton National Park, 112
Tetr 3 (Test and Training Satellite), 269, 365
Texas, 370
Texas A&M Univ., 111
Texas Instruments, Inc., 170
Texas, Univ. of, 144, 193
Textron, Inc.
  Bell Aerospace Co., div., 200, 290, 361
  Bell Helicopter Co., div., 357
TF-8A (jet research aircraft), 122, 356
  contract, 345–346
  test flight, 65–66, 77, 94–95, 101, 114–115, 140, 145, 163, 165, 192, 233, 240–241, 252, 256, 270–271, 272, 278–279, 287, 346, 349, 365–366
Thackray, Arnold, 187–188
Thant, U, U.N. Secretary General, 153–154, 222, 239, 252
Tharsis (Mars), 321
Thayer, Sylvanus, Award, 121
Thermoelectric Outer Planet Spacecraft (TOPS), 179

Thermonuclear fusion, 213
T'ry, Paul, 309
This Island Earth (NASA SP-250), 48
This New Ocean: A History of Project Mercury, 234
Thomas, Dr. Gary, 2, 93
Thompson-CSF, 3
Thompson, Gerald M., 161
Thompson Trophy Race, 382
Thor-Able-Star (booster), 180
Thor-Agena (booster), 8, 48
  Thorad-Agena, 80–81, 195, 254, 289, 350
Thor-Burner II (booster), 48, 157, 287
Thor-Delta (booster), 120, 269, 300–301
  long-tank, thrust-augmented, 294–295, 300–301
  thrust-augmented, 72–73, 106
Thorpe, Day, 5
3C-279 (quasar), 102
Thumba Equatorial Rocket Launching Station (TERLS), India, 18–19, 105, 111, 115
Tidbinbilla, Australia, 69
Tilt Rotor Research Aircraft Technology Program, 272
Time magazine, 281
Tischler, Adelbert O., 291
Titan (booster), 153
  Titan IIIB-Agena, 11
  Titan IIIB-Agena D, 79
  Titan IIIC, 95, 122, 306, 310, 313, 329
  Titan IIID, 163
Titan (missile), 224–225
Tito, President Josip Broz (Yugoslavia), 286
Tito, Mrs. Josip Broz, 286
TLI: translunar injection.
Tobias, Dr. Cornelius A., 63
Toksoz, Dr. Mehmet Nafi, 345
Tokyo, Japan, 133, 138, 257
Tokyo, Univ. of, 83–84
  Institute of Space and Aeronautical Science, 9, 48
Topaz (U.S.S.R. nuclear reactor), 250
TOPS. See Thermoelectric Outer Planet Spacecraft.
Torell, Bruce N., 178, 218
Toro (asteroid), 281–282, 232–233
Toronto, Univ. of, 315
Toulouse, France, 150, 242, 248
Tournesol (French satellite), 103, 369
Townsend, John W., Jr., 19, 302
Tracked air cushion research vehicle, 182, 348
Tracked Hovercraft, Ltd., 348
Tracking, 209
  Apollo 14 mission, 30
  balloon, 250–251, 254
  budget, 21
  cooperation, international, 106, 369–370
  deep space (DSN) antenna, 69, 102, 313, 318
  military, 357–358
  MSFN, 27, 69, 269

satellite, 102, 290
ship, 350, 357-358
sounding rocket, 64
station, 97
   Australia, 69
   South Africa, 267
   Spain, 69
Tracking and data relay satellite system (TDRSS), 173
Tracking ship, 350
Trans World Airlines, Inc. (TWA), 71
Transit 4–A (navigation satellite), 180
Transonic Aircraft Technology Program (TACT), 165, 272
TRANSPO '72, 309
Transportation, Dept. of (DOT) (see also Federal Aviation Administration), 288, 307
   air traffic control, 85
   automated transit system, 47, 55
   award, 75
   budget, 21, 211, 226
   contract, 19, 47, 85, 182
   cooperation, 3, 39, 53-54, 121-122, 187, 218-219, 242, 290, 369
   facilities, 194, 247, 264
   general-aviation safety, 276-277
   noise abatement, 21, 122-123, 242, 290
   organization, 81-82
   personnel, 45, 107, 123-124, 193, 253, 323
   R&D, 82, 122-123, 369
   STOL aircraft, 19, 39, 187, 218-219, 290
   supersonic transport, 21, 60, 74, 75, 81, 82, 84-85, 88, 129-130, 211, 323
   tracked air cushion research vehicle, 183
Transportation Systems Center, 82
Treasury Dept., 266
Treaty, 45, 47, 55, 88, 153-154, 162, 173, 313-314
Tribus, Myron, 11
Trubshaw, Brian, 48
Truman, President Harry S., 248
Truszynski, Gerald M., 69, 97
Truxal, Dr. John G., 16-17
TRW Inc., 89, 117, 174, 282, 306, 327, 345-346
   Systems Group, 329, 335-336
Tsiolkovsky, Konstantin, 234
Tu-104 (U.S.S.R. jet airliner), 140
Tu-134 (U.S.S.R. jet airliner), 140
Tu-144 (U.S.S.R. supersonic transport), 77, 79, 85, 126, 127, 147, 150, 157, 214, 232, 249, 263, 264, 314
Tu-154 (U.S.S.R. jet airliner), 140
Tucker, Dr. Wallace, 113
Tucson, Ariz., 40, 281
Tullahoma, Tenn., 60
Tumulty, William T., Jr., 269
Tupolev, Aleksey A., 127, 148, 264
Tupolev, Andrey N., 260, 263, 264
Turbine-alternator compressors (TAC), 147
Turbo-Mallard (amphibious aircraft), 288
Turcat, André, 249
TWA. See Trans World Airlines, Inc.
27th Tactical Fighter Wing, 314
Tycho Crater (moon), 245

U

U-2 (reconnaissance aircraft), 95, 115, 120, 158, 281, 329
Uchinoura Space Center, Japan, 9, 48, 200, 268
UH-ID (helicopter), 74
UHF: ultrahigh frequency.
Uhuru (*Explorer 42*) (Small Astronomy Satellite), 89, 113, 131, 319
U.K. See United Kingdom.
Ulan Bator, Mongolia, 238
ULMS. See Underwater long-range missile system.
Ultraviolet (UV), 8, 14, 23, 60, 119, 135, 148, 189, 220, 242, 302, 318, 327, 366
Ultraviolet panorama (S-138) experiment, 38
U.N. See United Nations.
Underground nuclear tests, 300, 303
Underwater long-range missile system (ULMS), 335
Unemployment Rates for Scientists, Spring 1971 (NSF report), 188
Unidentified satellite, 11, 49, 79, 80, 107, 122, 157, 163, 195, 227, 254, 287, 296, 341, 350
Union of Concerned Scientists, 209
United Aircraft Corp.
   Hamilton Standard Div., 363
   Pratt & Whitney Div., 60, 96, 131, 178, 198, 218, 232, 240, 246, 300, 354, 356
   Sikorsky Aircraft Div., 150, 357
United Airlines, 211
United Automobile, Aerospace, and Agricultural Implementation Workers of America, 343
United Cerebral Palsy Research Foundation, 253
United Kingdom (U.K.), 171, 249, 287, 312, 331
   aerospace industry, 88, 173, 329
   aircraft, 80, 224, 300, 329, 346
      Concorde, 157, 214, 346
   aircraft carrier, 224
   Apollo 14 mission, reaction, 43
   booster, 211, 300, 310
   computer use, 122
   cooperation, space, 3, 16, 19, 37, 57-58, 96, 291, 310, 348, 369
   Defence Ministry, 303, 329
   hovertrain, 348
   launch, 365, 368
      Ariel 4 (UK-4) (scientific satellite), 348
      Prospero (X-3) (technology satellite), 300
   missile-firing submarines, 224

nuclear power, 252
research and applications modules (RAM), 3
Science Research Council, 348, 356
space shuttle, 16, 37, 291
United Nations (U.N.), 70, 104, 110, 120, 166, 171, 250, 361, 370
   Apollo 15 message, 222
   astronaut visit, 65, 238, 239
   Committee on the Peaceful Uses of Outer Space, 173, 182, 334
   General Assembly, 153, 252, 254, 257, 313, 334
   lunar law treaty draft, 153
   Scientific Subcommittee on Space, 198
United Nations Day, 191
United Nations Peace Medal, 318
United States (U.S.) (see also appropriate agencies and Congress)
   atlas, 11
   award, 38–39, 60, 102, 139, 282, 307, 314
   budget, 17, 20–21, 118
   communications, 3, 71
   defense, 20, 21, 103–104, 119, 124, 170, 194, 201, 224–225, 284, 288, 319–320, 360, 361
   disarmament, 74, 85, 136
   economy, 71–72, 160, 295
   education, 146, 161, 260
   employment, 3–4, 20, 63, 76, 104–105, 118, 119, 124, 141–142, 188, 252–253, 308, 309
   energy sources, 129, 154
   engineering, 308
   environmental program (see also Air pollution; Noise, aircraft; and Water pollution), 2, 13, 42, 76, 78, 96, 170, 180, 231
   foreign policy, 54–55, 117
   Government reorganization, proposed, 81–82, 154
   international cooperation, 17, 104, 190, 191
   international cooperation, space. See International cooperation, space.
   meteorology, 2, 103, 180, 303–304
   metric system, 212
   nuclear nonproliferation treaty. See Nuclear nonproliferation treaty.
   oceanography, 12, 19, 24, 45
   patent policy, 238–239
   research and development, 13, 20, 33, 116, 118, 122–123, 124–125, 175–176, 201, 251, 291, 303, 305, 331–332, 336, 363
   science and technology, 11, 17, 18, 20, 30, 40, 58, 69, 76, 86, 89, 99, 104, 108–109, 113, 118, 126, 127–128, 147, 160, 167, 173, 187–188, 197, 248, 260, 291, 296, 303, 331–332, 363–364
   space program. See Space program, national; Space race; Space results.
   transportation (see also Supersonic transport), 36, 47, 53, 82, 128–129
   Vietnam war. See Vietnam war.

United States and Soviet Progress in Space: Some New Contrasts, 7–8
United States Space Science Program, 167
Universe, 132
Universities, 68, 85, 146, 161, 210, 248, 306, 308
Uranus (planet), 22
Urey, Dr. Harold C., 6, 109
U.S. Air Force (USAF) (see also individual bases, centers, and commands, such as Air Force Academy, Air Force Systems Command, Air Defense Command, Arnold Engineering Development Center, Edwards AFB)
   aircraft (see also individual aircraft, such as C-5A, F-15, F-111, Helicopter, YF-12), 1, 21, 70, 87, 94, 95, 112, 168, 239, 248, 255, 263, 279, 300, 302, 344
   anniversary, 94
   award, 72, 178, 266, 354
   budget, 94, 263
   contract, 2, 8, 12, 94, 294, 300, 310, 312, 341
   cooperation, 65, 70, 87, 95, 130, 142, 165, 173–174, 199, 210, 218–219, 225, 237, 272–276, 278, 281, 290, 328, 351, 365
   FRUSA (flexible rolled-up solar array), 8
   launch
      satellite, 11, 49, 79, 80, 107, 122, 157, 163, 195–196, 220–221, 227, 254, 287, 289, 296, 306, 350, 365
      failure, 49, 342
   lifting body, 11, 39, 49, 65, 85, 130, 142, 153, 210–211, 351, 365
   missile program, 77–78, 95, 199, 224–225, 262, 266, 312, 346
   personnel, 11, 36, 42, 180–181, 186, 191, 193, 320
   reentry vehicle, 220–221
   satellite, 243
   space program (see also Defense, Dept. of), 94, 328
   space shuttle, 276
   Transonic Aircraft Technology (TACT) Program, 165, 272
U.S. Arms Control and Disarmament Agency, 74
U.S. Army (USA), 43, 70, 74, 182, 186, 189, 272, 288, 294, 351, 357, 361, 365
U.S. Army (USA) Air Mobility Research and Development Laboratory, 233
U.S. Coast Guard, 21
U.S. Conference of Mayors, 177
U.S. Court of Appeals, District of Columbia, 300
U.S. Forest Service, 129
U.S. Geological Survey (USGS), 11, 104, 120, 171, 224, 245, 279, 281, 339–340, 348, 351

U.S. Marine Corps (USMC), 80
U.S. Military Academy, 11
U.S. Military Academy Assn. of Graduates, 121
U.S. Naval Aerospace Medical Institute, 182
U.S. Navy (USN), 18, 241
   aircraft, 11–12, 17, 18, 41, 46–47, 70, 112, 124, 141, 152, 163, 210, 211, 212, 219, 246, 252, 255, 271, 272, 284, 286, 301, 322, 323
   award, 72
   contract, 329, 335
   cooperation, 119–120, 244, 369–370
   missile, 52, 335
   missile-firing ship, 125
   personnel, 47, 345
   satellite, 95, 180
   Sealab Project, 292–293
   space shuttle, 244
   spacecraft recovery force, 349
   surface effect ship, 18, 199–200
U.S. Postal Service, 232
U.S. Supreme Court, 49, 65
USA. See U.S. Army.
USAF. See U.S. Air Force.
"USAF in Space" (exhibit), 153
USGS. See U.S. Geological Survey.
USMC. See U.S. Marine Corps.
USN. See U.S. Navy.
U.S.S. *Nathaniel Greene* (submarine), 309
U.S.S. *New Orleans*, 30, 42
U.S.S. *Observation Island*, 125
U.S.S. *Okinawa*, 207, 222
U.S.S.R. (Union of Soviet Socialist Republics) (see also Soviet Academy of Sciences, etc.), 330, 362
   aircraft, 12, 17, 38, 78, 79, 85, 103, 124, 126, 127, 140, 147, 148, 150, 157, 211, 214, 232, 248, 249, 260, 263, 264, 280, 286, 312, 314, 346, 360
   anniversary, 100, 152, 167
   Apollo 11 mission reaction, 162
   Apollo 14 mission reaction, 35, 43
   Apollo 15 mission reaction, 222
   astronomy, 357
   award, 180, 191, 218, 266, 311
   booster, 115, 246, 249
   communications satellite system, 238, 330
   cooperation, 54–55, 117, 190, 197, 287, 299
   cooperation, space, 10, 13–14, 16, 19–20, 36, 43, 54–55, 62, 68, 77, 101, 133–134, 139, 153–154, 159, 171, 177, 185, 196, 197, 198, 216–217, 235, 237, 238, 243, 248, 249, 275, 284, 290, 293, 299, 319, 321, 323, 325, 330, 332, 334, 337, 340, 347, 358, 369
   cosmonaut. See Cosmonaut.
   Council of Ministers, 216–217
   defector, 147, 168, 171, 176, 190, 329, 331
   disarmament (see also U.S.S.R., SALT talks), 54–55, 74, 137, 148
   five-year plan, 58
   launch, 366–368
   probe
      Luna 18, 246
      Luna 19, 267–268
      Mars 2, 134
      Mars 3, 147
   Salyut 1, 105
   satellite
      Cosmos, 7, 9, 11, 15, 42, 49, 54, 61, 78, 84, 93, 94, 96, 98, 102, 108, 110, 115, 124, 125, 128, 133, 135, 139, 146, 148, 153, 160, 174, 197, 200, 212, 219, 222, 227, 240, 250, 254, 255, 262, 265, 268, 270, 280, 284–285, 287, 291, 306, 323, 325, 327, 333, 334, 337, 339, 343, 347–348, 351, 352, 353, 358, 359
   Intercosmos 5, 337
   Meteor 7, 11
   Meteor 8, 105
   Meteor 9, 195
   Meteor 10, 360
   Molniya I–18, 210
   Molniya I–19, 355
   Molniya II–1, 330
   Oreol, 359
   sounding rocket, 237, 252, 304
   Soyuz 10, 107–108
   Soyuz 11, 154–155
   Lunokhod 1, 4, 8, 9–10, 20, 41, 43, 60, 66, 73, 98, 100, 133, 139, 145, 152, 168, 188, 196, 209, 220, 221, 246, 256, 272, 278, 306–307, 315, 322, 368
   Ministry of Health, 36
   missile and rocket program, 84, 103–104, 107, 109, 117, 122, 129, 135, 137, 146, 152, 194, 243, 248, 286, 303, 304, 320, 360
   missile threat to, 224–225
   nuclear energy facilities, U.S. visit to, 219
   nuclear test, 331
   Paris Air Show exhibit, 152, 168
   probe, 4, 9, 15, 20, 209, 293
   research and development, 86, 124–125, 201
   SALT talks, 54–55, 74, 148, 189, 190, 194–195, 252, 264–265, 266
   science and technology, 100–101, 110, 124–125, 140, 163, 172, 182, 183, 190, 210, 262, 290–291, 303–304, 309, 311, 315, 323
   Soyuz 11 cosmonauts funeral and burial, 185, 186
   space encyclopedia, 91
   space program, 7–8, 58, 62, 66, 73, 82, 86, 94, 100–101, 102, 110, 111, 152, 154–156, 157, 162–163, 166,

167, 174–175, 176, 183, 201, 218, 220, 221, 237, 283–284, 299–300, 329, 331, 353, 366–368
  military use, 7–8, 24, 115–116, 158, 176, 296–297, 339
  space station, 105, 107–108, 110–111, 112, 115, 119, 122, 127, 154–156, 174–175, 186, 198, 282–283, 368
  space tracking ship, 350
  spacecraft. See U.S.S.R., launch; and individual spacecraft, such as *Luna 16, Mars 2, Molniya 1–18, Salyut 1, Soyuz 10.*
  supersonic transport, 77, 79, 126, 127, 147, 148, 157
  treaty, 45, 47, 153–154, 177, 313–314
  weapons, 39, 104, 115–116, 129, 201
Ustinov, G., 314
Utah State Medical Assn., 256
Utah, Univ. of, 49, 57, 72, 85, 368
uv. See Ultraviolet.

V

vab. See Vehicle Assembly Building.
Van Allen radiation belts, 261
Vandenberg afb, Calif. (see also Western Test Range), 8
  missile target, 78
  satellite launch vehicle
    Agena, 350
    Atlas, 220
    Thor-Agena, 49, 195, 254
    Thor-Burner II, 48, 157
    Thorad-Agena, 80, 195, 254, 289, 350
    Titan IIIB-Agena, 11, 79, 107, 227, 296
    Titan IIID, 163
Vanderbilt Univ., 282
Vanguard (satellite), 75, 135, 362
Vanguard—A History, 75
Van Reeth, G., 352
Vecchietti, George J., 166
Vehicle Assembly Building (vab) (ksc), 271
Vehicle Technology for Civil Aviation: The 70s and Beyond (conference), 307
Vela (nuclear detection satellite), 153
Venus (planet, 14, 20, 22, 50, 62, 67, 174, 186, 218, 333
Venus 7 (U.S.S.R. interplanetary probe), 15, 20
Vernadsky Institute for Analytical Chemistry, 6
Veronique (French rocket), 235
Vertikal-2 (U.S.S.R. sounding rocket), 237, 252
Very-large-array (vla) radiotelescope, 280
vhf: very high frequency.
Victoria, Australia, 167
Vienna, Austria, 148, 199, 253, 265
Vietnam war, 220, 320, 350, 358
Viking (program), 22, 61, 67, 116, 164 339

Viking Orbiter (spacecraft), 211, 344
Viking, Project, 247, 264, 275
Vinogradov, Dr. Aleksander P., 6, 159, 217
Viper Dart (sounding rocket), 234, 236
Virgin Islands, 98, 133, 134, 308
Virginia, 97, 224
Virginia Institute of Marine Science, 292
Virginians for Dulles, 271
Virgo (galaxy), 132
vla. See Very-large-array telescope.
Vladimirov, Leonid, 329, 331
Volgograd, U.S.S.R., 304
Volkov, Vladislav N., 154, 180, 181, 186, 192, 200, 283
Volpe, Secretary of Transportation John A., 9, 19, 107, 109, 182, 194, 276–277, 295
von Braun, Dr. Wernher, 191, 298
  awards and honors, 266
  global resources management system, 32–33, 321
  science and technology, 147
  space program, national, 62–63, 68–69, 147, 249, 281, 283, 329, 342
  space shuttle, 62–63
von Kármán Lecture, 298
von Opel, Fritz, 99
Voskhod 1 mission, 100, 262
Vostok 1 (U.S.S.R. spacecraft), 152
Vostok 4 mission, 152
v/stol (vertical or short takeoff and landing) aircraft, 18, 65, 80, 298, 300, 329, 365
v/stolport, 287
vtol (vertical takeoff and landing) aircraft, 12, 262, 298, 300

W

Wagner, Charles S., 32, 44, 307
Wakelin, James H., Jr., 11
Walker, R. M., 83
Wallops Island, Va., 176
Wallops Station (nasa)
  bird migration cooperative experiment, 160–161
  Central Atlantic Regional Ecological Test Site (carets) project, 224
  contract, 126
  launch
    barium-ion-cloud (bic) cooperative probe, 259
    Planetary Atmosphere Experiments Test (paet), 168–169
  sounding rocket
    Aerobee 350, 52, 159
    Arcas, 218
    Black Brant VB, 54
    Black Brant VC, 199
    Boosted Arcas II, 22
    Javelin, 176
    Nike-Apache, 68, 234–235, 236–237, 257

Nike-Cajun, 3, 8, 68, 174, 187, 192, 234–235, 236
Nike-Tomahawk, 193
Viper Dart, 234–235, 236
meeting, 97
open house, 105
runway research program, 278
Wapakoneta, Ohio, 198
Washington, Bradford, award, 327
Washington, D.C., 41, 129, 187, 219, 222, 268, 280, 296, 301, 308, 314, 323
astronauts in, 59, 141
awards presented in, 135, 139, 178, 194, 263, 298–299, 302, 326
comsat communications system, 252, 265, 271, 272
meetings, 15, 18, 36, 38–39, 53, 68–69, 88, 112–113, 146–147, 166, 238, 281, 297, 306, 335
press conference, 59–60, 71, 102, 114, 122, 125–126, 141, 152, 156–157, 191–192
Washington National Airport, 165, 209, 271
Washington Univ., 83, 296
Washington, Univ. of, 13, 63
Wasserburg, Dr. Gerald J., 7, 167, 235–236
Water pollution, 293, 322
Water-vapor electrolysis system, 308
Watkins, Allen H., 139, 171
Watson, Ambassador Arthur K., 350
Watt, James, International Medal, 84
Weapon systems, 39, 40, 72, 78, 84, 303
Weather modification, 62, 320
Weaver, Kenneth F., 70
Webb, James E., 69
Webb, John A., Jr., 106
Weicker, Sen. Lowell P., 24
Weightlessness, effects, 127, 132, 153, 163, 181, 342
Weizmann Institute of Science, 18
Welch, Dr. William J., 299
Weliachew, Dr. Leonid N., 194
Wertz, Carl C., 178
Wesleyan Univ., 145
Wesselski, Clarence J., 251
West Point, N.Y., 121
Westerbork, The Netherlands, 280
Western Test Range (WTR), 88, 348
Wetherill, Dr. George W., 212
Wetport (offshore airport), 137–138
Wheeler, Dr. John A., 18
Whirlpool galaxy, 279–280
Whitcomb, Gov. Edgar H., 199
Whitcomb, Dr. Richard T., 299
White, L/C Edward H., II (USAF), 10, 357
White, Robert M., 19
White House, 9, 42–43, 124, 198, 212, 265, 282
appointments, 114, 253
astronauts, 59, 68, 356
awards presented at, 59, 138
space program, 37, 63, 272

White Sands Missile Range (WSMR), N. Mex.
atmospheric data experiment, 236
launch
Aerobee 150
atmospheric data, 171
micrometeorite study, 54
solar astronomy, 161–162
stellar data, 13
ultraviolet astronomy, 14, 23, 119
x-ray astronomy, 132
Aerobee 170
airglow, 41, 209
solar astronomy, 228
stellar data, 226
test and support mission, 292
ultraviolet astronomy, 79, 119, 226, 296
Aerobee 170B, 190–191
Aerobee 174, 350
Boosted Arcas II, 226, 285–286
Nike-Apache
airglow, 108
micrometeorite collection, 108
Nike-Cajun
airglow, 130, 131
atmospheric data, 8
ion composition study, 285
White Sands Test Facility, N. Mex., 131
White, Thomas D., National Defense Award, 150
White, Gen. Thomas D., Space Trophy, 178
Whittaker Corp., 200
Wichita, Kans., 82, 121
Wicker, Tom, 217
Wideroe, Turi, 135
Wiesner, Dr. Jerome B., 282
Wilford, John N., 343–344
Williams, Squadron Leader Graham, 135
Williamsburg, Va., 136, 307
Wilson, George C., 66, 115–116
Wind tunnel, 9, 70, 153, 365–366
Window Crater (moon), 205
Wing, aircraft
supercritical, 70, 122, 345, 356
flight test, 65, 77, 94, 101, 114, 141, 145, 163, 165, 192, 233, 240, 251–252, 255–256, 269, 272, 279, 287, 346, 349, 365–366
swing, 248
test, static, 255–256
Winstein, Dr. Saul, 18
Wisconsin, Univ. of, 15, 60, 79, 239, 279
Educational Satellite Center, 117
Witkin, Richard, 131
Wollin, Goesta, 302
Wood, Dr. John A., 7
Woomera, Australia, 356
Woomera Test Range, Australia, 300
Worden, L/C Alfred M. (USAF)
Apollo 15 mission
celebrations for, 222–223, 238–239
Congress, report to, 254
extravehicular activity, 141, 207–208
flight, 202–208

Kennedy Space Center visit, 271
Marshall Space Center visit, 272
medical aspects, 207, 222, 225, 228, 261, 289
New York Times article, 200–201, 229
Nixon, President Richard M., 201, 208–209, 221–222, 223, 256
photography, 206–207
preparations for, 141
press conference, 141, 227
tribute to, 221–222
U.N. visit, 239
White House visit, 256
Apollo 17 mission, 228–229
awards and honors, 239, 260, 262, 266, 346–347
European visit, 311
Working Group on Philosophy, Science, and Technology, 126
Working Group on Remote Measurement of Pollution, 231
World Administrative Radio Conference for Space Communications, 132
World Law Day, 166
World Weather Program, 103, 230
World Weather Program, Plan for Fiscal Year 1972, 103
World Weather Watch, 180
Wright Brothers Memorial Trophy, 263
Wright 1909 Military Flyer, 248
Wright-Patterson AFB, Ohio, 247, 268
WSMR. See White Sands Missile Range.
WTR. See Western Test Range.
Wydeven, Dr. Theodore, Jr., 308
Wydler, Rep. John W., 38
Wyle Laboratories, 49
Wynn, Rep. Larry, Jr., 38
Wyoming, 128

X

X–1 (rocket research aircraft), 239
X–3 (satellite). See *Prospero*.
X–14A (VTOL aircraft), 262
X–14B, 262
X–15 (rocket research aircraft), 239
X–24A (lifting body), 210, 351, 356, 365
test flight, 130, 153, 351
glide, 39, 210
powered, 11, 49, 65, 85, 142, 210
X–24B, 210, 351, 356, 365
XE (nuclear rocket engine), 53
XLR–129 (rocket engine), 87, 240, 276
X-ray, 67, 99
source, 50, 52, 63, 73, 89, 119, 132, 174, 189, 197, 226, 296, 356
weapon detection device, 278

Y

Yak-40 (U.S.S.R. jet transport), 140
Yangel, Mikhail K., 73, 297
Yardley, John F., 299
Yegorov, A. D., 167
Yegorov, Dr. Boris B., 111, 119, 262
Yeliseyev, Dr. Aleksey S., 107, 110–111, 282–283
Yellowstone National Park, 112
Yerevan, U.S.S.R., 277
YF–12 (jet interceptor), 174
YF–12A, 115
Young, Capt. John W. (USN), 61, 171, 323
Yugoslavia, 43, 242–243, 286, 305–306

Z

Zagreb, Yugoslavia, 43
Zhulin, Dr. I. A., 321, 340
Ziegler, Henri, 351
Ziegler, Ronald L., 135
Znaniye (Knowledge) (All Union Znaniye Society, U.S.S.R.), 177
Zoller, Lowell K., 196
Zond (program), 235
Zond (U.S.S.R. spacecraft), 268
Zond 2, 134
Zuckerman, Dr. Benjamin M., 299
Zumwalt, Adm. Elmo R., Jr. (USN), 18

NASA HISTORICAL PUBLICATIONS

HISTORIES

- Robert L. Rosholt, *An Administrative History of NASA, 1958–1963*, NASA SP-4101, 1966, GPO, $4.00.*
- Loyd S. Swenson, James M. Grimwood, and Charles C. Alexander, *This New Ocean: A History of Project Mercury*, NASA SP-4201, 1966, GPO, $5.50.
- Constance McL. Green and Milton Lomask, *Vanguard—A History*, NASA SP-4202, 1970; also Washington: Smithsonian Institution Press, 1971, $12.50.
- Alfred Rosenthal, *Venture into Space: Early Years of Goddard Space Flight Center*, NASA SP-4301, 1968, GPO, $2.50.
- Edwin P. Hartman, *Adventures in Research: A History of the Ames Research Center, 1940–1965*, NASA SP-4302, 1970, GPO, $4.75.

HISTORICAL STUDIES

- Eugene M. Emme (ed.), *History of Rocket Technology*, Detroit: Wayne State University, 1964, out of print.
- Mae Mills Link, *Space Medicine in Project Mercury*, NASA SP-4003, 1965, NTIS, $6.00.**
- *Historical Sketch of NASA*, NASA EP-29, 1965 and 1966, NTIS, $6.00.
- Katherine M. Dickson (Library of Congress), *History of Aeronautics and Astronautics: A Preliminary Bibliography*, NASA HHR-29, NTIS, $6.00.
- Eugene M. Emme (ed.), *Statements by the Presidents of the United States on International Cooperation in Space*, Senate Committee on Aeronautical and Space Sciences, Sen. Doc. 92-40, 1971, GPO, $0.55.
- William R. Corliss, *NASA Sounding Rockets, 1958–1968: A Historical Summary*, NASA SP-4401, 1971, GPO, $1.75.
- Helen T. Wells with Susan Whiteley, *Origins of NASA Names*, NASA SP-4402 (1973).
- Jane Van Nimmen and Leonard C. Bruno with Robert L. Rosholt, *NASA Historical Data Book, 1958–1968*, NASA SP-4012 (1973).

CHRONOLOGIES

- *Aeronautics and Astronautics: An American Chronology of Science and Technology in the Exploration of Space, 1915–1960,* compiled by Eugene M. Emme, Washington: NASA, 1961, NTIS, $6.00.
- *Aeronautical and Astronautical Events of 1961,* published by the House Committee on Science and Astronautics, 1962, NTIS, $6.00.
- *Astronautical and Aeronautical Events of 1962,* published by the House Committee on Science and Astronautics, 1963, NTIS, $6.00.
- *Astronautics and Aeronautics,* 1963, NASA SP–4004, 1964, NTIS, $6.00.
- ———, 1964, NASA SP–4005, 1965, NTIS, $6.00.
- ———, 1965, NASA SP–4006, 1966, NTIS, $6.00.
- ———, 1966, NASA SP–4007, 1967, NTIS, $6.00.
- ———, 1967, NASA SP–4008, 1968, GPO, $2.25.
- ———, 1968, NASA SP–4010, 1969, GPO, $2.00.
- ———, 1969, NASA SP–4014, 1970, GPO, $2.25.
- ———, 1970, NASA SP–4015 (1972).
- James M. Grimwood, *Project Mercury: A Chronology,* NASA SP–4001, 1963, NTIS, $6.00.
- James M. Grimwood and Barton C. Hacker, with Peter J. Vorzimmer, *Project Gemini Technology and Operations: A Chronology,* NASA SP–4002, 1969, GPO, $2.75.
- Ivan D. Ertel and Mary Louise Morse, *The Apollo Spacecraft: A Chronology,* Vol. I, *Through November 7, 1962,* NASA SP–4009, 1969, GPO, $2.50.
- Mary Louise Morse and Jean Kernahan Bays, *The Apollo Spacecraft: A Chronology,* Vol. II, *November 8, 1962–September 30, 1964,* NASA SP–4009 (1973).
- R. Cargill Hall, *Project Ranger: A Chronology,* JPL/HR–2, 1971, NTIS, $6.00.

---

* GPO: Titles may be ordered from the Superintendent of Documents, Government Printing Office, Washington, D.C. 20402.
** NTIS: Titles may be ordered from National Technical Information Service, Springfield, Va. 22151.

www.ingramcontent.com/pod-product-compliance
Lightning Source LLC
Chambersburg PA
CBHW081715170526
45167CB00009B/3581